$48.50

NATIONAL REPAIR & REMODELING ESTIMATOR

by Albert S. Paxton

Edited by J. A. O'Grady

Includes inside the back cover:

- An estimating CD with all the costs in this book, plus,
- An estimating program that makes it easy to use these costs,
- An interactive video guide to the National Estimator program,
- A program that converts your estimates into invoices,
- A program that exports your estimates to QuickBooks Pro.

The Show Me video guide to National Estimator starts a few seconds after you insert the CD in your CD-ROM drive.

To Mickey Green
Welcome to the
Nisqually EQ
CAT team
Al Paxton
4/26/2001

Craftsman Book Company
6058 Corte del Cedro / P.O. Box 6500 / Carlsbad, CA 92018

Preface

The author has corresponded with manufacturers and wholesalers of building material supplies and surveyed retail pricing services. From these sources, he has developed Average Material Unit Costs which should apply in most parts of the country.

Wherever possible, the author has listed Average Labor Unit Costs which are derived from the Average Manhours per Unit, the Crew Size, and the Wage Rates used in this book. Please read How to Use This Book for a more in-depth explanation of the arithmetic.

If you prefer, you can develop your own local labor unit costs. You can do this by simply multiplying the Average Manhours per Unit by your local crew wage rates per hour. Using your actual local labor wage rates for the trades will make your estimate more accurate.

What is a realistic labor unit cost to one reader may well be low or high to another reader, because of variations in labor efficiency. The Average Manhours per Unit figures were developed by time studies at job sites around the country. To determine the daily production rate for the crew, divide the total crew manhours per day by the Average Manhours per Unit.

The subject topics in this book are arranged in alphabetical order, A to Z. To help you find specific construction items, there is a complete alphabetical index at the end of the book, and a main subject index at the beginning of the book.

This manual shows crew, manhours, material, labor and equipment cost estimates based on Small Volume work, then a total cost and a total including overhead and profit. Only total cost and total including overhead and profit are shown for Large Volume. For those readers who need it, the breakdown for Large Volume is included on the National Estimator CD in the back of the book. No single price fits all repair and remodeling jobs. Generally, work done on smaller jobs costs more per unit installed and work on larger jobs costs less. The estimates in this book reflect that simple fact. The two estimates you find for each work item show the author's opinion of the likely range of costs for most contractors and for most jobs. So, which cost do you use, Low Volume or High Volume?

The only right price is the one that gets the job and earns a reasonable profit. Finding that price always requires estimating judgment. Use Small Volume cost estimates when some or most of the following conditions are likely:

- The crews won't work more than a few days on site.
- Better quality work is required.
- Productivity will probably be below average.
- Volume discounts on materials aren't available.
- Bidding is less competitive.
- Your overhead is higher than most contractors.

When few or none of those conditions apply, use Large Volume cost estimates.

Credits and Acknowledgments

This book has over 12,000 cost estimates for 2001. To develop these estimates, the author and editors relied on information supplied by hundreds of construction cost authorities. We offer our sincere thanks to the contractors, engineers, design professionals, construction estimators, material suppliers and manufacturers who, in the spirit of cooperation, have assisted in the preparation of this Twenty-fourth Edition of the National Repair & Remodeling Estimator. Our appreciation is extended to those listed below.

American Standard Products
DAP Products
Outwater Plastic Industries
Con-Rock Concrete
Georgia Pacific Products
Kohler Products
Wood Mode Cabinets
Transit Mixed Concrete
U.S. Gypsum Products
Henry Roofing Products

Special thanks to the following:

Dal-Tile Corporation, 1713 Stewart, Santa Monica, California

About the Author

Albert Paxton is president of Professional Construction Analysts, Inc., located at 28990 Pacific Coast Highway, Suite 205, Malibu, CA 90265-3944; (800) PCA-9690; fax (310) 589-0400. Mr. Paxton is a California licensed General Contractor (B1-425946). PCA's staff is comprised of estimators, engineers and project managers who are also expert witnesses, building appraisers and arbitrators operating throughout the United States.

PCA clients include property insurance carriers, financial institutions, self-insureds, and private individuals. The expertise of PCA is in both new and repair-remodel construction, both commercial and residential structures. In addition to individual structures, ***PCA*** assignments have included catastrophic events such as Hurricane Hugo, the Oakland Hills firestorm, Hurricane Andrew, Hurricane Iniki, and the Northridge earthquake.

 ISBN 1-57218-098-6

Cover photos by Fritz Light Photography, (760) 649-6496

Main Subject Index

How to Use This Book

1	2	3	4	5	6	7	8	9	10	11	12
			Costs Based on Small Volume							Large Volume	
Description	Oper	Unit	Crew Size	Man-Hours Per Unit	Avg Mat'l Unit Cost	Avg Labor Unit Cost	Avg Equip Unit Cost	Avg Total Unit Cost	Avg Price Incl O&P	Avg Total Unit Cost	Avg Price Incl O&P

The descriptions and cost data in this book are arranged in a series of columns, which are described below. The cost data is divided into two categories: Costs Based On Large Volume and Costs Based On Small Volume. These two categories provide the estimator with a pricing range for each construction topic.

The Description column (1) contains the pertinent, specific information necessary to make the pricing information relevant and accurate.

The Operation column (2) contains a description of the construction repair or remodeling operation being performed. Generally the operations are Demolition, Install, and Reset.

The Unit column (3) contains the unit of measurement or quantity which applies to the item described.

The Crew Size column (4) contains a description of the trade that usually installs or labors on the specified item. It includes information on the labor trade that installs the material and the typical crew size. Letters and numbers are used in the abbreviations in the crew size column. Full descriptions of these abbreviations are in the Crew Compositions and Wage Rates table, beginning on page 15.

The Manhours Per Unit column (5) is for the listed operation and listed crew.

The units per day in this book don't take into consideration unusually large or small quantities. But items such as travel, accessibility to work, experience of workers, and protection of undamaged property, which can favorably or adversely affect productivity, have been considered in developing Average Manhours per Unit. For further information about labor, see "Notes — Labor" in the Notes Section of some specific items.

The Average Material Unit Cost column (6) contains an average material cost for products (including, in many cases, the by-products used in installing the products) for small volume. It doesn't include an allowance for sales tax, delivery charges, overhead and profit. Percentages for waste, shrinkage, or coverage have been taken into consideration unless indicated. For other information, see "Dimensions" or "Installation" in the Notes Section.

If the item described has many or very unusual by-products which are essential to determining the Average Material Unit Cost, the author has provided examples of material pricing. These examples are placed throughout the book in the Notes Section.

Average Daily Production and Average Material Unit Cost should assist the estimator in:

1. Checking prices quoted by others.
2. Developing local prices.

You should verify labor rates and material prices locally. Though the prices in this book are average material prices, prices vary from locality to locality. A local hourly wage rate should normally include taxes, benefits, and insurance. Some contractors may also include overhead and profit in the hourly rate.

The Average Labor Unit Cost column (7) contains an average labor cost for small volume based on the Average Manhours per Unit and the Crew Compositions and Wage Rates table. The average labor unit cost equals the Average Manhours per Unit multiplied by the Average Crew Rate per hour. The rates include fringe benefits, taxes, and insurance. Examples that show how to determine the average labor unit cost are provided in the Notes Section.

The Average Equipment Unit Cost column (8) contains an average equipment cost for small volume based on both the average daily rental and the cost per day if owned and depreciated. The costs of daily maintenance and the operator are included.

The Average Total Unit Cost column for Small Volume (9) includes the sum of the Material, Equipment, and Labor Cost columns. It doesn't include an allowance for overhead and profit. The Average Total Unit Cost column for Large Volume (11) has the same information, although you don't see those columns in the book. The *National Estimator* CD in the back of the book does include the Material, Equipment, and Labor Cost columns for Large Volume. Check it if you want to see how we arrived at the Total Unit Cost and Price Including Overhead & Profit for Large Volume.

The Average Total Price Including Overhead and Profit columns (10 and 12) result from adding an overhead and profit allowance to Total Cost. This allowance reflects the author's interpretation of average fixed and variable overhead expenses and the labor intensiveness of the operation vs. the costs of materials for the operation. This allowance factor varies throughout the book, depending on the operation. Each contractor interprets O&P differently. The range can be from 15% to 80% of the Average Total Unit Cost.

Estimating Techniques

Estimating Repair/Remodeling Jobs: The unforeseen, unpredictable, or unexpected can ruin you.

Each year, the residential repair and remodeling industry grows. It's currently outpacing residential new construction due to increases in land costs, labor wage rates, interest rates, material costs, and economic uncertainty. When people can't afford a new home, they tend to remodel their old one. And there are always houses that need repair, from natural disasters or accidents like fire. The professional repair and remodeling contractor is moving to the forefront of the industry.

Repair and remodeling spawns three occupations: the contractor and his workers, the insurance company property claims adjuster, and the property damage appraiser. Each of these professionals shares common functions, including estimating the cost of the repair or remodeling work.

Estimating isn't an exact science. Yet the estimate determines the profit or loss for the contractor, the fairness of the claim payout by the adjuster, and the amount of grant or loan by the appraiser. Quality estimating must be uppermost in the mind of each of these professionals. And accurate estimates are possible only when you know exactly what materials are needed and the number of manhours required for demolition, removal, and installation. Remember, profits follow the professional. To be profitable you must control costs — and cost control is directly related to accurate, professional estimates.

There are four general types of estimates, each with a different purpose and a corresponding degree of accuracy:

- The guess method: "All bathrooms cost $5,000." or "It looks like an $8,000 job to me."
- The per measure method: (I like to call it the surprise package.) "Remodeling costs $60 per SF, the job is 500 SF, so the price is $30,000."

These two methods are the least accurate and accomplish little for the adjuster or the appraiser. The contractor might use the methods for qualifying customers (e.g., "I thought a bathroom would only cost $2,000."), but never as the basis for bidding or negotiating a price.

- The *piece estimate* or *stick-by-stick* method.
- The *unit cost estimate* method.

These two methods yield a detailed estimate itemizing all of the material quantities and costs, the labor manhours and wage rates, the subcontract costs, and the allowance for overhead and profit.

Though time-consuming, the detailed estimate is the most accurate and predictable. It's a very satisfactory tool for negotiating either the contract price or the adjustment of a building loss. The piece estimate and the unit cost estimate rely on historical data, such as manhours per specific job operation and recent material costs. The successful repair and remodeling contractor, or insurance/appraisal company, maintains records of previous jobs detailing allocation of crew manhours per day and materials expended.

While new estimators don't have historical data records, they can rely on reference books, magazines, and newsletters to estimate manhours and material costs. It is important to remember that **the reference must pertain to repair and remodeling**. This book is designed *specifically* to meet this requirement.

The reference material must specialize in repair and remodeling work because there's a large cost difference between new construction and repair and remodeling. Material and labor construction costs vary radically with the size of the job or project. Economies of scale come into play. The larger the quantity of materials, the better the purchase price should be. The larger the number of units to be installed, the greater the labor efficiency.

Repair and remodeling work, compared to new construction, is more expensive due to a normally smaller volume of work. Typical repair work involves only two or three rooms of a house, or one roof. In new construction, the job size may be three to five complete homes or an entire development. And there's another factor: a lot of repair and remodeling is done with the house occupied, forcing the crew to work around the normal, daily activities of the occupants. In new construction, the approach is systematic and logical — work proceeds from the ground up to the roof and to the inside of the structure.

Since the jobs are small, the repair and remodeling contractor doesn't employ trade specialists. Repairers employ the "jack-of-all-trades" who is less specialized and therefore less efficient. This isn't to say the repairer is less professional than the trade specialist. On the contrary, the repairer must know about many more facets of construction: not just framing, but painting, finish carpentry, roofing, and electrical as well. But because the repairer has to spread his expertise over a greater area, he will be less efficient than the specialist who repeats the same operation all day long.

Another factor reducing worker efficiency is poor access to the work area. With new construction, where building is an orderly "from the ground up" approach, workers have easy access to the work area for any given operation. The workers can spread out as much as needed, which facilitates efficiency and minimizes the manhours required to perform a given operation.

The opposite situation exists with repair and remodeling construction. Consider an example where the work area involves fire damage on the second floor. Materials either go up through the interior stairs or through a second story window. Neither is easy when the exterior and interior walls have a finished covering such as siding and drywall. That results in greater labor costs with repair and remodeling because it takes more manhours to perform many of the same tasks.

If, as a professional estimator, you want to start collecting historical data, the place to begin is with daily worker time sheets that detail:

1. total hours worked by each worker per day
2. what specific operations each worker performed that day
3. how many hours (to the nearest tenth) each worker used in each operation performed that day.

Second, you must catalog all material invoices daily, being sure that quantities and unit costs per item are clearly indicated.

Third, maintain a record of overhead expenses attributable to the particular project. Then, after a number of jobs, you'll be able to calculate an average percentage of the job's gross amount that's attributable to overhead. Many contractors add 45% for overhead and profit to their total direct costs (direct labor, direct material, and direct subcontract costs). But that figure may not be right for your jobs.

Finally, each week you should reconcile in a job summary file the actual costs versus the estimated costs, and determine why there is any difference. This information can't immediately help you on this job since the contract has been signed, but it will be invaluable to you on your next job.

Up to now I've been talking about general estimating theory. Now let's be more specific. On page 8 is a Building Repair Estimate form. Each line is keyed to an explanation. A filled-out copy of the form is also provided, and on page 10, a blank, full-size copy that you can reproduce for your own use.

You can adapt the Building Repair Estimate form, whether you're a contractor, adjuster, or appraiser. Use of the form will yield a detailed estimate that will identify:

- The room or area involved, including sizes, dimensions and measurements.
- The kind and quality of material to be used.
- The quantities of materials to be used and verification of their prices.
- The type of work to be performed (demolish, remove, install, remove and reset) by what type of crew.
- The crew manhours per job operation and verification of the hourly wage scale.
- All arithmetical calculations that can be verified.
- Areas of difference between your estimate and others.
- Areas that will be a basis for negotiation and discussion of details.

Each job estimate begins with a visual inspection of the work site. If it's a repair job, you've got to see the damage. Without a visual inspection, you can't select a method of repair and you can't properly evaluate the opinions of others regarding repair or replacement. With either repair or remodeling work, the visual inspection is essential to uncover the "hiders" — the unpredictable, unforeseen, and unexpected problems that can turn profit into loss, or simplicity into nightmare. You're looking for the many variables and unknowns that exist behind an exterior or interior wall covering.

Along with the Building Repair Estimate form, use this checklist to make sure you're not forgetting anything.

Checklist

- Site accessibility: Will you store materials and tools in the garage? Is it secure? You can save a half hour to an hour each day by storing tools in the garage. Will the landscaping prevent trucks from reaching the work site? Are wheelbarrows or concrete pumpers going to be required?
- Soil: What type and how much water content? Will the soil change your excavation estimate?
- Utility lines: What's under the soil and where? Should you schedule the utilities to stake their lines?
- Soundness of the structure: If you're going to remodel, repair or add on, how sound is that portion of the house that you're going to have to work around? Where are the load-bearing walls? Are you going to remove and reset any walls? Do the floor joists sag?
- Roof strength: Can the roof support the weight of another layer of shingles. (Is four layers of composition shingles already too much?)
- Electrical: Is another breaker box required for the additional load?

This checklist is by no means complete, but it is a start. Take pictures! A Polaroid camera will quickly pay for itself. When you're back at the office, the picture helps reconstruct the scene. Before and after pictures are also a sales tool representing your professional expertise.

During the visual inspection always be asking yourself "what if" this or that happened. Be looking for potential problem areas that would be extremely labor intensive or expensive in material to repair or replace.

Also spend some time getting to know your clients and their attitudes. Most of repair and remodeling work occurs while the house is occupied. If the work will be messy, let the homeowners know in advance. Their satisfaction is your ultimate goal — and their satisfaction will provide you a pleasant working atmosphere. You're there to communicate with them. At the end of an estimate and visual inspection, the homeowner should have a clear idea of what you can or can't do, how it will be done, and approximately how long it will take. Don't discuss costs now! Save the estimating for your quiet office with a print-out calculator and your cost files or reference books.

What you create on your estimate form during a visual inspection is a set of rough notes and diagrams that make the estimate speak. To avoid duplications and omissions, estimate in a *systematic sequence of inspection.* There are two questions to consider. First, where do you start the estimate? Second, in what order will you list the damaged or replaced items? It's customary to start in the room having either the most damage or requiring the most extensive remodeling. The sequence of listing is important. Start with either the floor or the ceiling. When starting with the floor, you might list items in the following sequence: Joists, subfloor, finish floor, base — listing from bottom to top. When starting with the ceiling, you reverse, and list from top to bottom. The important thing is to be *consistent* as you go from room to room! It's a good idea to figure the roof and foundation separately, instead of by the room.

After completing your visual inspection, go back to your office to cost out the items. Talk to your material supply houses and get unit costs for the quantity involved. Consult your job files or reference books and assign crew manhours to the different job operations.

There's one more reason for creating detailed estimates. Besides an estimate, what else have your notes given you? A material take-off sheet, a lumber list, a plan and specification sheet — the basis for writing a job summary for comparing estimated costs and profit versus actual costs and profit — and a project schedule that minimizes down time.

Here's the last step: Enter an amount for overhead and profit. No matter how small or large your work volume is, be realistic — everyone has overhead. An office, even in your home, costs money to operate. If family members help out, pay them. Everyone's time is valuable!

Don't forget to charge for performing your estimate. A professional expects to be paid. You'll render a better product if you know you're being paid for your time. If you want to soften the blow to the owner, say the first hour is free or that the cost of the estimate will be deducted from the job price if you get the job.

In conclusion, whether you're a contractor, adjuster, or appraiser, you're selling your personal service, your ideas, and your reputation. To be successful you must:

- Know yourself and your capabilities.
- Know what the job will require by ferreting out the "hiders."
- Know your products and your work crew.
- Know your productivity and be able to deliver in a reasonable manner and within a reasonable time frame.
- Know your client and make it clear that all change orders, no matter how large or small, will cost money.

National Estimator '01 Inside the back cover of this book you'll find an envelope with a compact disk. The disk has National Estimator, an easy-to-use estimating program with all the cost estimates in this book. Insert the CD in your computer and wait a few seconds. Installation should begin automatically. (If not, click Start, Settings, Control Panel, double-click Add/Remove Programs and Install.) Select Show Me from the installation menu and Julie will show you how to use National Estimator. When Show Me is complete, select Install Program. When the National Estimator program has been installed, click Help on the menu bar, click Contents, click Print all Topics, click File and click Print Topic to print a 28-page instruction manual for National Estimator.

Building Repair Estimate

Date (33)

Insured (1)	Claim or Policy No. (2) | Page of (3)
Loss Address (4)	Home Ph. | Cause of Loss (5)
City	Bus. Ph. | Other Ins. Y N (6)
Bldg. R.C.V. (7) Bldg. A.C.V. (8)	Insurance Amount (9)
Insurance Required R.C.V. ((10) %) A.C.V.((11))%	

Description of Item	Unit Cost or Material Price Only (12)			Labor Price Only (16)		
	Unit	Unit Price	Total (Col. A)	Hours	Rate	Total Col. B)
(20)	(13)	(14)	(15)	(17)	(18)	(19)
This is not an order to Repair Totals			(21)			(22)

The undersigned agrees to complete and guarantee repairs at a total of $ (29)	Total Column A	
Repairer (30)		
Street (31)	(23)	(24)
City Phone	(25)	
By (32)		(26)
Adjuster (34) Date of A/P (35)	(27)	
Adj. License No. (If Any)	(28) Grand Total	
Service Office Name (36)		

Note: This form does not replace the need for field notes, sketches and measurements.

Building Repair Estimate

Date

Insured John Q. Smith	Claim or Policy No. DP 0029 | Page 1 of 2
Loss Address 123 A. Main St.	Home Ph. 555-1241 | Cause of Loss Fire
City Anywhere, Anystate 00010	Bus. Ph. 555-1438 | Other Ins. Y (N)
Bldg. R.C.V. 10,000 Bldg. A.C.V. 90,000	Insurance Amount $100,000
Insurance Required R.C.V. (80%) A.C.V.(80%%)	

Description of Item	Unit Cost or Material Price Only			Labor Price Only		
	Unit	Unit Price	Total (Col. A)	Hours	Rate	Total Col. B)
Install 1/2" sheetrock (standard,						
on walls, including tape						
and finish 400 *(page 92)*	400	0.23	92.00	9.6	40.98	393.41
Paint walls, roller, smooth finish						
1 coat sealer 600 *(page 187)*	600	0.06	36.00	4.2	42.15	177.03
2 coats latex flat 600 *(page 189)*	600	0.14	84.00	7.2	42.15	303.48
This is not an order to Repair Totals			212.00			873.92

The undersigned agrees to complete and guarantee repairs at a total of $	Total Column A	212.00
Repairer ABC Construction		1085.92
Street 316 E. 2nd Street	6% Tax	65.16
City Anywhere Phone		1151.08
By Jack Williams	10% Overhead	115.11
Adjuster Stan Jones Date of A/P N/A	10% Profit	115.11
Adj. License No. (If Any) 561-84	Grand Total	1381.30
Service Office Name Phoenix		

Note: This form does not replace the need for field notes, sketches and measurements.

Keyed Explanations of the Building Repair Estimate Form

1. Insert name of insured(s).
2. Insert claim number or, if claim number is not available, insert policy number or binder number.
3. Insert the page number and the total number of pages.
4. Insert street address, city and state where loss or damage occurred.
5. Insert type of loss (wind, hail, fire, water, etc.)
6. Check YES if there is other insurance, whether collectible or not. Check NO if there's only one insurer.
7. Insert the present replacement cost of the building. What would it cost to build the structure today?
8. Insert present actual cash value of the building.
9. Insert the amount of insurance applicable. If there is more than one insurer, insert the total amount of applicable insurance provided by all insurers.
10. If the amount of insurance required is based on replacement cost value, circle RCV and insert the percent required by the policy, if any.
11. If the amount of insurance required is based on actual cash value, circle ACV and insert the percent required by the policy, if any.

    Note: (regarding 10 and 11) if there is a non-concurrency, i.e., one insurer requires insurance to 90% of value while another requires insurance to 80% of value, make a note here. Comment on the non-concurrency in the settlement report.
12. The installed price and/or material price only, as expressed in columns 13 through 15, may include any of the following (expressed in units and unit prices):

    Material only (no labor)

    Material and labor to replace

    Material and labor to remove and replace

    Unit Cost is determined by dividing dollar cost by quantity. The term cost, as used in unit cost, is not intended to include any allowance, percentage or otherwise, for overhead or profit. Usually, overhead and profit are expressed as a percentage of cost. Cost must be determined first. Insert a line or dash in a space(s) in columns 13, 14, 15, 17, 18 or 19 if the space is not to be used.
13. The *units* column includes both the quantity and the unit of measure, i.e., 100 SF, 100 BF, 200 CF, 100 CY, 20 ea., etc.
14. The *unit price* may be expressed in dollars, cents or both. If the units column has 100 SF and if the unit price column has $.10, this would indicate the price to be $.10 per SF.
15. The *total* column is merely the dollar product of the quantity (in column 13) times the price per unit measure (in column 14).

16-19. These columns are normally used to express labor as follows: hours times rate per hour. However, it is possible to express labor as a unit price, i.e., 100 SF in column 13, a dash in column 17, $.05 in column 18 and $5.00 in column 19.

20. Under *description of item*, the following may be included:

    Description of item to be repaired or replaced (studs 2" x 4" 8'0" #2 Fir, Sheetrock 1/2", etc.)

    Quantities or dimensions (20 pcs., 8'0" x 14'0", etc.)

    Location within a room or area (north wall, ceiling, etc.)

    Method of correcting damage (paint - 1 coat; sand, fill and finish; R&R; remove only; replace; resize; etc.)

21-22. Dollar totals of columns A and B respectively.

23-27. Spaces provided for items not included in the body of the estimate (subtotals, overhead, profit, sales tax, etc.)

28. Total cost of repair.
29. Insert the agreed amount here. The agreement may be between the claim representative and the insured or between the claim rep and the repairer. If the agreed price is different from the grand total, the reason(s) for the difference should be itemized on the estimate sheet. If there is no room, attach an additional estimate sheet.
30. PRINT the name of the insured or the repairer so that it is legible.
31. PRINT the address of the insured or repairer legibly. Include phone number.
32. The insured or a representative of the repairer should sign here indicating agreement with the claim rep's estimate.
33. Insured or representative of the repairer should insert date here.
34. Claim rep should sign here.
35. Claim rep should insert date here.
36. Insert name of service office here.

Building Repair Estimate

	Date			
Insured	Claim or Policy No.		Page of	
Loss Address	Home Ph.		Cause of Loss	
City	Bus. Ph.		Other Ins. Y N	
Building. R.C.V. Bldg. A.C.V.	Insurance Amount			
Insurance Required R.C.V.(%) A.C.V.(%)	Unit Cost or Material Price Only		Labor Price Only	

Description of Item	Unit	Unit Price	Total (Col. A)	Hours	Rate	Total Col. B)
THIS IS NOT AN ORDER TO REPAIR TOTALS						

The undersigned agrees to complete and guarantee repairs at a total of $	**Total Column A**	
Repairer		
Street		
City Phone		
By		
Adjuster Date of A/P		
Adj. License No. (If Any)	**Grand Total**	
Service Office Name		

Note: This form does not replace the need for field notes, sketches and measurements.

Wage Rates Used in This Book

Wage rates listed here and used in this book were compiled in the fall of 2000 and projected to mid-2001. Wage rates are in dollars per hour.

"Base Wage Per Hour" (Col. 1) includes items such as vacation pay and sick leave which are normally taxed as wages. Nationally, these benefits average 5.15% of the Base Wage Per Hour. This amount is paid by the Employer in addition to the Base Wage Per Hour.

"Liability Insurance and Employer Taxes" (Cols. 3 & 4) include national averages for state unemployment insurance (4.00%), federal unemployment insurance (0.80%), Social Security and Medicare tax (7.65%), liability insurance (2.29%), and Workers' Compensation Insurance which varies by trade. This total percentage (Col. 3) is applied to the sum of Base Wage Per Hour and Taxable Fringe Benefits (Col. 1 + Col. 2) and is listed in Dollars (Col. 4). This amount is paid by the Employer in addition to the Base Wage Per Hour and the Taxable Fringe Benefits.

"Non-Taxable Fringe Benefits" (Col. 5) include employer-sponsored medical insurance and other benefits, which nationally average 4.55% of the Base Wage Per Hour.

"Total Hourly Cost Used In This Book" is the sum of Columns 1, 2, 4, & 5.

	1	2	3	4	5	6
Trade	Base Wage Per Hour	Taxable Fringe Benefits (5.15% of Base Wage)	Liability Insurance & Employer Taxes %	$	Non-Taxable Fringe Benefits (4.55% of Base Wage)	Total Hourly Cost Used This Book
Air Tool Operator	$17.48	$0.90	33.05%	$6.07	$0.80	$25.25
Bricklayer or Stone Mason	$19.54	$1.01	31.72%	$6.52	$0.89	$27.96
Bricktender	$15.04	$0.77	31.72%	$5.01	$0.68	$21.50
Carpenter	$18.52	$0.95	34.14%	$6.65	$0.84	$26.96
Cement Mason	$18.75	$0.97	28.12%	$5.55	$0.85	$26.12
Electrician, Journeyman Wireman	$21.73	$1.12	21.88%	$5.00	$0.99	$28.84
Equipment Operator	$21.77	$1.12	31.40%	$7.19	$0.99	$31.07
Fence Erector	$19.70	$1.01	34.14%	$7.07	$0.90	$28.68
Floorlayer: Carpet, Linoleum, Soft Tile	$18.12	$0.93	32.55%	$6.20	$0.82	$26.07
Floorlayer: Hardwood	$19.03	$0.98	31.45%	$6.29	$0.87	$27.17
Glazier	$17.91	$0.92	29.03%	$5.47	$0.81	$25.11
Laborer, General Construction	$15.20	$0.78	33.05%	$5.28	$0.69	$21.95
Lather	$19.52	$1.01	26.12%	$5.36	$0.89	$26.78
Marble and Terrazzo Setter	$18.49	$0.95	24.56%	$4.77	$0.84	$25.05
Painter, Brush	$19.77	$1.02	30.84%	$6.41	$0.90	$28.10
Painter, Spray-Gun	$20.36	$1.05	30.84%	$6.60	$0.93	$28.94
Paperhanger	$20.76	$1.07	30.84%	$6.73	$0.94	$29.50
Plasterer	$19.40	$1.00	28.94%	$5.90	$0.88	$27.18
Plumber	$22.22	$1.14	26.08%	$6.09	$1.01	$30.46
Reinforcing Ironworker	$18.38	$0.95	27.15%	$5.25	$0.84	$25.42
Roofer, Foreman	$20.82	$1.07	48.29%	$10.57	$0.95	$33.41
Roofer, Journeyman	$18.93	$0.97	48.29%	$9.61	$0.86	$30.37
Roofer, Hot Mop Pitch	$19.50	$1.00	48.29%	$9.90	$0.89	$31.29
Roofer, Wood Shingles	$19.88	$1.02	48.29%	$10.09	$0.90	$31.89
Sheet Metal Worker	$21.41	$1.10	27.88%	$6.28	$0.97	$29.76
Tile Setter	$19.24	$0.99	24.56%	$4.97	$0.88	$26.08
Tile Setter Helper	$14.74	$0.76	24.56%	$3.81	$0.67	$19.98
Truck Driver	$15.62	$0.80	33.15%	$5.44	$0.71	$22.57

Area Modification Factors

Construction costs are higher in some cities than in other cities. Add or deduct the percentage shown on pages 12 to 14 to adapt the costs in this book to your job site. Adjust your estimated total project cost by the percentage shown for the appropriate city in this table to find your total estimated cost. Where 0% is shown it means no modification is required. Factors for Canada adjust to Canadian dollars.

These percentages were compiled by comparing the construction cost of buildings in nearly 600 communities throughout North America. Because these percentages are based on completed projects, they consider all construction cost variables, including labor, equipment and material cost, labor productivity, climate, job conditions and markup.

Modification factors are listed alphabetically by state. Areas within each state are listed by the first three digits of the postal zip code. For convenience, one representative city is identified in each zip code or range of zip codes.

These percentages are composites of many costs and will not necessarily be accurate when estimating the cost of any particular part of a building. But when used to modify costs for an entire structure, they should improve the accuracy of your estimates.

Zip	City	Factor
Alabama		**-14%**
350-352	Birmingham	-13%
354	Tuscaloosa	-13%
355	Jasper	-13%
356	Sheffield	-16%
357	Scottsboro	-16%
358	Huntsville	-13%
359	Gadsden	-13%
360-361	Montgomery	-13%
362	Anniston	-13%
363	Dothan	-16%
364	Evergreen	-16%
365-366	Mobile	-11%
367	Selma	-13%
368	Auburn	-13%
369	Bellamy	-13%
Alaska		**43%**
995	Anchorage	30%
996	King Salmon	51%
997	Fairbanks	45%
998	Juneau	44%
999	Ketchikan	46%
Arizona		**1%**
850	Phoenix	-1%
852-853	Mesa	-1%
855	Douglas	-1%
856-857	Tucson	-2%
859	Show Low	4%
860	Flagstaff	-1%
863	Prescott	3%
864	Kingman	3%
865	Chambers	3%
Arkansas		**-12%**
716	Pine Bluff	-11%
717	Camden	-12%
718	Hope	-12%
719	Hot Springs	-12%
720-722	Little Rock	-12%
723	W. Memphis	-12%
724	Jonesboro	-12%
725	Batesville	-12%
726	Harrison	-12%
727	Fayetteville	-11%
728	Russellville	-9%
729	Fort Smith	-9%
California		**14%**
900-901	Los Angeles	14%
902-905	Inglewood	13%
906	Whittier	13%
907-908	Long Beach	14%
910-912	Pasadena	13%
913-916	Van Nuys	13%
917-918	Alhambra	13%
919-921	San Diego	10%
922	El Centro	10%
923-925	San Bernardino	13%
926-928	Anaheim	13%
930	Oxnard	11%
931	Santa Barbara	11%
932-933	Bakersfield	13%
934	Lompoc	11%
935	Mojave	10%
936-938	Fresno	13%
939	Salinas	14%
940	Sunnyvale	17%
941	San Francisco	24%
942	Sacramento	10%
943-944	San Mateo	16%
945-947	Oakland	16%
948	Richmond	22%
949	Novato	22%
950-951	San Jose	17%
952-953	Stockton	11%
954	Santa Rosa	22%
955	Eureka	16%
956-958	Rancho Cordova	11%
959	Marysville	11%
960	Redding	12%
961	Herlong	18%
Colorado		**-1%**
800-801	Aurora	-3%
802-804	Denver	0%
805	Longmont	-3%
806	Greeley	-3%
807	Fort Morgan	-3%
808-809	Colorado Springs	1%
810	Pueblo	-4%
811	Pagosa Springs	-3%
812	Salida	-3%
813	Durango	-3%
814-815	Grand Junction	-3%
816	Glenwood Spr.	-3%
	Ski resorts	13%
Connecticut		**9%**
060-061	Hartford	9%
062	West Hartford	9%
063	Norwich	8%
064	Fairfield	10%
065	New Haven	10%
066	Bridgeport	8%
067	Waterbury	11%
068-069	Stamford	11%
Delaware		**3%**
197	Newark	3%
198	Wilmington	4%
199	Dover	3%
District of Columbia		**2%**
200-201	Washington	2%
202-205	Washington	2%
Florida		**-8%**
320	St. Augustine	-10%
321	Daytona Bch.	-10%
322	Jacksonville	-3%
323	Tallahassee	-14%
324	Panama City	-13%
325	Pensacola	-13%
326	Gainesville	-10%
327	Altamonte Spr.	-1%
328	Orlando	-1%
329	Melbourne	-1%
330-332	Miami	-6%
333	Ft. Lauderdale	-1%
334	W. Palm Beach	-1%
335-336	Tampa	-11%
337	St. Petersburg	-11%
338	Lakeland	-10%
339	Fort Myers	-10%
342	Bradenton	-10%
344	Ocala	-10%
346	Brooksville	-10%
347	Saint Cloud	-10%
349	Fort Pierce	-10%
Georgia		**-11%**
300-302	Marietta	-1%
303	Atlanta	-1%
304	Statesboro	-11%
305	Buford	-11%
306	Athens	-11%
307	Calhoun	-11%
308-309	Augusta	-17%
310	Dublin	-11%
311	Fort Valley	-11%
312	Macon	-11%
313	Hinesville	-13%
314	Savannah	-13%
315	Kings Bay	-12%
316	Valdosta	-11%
317	Albany	-13%
318-319	Columbus	-16%
Hawaii		**34%**
967	Ewa	34%
967	Halawa Heights	34%
967	Hilo	34%
967	Lualualei	34%
967	Mililani Town	34%
967	Pearl City	34%
967	Wahiawa	34%
967	Waianae	34%
967	Wailuku (Maui)	34%
968	Alimanu	34%
968	Honolulu	34%
968	Kailua	35%
Idaho		**-6%**
832	Pocatello	-6%
833	Sun Valley	-5%
834	Idaho Falls	-6%
835	Lewiston	-8%
836	Meridian	-8%
837	Boise	-6%
838	Coeur d'Alene	-2%
Illinois		**6%**
600	Arlington Hts.	6%
601	Carol Stream	10%
602	Quincy	6%
603	Oak Park	6%
604	Joliet	6%
605	Aurora	6%
606-608	Chicago	10%
609	Kankakee	5%
610-611	Rockford	2%
612	Green River	5%
613	Peru	6%
614	Galesburg	5%
615-616	Peoria	6%
617	Bloomington	5%
618-619	Urbana	6%
620	Granite City	5%
622	Belleville	3%
623	Decatur	5%
624	Lawrenceville	6%
625-627	Springfield	6%
628	Centralia	5%
629	Carbondale	5%
Indiana		**3%**
460-462	Indianapolis	3%
463-464	Gary	4%
465-466	South Bend	4%
467-468	Fort Wayne	4%
469	Kokomo	3%
470	Aurora	3%
471	Jeffersonville	3%
472	Columbus	4%
473	Muncie	4%

474 Bloomington 4%
475 Jasper 1%
476-477 Evansville 1%
478 Terre Haute 4%
479 Lafayette 4%

Iowa . . . .-4%

500-503 Des Moines 1%
504 Mason City . . . .-4%
505 Fort Dodge . . . .-4%
506-507 Waterloo . . . . . .-4%
508 Creston . . . .-4%
510 Cherokee . . . . .-4%
511 Sioux City . . . . .-4%
512 Sheldon . . . . . .-4%
513 Spencer . . . . . .-4%
514 Carroll . . . .-5%
515 Council Bluffs . .-4%
516 Shenandoah . . .-5%
520 Dubuque . . . . . .-4%
522-524 Cedar Rapids . .-4%
521 Decorah . . . . . .-4%
525 Ottumwa . . . . . .-4%
526 Burlington . . . . .-5%
527-528 Davenport . . . . .-2%

Kansas -5%

660-662 Kansas City . . .-5%
664-666 Topeka . . . .-5%
667 Fort Scott . . . . .-5%
668 Emporia . . . . . .-6%
669 Concordia . . . . .-5%
670-672 Wichita . . . .-4%
673 Independence . .-5%
674 Salina . . . .-5%
675 Hutchinson . . . .-6%
676 Hays . . . .-5%
677 Colby . . . .-5%
678 Dodge City . . . .-5%
679 Liberal . . . .-5%

Kentucky -4%

400-402 Louisville . . . . . .-5%
403-406 Lexington . . . . .-5%
407-409 London . . . .-7%
410 Covington . . . . .-5%
411-412 Ashland . . . . . .-5%
413-414 Campton . . . . . .-5%
415-416 Pikeville . . . . . .-5%
417-418 Hazard . . . .-5%
420 Paducah . . . . . .-5%
421 Bowling Green .-5%
422 Hopkinsville . . . .-4%
423 Owensboro . . . .-5%
424 White Plains . . .-3%
425-426 Somerset . . . . .-3%
427 Elizabethtown . .-3%

Louisiana -4%

700-701 New Orleans . . .1%
703 Houma 1%
704 Mandeville . . . . .-6%
705 Lafayette . . . . . .-5%
706 Lake Charles . . .-5%
707-708 Baton Rouge . . .-5%
710 Minden . . . .-8%
711 Shreveport . . . .-6%
712 Monroe . . . .-6%
713-714 Alexandria . . . . .-6%

Maine -4%

039-040 Brunswick . . . . .-2%
041 Portland . . . . . .-2%
042 Auburn . . . .-4%
043 Augusta . . . . . .-4%
044 Bangor . . . .-6%
045 Bath . . . .-4%
046 Cutler . . . .-4%
047 Northern Area . .-4%
048 Camden . . . . . .-5%
049 Dexter . . . .-4%

Maryland 0%

206-207 Laurel 2%
208-209 Bethesda 0%
210-212 Baltimore 0%
214 Annapolis 1%
215 Cumberland 0%
216 Church Hill 0%
217 Frederick 0%
218 Salisbury 0%
219 Elkton 0%

Massachusetts 15%

015-016 Ayer 15%
010 Chicopee 15%
011 Springfield 15%
012 Pittsfield 15%
013 Northfield 15%
014 Fitchburg 14%
017 Bedford 15%
018 Lawrence 15%
019 Dedham 15%
020 Hingham 16%
021-022 Boston 16%
023-024 Brockton 15%
025 Nantucket 15%
026 Centerville 15%
027 New Bedford . .15%

Michigan 6%

480 Royal Oak 7%
481-482 Detroit 6%
483 Pontiac 6%
484-485 Flint 6%
486-487 Saginaw 1%
488-489 Lansing 6%
490-491 Battle Creek 7%
492 Jackson 6%
493-495 Grand Rapids . . .7%
496 Traverse City . . .6%
497 Grayling 6%
498-499 Marquette 1%

Minnesota 5%

550-551 St Paul 9%
553-555 Minneapolis 9%
556-558 Duluth 3%
559 Rochester 5%
560 Mankato 5%
561 Magnolia 5%
562 Willmar 5%
563 St Cloud 5%
564 Brainerd 5%
565 Fergus Falls 5%
566 Bemidji 5%
567 Thief River Falls .5%

Mississippi -13%

386 Clarksdale . . . .-14%
387 Greenville . . . . .-5%
388 Tupelo . . . .-14%
389 Greenwood . . .-14%
390-392 Jackson . . . . . .-14%
393 Meridian . . . . .-14%
394 Laurel . . . .-14%
395 Gulfport . . . . . .-12%
396 McComb . . . . .-12%
397 Columbus . . . .-16%

Missouri 1%

630-631 St Louis 5%
633 Saint Charles . . .5%
634 Hannibal 1%
635 Kirksville 1%
636 Farmington 1%
637 Cape Girardeau .1%
638 Caruthersville . . .1%
639 Poplar Bluff 0%
640-641 Independence . .1%
644-645 Saint Joseph . . .1%
646 Chillicothe 1%
647 East Lynne 1%
648 Joplin 1%
650-651 Jefferson City . .-1%
652 Columbia 0%
653 Knob Noster . . .-1%
654-655 Lebanon 2%
656-658 Springfield 1%

Montana 1%

590-591 Billings 1%
592 Fairview 1%
593 Miles City 1%
594 Great Falls 2%
595 Havre 1%
596 Helena 1%
597 Butte 1%
598 Missoula 1%
599 Kalispell 1%

Nebraska -11%

680-681 Omaha . . . .-8%
683-685 Lincoln . . . . . .-11%
686 Columbus . . . .-11%
687 Norfolk . . . . . .-11%
688 Grand Island . .-11%
689 Hastings . . . . .-11%
690 McCook . . . . .-11%
691 North Platte . . .-11%
692 Valentine . . . . .-11%
693 Alliance . . . . . .-11%

Nevada 5%

889-891 Las Vegas 4%
893 Ely 4%
894 Fallon 7%
895 Reno 4%
897 Carson City 6%
898 Elko 6%

New Hampshire 2%

030-031 New Boston 3%
032-033 Manchester 2%
034 Concord 3%
035 Littleton 2%
036 Charlestown 2%
037 Lebanon 2%
038 Dover 2%

New Jersey 9%

070-073 Newark 9%
074-075 Paterson 9%
076 Hackensack 9%
077 Monmouth 8%
078 Dover 9%
079 Summit 9%
080-084 Atlantic City 9%
085 Princeton 9%
086 Trenton 8%
087 Brick 9%
088-089 Edison 9%

New Mexico -5%

870-871 Albuquerque . . .-4%
873 Gallup . . . .-4%
874 Farmington . . . .-4%
875 Santa Fe . . . . . .-4%
877 Holman . . . .-4%
878 Socorro . . . .-5%
879 Truth or Cons. . .-5%
880 Las Cruces . . . .-5%
881 Clovis . . . .-6%
882 Fort Sumner . . .-4%
883 Alamogordo . . .-5%
884 Tucumcari . . . . .-4%

New York 9%

100 NY (Man.) 26%
100 New York City . .13%
103 Staten Island . . .9%
104 Bronx 9%
105-108 White Plains 9%
109 West Point 9%
110 Queens 16%
111 Long Island . . .16%
112 Brooklyn 9%
113 Flushing 9%
114 Jamaica 9%
115 Garden City 9%
116 Rockaway 9%
117 Amityville 9%
118 Hicksville 9%
119 Montauk 9%
120-123 Albany 6%
124 Kingston 7%
125-126 Poughkeepsie . . .7%
127 Stewart 7%
128 Newcomb 6%
129 Plattsburgh 7%
130-132 Syracuse 8%
133-135 Utica 8%
136 Watertown 6%
137-139 Binghampton . . .8%
140 Batavia 8%
141 Tonawanda 8%
142 Buffalo 8%
143 Niagara Falls . . .8%
144-146 Rochester 7%
147 Jamestown 5%
148-149 Ithaca 5%

North Carolina -16%

270-274 Winston-Salem -17%
275-277 Raleigh . . . . . .-17%
278 Rocky Mount . .-17%
279 Elizabeth City .-17%
280-282 Charlotte . . . . .-17%
283 Fayetteville . . .-11%
284 Wilmington . . .-11%
285 Kinston . . . . . .-17%
286 Hickory . . . . . .-17%
287-289 Asheville . . . . .-17%

North Dakota -4%

580-581 Fargo . . . .-4%
582 Grand Forks . . .-4%
583 Nekoma . . . . . .-4%
584 Jamestown . . . .-4%
585 Bismarck . . . . . .-4%
586 Dickinson . . . . .-4%
587 Minot . . . .-3%
588 Williston . . . . . .-4%

Ohio 1%

430-431 Newark 1%
432 Columbus 1%
433 Marion 0%
434-436 Toledo 2%

437-438	Zanesville	0%
439	Stubenville	1%
440-441	Cleveland	3%
442-443	Akron	1%
444-445	Youngstown	1%
446-447	Canton	1%
448-449	Sandusky	0%
450-452	Cincinnati	0%
453-455	Dayton	0%
456	Chillicothe	0%
457	Marietta	1%
458	Lima	2%

Oklahoma		**-8%**
730-731	Oklahoma City	-7%
734	Ardmore	-8%
735	Lawton	-8%
736	Clinton	-8%
737	Enid	-8%
738	Woodward	-8%
739	Adams	-8%
740-741	Tulsa	-8%
743	Pryor	-8%
744	Muskogee	-6%
745	McAlester	-12%
746	Ponca City	-8%
747	Durant	-8%
748	Shawnee	-8%
749	Poteau	-8%

Oregon		**6%**
970-972	Portland	6%
973	Salem	7%
974	Eugene	7%
975	Grants Pass	7%
976	Klamath Falls	7%
977	Bend	7%
978	Pendleton	6%
979	Adrian	6%

Pennsylvania		**5%**
150-151	Warrendale	2%
152	Pittsburgh	2%
153	Washington	2%
154	Uniontown	2%
155	Somerset	2%
156	Greensburg	2%
157	Punxsutawney	5%
158	DuBois	5%
159	Johnstown	5%
160	Butler	5%
161	New Castle	5%
162	Kittanning	5%
163	Meadville	5%
164-165	Erie	5%
166	Altoona	5%
167	Bradford	5%
168	Clearfield	5%
169	Genesee	5%
170-171	Harrisburg	5%
172	Chambersburg	2%
173-174	York	5%
175-176	Lancaster	5%
177	Williamsport	5%
178	Beaver Springs	5%
179	Pottsville	5%
180-181	Allentown	5%
182	Hazleton	5%
183	E. Stroudsburg	5%
184-185	Scranton	4%
186-187	Wilkes Barre	5%
188	Montrose	5%
189	Warminster	8%
190-191	Philadelphia	9%
195-196	Reading	5%
193	Southeastern	5%
194	Valley Forge	5%

Rhode Island		**10%**
028	Bristol	9%
028	Coventry	9%
028	Davisville	9%
028	Narragansett	9%
028	Newport	10%
028	Warwick	10%
029	Cranston	10%
029	Providence	10%

South Carolina		**-13%**
290-292	Columbia	-15%
293	Spartanburg	-15%
294	Charleston	-13%
295	Myrtle Beach	-13%
296	Greenville	-15%
297	Rock Hill	-15%
298	Aiken	-9%
299	Beaufort	-9%

South Dakota		**-8%**
570-571	Sioux Falls	-10%
572	Watertown	-8%
573	Mitchell	-8%
574	Aberdeen	-7%
575	Pierre	-7%
576	Mobridge	-8%
577	Rapid City	-8%

Tennessee		**-8%**
370-372	Nashville	-9%
373	Cleveland	-9%
374	Chattanooga	-8%
376	Kingsport	-8%
377-379	Knoxville	-8%
380-381	Memphis	-7%
382	McKenzie	-8%
383	Jackson	-8%
384	Columbia	-8%
385	Cookeville	-8%

Texas		**-9%**
750	Plano	-5%
751-753	Dallas	-5%
754	Greenville	-11%
755	Texarkana	-8%
756	Longview	-11%
757	Tyler	-11%
758	Palestine	-11%
759	Lufkin	-11%
760	Arlington	-11%
761-762	Ft Worth	-5%
763	Wichita Falls	-11%
764	Woodson	-11%
765-767	Waco	-11%
768	Brownwood	-12%
769	San Angelo	-12%
770-772	Houston	-10%
773	Huntsville	-9%
774	Bay City	-10%
775	Galveston	-11%
776-777	Beaumont	-11%
778	Bryan	-11%
779	Victoria	-10%
780-782	San Antonio	-8%
783-784	Corpus Chrsti	-11%
785	McAllen	-11%
786-787	Austin	-11%
788	Del Rio	-2%
789	Giddings	-2%
790-791	Amarillo	-11%
792	Childress	-9%
793-794	Lubbock	-8%
795-796	Abilene	-11%
797	Midland	-10%
798-799	El Paso	-11%

Utah		**-7%**
840	Clearfield	-7%
841	Salt Lake City	-8%
843-844	Ogden	-6%
845	Green River	-7%
846-847	Provo	-8%

Vermont		**-4%**
050	White River Jct.	-4%
051	Spingfield	-4%
052	Bennington	-4%
053	Battleboro	-4%
054	Burlington	-3%
056	Montpelier	-5%
057	Rutland	-4%
058	Albany	-4%
059	Beecher Falls	-5%

Virginia		**-10%**
220-223	Alexandria	-11%
224-225	Fredericksburg	-15%
226	Winchester	-10%
227	Culpeper	-10%
228	Harrisonburg	-10%
229	Charlottesville	-10%
230-231	Williamsburg	-7%
232	Richmond	-10%
233-237	Norfolk	-7%
238	Petersburg	-9%
239	Farmville	-8%
240-241	Radford	-10%
242	Abingdon	-10%
243	Galax	-10%
244	Staunton	-10%
245	Lynchburg	-10%
246	Tazewell	-10%

Washington		**8%**
980-981	Seattle	8%
982	Everett	8%
983-984	Tacoma	8%
985	Olympia	8%
986	Vancouver	8%
988	Wenatchee	7%
989	Yakima	6%
990-992	Spokane	10%
993	Pasco	8%
994	Clarkston	8%

West Virginia		**-1%**
247-248	Bluefield	-3%
249	Lewisburg	-1%
250-253	Charleston	-1%
254	Martinsburg	-1%
255-257	Huntington	-1%
258-259	Beckley	-1%
260	Wheeling	-1%
261	Parkersburg	-1%
262	New Martinsville	-1
263-264	Clarksburg	-1
265	Morgantown	-1
266	Fairmont	-1
267	Romney	-1
268	Sugar Grove	-3

Wisconsin		**-2**
530-534	Milwaukee	4
535	Beloit	-2
537	Madison	-6
538	Prairie du Chien	-6
539	Portage	-7
540	Amery	-2
541-543	Green Bay	-1
544	Wausau	-2
545	Clam Lake	-2
546	La Crosse	-2
547	Eau Claire	-2
548	Ladysmith	-2
549	Oshkosh	-2

Wyoming		**-11**
820	Cheyenne	-10
821	Laramie	-11
822	Wheatland	-11
823	Rawlins	-10
824	Powell	-11
825	Riverton	-11
826	Casper	-11
827	Gillette	-11
828	Sheridan	-11
829-831	Rock Springs	-10

Canada

Alberta	**53.0**
Calgary	49.3
Edmonton	54.1
Ft. McMurray	55.7

British Columbia	**51.6**
Fraser Valley	50.9
Okanagan HRCC	45.6
Vancouver	62.5

Manitoba	**57.3**
North Manitoba	67.7
South Manitoba	57.0
Selkirk	49.7
Winnipeg	54.6

New Brunswick	**49.1**
Moncton	49.1

Nova Scotia	**57.8**
Amherst	56.9
Nova Scotia	52.7
Sydney	63.9

Newfoundland	**49.8**
Newfoundland & Laborador	49.8

Ontario	**51.1**
London	59.1
Thunder Bay	48.3
Toronto	59.1

Quebec	**59.5**
Montreal	63.4
Quebec	55.4

Saskatchewan	**58.6**
La Ronge	62.3
Prince Albert	57.7
Saskatoon	55.7

Crew Compositions & Wage Rates

Crew Code	Manhours per day	Total costs $/Hr.	$/Day	Average Crew Rate (ACR) Per Hour	Average Crew Rate (ACROP) per Hour w/O&P
AB					
1 Air tool operator	8.00	$25.25	$202.00		
1 Laborer	8.00	$21.95	$175.60		
TOTAL	**16.00**		**$377.60**	**$23.60**	**$35.40**
AD					
2 Air tool operators	16.00	$25.25	$404.00		
1 Laborer	8.00	$21.95	$175.60		
TOTAL	**24.00**		**$579.60**	**$24.15**	**$36.23**
BD					
3 Bricklayers	24.00	$27.96	$671.04		
2 Bricktenders	16.00	$21.50	$344.00		
TOTAL	**40.00**		**$1,015.04**	**$25.38**	**$38.32**
BK					
1 Bricklayer	8.00	$27.96	$223.68		
1 Bricktender	8.00	$21.50	$172.00		
TOTAL	**16.00**		**$395.68**	**$24.73**	**$37.34**
BO					
2 Bricklayers	16.00	$27.96	$447.36		
2 Bricktenders	16.00	$21.50	$344.00		
TOTAL	**32.00**		**$791.36**	**$24.73**	**$37.34**
2C					
2 Carpenters	16.00	$26.96	$431.36	$26.96	$40.98
CA					
1 Carpenter	8.00	$26.96	$215.68	$26.96	$40.98
CH					
1 Carpenter	8.00	$26.96	$215.68		
1/2 Laborer	4.00	$21.95	$87.80		
TOTAL	**12.00**		**$303.48**	**$25.29**	**$38.44**
CJ					
1 Carpenter	8.00	$26.96	$215.68		
1 Laborer	8.00	$21.95	$175.60		
TOTAL	**16.00**		**$391.28**	**$24.46**	**$37.17**
CN					
2 Carpenters	16.00	$26.96	$431.36		
1/2 Laborer	4.00	$21.95	$87.80		
TOTAL	**20.00**		**$519.16**	**$25.96**	**$39.46**
CS					
2 Carpenters	16.00	$26.96	$431.36		
1 Laborer	8.00	$21.95	$175.60		
TOTAL	**24.00**		**$606.96**	**$25.29**	**$38.44**
CU					
4 Carpenters	32.00	$26.96	$862.72		
1 Laborer	8.00	$21.95	$175.60		
TOTAL	**40.00**		**$1,038.32**	**$25.96**	**$39.46**
CW					
2 Carpenters	16.00	$26.96	$431.36		
2 Laborers	16.00	$21.95	$351.20		
TOTAL	**32.00**		**$782.56**	**$24.46**	**$37.17**

Crew Code	Manhours per day	Total costs $/Hr.	$/Day	Average Crew Rate (ACR) Per Hour	Average Crew Rate (ACROP) per Hour w/O&P
CX					
4 Carpenters	32.00	$26.96	$862.72	$26.96	$40.98
CY					
3 Carpenters	24.00	$26.96	$647.04		
2 Laborers	16.00	$21.95	$351.20		
1 Equipment operator	8.00	$31.07	$248.56		
1 Laborer	8.00	$21.95	$175.60		
TOTAL	**56.00**		**$1,422.40**	**$25.40**	**$38.61**
CZ					
4 Carpenters	32.00	$26.96	$862.72		
3 Laborers	24.00	$21.95	$526.80		
1 Equipment operator	8.00	$31.07	$248.56		
1 Laborer	8.00	$21.95	$175.60		
TOTAL	**72.00**		**$1,813.68**	**$25.19**	**$38.29**
DD					
2 Cement masons	16.00	$26.12	$417.92		
1 Laborer	8.00	$21.95	$175.60		
TOTAL	**24.00**		**$593.52**	**$24.73**	**$37.10**
DF					
3 Cement masons	24.00	$26.12	$626.88		
5 Laborers	40.00	$21.95	$878.00		
TOTAL	**64.00**		**$1,504.88**	**$23.51**	**$35.27**
EA					
1 Electrician	8.00	$28.84	$230.72	$28.84	$42.68
EB					
2 Electricians	16.00	$28.84	$461.44	$28.84	$42.68
ED					
1 Electrician	8.00	$28.84	$230.72		
1 Carpenter	8.00	$26.96	$215.68		
TOTAL	**16.00**		**$446.40**	**$27.90**	**$41.29**
FA					
1 Floorlayer	8.00	$26.07	$208.56	$26.07	$38.84
FB					
2 Floorlayers	16.00	$26.07	$417.12		
1/4 Laborer	2.00	$21.95	$43.90		
TOTAL	**18.00**		**$461.02**	**$25.61**	**$38.16**
FC					
1 Floorlayer (hardwood)	8.00	$27.17	$217.33	$27.17	$40.48
FD					
2 Floorlayers (hardwood)	16.00	$27.17	$434.66		
1/4 Laborer	2.00	$21.95	$43.90		
TOTAL	**18.00**		**$478.56**	**$26.59**	**$39.61**
GA					
1 Glazier	8.00	$25.11	$200.88	$25.11	$37.67
GB					
2 Glaziers	16.00	$25.11	$401.76	$25.11	$37.67
GC					
3 Glaziers	24.00	$25.11	$602.64	$25.11	$37.67
HA					
1 Fence erector	8.00	$28.68	$229.42	$28.68	$43.30
HB					
2 Fence erectors	16.00	$28.68	$458.84		
1 Laborer	8.00	$21.95	$175.60		
TOTAL	**24.00**		**$634.44**	**$26.43**	**$39.92**

Crew Code	Manhours per day	Total costs $/Hr.	$/Day	Average Crew Rate (ACR) Per Hour	Average Crew Rate (ACROP) per Hour w/O&P
1L					
1 Laborer	8.00	$21.95	$175.60	$21.95	$33.14
LB					
2 Laborers	16.00	$21.95	$351.20	$21.95	$33.14
LC					
2 Laborers	16.00	$21.95	$351.20		
1 Carpenter	8.00	$26.96	$215.68		
TOTAL	**24.00**		**$566.88**	**$23.62**	**$35.67**
LD					
2 Laborers	16.00	$21.95	$351.20		
2 Carpenters	16.00	$26.96	$431.36		
TOTAL	**32.00**		**$782.56**	**$24.46**	**$36.93**
LG					
5 Laborers	40.00	$21.95	$878.00		
1 Carpenter	8.00	$26.96	$215.68		
TOTAL	**48.00**		**$1,093.68**	**$22.79**	**$34.41**
LH					
3 Laborers	24.00	$21.95	$526.80	$21.95	$33.14
LJ					
4 Laborers	32.00	$21.95	$702.40	$21.95	$33.14
LK					
2 Laborers	16.00	$21.95	$351.20		
2 Carpenters	16.00	$26.96	$431.36		
1 Equipment operator	8.00	$31.07	$248.56		
1 Laborer	8.00	$21.95	$175.60		
TOTAL	**48.00**		**$1,206.72**	**$25.14**	**$37.96**
LL					
3 Laborers	24.00	$21.95	$526.80		
1 Carpenters	8.00	$26.96	$215.68		
1 Equipment operator	8.00	$31.07	$248.56		
1 Laborer	8.00	$21.95	$175.60		
TOTAL	**48.00**		**$1,166.64**	**$24.31**	**$36.70**
LR					
1 Lather	8.00	$26.78	$214.24	$26.78	$40.44
ML					
2 Bricklayers	16.00	$27.96	$447.36		
1 Bricktender	8.00	$21.50	$172.00		
TOTAL	**24.00**		**$619.36**	**$25.81**	**$38.97**
NA					
1 Painter (brush)	8.00	$28.10	$224.80	$28.10	$42.15
NC					
1 Painter (spray)	8.00	$28.94	$231.54	$28.94	$43.41
P3					
2 Plasterers	16.00	$27.18	$434.88		
1 Laborer	8.00	$21.95	$175.60		
TOTAL	**24.00**		**$610.48**	**$25.44**	**$38.41**
PE					
3 Plasterers	24.00	$27.18	$652.32		
2 Laborers	16.00	$21.95	$351.20		
TOTAL	**40.00**		**$1,003.52**	**$25.09**	**$37.88**

Crew Code	Manhours per day	Total costs $/Hr.	$/Day	Average Crew Rate (ACR) Per Hour	Average Crew Rate (ACROP) per Hour w/O&P
QA					
1 Paperhanger	8.00	$29.50	$235.99	$29.50	$44.25
2R					
2 Roofers (composition)	16.00	$30.37	$485.92	$30.37	$47.68
RG					
2 Roofers (composition)	16.00	$30.37	$485.92		
1 Laborer	8.00	$21.95	$175.60		
TOTAL	**24.00**		**$661.52**	**$27.56**	**$43.27**
RJ					
2 Roofers (wood shingles)	16.00	$31.89	$510.18	$31.89	$50.06
RL					
2 Roofers (composition)	16.00	$30.37	$485.92		
1/2 Laborer	4.00	$21.95	$87.80		
TOTAL	**20.00**		**$573.72**	**$28.69**	**$45.04**
RM					
2 Roofers (wood shingles)	16.00	$31.89	$510.18		
1/2 Laborer	4.00	$21.95	$87.80		
TOTAL	**20.00**		**$597.98**	**$29.90**	**$46.94**
RQ					
2 Roofers (wood shingles)	16.00	$31.89	$510.18		
7/8 Laborer	7.00	$21.95	$153.65		
TOTAL	**23.00**		**$663.83**	**$28.86**	**$45.31**
RS					
1 Roofer (foreman)	8.00	$33.41	$267.30		
3 Roofers (pitch)	24.00	$31.29	$750.91		
2 Laborers	16.00	$21.95	$351.20		
TOTAL	**48.00**		**$1,369.41**	**$28.53**	**$44.79**
RT					
2 Roofers (pitch)	16.00	$31.29	$500.61		
1 Laborer	8.00	$21.95	$175.60		
TOTAL	**24.00**		**$676.21**	**$28.18**	**$44.24**
SA					
1 Plumber	8.00	$30.46	$243.68	$30.46	$45.39
SB					
1 Plumber	8.00	$30.46	$243.68		
1 Laborer	8.00	$21.95	$175.60		
TOTAL	**16.00**		**$419.28**	**$26.21**	**$39.05**
SC					
1 Plumber	8.00	$30.46	$243.68		
1 Electrician	8.00	$28.84	$230.72		
TOTAL	**16.00**		**$474.40**	**$29.65**	**$44.18**
SD					
1 Plumber	8.00	$30.46	$243.68		
1 Laborer	8.00	$21.95	$175.60		
1 Electrician	8.00	$28.84	$230.72		
TOTAL	**24.00**		**$650.00**	**$27.08**	**$40.35**
SE					
2 Plumbers	16.00	$30.46	$487.36		
1 Laborer	8.00	$21.95	$175.60		
1 Electrician	8.00	$28.84	$230.72		
TOTAL	**32.00**		**$893.68**	**$27.93**	**$41.61**

Crew Code	Manhours per day	Total costs $/Hr.	$/Day	Average Crew Rate (ACR) Per Hour	Average Crew Rate (ACROP) per Hour w/O&P
SF					
2 Plumbers	16.00	$30.46	$487.36		
1 Laborer	8.00	$21.95	$175.60		
TOTAL	**24.00**		**$662.96**	**$27.62**	**$41.16**
SG					
3 Plumbers	24.00	$30.46	$731.04		
1 Laborer	8.00	$21.95	$175.60		
TOTAL	**32.00**		**$906.64**	**$28.33**	**$42.22**
TB					
1 Tile setter (ceramic)	8.00	$26.08	$208.64		
1 Tile setter's helper (ceramic)	8.00	$19.98	$159.87		
TOTAL	**16.00**		**$368.51**	**$23.03**	**$34.32**
UA					
1 Sheet metal worker	8.00	$29.76	$238.08	$29.76	$44.64
UB					
2 Sheet metal workers	16.00	$29.76	$476.16	$29.76	$44.64
UC					
2 Sheet metal workers	16.00	$29.76	$476.16		
2 Laborers	16.00	$21.95	$351.20		
TOTAL	**32.00**		**$827.36**	**$25.86**	**$38.78**
UD					
1 Sheet metal worker	8.00	$29.76	$238.08		
1 Laborer	8.00	$21.95	$175.60		
TOTAL	**16.00**		**$413.68**	**$25.86**	**$38.78**
UE					
1 Sheet metal worker	8.00	$29.76	$238.08		
1 Laborer	8.00	$21.95	$175.60		
1/2 Electrician	4.00	$28.84	$115.36		
TOTAL	**20.00**		**$529.04**	**$26.45**	**$39.68**
UF					
2 Sheet metal workers	16.00	$29.76	$476.16		
1 Laborer	8.00	$21.95	$175.60		
TOTAL	**24.00**		**$651.76**	**$27.16**	**$40.74**
VB					
1 Equipment operator	8.00	$31.07	$248.56		
1 Laborer	8.00	$21.95	$175.60		
TOTAL	**16.00**		**$424.16**	**$26.51**	**$39.77**

Abbreviations Used in This Book

ABS	acrylonitrile butadiene styrene
ACR	average crew rate
AGA	American Gas Association
AMP	ampere
Approx.	approximately
ASME	American Society of Mechanical Engineers
auto.	automatic
Avg.	average
Bdle.	bundle
BTU	British thermal unit
BTUH	British thermal unit per hour
C	100
cc	center to center or cubic centimeter
CF	cubic foot
CFM	cubic foot per minute
CLF	100 linear feet
Const.	construction
Corr.	corrugated
CSF	100 square feet
CSY	100 square yards
Ctn	carton
CWT	100 pounds
CY	cubic yard
Cu.	cubic
d	penny
D	deep or depth
Demo	demolish
dia.	diameter
D.S.A.	double strength, A grade
D.S.B.	double strength, B grade
Ea	each
e.g.	for example
etc.	et cetera
exp.	exposure
FAS	First and Select grade
F.H.A.	Federal Housing Administration
fl. oz.	fluid ounce
flt	flight
ft.	foot
ga.	gauge
gal	gallon
galv.	galvanized
GFI	ground fault interrupter
GPH	gallons per hour
GPM	gallons per minute
H	height or high
HP, hp	horsepower
Hr.	hour
HVAC	heating, ventilating, air conditioning
i.d.	inside diameter
i.e.	that is
Inst	install
I.P.S.	iron pipe size
KD	knocked down
KW, kw	kilowatts
L	length or long
lb, lbs.	pound(s)
LF	linear feet
LS	lump sum
M	1000
Mat'l	material
Max.	maximum
MBF	1000 board feet
MBHP	1000 boiler horsepower
Mi	miles
Min.	minimum
MSF	1000 square feet
O.B.	opposed blade
oc	on center
o.d.	outside dimension
oz.	ounce
pcs.	pieces
pkg.	package
PSI	per square inch
PVC	polyvinyl chloride
Qt.	quart
R.E.	rounded edge
R/L	random length
RS	rapid start (lamps)
R/W/L	random width and length
S4S	surfaced-four-sides
SF	square foot
SL	slimline (lamps)
Sq.	1 square or 100 square feet
S.S.B.	single strength, B quality
std.	standard
SY	square yard
T	thick
T&G	tongue and groove
U	thermal conductivity
U.I.	united inches
UL	Underwriters Laboratories
U.S.G.	United States Gypsum
VLF	vertical linear feet
W	width or wide
yr.	year

Symbols

/	per
%	percent
"	inches
'	foot or feet
x	by
o	degree
#	number or pounds
$	dollar
+/-	plus or minus

For crew abbreviations, please see Crew Compositions & Wage Rates chart, pages 15 to 19.

Acoustical and insulating tile

1. **Dimensions**
   a. Acoustical tile. 1/2" thick x 12" x 12", 24".
   b. Insulating tile, decorative. 1/2" thick x 12" x 12", 24"; 1/2" thick x 16" x 16", 32".

2. **Installation.** Tile may be applied to existing plaster (if joist spacing is suitable) or to wood furring strips. Tile may have a square edge or flange. Depending on the type and shape of the tile, you can use adhesive, staples, nails or clips to attach the tile.

3. **Estimating Technique.** Determine area and add 5% to 10% for waste.

4. **Notes on Material Pricing.** A material price of $17.00 a gallon for adhesive was used to compile the Average Material Cost/Unit on the following pages. Here are the coverage rates:

12" x 12"	1.25 Gal/CSF
12" x 24"	0.95 Gal/CSF
16" x 16"	0.75 Gal/CSF
16" x 32"	0.55 Gal/CSF

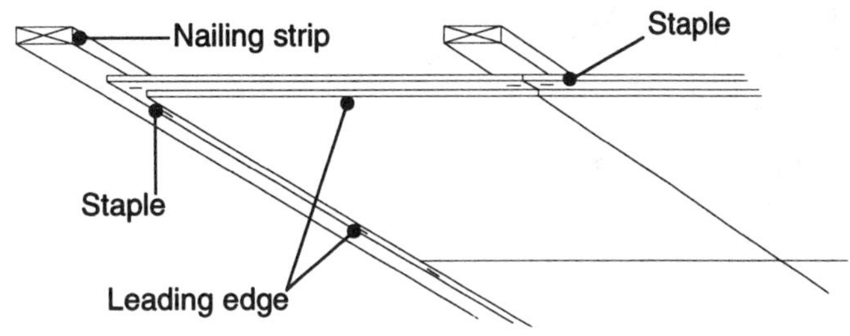

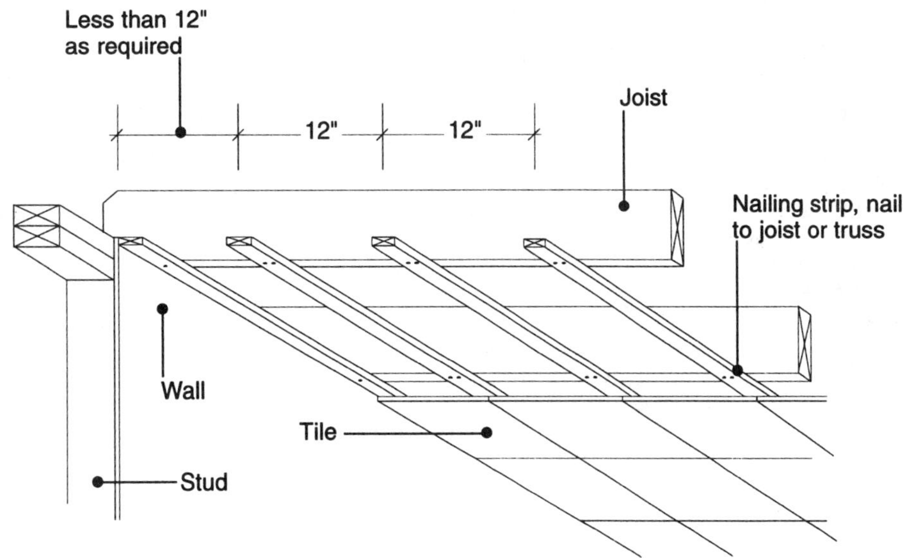

Description	Oper	Unit	Crew Size	Man-Hours Per Unit	Avg Mat'l Unit Cost	Avg Labor Unit Cost	Avg Equip Unit Cost	Avg Total Unit Cost	Avg Price Incl O&P	Large Volume Avg Total Unit Cost	Large Volume Avg Price Incl O&P
Acoustical treatment											
See also Suspended ceiling systems, page 249											
Ceiling and wall tile											
Adhesive set											
Tile only, no grid system	Demo	SF	LB	.018	---	.40	---	.40	**.60**	.26	**.40**
Tile on furring strips	Demo	SF	LB	.013	---	.29	---	.29	**.43**	.20	**.30**
Mineral fiber, vinyl coated, tile only											
Applied in square pattern by adhesive to solid backing; 5% tile waste											
1/2" thick, 12" x 12" or 12" x 24"											
Economy, mini perforated	Inst	SF	2C	.026	.95	.70	---	1.65	**2.11**	1.36	**1.69**
Standard, random perforated	Inst	SF	2C	.026	1.32	.70	---	2.02	**2.52**	1.70	**2.07**
Designer, swirl perforation	Inst	SF	2C	.026	2.12	.70	---	2.82	**3.40**	2.44	**2.88**
Deluxe, sculptured face	Inst	SF	2C	.026	2.42	.70	---	3.12	**3.73**	2.71	**3.18**
5/8" thick, 12" x 12" or 12" x 24"											
Economy, mini perforated	Inst	SF	2C	.026	1.06	.70	---	1.76	**2.23**	1.46	**1.80**
Standard, random perforated	Inst	SF	2C	.026	1.49	.70	---	2.19	**2.70**	1.85	**2.23**
Designer, swirl perforation	Inst	SF	2C	.026	2.41	.70	---	3.11	**3.72**	2.70	**3.17**
Deluxe, sculptured face	Inst	SF	2C	.026	2.75	.70	---	3.45	**4.09**	3.02	**3.52**
3/4" thick, 12" x 12" or 12" x 24"											
Economy, mini perforated	Inst	SF	2C	.026	1.18	.70	---	1.88	**2.36**	1.57	**1.93**
Standard, random perforated	Inst	SF	2C	.026	1.67	.70	---	2.37	**2.90**	2.03	**2.43**
Designer, swirl perforation	Inst	SF	2C	.026	2.73	.70	---	3.43	**4.07**	2.99	**3.49**
Deluxe, sculptured face	Inst	SF	2C	.026	3.12	.70	---	3.82	**4.50**	3.36	**3.89**
Applied by adhesive to furring strips											
ADD	Inst	SF	2C	.002	---	.05	---	.05	**.08**	.05	**.08**
Stapled											
Tile only, no grid system	Demo	SF	LB	.020	---	.44	---	.44	**.66**	.31	**.46**
Tile on furring strips	Demo	SF	LB	.015	---	.33	---	.33	**.50**	.22	**.33**
Mineral fiber, vinyl coated, tile only											
Applied in square pattern by staples, nails, or clips; 5% tile waste											
1/2" thick, 12" x 12" or 12" x 24"											
Economy, mini perforated	Inst	SF	2C	.024	.81	.65	---	1.46	**1.87**	1.21	**1.52**
Standard, random perforated	Inst	SF	2C	.024	1.21	.65	---	1.86	**2.31**	1.57	**1.92**
Designer, swirl perforation	Inst	SF	2C	.024	1.94	.65	---	2.59	**3.12**	2.24	**2.65**
Deluxe, sculptured face	Inst	SF	2C	.024	2.20	.65	---	2.85	**3.40**	2.48	**2.92**
5/8" thick, 12" x 12" or 12" x 24"											
Economy, mini perforated	Inst	SF	2C	.024	.97	.65	---	1.62	**2.05**	1.35	**1.68**
Standard, random perforated	Inst	SF	2C	.024	1.35	.65	---	2.00	**2.47**	1.70	**2.06**
Designer, swirl perforation	Inst	SF	2C	.024	2.20	.65	---	2.85	**3.40**	2.48	**2.92**
Deluxe, sculptured face	Inst	SF	2C	.024	2.50	.65	---	3.15	**3.73**	2.75	**3.22**
3/4" thick, 12" x 12" or 12" x 24"											
Economy, mini perforated	Inst	SF	2C	.024	1.08	.65	---	1.73	**2.17**	1.45	**1.79**
Standard, random perforated	Inst	SF	2C	.024	1.52	.65	---	2.17	**2.66**	1.85	**2.23**
Designer, swirl perforation	Inst	SF	2C	.024	2.48	.65	---	3.13	**3.71**	2.73	**3.19**
Deluxe, sculptured face	Inst	SF	2C	.024	2.84	.65	---	3.49	**4.11**	3.07	**3.57**
Applied by staples, nails or clips to furring strips											
ADD	Inst	SF	2C	.024	---	.65	---	.65	**.98**	.46	**.70**

Description	Oper	Unit	Crew Size	Man-Hours Per Unit	Avg Mat'l Unit Cost	Avg Labor Unit Cost	Avg Equip Unit Cost	Avg Total Unit Cost	Avg Price Incl O&P	Large Volume Avg Total Unit Cost	Large Volume Avg Price Incl O&P
			Costs Based On Small Volume							Large Volume	
Tile patterns, effect on labor											
Herringbone, Reduce output	Inst	%	2C	-25.0	---	---	---	---	**---**	---	**---**
Diagonal, Reduce output	Inst	%	2C	-20.0	---	---	---	---	**---**	---	**---**
Ashlar, Reduce output	Inst	%	2C	-30.0	---	---	---	---	**---**	---	**---**
Furring strips, 8% waste included											
Over wood											
1" x 4", 12" oc	Inst	SF	2C	.014	.25	.38	---	.63	**.85**	.50	**.66**
1" x 4", 16" oc	Inst	SF	2C	.012	.22	.32	---	.54	**.73**	.42	**.55**
Over plaster											
1" x 4", 12" oc	Inst	SF	2C	.018	.25	.49	---	.74	**1.01**	.58	**.79**
1" x 4", 16" oc	Inst	SF	2C	.014	.22	.38	---	.60	**.82**	.47	**.63**

Adhesives

Better quality, gun applied in continuous bead to wood or metal framing or furring members. Per 100 SF of surface area including 6% waste.

Panel adhesives

Description	Oper	Unit	Crew Size	Man-Hours Per Unit	Avg Mat'l Unit Cost	Avg Labor Unit Cost	Avg Equip Unit Cost	Avg Total Unit Cost	Avg Price Incl O&P	Large Volume Avg Total Unit Cost	Large Volume Avg Price Incl O&P
Subfloor adhesive, on floors											
12" oc members											
1/8" diameter (11 fl.oz./CSF)	Inst	CSF	CA	.108	2.65	2.91	---	5.56	**7.34**	4.33	**5.61**
1/4" diameter (43 fl.oz./CSF)	Inst	CSF	CA	.108	10.60	2.91	---	13.51	**16.10**	11.26	**13.20**
3/8" diameter (97 fl.oz./CSF)	Inst	CSF	CA	.108	23.90	2.91	---	26.81	**30.70**	22.82	**26.00**
1/2" diameter (172 fl.oz./CSF)	Inst	CSF	CA	.108	42.30	2.91	---	45.21	**50.90**	38.92	**43.60**
16" oc members											
1/8" diameter (9 fl.oz./CSF)	Inst	CSF	CA	.080	2.12	2.16	---	4.28	**5.61**	3.36	**4.33**
1/4" diameter (34 fl.oz./CSF)	Inst	CSF	CA	.080	8.47	2.16	---	10.63	**12.60**	8.90	**10.40**
3/8" diameter (78 fl.oz./CSF)	Inst	CSF	CA	.080	19.10	2.16	---	21.26	**24.30**	18.21	**20.60**
1/2" diameter (137 fl.oz./CSF)	Inst	CSF	CA	.080	33.80	2.16	---	35.96	**40.50**	31.01	**34.70**
24" oc members											
1/8" diameter (7 fl.oz./CSF)	Inst	CSF	CA	.074	1.60	2.00	---	3.60	**4.79**	2.79	**3.66**
1/4" diameter (26 fl.oz./CSF)	Inst	CSF	CA	.074	6.35	2.00	---	8.35	**10.00**	6.94	**8.22**
3/8" diameter (58 fl.oz./CSF)	Inst	CSF	CA	.074	14.30	2.00	---	16.30	**18.80**	13.90	**15.90**
1/2" diameter (103 fl.oz./CSF)	Inst	CSF	CA	.074	25.40	2.00	---	27.40	**30.90**	23.50	**26.50**
Wall sheathing or shear panel wall adhesive on walls, floors or ceilings											
12" oc members											
1/8" diameter (11 fl.oz./CSF)	Inst	CSF	CA	.143	3.41	3.86	---	7.27	**9.61**	5.68	**7.38**
1/4" diameter (43 fl.oz./CSF)	Inst	CSF	CA	.143	13.60	3.86	---	17.46	**20.80**	14.60	**17.20**
3/8" diameter (97 fl.oz./CSF)	Inst	CSF	CA	.143	30.70	3.86	---	34.56	**39.60**	29.50	**33.60**
1/2" diameter (172 fl.oz./CSF)	Inst	CSF	CA	.143	54.30	3.86	---	58.16	**65.60**	50.10	**56.20**
16" oc members											
1/8" diameter (9 fl.oz./CSF)	Inst	CSF	CA	.130	2.73	3.50	---	6.23	**8.33**	4.83	**6.35**
1/4" diameter (34 fl.oz./CSF)	Inst	CSF	CA	.130	10.90	3.50	---	14.40	**17.30**	11.95	**14.20**
3/8" diameter (78 fl.oz./CSF)	Inst	CSF	CA	.130	24.60	3.50	---	28.10	**32.30**	23.85	**27.30**
1/2" diameter (137 fl.oz./CSF)	Inst	CSF	CA	.130	43.50	3.50	---	47.00	**53.10**	40.35	**45.40**
24" oc members											
1/8" diameter (7 fl.oz./CSF)	Inst	CSF	CA	.120	2.05	3.24	---	5.29	**7.17**	4.05	**5.41**
1/4" diameter (26 fl.oz./CSF)	Inst	CSF	CA	.120	8.16	3.24	---	11.40	**13.90**	9.38	**11.30**
3/8" diameter (58 fl.oz./CSF)	Inst	CSF	CA	.120	18.40	3.24	---	21.64	**25.20**	18.36	**21.10**
1/2" diameter (103 fl.oz./CSF)	Inst	CSF	CA	.120	32.60	3.24	---	35.84	**40.80**	30.66	**34.70**

Description	Oper	Unit	Crew Size	Man-Hours Per Unit	Avg Mat'l Unit Cost	Avg Labor Unit Cost	Avg Equip Unit Cost	Avg Total Unit Cost	Avg Price Incl O&P	Avg Total Unit Cost (Large Volume)	Avg Price Incl O&P (Large Volume)
Polystyrene or polyurethane foam panel adhesive, on walls											
12" oc members											
1/8" diameter (11 fl.oz./CSF)	Inst	CSF	CA	.143	4.29	3.86	---	8.15	**10.60**	6.44	**8.21**
1/4" diameter (43 fl.oz./CSF)	Inst	CSF	CA	.143	17.10	3.86	---	20.96	**24.70**	17.60	**20.50**
3/8" diameter (97 fl.oz./CSF)	Inst	CSF	CA	.143	38.60	3.86	---	42.46	**48.30**	36.40	**41.20**
1/2" diameter (172 fl.oz./CSF)	Inst	CSF	CA	.143	68.30	3.86	---	72.16	**81.00**	62.30	**69.70**
16" oc members											
1/8" diameter (9 fl.oz./CSF)	Inst	CSF	CA	.130	3.43	3.50	---	6.93	**9.10**	5.45	**7.03**
1/4" diameter (34 fl.oz./CSF)	Inst	CSF	CA	.130	13.70	3.50	---	17.20	**20.40**	14.35	**16.90**
3/8" diameter (78 fl.oz./CSF)	Inst	CSF	CA	.130	30.90	3.50	---	34.40	**39.30**	29.35	**33.40**
1/2" diameter (137 fl.oz./CSF)	Inst	CSF	CA	.130	54.60	3.50	---	58.10	**65.40**	50.15	**56.20**
24" oc members											
1/8" diameter (7 fl.oz./CSF)	Inst	CSF	CA	.129	2.57	3.48	---	6.05	**8.11**	4.51	**5.92**
1/4" diameter (26 fl.oz./CSF)	Inst	CSF	CA	.129	10.30	3.48	---	13.78	**16.60**	11.22	**13.30**
3/8" diameter (58 fl.oz./CSF)	Inst	CSF	CA	.129	23.20	3.48	---	26.68	**30.80**	22.46	**25.70**
1/2" diameter (103 fl.oz./CSF)	Inst	CSF	CA	.129	41.00	3.48	---	44.48	**50.30**	38.06	**42.80**
Gypsum drywall adhesive, on ceilings											
12" oc members											
1/8" diameter (11 fl.oz./CSF)	Inst	CSF	CA	.143	3.82	3.86	---	7.68	**10.10**	6.03	**7.76**
1/4" diameter (43 fl.oz./CSF)	Inst	CSF	CA	.143	15.20	3.86	---	19.06	**22.60**	16.00	**18.70**
3/8" diameter (97 fl.oz./CSF)	Inst	CSF	CA	.143	34.30	3.86	---	38.16	**43.60**	32.70	**37.00**
1/2" diameter (172 fl.oz./CSF)	Inst	CSF	CA	.143	60.70	3.86	---	64.56	**72.70**	55.70	**62.40**
16" oc members											
1/8" diameter (9 fl.oz./CSF)	Inst	CSF	CA	.130	3.06	3.50	---	6.56	**8.69**	5.12	**6.67**
1/4" diameter (34 fl.oz./CSF)	Inst	CSF	CA	.130	12.20	3.50	---	15.70	**18.70**	13.05	**15.40**
3/8" diameter (78 fl.oz./CSF)	Inst	CSF	CA	.130	27.50	3.50	---	31.00	**35.50**	26.45	**30.10**
1/2" diameter (137 fl.oz./CSF)	Inst	CSF	CA	.130	48.60	3.50	---	52.10	**58.80**	44.85	**50.40**
24" oc members											
1/8" diameter (7 fl.oz./CSF)	Inst	CSF	CA	.120	2.29	3.24	---	5.53	**7.44**	4.26	**5.64**
1/4" diameter (26 fl.oz./CSF)	Inst	CSF	CA	.120	9.13	3.24	---	12.37	**15.00**	10.23	**12.20**
3/8" diameter (58 fl.oz./CSF)	Inst	CSF	CA	.120	20.60	3.24	---	23.84	**27.60**	20.26	**23.20**
1/2" diameter (103 fl.oz./CSF)	Inst	CSF	CA	.120	36.40	3.24	---	39.64	**45.00**	34.06	**38.40**
Gypsum drywall adhesive, on walls											
12" oc members											
1/8" diameter (11 fl.oz./CSF)	Inst	CSF	CA	.143	3.82	3.86	---	7.68	**10.10**	6.03	**7.76**
1/4" diameter (43 fl.oz./CSF)	Inst	CSF	CA	.143	15.20	3.86	---	19.06	**22.60**	16.00	**18.70**
3/8" diameter (97 fl.oz./CSF)	Inst	CSF	CA	.143	34.30	3.86	---	38.16	**43.60**	32.70	**37.00**
1/2" diameter (172 fl.oz./CSF)	Inst	CSF	CA	.143	60.70	3.86	---	64.56	**72.70**	55.70	**62.40**
16" oc members											
1/8" diameter (9 fl.oz./CSF)	Inst	CSF	CA	.130	3.06	3.50	---	6.56	**8.69**	5.12	**6.67**
1/4" diameter (34 fl.oz./CSF)	Inst	CSF	CA	.130	12.20	3.50	---	15.70	**18.70**	13.05	**15.40**
3/8" diameter (78 fl.oz./CSF)	Inst	CSF	CA	.130	27.50	3.50	---	31.00	**35.50**	26.45	**30.10**
1/2" diameter (137 fl.oz./CSF)	Inst	CSF	CA	.130	48.60	3.50	---	52.10	**58.80**	44.85	**50.40**
24" oc members											
1/8" diameter (7 fl.oz./CSF)	Inst	CSF	CA	.120	2.29	3.24	---	5.53	**7.44**	4.26	**5.64**
1/4" diameter (26 fl.oz./CSF)	Inst	CSF	CA	.120	9.13	3.24	---	12.37	**15.00**	10.23	**12.20**
3/8" diameter (58 fl.oz./CSF)	Inst	CSF	CA	.120	20.60	3.24	---	23.84	**27.60**	20.26	**23.20**
1/2" diameter (103 fl.oz./CSF)	Inst	CSF	CA	.120	36.40	3.24	---	39.64	**45.00**	34.06	**38.40**

Description	Oper	Unit	Crew Size	Man-Hours Per Unit	Avg Mat'l Unit Cost	Avg Labor Unit Cost	Avg Equip Unit Cost	Avg Total Unit Cost	Avg Price Incl O&P	Large Volume Avg Total Unit Cost	Large Volume Avg Price Incl O&P
Hardboard or plastic panel adhesive, on walls											
12" oc members											
1/8" diameter (11 fl.oz./CSF)	Inst	CSF	CA	.143	3.82	3.86	---	7.68	**10.10**	6.03	**7.76**
1/4" diameter (43 fl.oz./CSF)	Inst	CSF	CA	.143	15.20	3.86	---	19.06	**22.60**	16.00	**18.70**
3/8" diameter (97 fl.oz./CSF)	Inst	CSF	CA	.143	34.30	3.86	---	38.16	**43.60**	32.70	**37.00**
1/2" diameter (172 fl.oz./CSF)	Inst	CSF	CA	.143	60.70	3.86	---	64.56	**72.70**	55.70	**62.40**
16" oc members											
1/8" diameter (9 fl.oz./CSF)	Inst	CSF	CA	.130	3.06	3.50	---	6.56	**8.69**	5.12	**6.67**
1/4" diameter (34 fl.oz./CSF)	Inst	CSF	CA	.130	12.20	3.50	---	15.70	**18.70**	13.05	**15.40**
3/8" diameter (78 fl.oz./CSF)	Inst	CSF	CA	.130	27.50	3.50	---	31.00	**35.50**	26.45	**30.10**
1/2" diameter (137 fl.oz./CSF)	Inst	CSF	CA	.130	48.60	3.50	---	52.10	**58.80**	44.85	**50.40**
24" oc members											
1/8" diameter (7 fl.oz./CSF)	Inst	CSF	CA	.120	2.29	3.24	---	5.53	**7.44**	4.26	**5.64**
1/4" diameter (26 fl.oz./CSF)	Inst	CSF	CA	.120	9.13	3.24	---	12.37	**15.00**	10.23	**12.20**
3/8" diameter (58 fl.oz./CSF)	Inst	CSF	CA	.120	20.60	3.24	---	23.84	**27.60**	20.26	**23.20**
1/2" diameter (103 fl.oz./CSF)	Inst	CSF	CA	.120	36.40	3.24	---	39.64	**45.00**	34.06	**38.40**

Air conditioning and ventilating systems

System components

Condensing units

Description	Oper	Unit	Crew Size	Man-Hours Per Unit	Avg Mat'l Unit Cost	Avg Labor Unit Cost	Avg Equip Unit Cost	Avg Total Unit Cost	Avg Price Incl O&P	Large Volume Avg Total Unit Cost	Large Volume Avg Price Incl O&P
Air cooled, compressor, standard controls											
1.0 ton	Inst	Ea	SB	10.0	730.00	262.00	---	992.00	**1120.00**	850.00	**1050.00**
1.5 ton	Inst	Ea	SB	11.4	895.00	299.00	---	1194.00	**1340.00**	1025.00	**1260.00**
2.0 ton	Inst	Ea	SB	13.3	1030.00	349.00	---	1379.00	**1550.00**	1180.00	**1450.00**
2.5 ton	Inst	Ea	SB	16.0	1230.00	419.00	---	1649.00	**1850.00**	1415.00	**1740.00**
3.0 ton	Inst	Ea	SB	20.0	1450.00	524.00	---	1974.00	**2230.00**	1699.00	**2090.00**
4.0 ton	Inst	Ea	SB	26.7	1910.00	700.00	---	2610.00	**2950.00**	2238.00	**2760.00**
5.0 ton	Inst	Ea	SB	40.0	2340.00	1050.00	---	3390.00	**3900.00**	2889.00	**3610.00**
Minimum Job Charge	Inst	Job	SB	6.67	---	175.00	---	175.00	**260.00**	140.00	**208.00**

Diffusers

Description	Oper	Unit	Crew Size	Man-Hours Per Unit	Avg Mat'l Unit Cost	Avg Labor Unit Cost	Avg Equip Unit Cost	Avg Total Unit Cost	Avg Price Incl O&P	Large Volume Avg Total Unit Cost	Large Volume Avg Price Incl O&P
Aluminum, opposed blade damper, unless otherwise noted											
Ceiling or sidewall, linear											
2" wide	Inst	LF	UA	.333	26.80	9.91	---	36.71	**41.70**	31.45	**38.90**
4" wide	Inst	LF	UA	.370	35.90	11.00	---	46.90	**52.40**	40.31	**49.40**
6" wide	Inst	LF	UA	.417	44.50	12.40	---	56.90	**63.10**	48.91	**59.70**
8" wide	Inst	LF	UA	.500	51.90	14.90	---	66.80	**74.20**	57.40	**70.20**
Perforated, 24" x 24" panel size											
6" x 6"	Inst	Ea	UA	.625	73.00	18.60	---	91.60	**101.00**	78.90	**95.90**
8" x 8"	Inst	Ea	UA	.714	75.20	21.30	---	96.50	**107.00**	83.00	**101.00**
10" x 10"	Inst	Ea	UA	.833	76.40	24.80	---	101.20	**114.00**	86.90	**107.00**
12" x 12"	Inst	Ea	UA	1.00	79.20	29.80	---	109.00	**124.00**	93.30	**116.00**
18" x 18"	Inst	Ea	UA	1.25	105.00	37.20	---	142.20	**161.00**	122.30	**151.00**
Rectangular, one- to four-way blow											
6" x 6"	Inst	Ea	UA	.625	42.80	18.60	---	61.40	**70.70**	52.40	**65.50**
12" x 6"	Inst	Ea	UA	.714	54.70	21.30	---	76.00	**86.60**	65.00	**80.70**
12" x 9"	Inst	Ea	UA	.833	64.40	24.80	---	89.20	**102.00**	76.40	**94.80**
12" x 12"	Inst	Ea	UA	1.00	75.20	29.80	---	105.00	**120.00**	89.80	**112.00**
24" x 12"	Inst	Ea	UA	1.25	128.00	37.20	---	165.20	**183.00**	141.80	**173.00**

Description	Oper	Unit	Crew Size	Man-Hours Per Unit	Avg Mat'l Unit Cost	Avg Labor Unit Cost	Avg Equip Unit Cost	Avg Total Unit Cost	Avg Price Incl O&P	Avg Total Unit Cost	Avg Price Incl O&P
			Costs Based On Small Volume							Large Volume	
T-bar mounting, 24" x 24" lay in frame											
6" x 6"	Inst	Ea	UA	.625	69.00	18.60	---	87.60	**96.90**	75.40	**91.90**
9" x 9"	Inst	Ea	UA	.714	77.00	21.30	---	98.30	**109.00**	84.50	**103.00**
12" x 12"	Inst	Ea	UA	.833	101.00	24.80	---	125.80	**138.00**	108.40	**132.00**
15" x 15"	Inst	Ea	UA	1.00	105.00	29.80	---	134.80	**150.00**	115.80	**142.00**
18" x 18"	Inst	Ea	UA	1.25	138.00	37.20	---	175.20	**194.00**	150.80	**184.00**
Minimum Job Charge	Inst	Job	UA	2.86	---	85.10	---	85.10	**128.00**	68.20	**102.00**
Ductwork											
Fabricated rectangular, includes fittings, joints, supports											
Aluminum alloy											
Under 100 lbs	Inst	Lb	UF	.429	3.52	11.70	---	15.22	**21.00**	12.41	**17.50**
100 to 500 lbs	Inst	Lb	UF	.333	2.63	9.04	---	11.67	**16.20**	9.56	**13.50**
500 to 1,000 lbs	Inst	Lb	UF	.273	1.81	7.41	---	9.22	**12.90**	7.51	**10.70**
Over 1,000 lbs	Inst	Lb	UF	.231	1.38	6.27	---	7.65	**10.80**	6.23	**8.93**
Galvanized steel											
Under 400 lbs	Inst	Lb	UF	.150	.74	4.07	---	4.81	**6.85**	3.91	**5.64**
400 to 1,000 lbs	Inst	Lb	UF	.140	.63	3.80	---	4.43	**6.33**	3.59	**5.20**
1,000 to 2,000 lbs	Inst	Lb	UF	.130	.54	3.53	---	4.07	**5.84**	3.29	**4.78**
2,000 to 5,000 lbs	Inst	Lb	UF	.125	.46	3.40	---	3.86	**5.55**	3.12	**4.53**
Over 10,000 lbs	Inst	Lb	UF	.120	.40	3.26	---	3.66	**5.29**	2.96	**4.31**
Flexible, coated fabric on spring steel, aluminum, or corrosion resistant metal											
Non-insulated											
3" diameter	Inst	LF	UD	.073	.89	1.89	---	2.78	**3.72**	2.28	**3.15**
5" diameter	Inst	LF	UD	.089	1.04	2.30	---	3.34	**4.49**	2.75	**3.80**
6" diameter	Inst	LF	UD	.100	1.16	2.59	---	3.75	**5.04**	3.09	**4.28**
7" diameter	Inst	LF	UD	.114	1.53	2.95	---	4.48	**5.95**	3.69	**5.07**
8" diameter	Inst	LF	UD	.133	1.58	3.44	---	5.02	**6.74**	4.16	**5.75**
10" diameter	Inst	LF	UD	.160	2.03	4.14	---	6.17	**8.23**	5.09	**7.01**
12" diameter	Inst	LF	UD	.200	2.41	5.17	---	7.58	**10.20**	6.25	**8.63**
Insulated											
3" diameter	Inst	LF	UD	.080	1.53	2.07	---	3.60	**4.63**	3.00	**4.02**
4" diameter	Inst	LF	UD	.089	1.53	2.30	---	3.83	**4.98**	3.18	**4.29**
5" diameter	Inst	LF	UD	.100	1.79	2.59	---	4.38	**5.67**	3.64	**4.91**
6" diameter	Inst	LF	UD	.114	1.97	2.95	---	4.92	**6.39**	4.08	**5.52**
7" diameter	Inst	LF	UD	.133	2.41	3.44	---	5.85	**7.57**	4.88	**6.58**
8" diameter	Inst	LF	UD	.160	2.46	4.14	---	6.60	**8.66**	5.47	**7.45**
10" diameter	Inst	LF	UD	.200	3.20	5.17	---	8.37	**11.00**	6.95	**9.44**
12" diameter	Inst	LF	UD	.267	3.73	6.90	---	10.63	**14.10**	8.78	**12.00**
14" diameter	Inst	LF	UD	.400	4.51	10.30	---	14.81	**20.00**	12.24	**17.00**
Duct accessories											
Air extractors											
12" x 4"	Inst	Ea	UA	.417	14.00	12.40	---	26.40	**32.60**	22.21	**29.00**
8" x 6"	Inst	Ea	UA	.500	14.00	14.90	---	28.90	**36.30**	24.20	**31.90**
20" x 8"	Inst	Ea	UA	.625	31.40	18.60	---	50.00	**59.30**	42.40	**54.00**
18" x 10"	Inst	Ea	UA	.714	30.80	21.30	---	52.10	**62.70**	44.00	**56.50**
24" x 12"	Inst	Ea	UA	1.00	42.80	29.80	---	72.60	**87.40**	61.30	**78.80**
Dampers, fire, curtain type, vertical											
8" x 4"	Inst	Ea	UA	.417	19.00	12.40	---	31.40	**37.60**	26.61	**34.00**
12" x 4"	Inst	Ea	UA	.455	19.00	13.50	---	32.50	**39.30**	27.50	**35.40**
16" x 14"	Inst	Ea	UA	.833	30.20	24.80	---	55.00	**67.40**	46.40	**60.30**
24" x 20"	Inst	Ea	UA	1.25	38.20	37.20	---	75.40	**94.00**	63.30	**83.20**

Description	Oper	Unit	Crew Size	Man-Hours Per Unit	Avg Mat'l Unit Cost	Avg Labor Unit Cost	Avg Equip Unit Cost	Avg Total Unit Cost	Avg Price Incl O&P	Large Volume Avg Total Unit Cost	Large Volume Avg Price Incl O&P
Dampers, multi-blade, opposed blade											
12" x 12"	Inst	Ea	UA	.625	27.90	18.60	---	46.50	**55.80**	39.40	**50.50**
12" x 18"	Inst	Ea	UA	.833	37.10	24.80	---	61.90	**74.20**	52.40	**67.20**
24" x 24"	Inst	Ea	UA	1.25	78.70	37.20	---	115.90	**134.00**	98.80	**124.00**
48" x 36"	Inst	Ea	UD	3.33	238.00	86.10	---	324.10	**367.00**	278.10	**344.00**
Dampers, variable volume modulating, motorized											
12" x 12"	Inst	Ea	UA	1.00	144.00	29.80	---	173.80	**188.00**	149.80	**181.00**
24" x 12"	Inst	Ea	UA	1.67	172.00	49.70	---	221.70	**247.00**	190.60	**233.00**
30" x 18"	Inst	Ea	UA	2.50	319.00	74.40	---	393.40	**431.00**	339.50	**411.00**
Thermostat, ADD	Inst	Ea	UA	1.67	34.20	49.70	---	83.90	**109.00**	69.60	**93.90**
Dampers, multi-blade, parallel blade											
8" x 8"	Inst	Ea	UA	.500	58.10	14.90	---	73.00	**80.50**	62.90	**76.50**
16" x 10"	Inst	Ea	UA	.625	61.00	18.60	---	79.60	**88.90**	68.40	**83.90**
18" x 12"	Inst	Ea	UA	.833	62.70	24.80	---	87.50	**99.90**	74.90	**93.00**
28" x 16"	Inst	Ea	UA	1.25	84.90	37.20	---	122.10	**141.00**	104.30	**130.00**
Mixing box, with electric or pneumatic motor, with silencer											
150 to 250 CFM	Inst	Ea	UD	2.00	519.00	51.70	---	570.70	**596.00**	496.40	**585.00**
270 to 600 CFM	Inst	Ea	UD	2.50	536.00	64.70	---	600.70	**633.00**	521.70	**618.00**
Fans											
Air conditioning or processed air handling											
Axial flow, compact, low sound 2.5" self-propelled											
3,800 CFM, 5 HP	Inst	Ea	UE	8.33	3790.00	220.00	---	4010.00	**4120.00**	3506.00	**4090.00**
15,600 CFM, 10 HP	Inst	Ea	UE	16.7	6610.00	442.00	---	7052.00	**7270.00**	6152.00	**7200.00**
In-line centrifugal, supply/exhaust booster, aluminum wheel/hub, disconnect switch, 1/4" self-propelled											
500 CFM, 10" dia. connect	Inst	Ea	UE	8.33	798.00	220.00	---	1018.00	**1130.00**	876.00	**1070.00**
1,520 CFM, 16" dia. connect	Inst	Ea	UE	12.5	1040.00	331.00	---	1371.00	**1540.00**	1180.00	**1450.00**
3,480 CFM, 20" dia. connect	Inst	Ea	UE	25.0	1370.00	661.00	---	2031.00	**2360.00**	1729.00	**2170.00**
Ceiling fan, right angle, extra quiet 0.10" self-propelled											
200 CFM	Inst	Ea	UE	1.56	170.00	41.30	---	211.30	**232.00**	182.10	**221.00**
900 CFM	Inst	Ea	UE	2.08	405.00	55.00	---	460.00	**487.00**	399.20	**475.00**
3,000 CFM	Inst	Ea	UE	3.13	770.00	82.80	---	852.80	**894.00**	741.10	**875.00**
Exterior wall or roof cap	Inst	Ea	UA	.833	112.00	24.80	---	136.80	**149.00**	118.40	**143.00**
Roof ventilators, corrosive fume resistant, plastic blades											
Roof ventilator, centrifugal V belt drive motor, 1/4" self-propelled											
250 CFM, 1/4 HP	Inst	Ea	UE	6.25	2370.00	165.00	---	2535.00	**2610.00**	2212.00	**2580.00**
900 CFM, 1/3 HP	Inst	Ea	UE	8.33	2570.00	220.00	---	2790.00	**2900.00**	2426.00	**2850.00**
1,650 CFM, 1/2 HP	Inst	Ea	UE	10.0	3050.00	265.00	---	3315.00	**3450.00**	2892.00	**3390.00**
2,250 CFM, 1 HP	Inst	Ea	UE	12.5	3190.00	331.00	---	3521.00	**3690.00**	3065.00	**3620.00**
Utility set centrifugal V belt drive motor, 1/4" self-propelled											
1,200 CFM, 1/4 HP	Inst	Ea	UE	6.25	2370.00	165.00	---	2535.00	**2610.00**	2212.00	**2580.00**
1,500 CFM, 1/3 HP	Inst	Ea	UE	8.33	2390.00	220.00	---	2610.00	**2720.00**	2276.00	**2680.00**
1,850 CFM, 1/2 HP	Inst	Ea	UE	10.0	2390.00	265.00	---	2655.00	**2790.00**	2312.00	**2730.00**
2,200 CFM, 3/4 HP	Inst	Ea	UE	12.5	2420.00	331.00	---	2751.00	**2920.00**	2395.00	**2840.00**
Direct drive											
320 CFM, 11" x 11" damper	Inst	Ea	UE	6.25	286.00	165.00	---	451.00	**534.00**	383.00	**487.00**
600 CFM, 11" x 11" damper	Inst	Ea	UE	6.25	315.00	165.00	---	480.00	**563.00**	408.00	**516.00**

Description	Oper	Unit	Crew Size	Man-Hours Per Unit	Avg Mat'l Unit Cost	Avg Labor Unit Cost	Avg Equip Unit Cost	Avg Total Unit Cost	Avg Price Incl O&P	Large Volume Avg Total Unit Cost	Large Volume Avg Price Incl O&P
Ventilation, residential											
Attic											
Roof type ventilators											
Aluminum dome, damper and curb											
6" diameter, 300 CFM	Inst	Ea	EA	.833	239.00	24.00	---	263.00	**275.00**	229.20	**270.00**
7" diameter, 450 CFM	Inst	Ea	EA	.909	261.00	26.20	---	287.20	**300.00**	250.00	**294.00**
9" diameter, 900 CFM	Inst	Ea	EA	1.00	422.00	28.80	---	450.80	**464.00**	393.10	**460.00**
12" diameter, 1,000 CFM	Inst	Ea	EA	1.25	259.00	36.10	---	295.10	**312.00**	255.80	**304.00**
16" diameter, 1,500 CFM	Inst	Ea	EA	1.43	312.00	41.20	---	353.20	**373.00**	306.90	**364.00**
20" diameter, 2,500 CFM	Inst	Ea	EA	1.67	382.00	48.20	---	430.20	**453.00**	373.40	**442.00**
26" diameter, 4,000 CFM	Inst	Ea	EA	2.00	462.00	57.70	---	519.70	**547.00**	451.10	**534.00**
32" diameter, 6,500 CFM	Inst	Ea	EA	2.50	638.00	72.10	---	710.10	**745.00**	617.70	**729.00**
38" diameter, 8,000 CFM	Inst	Ea	EA	3.33	946.00	96.00	---	1042.00	**1090.00**	907.00	**1070.00**
50" diameter, 13,000 CFM	Inst	Ea	EA	5.00	1370.00	144.00	---	1514.00	**1580.00**	1315.00	**1550.00**
Plastic ABS dome											
900 CFM	Inst	Ea	EA	.833	76.40	24.00	---	100.40	**112.00**	86.20	**106.00**
1,600 CFM	Inst	Ea	EA	1.00	114.00	28.80	---	142.80	**157.00**	123.10	**149.00**
Wall type ventilators, one speed, with shutter											
12" diameter, 1,000 CFM	Inst	Ea	EA	.833	141.00	24.00	---	165.00	**177.00**	143.20	**171.00**
14" diameter, 1,500 CFM	Inst	Ea	EA	1.00	172.00	28.80	---	200.80	**215.00**	174.10	**208.00**
16" diameter, 2,000 CFM	Inst	Ea	EA	1.25	259.00	36.10	---	295.10	**312.00**	255.80	**304.00**
Entire structure, wall type, one speed, with shutter											
30" diameter, 4,800 CFM	Inst	Ea	EA	2.00	445.00	57.70	---	502.70	**530.00**	436.10	**517.00**
36" diameter, 7,000 CFM	Inst	Ea	EA	2.50	485.00	72.10	---	557.10	**591.00**	482.70	**574.00**
42" diameter, 10,000 CFM	Inst	Ea	EA	3.33	593.00	96.00	---	689.00	**735.00**	597.00	**712.00**
48" diameter, 16,000 CFM	Inst	Ea	EA	5.00	775.00	144.00	---	919.00	**989.00**	795.00	**953.00**
Two speeds, ADD	Inst	Ea	---	---	55.30	---	---	55.30	**63.60**	48.50	**55.80**
Entire structure, lay-down type, one speed, with shutter											
30" diameter, 4,500 CFM	Inst	Ea	EA	1.67	485.00	48.20	---	533.20	**556.00**	463.40	**546.00**
36" diameter, 6,500 CFM	Inst	Ea	EA	2.00	530.00	57.70	---	587.70	**615.00**	511.10	**603.00**
42" diameter, 9,000 CFM	Inst	Ea	EA	2.50	627.00	72.10	---	699.10	**734.00**	607.70	**718.00**
48" diameter, 12,000 CFM	Inst	Ea	EA	3.33	827.00	96.00	---	923.00	**969.00**	802.00	**948.00**
Two speeds, ADD	Inst	Ea	---	---	13.60	---	---	13.60	**15.60**	11.90	**13.70**
12 hour timer, ADD	Inst	Ea	EA	.625	24.50	18.00	---	42.50	**51.20**	35.90	**46.10**
Minimum Job Charge	Inst	Job	EA	2.86	---	82.50	---	82.50	**122.00**	66.00	**97.70**
Grilles											
Aluminum											
Air return											
6" x 6"	Inst	Ea	UA	.417	11.30	12.40	---	23.70	**29.90**	19.81	**26.30**
10" x 6"	Inst	Ea	UA	.455	13.80	13.50	---	27.30	**34.10**	22.90	**30.20**
16" x 8"	Inst	Ea	UA	.500	20.10	14.90	---	35.00	**42.40**	29.50	**38.10**
12" x 12"	Inst	Ea	UA	.625	20.10	18.60	---	38.70	**48.00**	32.50	**42.60**
24" x 12"	Inst	Ea	UA	.714	35.90	21.30	---	57.20	**67.80**	48.50	**61.70**
48" x 24"	Inst	Ea	UA	1.00	138.00	29.80	---	167.80	**183.00**	144.80	**175.00**
Minimum Job Charge	Inst	Job	UA	2.86	---	85.10	---	85.10	**128.00**	68.20	**102.00**

			Costs Based On Small Volume							Large Volume	
Description	Oper	Unit	Crew Size	Man-Hours Per Unit	Avg Mat'l Unit Cost	Avg Labor Unit Cost	Avg Equip Unit Cost	Avg Total Unit Cost	Avg Price Incl O&P	Avg Total Unit Cost	Avg Price Incl O&P
Registers, air supply											
Ceiling/wall, O.B. damper, anodized aluminum											
One or two way deflection, adjustable curved face bars											
14" x 8"	Inst	Ea	UA	.625	23.90	18.60	---	42.50	**51.80**	35.90	**46.50**
Baseboard, adjustable damper, enameled steel											
10" x 6"	Inst	Ea	UA	.417	10.40	12.40	---	22.80	**29.00**	19.01	**25.30**
12" x 5"	Inst	Ea	UA	.455	11.30	13.50	---	24.80	**31.60**	20.70	**27.60**
12" x 6"	Inst	Ea	UA	.455	10.40	13.50	---	23.90	**30.70**	19.90	**26.70**
12" x 8"	Inst	Ea	UA	.500	15.10	14.90	---	30.00	**37.40**	25.20	**33.10**
14" x 6"	Inst	Ea	UA	.556	11.30	16.60	---	27.90	**36.10**	23.10	**31.20**
Minimum Job Charge	Inst	Job	UA	2.86	---	85.10	---	85.10	**128.00**	68.20	**102.00**
Ventilators											
Base, damper, screen; 8" neck diameter											
215 CFM @ 5 MPH wind	Inst	Ea	UD	3.33	63.30	86.10	---	149.40	**192.00**	124.60	**167.00**
System units complete											
Fan coil air conditioning											
Cabinet mounted, with filters											
Chilled water											
0.5 ton cooling	Inst	Ea	SB	5.00	821.00	131.00	---	952.00	**1020.00**	825.00	**984.00**
1 ton cooling	Inst	Ea	SB	6.67	929.00	175.00	---	1104.00	**1190.00**	955.00	**1150.00**
2.5 ton cooling	Inst	Ea	SB	10.0	1680.00	262.00	---	1942.00	**2070.00**	1690.00	**2010.00**
3 ton cooling	Inst	Ea	SB	20.0	1850.00	524.00	---	2374.00	**2630.00**	2049.00	**2490.00**
10 ton cooling	Inst	Ea	SF	15.0	2680.00	414.00	---	3094.00	**3300.00**	2681.00	**3200.00**
15 ton cooling	Inst	Ea	SF	20.0	3730.00	552.00	---	4282.00	**4560.00**	3722.00	**4420.00**
20 ton cooling	Inst	Ea	SF	40.0	4790.00	1100.00	---	5890.00	**6430.00**	5084.00	**6150.00**
30 ton cooling	Inst	Ea	SF	60.0	7100.00	1660.00	---	8760.00	**9570.00**	7560.00	**9130.00**
Minimum Job Charge	Inst	Job	SB	6.67	---	175.00	---	175.00	**260.00**	140.00	**208.00**
Heat pumps											
Air to air, split system, not including curbs or pads											
2 ton cooling, 8.5 MBHP heat	Inst	Ea	SB	20.0	2340.00	524.00	---	2864.00	**3120.00**	2469.00	**2980.00**
3 ton cooling, 13 MBHP heat	Inst	Ea	SB	40.0	3020.00	1050.00	---	4070.00	**4580.00**	3489.00	**4300.00**
7.5 ton cooling, 33 MBHP heat	Inst	Ea	SB	61.5	7320.00	1610.00	---	8930.00	**9730.00**	7700.00	**9280.00**
15 ton cooling, 64 MBHP heat	Inst	Ea	SB	80.0	13100	2100.00	---	15200	**16200**	13180	**15700**
Air to air, single package, not including curbs, pads, or plenums											
2 ton cooling, 6.5 MBHP heat	Inst	Ea	SB	16.0	2820.00	419.00		3239.00	**3450.00**	2815.00	**3350.00**
3 ton cooling, 10 MBHP heat	Inst	Ea	SB	26.7	3390.00	700.00	---	4090.00	**4430.00**	3538.00	**4250.00**
Water source to air, single package											
1 ton cooling, 13 MBHP heat	Inst	Ea	SB	11.4	1080.00	299.00	---	1379.00	**1530.00**	1190.00	**1450.00**
2 ton cooling, 19 MBHP heat	Inst	Ea	SB	13.3	1370.00	349.00	---	1719.00	**1890.00**	1480.00	**1800.00**
5 ton cooling, 29 MBHP heat	Inst	Ea	SB	26.7	2390.00	700.00	---	3090.00	**3440.00**	2658.00	**3250.00**
Minimum Job Charge	Inst	Job	SB	10.0	---	262.00	---	262.00	**391.00**	210.00	**312.00**
Roof top air conditioners											
Standard controls, curb, energy economizer											
Single zone, electric fired cooling, gas fired heating											
3 ton cooling, 60 MBHP heat	Inst	Ea	SB	16.7	4530.00	438.00	---	4968.00	**5180.00**	4329.00	**5090.00**
4 ton cooling, 95 MBHP heat	Inst	Ea	SB	20.0	4700.00	524.00	---	5224.00	**5480.00**	4549.00	**5370.00**
10 ton cooling, 200 MBHP heat	Inst	Ea	SF	75.0	10600	2070.00	---	12670	**13700**	10960	**13200**
30 ton cooling, 540 MBHP heat	Inst	Ea	SG	200	32600	5670.00	---	38270	**41000**	33130	**39600**
40 ton cooling, 675 MBHP heat	Inst	Ea	SG	246	42400	6970.00	---	49370	**52800**	42870	**51200**
50 ton cooling, 810 MBHP heat	Inst	Ea	SG	320	51400	9070.00	---	60470	**64900**	52660	**63100**

Description	Oper	Unit	Crew Size	Man-Hours Per Unit	Avg Mat'l Unit Cost	Avg Labor Unit Cost	Avg Equip Unit Cost	Avg Total Unit Cost	Avg Price Incl O&P	Large Volume Avg Total Unit Cost	Large Volume Avg Price Incl O&P
Single zone, gas fired cooling and heating											
3 ton cooling, 90 MBHP heat	Inst	Ea	SB	20.0	5640.00	524.00	---	6164.00	**6420.00**	5369.00	**6320.00**
Multizone, electric fired cooling, gas fired cooling											
20 ton cooling, 360 MBHP heat	Inst	Ea	SG	200	51000	5670.00	---	56670	**59400**	49230	**58200**
30 ton cooling, 540 MBHP heat	Inst	Ea	SG	286	63800	8100.00	---	71900	**75900**	62490	**74100**
70 ton cooling, 1,500 MBHP hea	Inst	Ea	SG	500	120000	14200	---	134200	**141000**	116300	**138000**
80 ton cooling, 1,500 MBHP hea	Inst	Ea	SG	571	137000	16200	---	153200	**161000**	132900	**157000**
105 ton cooling,1,500 MBHP heat	Inst	Ea	SG	800	180000	22700	---	202700	**213000**	176100	**208000**
Minimum Job Charge	Inst	Job	SB	6.25	---	164.00	---	164.00	**244.00**	131.00	**195.00**
Window unit air conditioners											
Semi-permanent installation, 3 speed fan, 125 volt GFI receptacle, energy efficient models											
6,000 BTUH (0.5 ton cooling)	Inst	Ea	EB	5.00	547.00	144.00	---	691.00	**761.00**	595.00	**723.00**
9,000 BTUH (0.75 ton cooling)	Inst	Ea	EB	5.00	604.00	144.00	---	748.00	**818.00**	645.00	**780.00**
12,000 BTUH (1 ton cooling)	Inst	Ea	EB	5.00	644.00	144.00	---	788.00	**858.00**	680.00	**820.00**
18,000 BTUH (1.5 ton cooling)	Inst	Ea	EB	6.67	735.00	192.00	---	927.00	**1020.00**	799.00	**969.00**
24,000 BTUH (2.0 ton cooling)	Inst	Ea	EB	6.67	804.00	192.00	---	996.00	**1090.00**	859.00	**1040.00**
36,000 BTUH (3.0 ton cooling)	Inst	Ea	EB	6.67	1040.00	192.00	---	1232.00	**1320.00**	1064.00	**1270.00**
Minimum Job Charge	Inst	Job	EA	3.33	---	96.00	---	96.00	**142.00**	77.00	**114.00**

Awnings. See Canopies, page 47

Backfill. See Excavation, page 96

Bath accessories.
For plumbing fixtures, see individual items.

The material cost of an item includes the fixture, water supply, and trim (includes fittings and faucets). The labor cost of an item includes installation of the fixture and connection of water supply and/or electricity, but no demolition or clean-up is included. Average rough-in of pipe, waste, and vent is an "add" item shown below each major grouping of fixtures, unless noted otherwise.

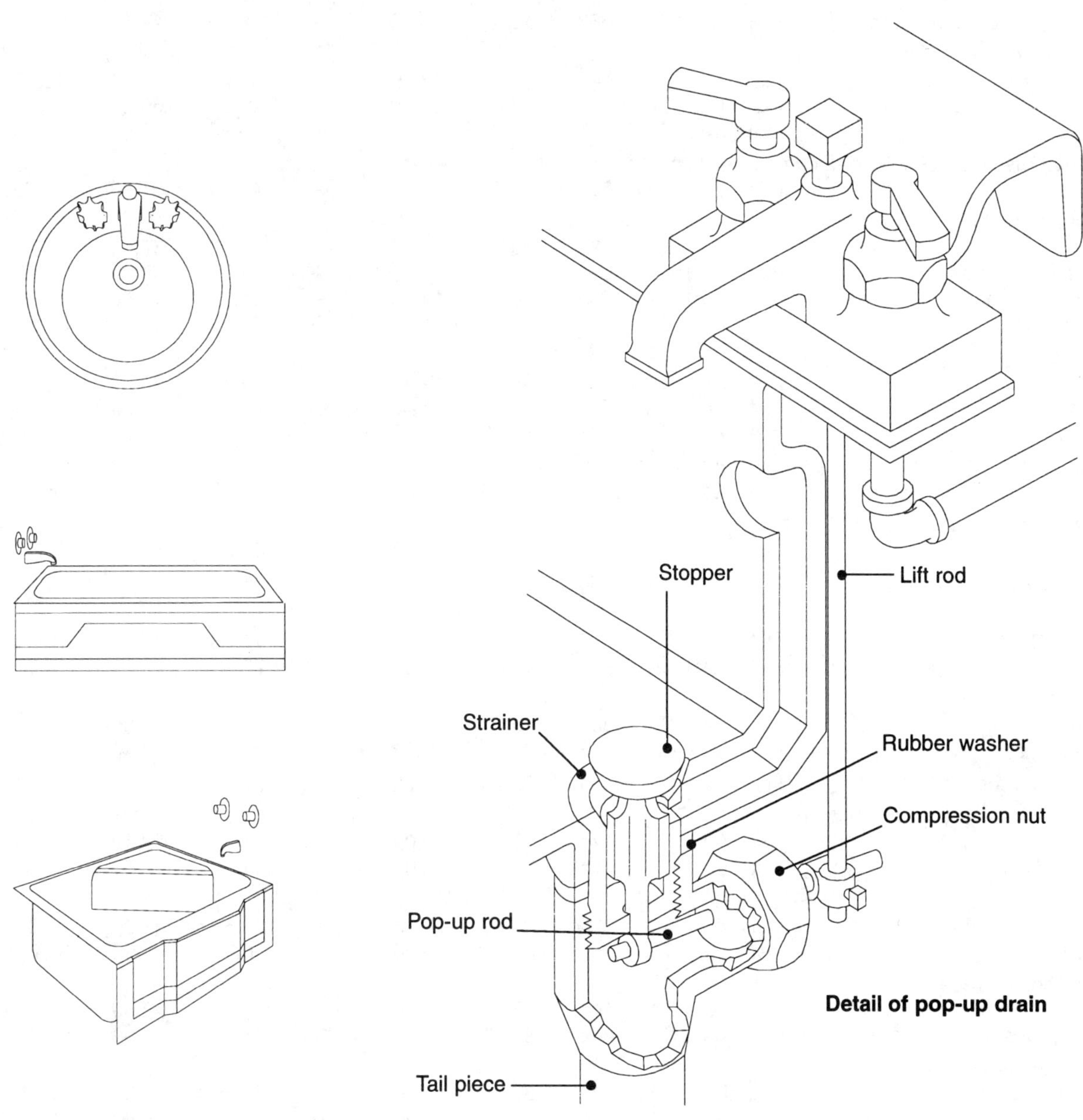

From: ***Basic Plumbing with Illustrations*** Craftsman Book Company

Bath accessories

Average quality, Nutone Products

Description	Oper	Unit	Crew Size	Man-Hours Per Unit	Avg Mat'l Unit Cost	Avg Labor Unit Cost	Avg Equip Unit Cost	Avg Total Unit Cost	Avg Price Incl O&P	Large Volume Avg Total Unit Cost	Large Volume Avg Price Incl O&P
Cup holder, chrome, surface mounted											
	Inst	Ea	CA	.444	10.40	12.00	---	22.40	**30.20**	18.09	**24.10**
Cup holder; chrome; recessed	Inst	Ea	CA	.667	18.40	18.00	---	36.40	**48.50**	29.60	**39.00**
Cup and toothbrush holder											
Antique brass	Inst	Ea	CA	.444	39.80	12.00	---	51.80	**64.00**	43.78	**53.70**
Brass	Inst	Ea	CA	.444	37.00	12.00	---	49.00	**60.70**	41.28	**50.80**
Chrome	Inst	Ea	CA	.444	27.20	12.00	---	39.20	**49.40**	32.78	**41.00**
Cup, toothbrush, and soapholder, chrome recessed											
	Inst	Ea	CA	.667	30.00	18.00	---	48.00	**61.80**	39.70	**50.60**
Glass shelf, 24" long											
Chrome	Inst	Ea	CA	.533	30.70	14.40	---	45.10	**57.10**	37.70	**47.30**
Antique brass	Inst	Ea	CA	.533	49.80	14.40	---	64.20	**79.20**	54.40	**66.50**
Polished chrome	Inst	Ea	CA	.533	40.00	14.40	---	54.40	**67.80**	45.80	**56.60**
Stainless steel	Inst	Ea	CA	.533	26.30	14.40	---	40.70	**52.10**	33.80	**42.90**
Grab bars, stainless steel											
16" long	Inst	Ea	CA	.533	31.10	14.40	---	45.50	**57.70**	38.00	**47.70**
24" long	Inst	Ea	CA	.533	34.30	14.40	---	48.70	**61.30**	40.80	**50.90**
32" long	Inst	Ea	CA	.533	36.90	14.40	---	51.30	**64.20**	43.00	**53.50**
Angle bar, 16" L x 32" H	Inst	Ea	CA	.533	74.10	14.40	---	88.50	**107.00**	75.70	**91.00**
Robe hooks, single or double	Inst	Ea	CA	.333	11.20	8.98	---	20.18	**26.50**	16.54	**21.50**
Shower curtain rod 1" dia x 5' 5" L with adjacent rod holder											
	Inst	Ea	CA	.333	15.70	8.98	---	24.68	**31.70**	20.44	**26.00**
Soap holder/dish, chrome	Inst	Ea	CA	.333	27.40	8.98	---	36.38	**45.20**	30.74	**37.90**
Soap holder with drain											
Chrome	Inst	Ea	CA	.333	10.60	8.98	---	19.58	**25.90**	16.05	**21.00**
Antique brass	Inst	Ea	CA	.333	14.00	8.98	---	22.98	**29.80**	19.04	**24.30**
Soap holder with tray											
Chrome	Inst	Ea	CA	.333	38.40	8.98	---	47.38	**57.80**	40.34	**48.90**
Antique brass	Inst	Ea	CA	.333	39.80	8.98	---	48.78	**59.40**	41.54	**50.30**
Soap holder, recessed, chrome	Inst	Ea	CA	.444	22.70	12.00	---	34.70	**44.30**	28.88	**36.50**
Soap holder and utility bar, recessed											
Chrome	Inst	Ea	CA	.444	19.00	12.00	---	31.00	**40.10**	25.68	**32.80**
Antique brass	Inst	Ea	CA	.444	29.10	12.00	---	41.10	**51.70**	34.48	**43.00**
Toilet roll holder											
Chrome	Inst	Ea	CA	.333	24.60	8.98	---	33.58	**42.00**	28.34	**35.00**
Antique brass	Inst	Ea	CA	.333	28.00	8.98	---	36.98	**45.90**	31.24	**38.40**
Toilet roll holder, recessed											
Chrome	Inst	Ea	CA	.444	56.00	12.00	---	68.00	**82.60**	57.98	**70.00**
Antique brass	Inst	Ea	CA	.444	61.60	12.00	---	73.60	**89.00**	62.88	**75.60**
Toilet roll holder with hood, recessed											
Chrome	Inst	Ea	CA	.444	67.20	12.00	---	79.20	**95.50**	67.78	**81.30**
Antique brass	Inst	Ea	CA	.444	84.00	12.00	---	96.00	**115.00**	82.48	**98.20**
Towel bars, round											
Chrome											
18"	Inst	Ea	CA	.333	41.20	8.98	---	50.18	**61.00**	42.74	**51.70**
24"	Inst	Ea	CA	.333	45.50	8.98	---	54.48	**65.90**	46.54	**56.00**
30"	Inst	Ea	CA	.333	44.50	8.98	---	53.48	**64.80**	45.74	**55.10**
36"	Inst	Ea	CA	.333	64.40	8.98	---	73.38	**87.70**	63.14	**75.10**

Description	Oper	Unit	Crew Size	Man-Hours Per Unit	Avg Mat'l Unit Cost	Avg Labor Unit Cost	Avg Equip Unit Cost	Avg Total Unit Cost	Avg Price Incl O&P	Large Volume Avg Total Unit Cost	Large Volume Avg Price Incl O&P
			Costs Based On Small Volume							Large Volume	
Polished brass											
18"	Inst	Ea	CA	.333	52.10	8.98	---	61.08	**73.50**	52.34	**62.70**
24"	Inst	Ea	CA	.333	55.00	8.98	---	63.98	**76.90**	54.84	**65.60**
30"	Inst	Ea	CA	.333	58.80	8.98	---	67.78	**81.30**	58.24	**69.40**
36"	Inst	Ea	CA	.333	60.80	8.98	---	69.78	**83.50**	59.94	**71.40**
Towel pin, chrome	Inst	Ea	CA	.296	11.20	7.98	---	19.18	**25.00**	15.79	**20.40**
Towel ring											
Chrome	Inst	Ea	CA	.296	33.30	7.98	---	41.28	**50.50**	35.19	**42.60**
Antique brass	Inst	Ea	CA	.296	43.10	7.98	---	51.08	**61.70**	43.69	**52.50**
Towel ladder, 4 arms											
Antique brass	Inst	Ea	CA	.444	119.00	12.00	---	131.00	**155.00**	112.98	**133.00**
Polished chrome	Inst	Ea	CA	.444	102.00	12.00	---	114.00	**136.00**	98.68	**117.00**
Towel supply shelf											
18" long	Inst	Ea	CA	.333	44.40	8.98	---	53.38	**64.70**	45.54	**54.90**
24" long	Inst	Ea	CA	.333	47.00	8.98	---	55.98	**67.70**	47.94	**57.60**
Towel three arm swing bar, chrome	Inst	Ea	CA	.296	70.30	7.98	---	78.28	**93.00**	67.49	**79.90**
Wall to floor angle bar with flange, bolt, washer and screws	Inst	Ea	CA	.571	81.20	15.40	---	96.60	**117.00**	83.10	**99.90**
Medicine cabinets											
No electrical work included; for wall outlet cost, see Electrical, page 93											
Surface mounting, no wall opening											
Swing door cabinets with reversible mirror door											
16" x 22"	Inst	Ea	CA	1.00	50.00	27.00	---	77.00	**98.50**	65.30	**83.10**
16" x 26"	Inst	Ea	CA	1.00	70.60	27.00	---	97.60	**122.00**	83.40	**104.00**
Swing door, corner cabinets with reversible mirror door											
16" x 36"	Inst	Ea	CA	1.00	123.00	27.00	---	150.00	**183.00**	129.60	**157.00**
Sliding door cabinets, bypassing mirror doors											
Toplighted, stainless steel											
20" x 20"	Inst	Ea	CA	1.00	112.00	27.00	---	139.00	**170.00**	119.60	**145.00**
24" x 20"	Inst	Ea	CA	1.00	130.00	27.00	---	157.00	**190.00**	134.60	**163.00**
28" x 20"	Inst	Ea	CA	1.00	151.00	27.00	---	178.00	**215.00**	153.60	**185.00**
Cosmetic box with framed mirror stainless steel, Builders series											
18" x 26"	Inst	Ea	CA	1.00	89.10	27.00	---	116.10	**143.00**	99.60	**122.00**
24" x 32"	Inst	Ea	CA	1.00	105.00	27.00	---	132.00	**161.00**	113.10	**138.00**
30" x 32"	Inst	Ea	CA	1.00	120.00	27.00	---	147.00	**179.00**	126.60	**153.00**
36" x 32"	Inst	Ea	CA	1.00	135.00	27.00	---	162.00	**197.00**	139.60	**169.00**
48" x 32"	Inst	Ea	CA	1.00	188.00	27.00	---	215.00	**257.00**	185.60	**222.00**
3-way mirror, tri-view											
30" x 30"											
Frameless, beveled mirror	Inst	Ea	CA	1.14	235.00	30.70	---	265.70	**317.00**	230.00	**273.00**
Natural oak	Inst	Ea	CA	1.14	269.00	30.70	---	299.70	**356.00**	259.00	**307.00**
White finish	Inst	Ea	CA	1.14	246.00	30.70	---	276.70	**330.00**	240.00	**284.00**
36" x 30"											
Frameless, beveled mirror	Inst	Ea	CA	1.14	269.00	30.70	---	299.70	**356.00**	259.00	**307.00**
Natural oak	Inst	Ea	CA	1.14	324.00	30.70	---	354.70	**419.00**	307.00	**362.00**
White finish	Inst	Ea	CA	1.14	276.00	30.70	---	306.70	**364.00**	265.00	**314.00**
48" x 30"											
Frameless, beveled mirror	Inst	Ea	CA	1.33	346.00	35.90	---	381.90	**453.00**	330.00	**389.00**
Natural oak	Inst	Ea	CA	1.33	351.00	35.90	---	386.90	**458.00**	334.00	**394.00**
White finish	Inst	Ea	CA	1.33	346.00	35.90	---	381.90	**453.00**	330.00	**389.00**

Description	Oper	Unit	Crew Size	Man-Hours Per Unit	Avg Mat'l Unit Cost	Avg Labor Unit Cost	Avg Equip Unit Cost	Avg Total Unit Cost	Avg Price Incl O&P	Large Volume Avg Total Unit Cost	Large Volume Avg Price Incl O&P
With matching light fixture											
with 2 lights	Inst	Ea	EA	2.67	57.10	77.00	---	134.10	**180.00**	107.70	**143.00**
with 3 lights	Inst	Ea	EA	2.67	93.00	77.00	---	170.00	**221.00**	139.00	**179.00**
with 4 lights	Inst	Ea	EA	2.67	101.00	77.00	---	178.00	**230.00**	145.90	**187.00**
with 6 lights	Inst	Ea	EA	2.67	134.00	77.00	---	211.00	**269.00**	175.70	**221.00**
With matching light fixture for beveled mirror cabinets only											
with 2 lights	Inst	Ea	EA	2.67	199.00	77.00	---	276.00	**343.00**	231.70	**286.00**
with 3 lights	Inst	Ea	EA	2.67	106.00	77.00	---	183.00	**236.00**	150.80	**192.00**
with 4 lights	Inst	Ea	EA	2.67	146.00	77.00	---	223.00	**281.00**	184.70	**232.00**
with 6 lights	Inst	Ea	EA	2.67	175.00	77.00	---	252.00	**315.00**	210.70	**261.00**
Recessed mounting, overall sizes											
Swing door with mirror											
Builders series											
14" x 18"											
Polished brass strip	Inst	Ea	CA	1.33	57.10	35.90	---	93.00	**120.00**	77.00	**98.50**
Polished edge strip	Inst	Ea	CA	1.33	82.90	35.90	---	118.80	**150.00**	99.50	**124.00**
Stainless steel strip	Inst	Ea	CA	1.33	69.40	35.90	---	105.30	**134.00**	87.80	**111.00**
14" x 24"											
Polished brass strip	Inst	Ea	CA	1.33	58.20	35.90	---	94.10	**121.00**	78.00	**99.60**
Polished edge strip	Inst	Ea	CA	1.33	84.00	35.90	---	119.90	**151.00**	100.50	**126.00**
Stainless steel strip	Inst	Ea	CA	1.33	70.60	35.90	---	106.50	**136.00**	88.70	**112.00**
Decorator series											
14" x 18"											
Frameless beveled mirror	Inst	Ea	CA	1.33	84.00	35.90	---	119.90	**151.00**	100.50	**126.00**
Natural oak	Inst	Ea	CA	1.33	123.00	35.90	---	158.90	**196.00**	135.00	**165.00**
White birch	Inst	Ea	CA	1.33	119.00	35.90	---	154.90	**191.00**	131.00	**160.00**
White finish	Inst	Ea	CA	1.33	153.00	35.90	---	188.90	**231.00**	161.00	**195.00**
14" x 24"											
Frameless beveled mirror	Inst	Ea	CA	1.33	86.20	35.90	---	122.10	**154.00**	102.50	**128.00**
Natural oak	Inst	Ea	CA	1.33	125.00	35.90	---	160.90	**199.00**	137.00	**167.00**
White birch	Inst	Ea	CA	1.33	121.00	35.90	---	156.90	**194.00**	133.00	**163.00**
White finish	Inst	Ea	CA	1.33	131.00	35.90	---	166.90	**205.00**	142.00	**173.00**
Oak framed cabinet with oval mirror											
20" x 36"	Inst	Ea	CA	1.33	195.00	35.90	---	230.90	**279.00**	198.00	**237.00**
Mirrors											
Decorator oval mirrors, antique gold											
16" x 24"	Inst	Ea	CA	.211	67.20	5.69	---	72.89	**85.90**	63.11	**74.20**
16" x 32"	Inst	Ea	CA	.211	84.00	5.69	---	89.69	**105.00**	77.81	**91.10**

Shower equipment. **See Shower stalls, page 219**

Vanity cabinets. **See Cabinets, page 46**

Description	Oper	Unit	Crew Size	Man-Hours Per Unit	Avg Mat'l Unit Cost	Avg Labor Unit Cost	Avg Equip Unit Cost	Avg Total Unit Cost	Avg Price Incl O&P	Large Volume Avg Total Unit Cost	Large Volume Avg Price Incl O&P
			Costs Based On Small Volume							Large Volume	

Bathtubs

Plumbing fixtures with good quality supply fittings and faucets included in material cost. Labor cost for installation only of fixture, supply fittings, and faucets. For rough-in, see Adjustments in this bathtub section, page 38

Free-standing

Kohler Products, enameled cast iron

Description	Oper	Unit	Crew Size	Man-Hours Per Unit	Avg Mat'l Unit Cost	Avg Labor Unit Cost	Avg Equip Unit Cost	Avg Total Unit Cost	Avg Price Incl O&P	Large Volume Avg Total Unit Cost	Large Volume Avg Price Incl O&P
Birthday Bath											
72" L x 37-1/2" W											
White	Inst	Ea	SB	12.3	4480.00	322.00	---	4802.00	**5630.00**	4100.00	**4790.00**
Colors	Inst	Ea	SB	12.3	5120.00	322.00	---	5442.00	**6360.00**	4660.00	**5430.00**
Premium Colors	Inst	Ea	SB	12.3	5600.00	322.00	---	5922.00	**6920.00**	5080.00	**5910.00**
Required accessories											
Antique ball-and-claw legs (four)	Inst	Set	SB	---	948.00	---	---	948.00	**1090.00**	824.00	**948.00**
Antique bath faucet without riser	Inst	Ea	SB	---	716.00	---	---	716.00	**823.00**	622.00	**716.00**
Antique drain, chain & stopper	Inst	Set	SB	---	273.00	---	---	273.00	**314.00**	238.00	**274.00**
Antique riser tubes	Inst	Set	SB	---	88.30	---	---	88.30	**102.00**	76.90	**88.40**
Vintage Bath											
72" L x 42" W x 22" H											
White	Inst	Ea	SB	12.3	4310.00	322.00	---	4632.00	**5430.00**	3960.00	**4620.00**
Colors	Inst	Ea	SB	12.3	5350.00	322.00	---	5672.00	**6630.00**	4860.00	**5670.00**
Required accessories											
Wood base	Inst	Ea	SB	---	1080.00	---	---	1080.00	**1240.00**	937.00	**1080.00**
Ceramic base	Inst	Ea	SB	---	716.00	---	---	716.00	**823.00**	623.00	**716.00**
Adjustable feet	Inst	Set	SB	---	34.90	---	---	34.90	**40.20**	30.40	**34.90**

Recessed

American Standard Products, enameled steel

Description	Oper	Unit	Crew Size	Man-Hours Per Unit	Avg Mat'l Unit Cost	Avg Labor Unit Cost	Avg Equip Unit Cost	Avg Total Unit Cost	Avg Price Incl O&P	Large Volume Avg Total Unit Cost	Large Volume Avg Price Incl O&P
60" L x 30" W x 16-1/8" H (Solar)											
White, slip resistant	Inst	Ea	SB	10.3	361.00	270.00	---	631.00	**818.00**	489.00	**622.00**
Colors, slip resistant	Inst	Ea	SB	10.3	375.00	270.00	---	645.00	**833.00**	501.00	**636.00**
60" L x 30" W x 15" H (Salem)											
White, slip resistant	Inst	Ea	SB	10.3	381.00	270.00	---	651.00	**840.00**	506.00	**641.00**
Colors, slip resistant	Inst	Ea	SB	10.3	412.00	270.00	---	682.00	**876.00**	534.00	**673.00**

Kohler Products, enameled cast iron

Description	Oper	Unit	Crew Size	Man-Hours Per Unit	Avg Mat'l Unit Cost	Avg Labor Unit Cost	Avg Equip Unit Cost	Avg Total Unit Cost	Avg Price Incl O&P	Large Volume Avg Total Unit Cost	Large Volume Avg Price Incl O&P
42" L x 36" W x 14" H (Standish)											
White	Inst	Ea	SB	10.3	1100.00	270.00	---	1370.00	**1660.00**	1130.00	**1360.00**
Colors	Inst	Ea	SB	10.3	1270.00	270.00	---	1540.00	**1870.00**	1285.00	**1540.00**
48" L x 42" W x 14" H (Bradford)											
White	Inst	Ea	SB	10.3	1530.00	270.00	---	1800.00	**2160.00**	1505.00	**1790.00**
Colors	Inst	Ea	SB	10.3	1810.00	270.00	---	2080.00	**2490.00**	1755.00	**2070.00**
54" L x 30" W x 14" H (Seaforth)											
White	Inst	Ea	SB	10.3	926.00	270.00	---	1196.00	**1470.00**	981.00	**1190.00**
Colors	Inst	Ea	SB	10.3	1080.00	270.00	---	1350.00	**1640.00**	1113.00	**1340.00**
60" L x 32" W x 16" H (Mendota)											
White	Inst	Ea	SB	10.3	839.00	270.00	---	1109.00	**1370.00**	905.00	**1100.00**
Colors	Inst	Ea	SB	10.3	996.00	270.00	---	1266.00	**1550.00**	1041.00	**1260.00**
Premium Colors	Inst	Ea	SB	10.3	1100.00	270.00	---	1370.00	**1660.00**	1130.00	**1360.00**

Description	Oper	Unit	Crew Size	Man-Hours Per Unit	Avg Mat'l Unit Cost	Avg Labor Unit Cost	Avg Equip Unit Cost	Avg Total Unit Cost	Avg Price Incl O&P	Large Volume Avg Total Unit Cost	Large Volume Avg Price Incl O&P
Bathtub with whirlpool											
Plumbing installation											
White	Inst	Ea	SB	20.5	2600.00	537.00	---	3137.00	**3790.00**	2609.00	**3120.00**
Colors	Inst	Ea	SB	20.5	2870.00	537.00	---	3407.00	**4100.00**	2849.00	**3390.00**
Electrical installation	Inst	Ea	EA	9.41	---	271.00	---	271.00	**402.00**	177.00	**262.00**
60" L x 30" W x 14" H (Villager)											
White	Inst	Ea	SB	10.3	628.00	270.00	---	898.00	**1120.00**	721.00	**889.00**
Colors	Inst	Ea	SB	10.3	722.00	270.00	---	992.00	**1230.00**	803.00	**983.00**
66" L x 32" W x 16" H (Dynametric)											
White	Inst	Ea	SB	10.3	1210.00	270.00	---	1480.00	**1800.00**	1235.00	**1470.00**
Colors	Inst	Ea	SB	10.3	1440.00	270.00	---	1710.00	**2060.00**	1425.00	**1700.00**
Sunken											
Kohler Products, enameled cast iron											
60" L x 32" W x 18" H (Tea for Two)											
White	Inst	Ea	SB	10.3	1360.00	270.00	---	1630.00	**1970.00**	1355.00	**1620.00**
Colors	Inst	Ea	SB	10.3	1580.00	270.00	---	1850.00	**2210.00**	1545.00	**1840.00**
Premium Colors	Inst	Ea	SB	10.3	1740.00	270.00	---	2010.00	**2400.00**	1685.00	**2000.00**
Bathtub with whirlpool											
Plumbing installation											
White	Inst	Ea	SB	20.5	2600.00	537.00	---	3137.00	**3790.00**	2609.00	**3120.00**
Colors	Inst	Ea	SB	20.5	2770.00	537.00	---	3307.00	**3990.00**	2759.00	**3290.00**
Premium Colors	Inst	Ea	SB	20.5	2870.00	537.00	---	3407.00	**4100.00**	2849.00	**3390.00**
Electrical installation	Inst	Ea	EA	9.41	---	271.00	---	271.00	**402.00**	177.00	**262.00**
60" L x 36" W x 20" H (Steeping Bath)											
White	Inst	Ea	SB	10.3	1620.00	270.00	---	1890.00	**2260.00**	1585.00	**1880.00**
Colors	Inst	Ea	SB	10.3	1920.00	270.00	---	2190.00	**2610.00**	1845.00	**2180.00**
Premium Colors	Inst	Ea	SB	10.3	2150.00	270.00	---	2420.00	**2880.00**	2045.00	**2410.00**
Bathtub with whirlpool											
Plumbing installation											
White	Inst	Ea	SB	20.5	4350.00	537.00	---	4887.00	**5810.00**	4139.00	**4870.00**
Colors	Inst	Ea	SB	20.5	4700.00	537.00	---	5237.00	**6210.00**	4439.00	**5220.00**
Premium Colors	Inst	Ea	SB	20.5	4890.00	537.00	---	5427.00	**6430.00**	4609.00	**5410.00**
Electrical installation	Inst	Ea	EA	9.41	---	271.00	---	271.00	**402.00**	177.00	**262.00**
66" L x 32" W x 18" H (Maestro)											
White	Inst	Ea	SB	10.3	1070.00	270.00	---	1340.00	**1630.00**	1106.00	**1330.00**
Colors	Inst	Ea	SB	10.3	1220.00	270.00	---	1490.00	**1810.00**	1235.00	**1480.00**
Premium Colors	Inst	Ea	SB	10.3	1330.00	270.00	---	1600.00	**1940.00**	1335.00	**1600.00**
Bathtub with whirlpool											
Plumbing installation											
White	Inst	Ea	SB	20.5	3100.00	537.00	---	3637.00	**4360.00**	3039.00	**3620.00**
Colors	Inst	Ea	SB	20.5	3340.00	537.00	---	3877.00	**4640.00**	3249.00	**3860.00**
Premium Colors	Inst	Ea	SB	20.5	3470.00	537.00	---	4007.00	**4790.00**	3369.00	**3990.00**
Electrical installation	Inst	Ea	EA	9.41	---	271.00	---	271.00	**402.00**	177.00	**262.00**
72" L x 36" W x 18" H (Caribbean)											
White	Inst	Ea	SB	10.3	1880.00	270.00	---	2150.00	**2560.00**	1815.00	**2140.00**
Colors	Inst	Ea	SB	10.3	2180.00	270.00	---	2450.00	**2910.00**	2075.00	**2450.00**
Premium Colors	Inst	Ea	SB	10.3	2410.00	270.00	---	2680.00	**3180.00**	2275.00	**2670.00**
Bathtub with whirlpool											
Plumbing installation											
White	Inst	Ea	SB	20.5	4100.00	537.00	---	4637.00	**5510.00**	3919.00	**4620.00**
Colors	Inst	Ea	SB	20.5	4450.00	537.00	---	4987.00	**5910.00**	4219.00	**4970.00**
Premium Colors	Inst	Ea	SB	20.5	4640.00	537.00	---	5177.00	**6130.00**	4389.00	**5160.00**
Electrical installation	Inst	Ea	EA	9.41	---	271.00	---	271.00	**402.00**	177.00	**262.00**

Description	Oper	Unit	Crew Size	Man-Hours Per Unit	Avg Mat'l Unit Cost	Avg Labor Unit Cost	Avg Equip Unit Cost	Avg Total Unit Cost	Avg Price Incl O&P	Large Volume Avg Total Unit Cost	Large Volume Avg Price Incl O&P
			Costs Based On Small Volume								
72" L x 36" W x 21" H (Tea for Two)											
White	Inst	Ea	SB	10.3	2150.00	270.00	---	2420.00	**2870.00**	2045.00	**2410.00**
Colors	Inst	Ea	SB	10.3	2560.00	270.00	---	2830.00	**3340.00**	2405.00	**2820.00**
Premium Colors	Inst	Ea	SB	10.3	2870.00	270.00	---	3140.00	**3700.00**	2665.00	**3130.00**
Bathtub with whirlpool											
Plumbing installation											
White	Inst	Ea	SB	20.5	3810.00	537.00	---	4347.00	**5180.00**	3659.00	**4330.00**
Colors	Inst	Ea	SB	20.5	4130.00	537.00	---	4667.00	**5550.00**	3939.00	**4650.00**
Premium Colors	Inst	Ea	SB	20.5	4300.00	537.00	---	4837.00	**5750.00**	4089.00	**4820.00**
Electrical installation	Inst	Ea	EA	9.41	---	271.00	---	271.00	**402.00**	177.00	**262.00**
72" L x 42" W x 22" H (Seawall Bath) with personal shower system											
White	Inst	Ea	SB	10.3	5130.00	270.00	---	5400.00	**6300.00**	4635.00	**5390.00**
Colors	Inst	Ea	SB	10.3	5700.00	270.00	---	5970.00	**6960.00**	5135.00	**5970.00**
Premium Colors	Inst	Ea	SB	10.3	6130.00	270.00	---	6400.00	**7460.00**	5515.00	**6400.00**
Bathtub with whirlpool											
Plumbing installation											
White	Inst	Ea	SB	20.5	9020.00	537.00	---	9557.00	**11200**	8199.00	**9540.00**
Colors	Inst	Ea	SB	20.5	9430.00	537.00	---	9967.00	**11600**	8559.00	**9960.00**
Premium Colors	Inst	Ea	SB	20.5	9660.00	537.00	---	10197	**11900**	8759.00	**10200**
Electrical installation	Inst	Ea	EA	9.41	---	271.00	---	271.00	**402.00**	177.00	**262.00**
Options											
Grab bar kit											
White	Inst	Ea	SB	---	216.00	---	---	216.00	**248.00**	188.00	**216.00**
Colors	Inst	Ea	SB	---	270.00	---	---	270.00	**311.00**	235.00	**270.00**
Drain kit											
White	Inst	Ea	SB	---	162.00	---	---	162.00	**186.00**	141.00	**162.00**
Colors	Inst	Ea	SB	---	216.00	---	---	216.00	**248.00**	188.00	**216.00**
Kohler Products, fiberglass											
60" L x 51" W x 25-3/4" H (Cape May Bath)											
White	Inst	Ea	SB	10.3	562.00	270.00	---	832.00	**1050.00**	664.00	**823.00**
Colors	Inst	Ea	SB	10.3	579.00	270.00	---	849.00	**1070.00**	679.00	**839.00**
60" L x 38" W x 25" H (Precedence)											
White	Inst	Ea	SB	10.3	3760.00	270.00	---	4030.00	**4720.00**	3445.00	**4020.00**
Colors	Inst	Ea	SB	10.3	3930.00	270.00	---	4200.00	**4920.00**	3595.00	**4190.00**
American Standard Products, acrylic											
60" L x 32" W x 21-1/2" H (Elisse Remodel)											
Bathtub with whirlpool											
Plumbing installation											
White	Inst	Ea	SB	24.6	2290.00	645.00	---	2935.00	**3590.00**	2409.00	**2920.00**
Colors	Inst	Ea	SB	24.6	2370.00	645.00	---	3015.00	**3690.00**	2479.00	**3000.00**
Electrical installation	Inst	Ea	EA	9.41	---	271.00	---	271.00	**402.00**	177.00	**262.00**
72" L x 44" W x 20" H (Elisse)											
White	Inst	Ea	SB	12.3	1130.00	322.00	---	1452.00	**1780.00**	1193.00	**1440.00**
Colors	Inst	Ea	SB	12.3	1210.00	322.00	---	1532.00	**1870.00**	1260.00	**1520.00**
Bathtub with whirlpool											
Plumbing installation											
White	Inst	Ea	SB	24.6	2630.00	645.00	---	3275.00	**3990.00**	2709.00	**3260.00**
Colors	Inst	Ea	SB	24.6	2710.00	645.00	---	3355.00	**4080.00**	2779.00	**3340.00**
Electrical installation	Inst	Ea	EA	9.41	---	271.00	---	271.00	**402.00**	177.00	**262.00**

Description	Oper	Unit	Crew Size	Man-Hours Per Unit	Avg Mat'l Unit Cost	Avg Labor Unit Cost	Avg Equip Unit Cost	Avg Total Unit Cost	Avg Price Incl O&P	Large Volume Avg Total Unit Cost	Large Volume Avg Price Incl O&P
72" L x 38" W x 20" H (Elisse Oval)											
White	Inst	Ea	SB	12.3	1020.00	322.00	---	1342.00	**1650.00**	1096.00	**1330.00**
Colors	Inst	Ea	SB	12.3	1100.00	322.00	---	1422.00	**1750.00**	1167.00	**1410.00**
Bathtub with whirlpool											
Plumbing installation											
White	Inst	Ea	SB	24.6	2410.00	645.00	---	3055.00	**3730.00**	2509.00	**3030.00**
Colors	Inst	Ea	SB	24.6	2490.00	645.00	---	3135.00	**3820.00**	2579.00	**3110.00**
Electric installation	Inst	Ea	EA	9.41	162.00	271.00	---	433.00	**588.00**	318.00	**425.00**
American Standard Products, Americast											
60" L x 32" W x 18" H (Cambridge)											
White	Inst	Ea	SB	12.3	735.00	322.00	---	1057.00	**1330.00**	849.00	**1050.00**
Colors	Inst	Ea	SB	12.3	880.00	322.00	---	1202.00	**1490.00**	975.00	**1190.00**
Bathtub with whirlpool											
Plumbing installation											
White	Inst	Ea	SB	24.6	2220.00	645.00	---	2865.00	**3510.00**	2349.00	**2840.00**
Colors	Inst	Ea	SB	24.6	2390.00	645.00	---	3035.00	**3710.00**	2499.00	**3010.00**
Electrical installation	Inst	Ea	EA	9.41	150.00	271.00	---	421.00	**574.00**	308.00	**413.00**
66" L x 34" W x 17" H (Princeton) w/ luxury ledge											
White	Inst	Ea	SB	12.3	839.00	322.00	---	1161.00	**1450.00**	940.00	**1150.00**
Colors	Inst	Ea	SB	12.3	944.00	322.00	---	1266.00	**1570.00**	1032.00	**1260.00**
66" L x 32" W x 18" H (Stratford)											
White	Inst	Ea	SB	12.3	1080.00	322.00	---	1402.00	**1730.00**	1152.00	**1400.00**
Colors	Inst	Ea	SB	12.3	1240.00	322.00	---	1562.00	**1900.00**	1290.00	**1550.00**
Bathtub with whirlpool											
Plumbing installation											
White	Inst	Ea	SB	24.6	2870.00	645.00	---	3515.00	**4260.00**	2909.00	**3490.00**
Colors	Inst	Ea	SB	24.6	3020.00	645.00	---	3665.00	**4430.00**	3049.00	**3650.00**
Electrical installation	Inst	Ea	EA	9.41	150.00	271.00	---	421.00	**574.00**	308.00	**413.00**
Adjustments											
Remove & reset tub	Reset	Ea	SB	8.21	55.00	215.00	---	270.00	**384.00**	187.90	**263.00**
Install rough-in	Inst	Ea	SB	15.4	85.00	404.00	---	489.00	**699.00**	336.00	**476.00**
Shower fixture over tub with mixer valve											
ADD	Inst	Ea	SA	3.08	115.00	93.80	---	208.80	**272.00**	160.90	**206.00**

Block, concrete. See Masonry, page 164

Brick. See Masonry, page 161

Cabinets

Top quality cabinets are built with the structural stability of fine furniture. Framing stock is kiln dried and a full 1" thick. Cabinets have backs, usually 5-ply $^{3}/_{16}$"-thick plywood, with all backs and interiors finished. Frames should be constructed of hardwood with mortise and tenon joints; corner blocks should be used on all four corners of all base cabinets. Doors are usually of select $^{7}/_{16}$" thick solid core construction using semi-concealed hinges. End panels are $^{1}/_{2}$" thick and attached to frames with mortise and tenon joints, glued and pinned under pressure. Panels should also be dadoed to receive the tops and bottoms of wall cabinets. Shelves are adjustable with veneer faces and front edges. The hardware includes magnetic catches, heavy duty die cast pulls and hinges, and ball-bearing suspension system. The finish is scratch and stain resistant, including a first coat of hand-wiped stain, a sealer coat, and a synthetic varnish with plastic laminate characteristics.

Average quality cabinets feature hardwood frame construction with plywood backs and veneered plywood end panels. Joints are glued mortise and tenon. Doors are solid core attached with exposed self-closing hinges. Shelves are adjustable, and drawers ride on a ball-bearing side suspension glide system. The finish is usually three coats including stain, sealer, and a mar-resistant top coat for easy cleaning.

Economy quality cabinets feature pine construction with joints glued under pressure. Doors, drawer fronts, and side or end panels are constructed of either $^{1}/_{2}$"-thick wood composition board or $^{1}/_{2}$"-thick veneered pine. Face frames are $^{3}/_{4}$"-thick wood composition board or $^{3}/_{4}$"-thick pine. Features include adjustable shelves, hinge straps, and a three-point suspension system on drawers (using nylon rollers). The finish consists of a filler coat, base coat, and final polyester top coat.

Kitchen Cabinet Installation Procedure

To develop a layout plan, measure and write down the following:

1. Floor space.
2. Height and width of all walls.
3. Location of electrical outlets.
4. Size and position of doors, windows, and vents.
5. Location of any posts or pillars. Walls must be prepared if chair rails or baseboards are located where cabinets will be installed.
6. Common height and depth of base cabinets (including 1" for countertops) and wall cabinets.

When you plan the cabinet placement, consider this Rule of Thumb:

Allow between 4$^{1}/_{2}$' and 5$^{1}/_{2}$' of counter surface between the refrigerator and sink. Allow between 3' and 4' between the sink and range.

1. What do you have to fit into the available space?
2. Is there enough counter space on both sides of all appliances and sinks? The kitchen has three work centers, each with a major appliance as its hub, and each needing adequate counter space. They are:

   a. Fresh and frozen food center — Refrigerator-freezer

   b. Clean-up center — Sink with disposal-dishwasher

   c. Cooking center — Range-oven
3. Will the sink workspace fit neatly in front of a window?
4. The kitchen triangle is the most efficient kitchen design; it means placing each major center at approximately equidistant triangle points. The ideal triangle is 22 feet total. It should never be less than 13 feet or more than 25 feet.
5. Where are the centers located? A logical working and walking pattern is from refrigerator to sink to range. The refrigerator should be at a triangle point near a door, to minimize the distance to bring in groceries and reduce traffic that could interfere with food preparation. The range should be at a triangle point near the serving and dining area. The sink is located between the two. The refrigerator should be located far enough from the range so that the heat will not affect the refrigerator's cooling efficiency.
6. Does the plan allow for lighting the range and sink work centers and for ventilating the range center?
7. Make sure that open doors (such as cabinet doors or entrance/exit doors) won't interfere with access to an appliance. To clear appliances and cabinets, a door opening should not be less than 30" from the corner, since such equipment is 24" to 28" in depth. A clearance of 48" is necessary when a range is next to a door.
8. Next locate the wall studs with a stud finder or hammer, since all cabinets attach to walls with screws, never nails. Also remove chair rails or baseboards where they conflict with cabinets.

Wall Cabinets: From the highest point on the floor, measure up approximately 84" to determine the top of wall cabinets. Using two #10 x 2$^{1}/_{2}$" wood screws, drill through hanging strips built into the cabinet backs at top and bottom. Use a level to assure that cabinets and doors are aligned, then tighten the screws.

Base Cabinets: Start with a corner unit and a base unit on each side of the corner unit. Place this combination in the corner and work out from both sides. Use "C" clamps when connecting cabinets to draw adjoining units into alignment. With the front face plumb and the unit level from front to back and across the front edge, attach the unit to wall studs by screwing through the hanging strips. To attach adjoining cabinets, drill two holes in the vertical side of one cabinet, inside the door (near top and bottom), and just into the stile of the adjoining cabinet.

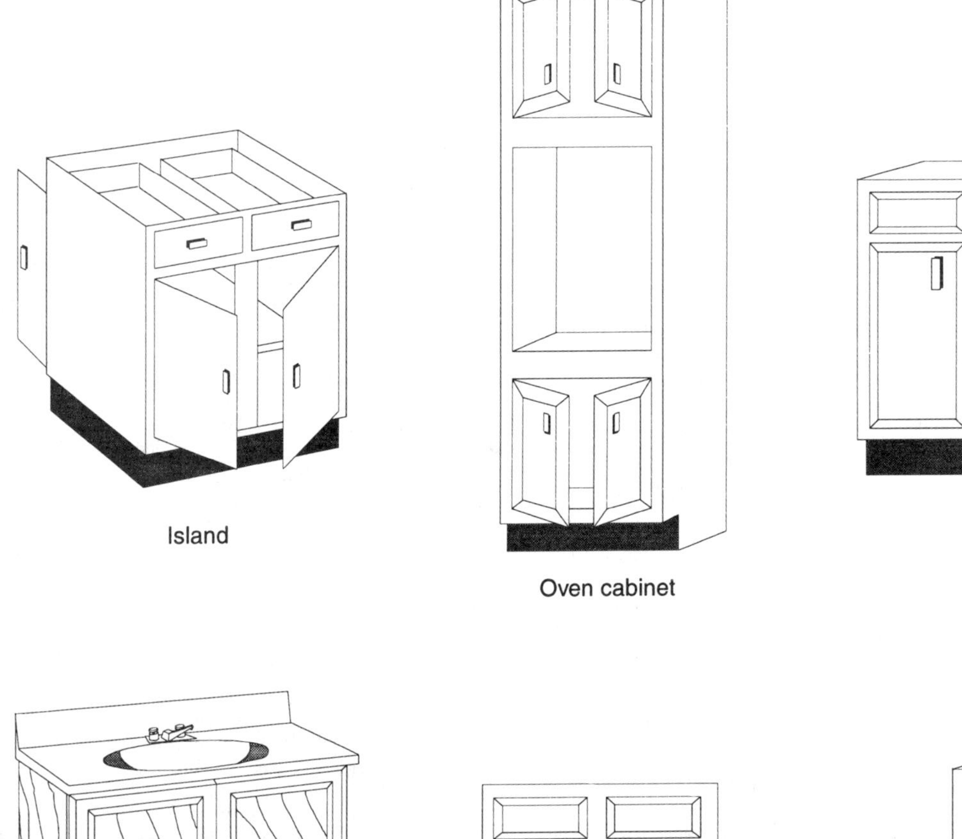

Island

Oven cabinet

Base

Vanity

Sink front

Base

Description	Oper	Unit	Crew Size	Man-Hours Per Unit	Avg Mat'l Unit Cost	Avg Labor Unit Cost	Avg Equip Unit Cost	Avg Total Unit Cost	Avg Price Incl O&P	Large Volume Avg Total Unit Cost	Large Volume Avg Price Incl O&P
			Costs Based On Small Volume							Large Volume	
Cabinets											
Labor costs include hanging and fitting of cabinets											
Kitchen											
3' x 4', wood; base, wall, or peninsula	Demo	Ea	LB	1.067	---	23.40	---	23.40	**35.40**	14.10	**21.20**
Kitchen; all hardware included											
Base cabinets, 35" H, 24" D; no tops											
12" W, 1 door, 1 drawer											
High quality workmanship	Inst	Ea	CJ	1.67	144.00	40.90	---	184.90	**235.00**	159.50	**199.00**
Good quality workmanship	Inst	Ea	CJ	1.11	117.00	27.20	---	144.20	**182.00**	126.30	**157.00**
Average quality workmanship	Inst	Ea	CJ	1.11	89.80	27.20	---	117.00	**149.00**	100.60	**126.00**
15" W, 1 door, 1 drawer											
High quality workmanship	Inst	Ea	CJ	1.67	152.00	40.90	---	192.90	**244.00**	167.50	**208.00**
Good quality workmanship	Inst	Ea	CJ	1.11	125.00	27.20	---	152.20	**192.00**	134.30	**166.00**
Average quality workmanship	Inst	Ea	CJ	1.11	95.00	27.20	---	122.20	**155.00**	105.60	**132.00**
18" W, 1 door, 1 drawer											
High quality workmanship	Inst	Ea	CJ	1.67	157.00	40.90	---	197.90	**251.00**	172.50	**214.00**
Good quality workmanship	Inst	Ea	CJ	1.11	129.00	27.20	---	156.20	**196.00**	138.30	**171.00**
Average quality workmanship	Inst	Ea	CJ	1.11	99.00	27.20	---	126.20	**160.00**	109.30	**136.00**
21" W, 1 door, 1 drawer											
High quality workmanship	Inst	Ea	CJ	1.93	174.00	47.20	---	221.20	**281.00**	192.40	**240.00**
Good quality workmanship	Inst	Ea	CJ	1.23	144.00	30.10	---	174.10	**218.00**	153.10	**190.00**
Average quality workmanship	Inst	Ea	CJ	1.23	111.00	30.10	---	141.10	**179.00**	122.10	**153.00**
24" W, 1 door, 1 drawer											
High quality workmanship	Inst	Ea	CJ	1.93	191.00	47.20	---	238.20	**301.00**	208.40	**259.00**
Good quality workmanship	Inst	Ea	CJ	1.23	157.00	30.10	---	187.10	**234.00**	166.10	**205.00**
Average quality workmanship	Inst	Ea	CJ	1.23	139.00	30.10	---	169.10	**212.00**	148.10	**184.00**
30" W, 2 doors, 2 drawers											
High quality workmanship	Inst	Ea	CJ	2.58	244.00	63.10	---	307.10	**389.00**	266.70	**333.00**
Good quality workmanship	Inst	Ea	CJ	1.74	211.00	42.60	---	253.60	**318.00**	223.40	**277.00**
Average quality workmanship	Inst	Ea	CJ	1.74	160.00	42.60	---	202.60	**256.00**	175.40	**219.00**
36" W, 2 doors, 2 drawers											
High quality workmanship	Inst	Ea	CJ	2.58	281.00	63.10	---	344.10	**433.00**	301.70	**374.00**
Good quality workmanship	Inst	Ea	CJ	1.74	232.00	42.60	---	274.60	**343.00**	243.40	**301.00**
Average quality workmanship	Inst	Ea	CJ	1.74	177.00	42.60	---	219.60	**277.00**	191.40	**238.00**
42" W, 2 doors, 2 drawers											
High quality workmanship	Inst	Ea	CJ	3.33	325.00	81.50	---	406.50	**513.00**	353.90	**440.00**
Good quality workmanship	Inst	Ea	CJ	2.22	269.00	54.30	---	323.30	**406.00**	285.50	**353.00**
Average quality workmanship	Inst	Ea	CJ	2.22	205.00	54.30	---	259.30	**328.00**	224.50	**280.00**
48" W, 4 doors, 2 drawers											
High quality workmanship	Inst	Ea	CJ	3.33	313.00	81.50	---	394.50	**499.00**	342.90	**427.00**
Good quality workmanship	Inst	Ea	CJ	2.22	259.00	54.30	---	313.30	**393.00**	275.50	**341.00**
Average quality workmanship	Inst	Ea	CJ	2.22	195.00	54.30	---	249.30	**317.00**	216.50	**270.00**
Base corner cabinet, blind, 35" H, 24" D, no tops											
36" W, 1 door, 1 drawer											
High quality workmanship	Inst	Ea	CJ	2.39	205.00	58.50	---	263.50	**334.00**	227.00	**284.00**
Good quality workmanship	Inst	Ea	CJ	1.65	164.00	40.40	---	204.40	**258.00**	178.20	**221.00**
Average quality workmanship	Inst	Ea	CJ	1.65	131.00	40.40	---	171.40	**218.00**	147.20	**184.00**

Description	Oper	Unit	Crew Size	Man-Hours Per Unit	Avg Mat'l Unit Cost	Avg Labor Unit Cost	Avg Equip Unit Cost	Avg Total Unit Cost	Avg Price Incl O&P	Large Volume Avg Total Unit Cost	Large Volume Avg Price Incl O&P
42" W, 1 door, 1 drawer											
High quality workmanship	Inst	Ea	CJ	3.02	226.00	73.90	---	299.90	**383.00**	256.50	**322.00**
Good quality workmanship	Inst	Ea	CJ	1.93	186.00	47.20	---	233.20	**295.00**	203.40	**253.00**
Average quality workmanship	Inst	Ea	CJ	1.93	141.00	47.20	---	188.20	**241.00**	161.40	**202.00**
36" x 36" (lazy Susan)											
High quality workmanship	Inst	Ea	CJ	2.39	305.00	58.50	---	363.50	**455.00**	321.00	**397.00**
Good quality workmanship	Inst	Ea	CJ	1.65	252.00	40.40	---	292.40	**364.00**	261.20	**321.00**
Average quality workmanship	Inst	Ea	CJ	1.65	193.00	40.40	---	233.40	**293.00**	205.20	**254.00**
Utility closets, 81" to 85" H, 24" W, 24" D											
High quality workmanship	Inst	Ea	CJ	3.33	467.00	81.50	---	548.50	**685.00**	487.90	**601.00**
Good quality workmanship	Inst	Ea	CJ	2.22	364.00	54.30	---	418.30	**520.00**	374.50	**460.00**
Average quality workmanship	Inst	Ea	CJ	2.22	277.00	54.30	---	331.30	**415.00**	292.50	**362.00**
Drawer base cabinets, 35" H, 24" D											
4 drawers, no tops											
18" W											
High quality workmanship	Inst	Ea	CJ	1.67	195.00	40.90	---	235.90	**297.00**	208.50	**257.00**
Good quality workmanship	Inst	Ea	CJ	1.11	168.00	27.20	---	195.20	**242.00**	173.30	**214.00**
Average quality workmanship	Inst	Ea	CJ	1.11	128.00	27.20	---	155.20	**195.00**	136.30	**169.00**
24" W											
High quality workmanship	Inst	Ea	CJ	1.93	219.00	47.20	---	266.20	**335.00**	234.40	**290.00**
Good quality workmanship	Inst	Ea	CJ	1.23	181.00	30.10	---	211.10	**263.00**	188.10	**231.00**
Average quality workmanship	Inst	Ea	CJ	1.23	139.00	30.10	---	169.10	**212.00**	148.10	**184.00**
Island cabinets											
Island base cabinets, 35" H, 24" D											
24" W, 1 door both sides											
High quality workmanship	Inst	Ea	CJ	2.05	341.00	50.10	---	391.10	**485.00**	350.10	**430.00**
Good quality workmanship	Inst	Ea	CJ	1.37	280.00	33.50	---	313.50	**387.00**	283.10	**346.00**
Average quality workmanship	Inst	Ea	CJ	1.37	215.00	33.50	---	248.50	**309.00**	222.10	**273.00**
30" W, 2 doors both sides											
High quality workmanship	Inst	Ea	CJ	2.67	360.00	65.30	---	425.30	**532.00**	378.10	**466.00**
Good quality workmanship	Inst	Ea	CJ	1.78	297.00	43.50	---	340.50	**423.00**	305.20	**375.00**
Average quality workmanship	Inst	Ea	CJ	1.78	226.00	43.50	---	269.50	**337.00**	238.20	**294.00**
36" W, 2 doors both sides											
High quality workmanship	Inst	Ea	CJ	2.67	379.00	65.30	---	444.30	**554.00**	395.10	**487.00**
Good quality workmanship	Inst	Ea	CJ	1.78	313.00	43.50	---	356.50	**442.00**	320.20	**392.00**
Average quality workmanship	Inst	Ea	CJ	1.78	238.00	43.50	---	281.50	**351.00**	249.20	**308.00**
48" W, 4 doors both sides											
High quality workmanship	Inst	Ea	CJ	3.56	407.00	87.10	---	494.10	**620.00**	434.10	**537.00**
Good quality workmanship	Inst	Ea	CJ	2.32	335.00	56.80	---	391.80	**489.00**	349.00	**430.00**
Average quality workmanship	Inst	Ea	CJ	2.32	255.00	56.80	---	311.80	**392.00**	273.00	**339.00**
Corner island base cabinets, 35" H, 24" D											
42" W, 4 doors, 2 drawers											
High quality workmanship	Inst	Ea	CJ	3.56	360.00	87.10	---	447.10	**565.00**	391.10	**485.00**
Good quality workmanship	Inst	Ea	CJ	2.32	294.00	56.80	---	350.80	**439.00**	311.00	**383.00**
Average quality workmanship	Inst	Ea	CJ	2.32	222.00	56.80	---	278.80	**352.00**	242.00	**302.00**
48" W, 6 doors, 2 drawers											
High quality workmanship	Inst	Ea	CJ	4.44	378.00	109.00	---	487.00	**618.00**	420.30	**525.00**
Good quality workmanship	Inst	Ea	CJ	2.96	312.00	72.40	---	384.40	**484.00**	336.50	**417.00**
Average quality workmanship	Inst	Ea	CJ	2.96	236.00	72.40	---	308.40	**394.00**	265.50	**333.00**

Description	Oper	Unit	Crew Size	Man-Hours Per Unit	Avg Mat'l Unit Cost	Avg Labor Unit Cost	Avg Equip Unit Cost	Avg Total Unit Cost	Avg Price Incl O&P	Large Volume Avg Total Unit Cost	Large Volume Avg Price Incl O&P
			Costs Based On Small Volume							Large Volume	
Hanging corner island wall cabinets, 18" H, 24" D											
18" W, 3 doors											
High quality workmanship	Inst	Ea	CJ	2.42	261.00	59.20	---	320.20	**404.00**	281.50	**349.00**
Good quality workmanship	Inst	Ea	CJ	1.62	191.00	39.60	---	230.60	**290.00**	203.70	**252.00**
Average quality workmanship	Inst	Ea	CJ	1.62	147.00	39.60	---	186.60	**236.00**	161.70	**201.00**
24" W, 3 doors											
High quality workmanship	Inst	Ea	CJ	2.42	269.00	59.20	---	328.20	**413.00**	288.50	**357.00**
Good quality workmanship	Inst	Ea	CJ	1.62	199.00	39.60	---	238.60	**299.00**	210.70	**261.00**
Average quality workmanship	Inst	Ea	CJ	1.62	150.00	39.60	---	189.60	**241.00**	164.70	**206.00**
30" W, 3 doors											
High quality workmanship	Inst	Ea	CJ	3.14	277.00	76.80	---	353.80	**449.00**	306.00	**382.00**
Good quality workmanship	Inst	Ea	CJ	2.11	207.00	51.60	---	258.60	**327.00**	225.80	**280.00**
Average quality workmanship	Inst	Ea	CJ	2.11	158.00	51.60	---	209.60	**269.00**	179.80	**225.00**
Hanging island cabinets, 18" H, 12" D											
30" W, 2 doors both sides											
High quality workmanship	Inst	Ea	CJ	2.86	277.00	70.00	---	347.00	**439.00**	301.60	**376.00**
Good quality workmanship	Inst	Ea	CJ	1.90	203.00	46.50	---	249.50	**315.00**	218.90	**272.00**
Average quality workmanship	Inst	Ea	CJ	1.90	154.00	46.50	---	200.50	**256.00**	172.90	**216.00**
36" W, 2 doors both sides											
High quality workmanship	Inst	Ea	CJ	2.86	285.00	70.00	---	355.00	**448.00**	309.60	**385.00**
Good quality workmanship	Inst	Ea	CJ	1.90	210.00	46.50	---	256.50	**322.00**	224.90	**279.00**
Average quality workmanship	Inst	Ea	CJ	1.90	160.00	46.50	---	206.50	**262.00**	177.90	**222.00**
Hanging island cabinets, 24" H, 12" D											
24" W, 2 doors both sides											
High quality workmanship	Inst	Ea	CJ	2.42	261.00	59.20	---	320.20	**404.00**	281.50	**349.00**
Good quality workmanship	Inst	Ea	CJ	1.62	182.00	39.60	---	221.60	**279.00**	194.70	**241.00**
Average quality workmanship	Inst	Ea	CJ	1.62	135.00	39.60	---	174.60	**222.00**	149.70	**188.00**
30" W, 2 doors both sides											
High quality workmanship	Inst	Ea	CJ	3.14	272.00	76.80	---	348.80	**443.00**	301.00	**376.00**
Good quality workmanship	Inst	Ea	CJ	2.11	197.00	51.60	---	248.60	**314.00**	215.80	**269.00**
Average quality workmanship	Inst	Ea	CJ	2.11	145.00	51.60	---	196.60	**253.00**	166.80	**211.00**
36" W, 2 doors both sides											
High quality workmanship	Inst	Ea	CJ	3.14	285.00	76.80	---	361.80	**459.00**	314.00	**391.00**
Good quality workmanship	Inst	Ea	CJ	2.11	210.00	51.60	---	261.60	**330.00**	227.80	**283.00**
Average quality workmanship	Inst	Ea	CJ	2.11	160.00	51.60	---	211.60	**270.00**	180.80	**227.00**
42" W, 2 doors both sides											
High quality workmanship	Inst	Ea	CJ	4.10	345.00	100.00	---	445.00	**566.00**	384.20	**480.00**
Good quality workmanship	Inst	Ea	CJ	2.71	253.00	66.30	---	319.30	**405.00**	277.90	**346.00**
Average quality workmanship	Inst	Ea	CJ	2.71	194.00	66.30	---	260.30	**334.00**	221.90	**279.00**
48" W, 4 doors both sides											
High quality workmanship	Inst	Ea	CJ	4.10	413.00	100.00	---	513.00	**648.00**	448.20	**557.00**
Good quality workmanship	Inst	Ea	CJ	2.71	305.00	66.30	---	371.30	**467.00**	325.90	**404.00**
Average quality workmanship	Inst	Ea	CJ	2.71	232.00	66.30	---	298.30	**380.00**	257.90	**322.00**
Oven cabinets											
81" to 85" H, 24" D, 27" W											
High quality workmanship	Inst	Ea	CJ	3.33	499.00	81.50	---	580.50	**723.00**	517.90	**637.00**
Good quality workmanship	Inst	Ea	CJ	2.22	388.00	54.30	---	442.30	**548.00**	397.50	**487.00**
Average quality workmanship	Inst	Ea	CJ	2.22	297.00	54.30	---	351.30	**439.00**	311.50	**384.00**

Description	Oper	Unit	Crew Size	Man-Hours Per Unit	Avg Mat'l Unit Cost	Avg Labor Unit Cost	Avg Equip Unit Cost	Avg Total Unit Cost	Avg Price Incl O&P	Avg Total Unit Cost	Avg Price Incl O&P
			Costs Based On Small Volume							Large Volume	
Sink/range											
Base cabinets, 35" H, 24" D, 2 doors, no tops included											
30" W											
High quality workmanship	Inst	Ea	CJ	2.22	162.00	54.30	---	216.30	**277.00**	185.50	**232.00**
Good quality workmanship	Inst	Ea	CJ	1.48	133.00	36.20	---	169.20	**215.00**	146.70	**183.00**
Average quality workmanship	Inst	Ea	CJ	1.48	102.00	36.20	---	138.20	**177.00**	117.20	**148.00**
36" W											
High quality workmanship	Inst	Ea	CJ	2.22	207.00	54.30	---	261.30	**331.00**	227.50	**283.00**
Good quality workmanship	Inst	Ea	CJ	1.48	172.00	36.20	---	208.20	**261.00**	182.70	**226.00**
Average quality workmanship	Inst	Ea	CJ	1.48	131.00	36.20	---	167.20	**212.00**	144.70	**180.00**
42" W											
High quality workmanship	Inst	Ea	CJ	2.91	230.00	71.20	---	301.20	**384.00**	258.60	**324.00**
Good quality workmanship	Inst	Ea	CJ	1.93	190.00	47.20	---	237.20	**300.00**	207.40	**257.00**
Average quality workmanship	Inst	Ea	CJ	1.93	144.00	47.20	---	191.20	**244.00**	163.40	**205.00**
48" W											
High quality workmanship	Inst	Ea	CJ	2.91	260.00	71.20	---	331.20	**420.00**	286.60	**358.00**
Good quality workmanship	Inst	Ea	CJ	1.93	214.00	47.20	---	261.20	**328.00**	229.40	**284.00**
Average quality workmanship	Inst	Ea	CJ	1.93	164.00	47.20	---	211.20	**268.00**	182.40	**228.00**
Sink/range front cabinets, 35" H, 2 doors											
30" W											
High quality workmanship	Inst	Ea	CJ	1.93	102.00	47.20	---	149.20	**194.00**	123.90	**158.00**
Good quality workmanship	Inst	Ea	CJ	1.33	84.50	32.50	---	117.00	**151.00**	99.00	**125.00**
Average quality workmanship	Inst	Ea	CJ	1.33	64.70	32.50	---	97.20	**127.00**	80.40	**103.00**
36" W											
High quality workmanship	Inst	Ea	CJ	1.93	137.00	47.20	---	184.20	**236.00**	157.40	**198.00**
Good quality workmanship	Inst	Ea	CJ	1.33	114.00	32.50	---	146.50	**186.00**	126.60	**158.00**
Average quality workmanship	Inst	Ea	CJ	1.33	88.40	32.50	---	120.90	**156.00**	102.70	**129.00**
42" W											
High quality workmanship	Inst	Ea	CJ	2.50	150.00	61.20	---	211.20	**274.00**	177.90	**226.00**
Good quality workmanship	Inst	Ea	CJ	1.67	125.00	40.90	---	165.90	**213.00**	142.50	**179.00**
Average quality workmanship	Inst	Ea	CJ	1.67	95.00	40.90	---	135.90	**176.00**	113.80	**144.00**
48" W											
High quality workmanship	Inst	Ea	CJ	2.50	158.00	61.20	---	219.20	**283.00**	185.90	**235.00**
Good quality workmanship	Inst	Ea	CJ	1.67	136.00	40.90	---	176.90	**225.00**	152.50	**190.00**
Average quality workmanship	Inst	Ea	CJ	1.67	103.00	40.90	---	143.90	**186.00**	121.20	**153.00**
Wall cabinets											
Refrigerator cabinets, 15" H, 12" D, 2 doors											
30" W											
High quality workmanship	Inst	Ea	CJ	2.58	137.00	63.10	---	200.10	**261.00**	166.70	**212.00**
Good quality workmanship	Inst	Ea	CJ	1.70	102.00	41.60	---	143.60	**185.00**	120.70	**153.00**
Average quality workmanship	Inst	Ea	CJ	1.70	77.90	41.60	---	119.50	**157.00**	98.40	**126.00**
33" W											
High quality workmanship	Inst	Ea	CJ	2.58	143.00	63.10	---	206.10	**267.00**	171.70	**218.00**
Good quality workmanship	Inst	Ea	CJ	1.70	104.00	41.60	---	145.60	**188.00**	123.20	**156.00**
Average quality workmanship	Inst	Ea	CJ	1.70	79.20	41.60	---	120.80	**158.00**	99.60	**128.00**
36" W											
High quality workmanship	Inst	Ea	CJ	2.58	150.00	63.10	---	213.10	**276.00**	178.70	**227.00**
Good quality workmanship	Inst	Ea	CJ	1.70	111.00	41.60	---	152.60	**196.00**	129.20	**163.00**
Average quality workmanship	Inst	Ea	CJ	1.70	84.50	41.60	---	126.10	**165.00**	104.60	**134.00**

Description	Oper	Unit	Crew Size	Man-Hours Per Unit	Avg Mat'l Unit Cost	Avg Labor Unit Cost	Avg Equip Unit Cost	Avg Total Unit Cost	Avg Price Incl O&P	Avg Total Unit Cost	Avg Price Incl O&P
			Costs Based On Small Volume							Large Volume	
Range cabinets, 21" H, 12" D											
24" W, 1 door											
High quality workmanship	Inst	Ea	CJ	2.50	117.00	61.20	---	178.20	**234.00**	146.90	**189.00**
Good quality workmanship	Inst	Ea	CJ	1.68	87.10	41.10	---	128.20	**167.00**	106.50	**136.00**
Average quality workmanship	Inst	Ea	CJ	1.68	66.00	41.10	---	107.10	**142.00**	86.70	**112.00**
30" W, 2 doors											
High quality workmanship	Inst	Ea	CJ	2.50	131.00	61.20	---	192.20	**250.00**	159.90	**203.00**
Good quality workmanship	Inst	Ea	CJ	1.68	96.40	41.10	---	137.50	**178.00**	115.20	**146.00**
Average quality workmanship	Inst	Ea	CJ	1.68	73.90	41.10	---	115.00	**151.00**	94.10	**121.00**
33" W, 2 doors											
High quality workmanship	Inst	Ea	CJ	2.67	136.00	65.30	---	201.30	**262.00**	167.10	**213.00**
Good quality workmanship	Inst	Ea	CJ	1.78	100.00	43.50	---	143.50	**187.00**	120.40	**153.00**
Average quality workmanship	Inst	Ea	CJ	1.78	76.60	43.50	---	120.10	**158.00**	98.10	**126.00**
36" W, 2 doors											
High quality workmanship	Inst	Ea	CJ	2.67	145.00	65.30	---	210.30	**273.00**	175.10	**223.00**
Good quality workmanship	Inst	Ea	CJ	1.78	106.00	43.50	---	149.50	**193.00**	125.40	**159.00**
Average quality workmanship	Inst	Ea	CJ	1.78	80.50	43.50	---	124.00	**163.00**	101.80	**131.00**
42" W, 2 doors											
High quality workmanship	Inst	Ea	CJ	2.67	158.00	65.30	---	223.30	**289.00**	188.10	**238.00**
Good quality workmanship	Inst	Ea	CJ	1.78	117.00	43.50	---	160.50	**207.00**	136.20	**172.00**
Average quality workmanship	Inst	Ea	CJ	1.78	89.80	43.50	---	133.30	**174.00**	110.50	**141.00**
48" W, 2 doors											
High quality workmanship	Inst	Ea	CJ	3.33	180.00	81.50	---	261.50	**339.00**	217.90	**277.00**
Good quality workmanship	Inst	Ea	CJ	2.22	133.00	54.30	---	187.30	**243.00**	157.50	**200.00**
Average quality workmanship	Inst	Ea	CJ	2.22	102.00	54.30	---	156.30	**204.00**	128.00	**164.00**
Range cabinets, 30" H, 12" D											
12" W, 1 door											
High quality workmanship	Inst	Ea	CJ	1.95	103.00	47.70	---	150.70	**196.00**	125.60	**160.00**
Good quality workmanship	Inst	Ea	CJ	1.30	75.20	31.80	---	107.00	**139.00**	89.80	**114.00**
Average quality workmanship	Inst	Ea	CJ	1.30	56.80	31.80	---	88.60	**116.00**	72.40	**93.00**
15" W, 1 door											
High quality workmanship	Inst	Ea	CJ	1.95	114.00	47.70	---	161.70	**209.00**	135.90	**172.00**
Good quality workmanship	Inst	Ea	CJ	1.30	83.20	31.80	---	115.00	**148.00**	97.20	**123.00**
Average quality workmanship	Inst	Ea	CJ	1.30	63.40	31.80	---	95.20	**124.00**	78.60	**100.00**
18" W, 1 door											
High quality workmanship	Inst	Ea	CJ	1.95	119.00	47.70	---	166.70	**215.00**	140.90	**178.00**
Good quality workmanship	Inst	Ea	CJ	1.30	87.10	31.80	---	118.90	**153.00**	100.90	**127.00**
Average quality workmanship	Inst	Ea	CJ	1.30	66.00	31.80	---	97.80	**128.00**	81.10	**103.00**
21" W, 1 door											
High quality workmanship	Inst	Ea	CJ	2.22	125.00	54.30	---	179.30	**233.00**	150.50	**191.00**
Good quality workmanship	Inst	Ea	CJ	1.48	92.40	36.20	---	128.60	**166.00**	108.50	**137.00**
Average quality workmanship	Inst	Ea	CJ	1.48	70.00	36.20	---	106.20	**139.00**	87.40	**112.00**
24" W, 1 door											
High quality workmanship	Inst	Ea	CJ	2.22	140.00	54.30	---	194.30	**250.00**	163.50	**207.00**
Good quality workmanship	Inst	Ea	CJ	1.48	103.00	36.20	---	139.20	**179.00**	118.40	**149.00**
Average quality workmanship	Inst	Ea	CJ	1.48	77.90	36.20	---	114.10	**148.00**	94.90	**121.00**
27" W, 2 doors											
High quality workmanship	Inst	Ea	CJ	2.22	154.00	54.30	---	208.30	**268.00**	177.50	**224.00**
Good quality workmanship	Inst	Ea	CJ	1.48	114.00	36.20	---	150.20	**191.00**	128.70	**161.00**
Average quality workmanship	Inst	Ea	CJ	1.48	87.10	36.20	---	123.30	**160.00**	103.50	**131.00**

Description	Oper	Unit	Crew Size	Man-Hours Per Unit	Avg Mat'l Unit Cost	Avg Labor Unit Cost	Avg Equip Unit Cost	Avg Total Unit Cost	Avg Price Incl O&P	Large Volume Avg Total Unit Cost	Large Volume Avg Price Incl O&P
30" W, 2 doors											
High quality workmanship	Inst	Ea	CJ	2.96	164.00	72.40	---	236.40	**306.00**	197.50	**251.00**
Good quality workmanship	Inst	Ea	CJ	1.98	121.00	48.40	---	169.40	**219.00**	143.10	**181.00**
Average quality workmanship	Inst	Ea	CJ	1.98	91.10	48.40	---	139.50	**183.00**	114.70	**147.00**
33" W, 2 doors											
High quality workmanship	Inst	Ea	CJ	2.96	173.00	72.40	---	245.40	**318.00**	205.50	**261.00**
Good quality workmanship	Inst	Ea	CJ	1.98	128.00	48.40	---	176.40	**227.00**	149.10	**189.00**
Average quality workmanship	Inst	Ea	CJ	1.98	97.70	48.40	---	146.10	**191.00**	120.90	**154.00**
36" W, 2 doors											
High quality workmanship	Inst	Ea	CJ	2.96	180.00	72.40	---	252.40	**325.00**	212.50	**269.00**
Good quality workmanship	Inst	Ea	CJ	1.98	133.00	48.40	---	181.40	**234.00**	154.10	**195.00**
Average quality workmanship	Inst	Ea	CJ	1.98	102.00	48.40	---	150.40	**196.00**	124.60	**159.00**
42" W, 2 doors											
High quality workmanship	Inst	Ea	CJ	3.81	195.00	93.20	---	288.20	**376.00**	240.00	**305.00**
Good quality workmanship	Inst	Ea	CJ	2.54	144.00	62.10	---	206.10	**267.00**	172.20	**219.00**
Average quality workmanship	Inst	Ea	CJ	2.54	110.00	62.10	---	172.10	**226.00**	140.20	**180.00**
48" W, 2 doors											
High quality workmanship	Inst	Ea	CJ	3.81	231.00	93.20	---	324.20	**419.00**	273.00	**346.00**
Good quality workmanship	Inst	Ea	CJ	2.54	170.00	62.10	---	232.10	**299.00**	197.20	**248.00**
Average quality workmanship	Inst	Ea	CJ	2.54	129.00	62.10	---	191.10	**250.00**	159.20	**202.00**
24" W, blind corner unit, 1 door											
High quality workmanship	Inst	Ea	CJ	2.81	125.00	68.70	---	193.70	**255.00**	159.10	**204.00**
Good quality workmanship	Inst	Ea	CJ	1.90	99.00	46.50	---	145.50	**189.00**	120.90	**154.00**
Average quality workmanship	Inst	Ea	CJ	1.90	96.40	46.50	---	142.90	**186.00**	118.40	**151.00**
24" x 24" angle corner unit, stationary											
High quality workmanship	Inst	Ea	CJ	2.81	180.00	68.70	---	248.70	**320.00**	210.10	**265.00**
Good quality workmanship	Inst	Ea	CJ	1.90	132.00	46.50	---	178.50	**229.00**	151.90	**191.00**
Average quality workmanship	Inst	Ea	CJ	1.90	103.00	46.50	---	149.50	**194.00**	124.60	**158.00**
24" x 24" angle corner unit, lazy Susan											
High quality workmanship	Inst	Ea	CJ	2.81	267.00	68.70	---	335.70	**424.00**	291.10	**363.00**
Good quality workmanship	Inst	Ea	CJ	1.90	216.00	46.50	---	262.50	**330.00**	230.90	**286.00**
Average quality workmanship	Inst	Ea	CJ	1.90	148.00	46.50	---	194.50	**248.00**	166.90	**209.00**

Vanity cabinets and sink top

Description	Oper	Unit	Crew Size	Man-Hours Per Unit	Avg Mat'l Unit Cost	Avg Labor Unit Cost	Avg Equip Unit Cost	Avg Total Unit Cost	Avg Price Incl O&P	Large Volume Avg Total Unit Cost	Large Volume Avg Price Incl O&P
Disconnect plumbing and remove to dumpster	Demo	Ea	LB	1.67	---	36.70	---	36.70	**55.30**	22.00	**33.10**
Remove old unit, replace with new unit, reconnect plumbing	Reset	Ea	SB	3.81	---	99.90	---	99.90	**149.00**	60.00	**89.40**

Vanity units, with marble tops, good quality fittings and faucets; hardware, deluxe, finished models

Stained ash and birch primed composition construction

Labor costs include fitting and hanging only of vanity units

For rough-in costs, see Adjustments in this cabinets section, page 47

2 door units

Description	Oper	Unit	Crew Size	Man-Hours Per Unit	Avg Mat'l Unit Cost	Avg Labor Unit Cost	Avg Equip Unit Cost	Avg Total Unit Cost	Avg Price Incl O&P	Large Volume Avg Total Unit Cost	Large Volume Avg Price Incl O&P
20" x 16"											
Ash	Inst	Ea	SB	8.42	298.00	221.00	---	519.00	**687.00**	411.00	**532.00**
Birch	Inst	Ea	SB	8.42	272.00	221.00	---	493.00	**655.00**	386.00	**502.00**
Composition construction	Inst	Ea	SB	8.42	206.00	221.00	---	427.00	**576.00**	324.00	**427.00**

Description	Oper	Unit	Crew Size	Man-Hours Per Unit	Avg Mat'l Unit Cost	Avg Labor Unit Cost	Avg Equip Unit Cost	Avg Total Unit Cost	Avg Price Incl O&P	Avg Total Unit Cost	Avg Price Incl O&P
			Costs Based On Small Volume							Large Volume	
25" x 19"											
Ash	Inst	Ea	SB	8.42	347.00	221.00	---	568.00	**745.00**	457.00	**587.00**
Birch	Inst	Ea	SB	8.42	317.00	221.00	---	538.00	**709.00**	429.00	**552.00**
Composition construction	Inst	Ea	SB	8.42	226.00	221.00	---	447.00	**600.00**	343.00	**450.00**
31" x 19"											
Ash	Inst	Ea	SB	8.42	413.00	221.00	---	634.00	**825.00**	519.00	**661.00**
Birch	Inst	Ea	SB	8.42	375.00	221.00	---	596.00	**779.00**	483.00	**618.00**
Composition construction	Inst	Ea	SB	8.42	268.00	221.00	---	489.00	**650.00**	383.00	**497.00**
35" x 19"											
Ash	Inst	Ea	SB	8.89	446.00	233.00	---	679.00	**883.00**	559.00	**711.00**
Birch	Inst	Ea	SB	8.89	404.00	233.00	---	637.00	**832.00**	519.00	**663.00**
Composition construction	Inst	Ea	SB	8.89	292.00	233.00	---	525.00	**697.00**	414.00	**537.00**
For drawers in any above unit, ADD per drawer	Inst	Ea	SB	---	39.60	---	---	39.60	**47.50**	37.20	**44.60**
2 door cutback units with 3 drawers											
36" x 19"											
Ash	Inst	Ea	SB	8.89	517.00	233.00	---	750.00	**968.00**	626.00	**791.00**
Birch	Inst	Ea	SB	8.89	465.00	233.00	---	698.00	**905.00**	576.00	**732.00**
Composition construction	Inst	Ea	SB	8.89	347.00	233.00	---	580.00	**764.00**	466.00	**599.00**
49" x 19											
Ash	Inst	Ea	SB	8.89	647.00	233.00	---	880.00	**1120.00**	748.00	**937.00**
Birch	Inst	Ea	SB	8.89	560.00	233.00	---	793.00	**1020.00**	666.00	**839.00**
Composition construction	Inst	Ea	SB	8.89	418.00	233.00	---	651.00	**849.00**	533.00	**680.00**
60" x 19"											
Ash	Inst	Ea	SB	11.4	816.00	299.00	---	1115.00	**1420.00**	941.00	**1180.00**
Birch	Inst	Ea	SB	11.4	677.00	299.00	---	976.00	**1260.00**	811.00	**1020.00**
Composition construction	Inst	Ea	SB	11.4	527.00	299.00	---	826.00	**1080.00**	670.00	**854.00**
Corner unit, 1 door											
22" x 22"											
Ash	Inst	Ea	SB	8.42	384.00	221.00	---	605.00	**790.00**	492.00	**628.00**
Birch	Inst	Ea	SB	8.42	363.00	221.00	---	584.00	**764.00**	472.00	**604.00**
Composition construction	Inst	Ea	SB	8.42	296.00	221.00	---	517.00	**684.00**	409.00	**529.00**
Adjustments											
To remove and reset											
Vanity units with tops	Reset	Ea	SA	5.71	---	174.00	---	174.00	**259.00**	101.00	**151.00**
Top only	Reset	Ea	SA	3.33	---	101.00	---	101.00	**151.00**	60.90	**90.80**
To install rough-in	Inst	Ea	SB	6.67	---	175.00	---	175.00	**260.00**	105.00	**156.00**

Canopies

Costs for awnings include all hardware

Residential prefabricated

Aluminum

Description	Oper	Unit	Crew Size	Man-Hours Per Unit	Avg Mat'l Unit Cost	Avg Labor Unit Cost	Avg Equip Unit Cost	Avg Total Unit Cost	Avg Price Incl O&P	Avg Total Unit Cost	Avg Price Incl O&P
Carport, freestanding											
16' x 8'	Inst	Ea	2C	8.21	2810.00	221.00	---	3031.00	**3570.00**	2644.00	**3090.00**
20' x 10'	Inst	Ea	2C	12.3	2970.00	332.00	---	3302.00	**3920.00**	2856.00	**3360.00**

Description	Oper	Unit	Crew Size	Man-Hours Per Unit	Avg Mat'l Unit Cost	Avg Labor Unit Cost	Avg Equip Unit Cost	Avg Total Unit Cost	Avg Price Incl O&P	Large Volume Avg Total Unit Cost	Large Volume Avg Price Incl O&P
			Costs Based On Small Volume							Large Volume	
Door canopies, 36" projection											
4' wide	Inst	Ea	CA	1.54	469.00	41.50	---	510.50	**602.00**	443.00	**520.00**
5' wide	Inst	Ea	CA	2.05	525.00	55.30	---	580.30	**688.00**	501.90	**591.00**
6' wide	Inst	Ea	CA	3.08	569.00	83.00	---	652.00	**780.00**	558.90	**663.00**
8' wide	Inst	Ea	2C	4.92	775.00	133.00	---	908.00	**1090.00**	774.30	**923.00**
10' wide	Inst	Ea	2C	8.21	888.00	221.00	---	1109.00	**1360.00**	932.00	**1120.00**
12' wide	Inst	Ea	2C	12.3	1060.00	332.00	---	1392.00	**1720.00**	1154.00	**1410.00**
Door canopies, 42" projection											
4' wide	Inst	Ea	CA	1.54	538.00	41.50	---	579.50	**681.00**	504.00	**590.00**
5' wide	Inst	Ea	CA	2.05	588.00	55.30	---	643.30	**760.00**	557.90	**654.00**
6' wide	Inst	Ea	CA	3.08	744.00	83.00	---	827.00	**982.00**	713.90	**841.00**
8' wide	Inst	Ea	2C	4.92	869.00	133.00	---	1002.00	**1200.00**	857.30	**1020.00**
10' wide	Inst	Ea	2C	8.21	1010.00	221.00	---	1231.00	**1490.00**	1038.00	**1250.00**
12' wide	Inst	Ea	2C	12.3	1180.00	332.00	---	1512.00	**1860.00**	1266.00	**1530.00**
Door canopies, 48" projection											
4' wide	Inst	Ea	CA	1.54	588.00	41.50	---	629.50	**739.00**	549.00	**641.00**
5' wide	Inst	Ea	CA	2.05	688.00	55.30	---	743.30	**875.00**	646.90	**757.00**
6' wide	Inst	Ea	CA	3.08	781.00	83.00	---	864.00	**1020.00**	747.90	**880.00**
8' wide	Inst	Ea	2C	4.92	1060.00	133.00	---	1193.00	**1420.00**	1030.30	**1220.00**
10' wide	Inst	Ea	2C	8.21	1190.00	221.00	---	1411.00	**1700.00**	1194.00	**1430.00**
12' wide	Inst	Ea	2C	12.3	1340.00	332.00	---	1672.00	**2040.00**	1406.00	**1690.00**
Door canopies, 54" projection											
4' wide	Inst	Ea	CA	1.54	675.00	41.50	---	716.50	**839.00**	626.00	**730.00**
5' wide	Inst	Ea	CA	2.05	719.00	55.30	---	774.30	**911.00**	673.90	**788.00**
6' wide	Inst	Ea	CA	3.08	894.00	83.00	---	977.00	**1150.00**	847.90	**995.00**
8' wide	Inst	Ea	2C	4.92	1220.00	133.00	---	1353.00	**1600.00**	1166.30	**1380.00**
10' wide	Inst	Ea	2C	8.21	1380.00	221.00	---	1601.00	**1920.00**	1364.00	**1620.00**
12' wide	Inst	Ea	2C	12.3	1560.00	332.00	---	1892.00	**2300.00**	1606.00	**1920.00**
Patio cover											
16' x 8'	Inst	Ea	2C	8.21	1500.00	221.00	---	1721.00	**2060.00**	1474.00	**1750.00**
20' x 10'	Inst	Ea	2C	12.3	2030.00	332.00	---	2362.00	**2840.00**	2016.00	**2400.00**
Window awnings, 3' high											
4' wide	Inst	Ea	CA	1.76	156.00	47.50	---	203.50	**252.00**	169.70	**206.00**
6' wide	Inst	Ea	CA	2.46	189.00	66.30	---	255.30	**318.00**	211.10	**258.00**
9' wide	Inst	Ea	2C	4.10	259.00	111.00	---	370.00	**466.00**	302.00	**374.00**
12' wide	Inst	Ea	2C	6.15	325.00	166.00	---	491.00	**626.00**	397.00	**496.00**
Window awnings, 4' high											
4' wide	Inst	Ea	CA	1.76	180.00	47.50	---	227.50	**279.00**	190.70	**231.00**
6' wide	Inst	Ea	CA	2.46	239.00	66.30	---	305.30	**375.00**	255.10	**309.00**
9' wide	Inst	Ea	2C	4.10	325.00	111.00	---	436.00	**542.00**	361.00	**441.00**
12' wide	Inst	Ea	2C	6.15	425.00	166.00	---	591.00	**741.00**	485.00	**598.00**
Window awnings, 6' high											
4' wide	Inst	Ea	CA	1.76	296.00	47.50	---	343.50	**413.00**	293.70	**349.00**
6' wide	Inst	Ea	CA	2.46	400.00	66.30	---	466.30	**561.00**	398.10	**474.00**
9' wide	Inst	Ea	2C	4.10	500.00	111.00	---	611.00	**743.00**	516.00	**620.00**
12' wide	Inst	Ea	2C	6.15	563.00	166.00	---	729.00	**899.00**	608.00	**738.00**
Roll-up											
3' wide	Inst	Ea	CA	1.54	123.00	41.50	---	164.50	**204.00**	136.00	**166.00**
4' wide	Inst	Ea	CA	1.76	158.00	47.50	---	205.50	**253.00**	170.70	**208.00**
6' wide	Inst	Ea	CA	2.46	195.00	66.30	---	261.30	**325.00**	216.10	**265.00**
9' wide	Inst	Ea	2C	4.10	283.00	111.00	---	394.00	**493.00**	323.00	**398.00**
12' wide	Inst	Ea	2C	6.15	346.00	166.00	---	512.00	**650.00**	415.00	**518.00**

Description	Oper	Unit	Crew Size	Man-Hours Per Unit	Avg Mat'l Unit Cost	Avg Labor Unit Cost	Avg Equip Unit Cost	Avg Total Unit Cost	Avg Price Incl O&P	Large Volume Avg Total Unit Cost	Large Volume Avg Price Incl O&P
Canvas											
Traditional fabric awning, with waterproof, colorfast acrylic duck, double stitched seams, tubular metal framing and all hardware included											
Window awning, 24" drop											
3' wide	Inst	Ea	CA	1.54	200.00	41.50	---	241.50	**293.00**	205.00	**245.00**
4' wide	Inst	Ea	CA	1.76	245.00	47.50	---	292.50	**354.00**	248.70	**297.00**
Window awning, 30" drop											
3' wide	Inst	Ea	CA	1.54	229.00	41.50	---	270.50	**326.00**	230.00	**275.00**
4' wide	Inst	Ea	CA	1.76	275.00	47.50	---	322.50	**388.00**	274.70	**328.00**
5' wide	Inst	Ea	CA	2.05	313.00	55.30	---	368.30	**443.00**	313.90	**374.00**
6' wide	Inst	Ea	CA	2.46	350.00	66.30	---	416.30	**503.00**	354.10	**423.00**
8' wide	Inst	Ea	2C	4.92	390.00	133.00	---	523.00	**650.00**	432.30	**529.00**
10' wide	Inst	Ea	2C	6.15	444.00	166.00	---	610.00	**762.00**	502.00	**617.00**
Minimum Job Charge	Inst	Ea	CA	3.08	---	83.00	---	83.00	**126.00**	53.90	**82.00**
Carpet											
Price includes consultation, measurement, pad, carpet, installation, and the use of tack strips and hot melt tape on seams.											
Nylon carpet with rebond pad											
Minimum quality, 25 to 35 oz.	Inst	SY	FA	.143	23.30	3.73	---	27.03	**32.30**	22.11	**26.30**
Medium quality, 35 to 50 oz.	Inst	SY	FA	.143	27.90	3.73	---	31.63	**37.60**	26.01	**30.80**
Better quality, 50 oz. plus	Inst	SY	FA	.143	37.20	3.73	---	40.93	**48.30**	33.81	**39.80**
Berber carpet, large loop	Inst	SY	FA	.143	65.10	3.73	---	68.83	**80.40**	57.21	**66.70**
ADD for box steps	Inst	Riser	FA	.088	2.33	2.29	---	4.62	**6.10**	3.57	**4.65**
ADD for wrapped steps, open riser, sewn	Inst	Step	FA	.152	4.19	3.96	---	8.15	**10.70**	6.30	**8.19**
ADD for sewn edge treatment	Inst	Riser	FA	.143	4.19	3.73	---	7.92	**10.40**	6.12	**7.92**
ADD for circular stair steps	Inst	Step	FA	.327	7.67	8.52	---	16.19	**21.50**	12.41	**16.30**
Caulking											
Material costs are typical costs for the listed bead diameters. Figures in parentheses, following bead diameter, indicate approximate coverage including 5% waste. Labor costs are per LF of bead length and assume good quality application on smooth to slightly irregular surfaces.											
Multi-purpose caulk, good quality											
1/8" (11.6 LF/fluid oz.)	Inst	LF	CA	.026	.01	.70	---	.71	**1.08**	.50	**.75**
1/4" (2.91 LF/fluid oz.)	Inst	LF	CA	.036	.08	.97	---	1.05	**1.57**	.74	**1.11**
3/8" (1.29 LF/fluid oz.)	Inst	LF	CA	.043	.18	1.16	---	1.34	**1.98**	.96	**1.41**
1/2" (.729 LF/fluid oz.)	Inst	LF	CA	.048	.31	1.29	---	1.60	**2.34**	1.15	**1.66**
Butyl flex caulk, premium quality											
1/8" (11.6 LF/fluid oz.)	Inst	LF	CA	.026	.04	.70	---	.74	**1.11**	.52	**.77**
1/4" (2.91 LF/fluid oz.)	Inst	LF	CA	.036	.14	.97	---	1.11	**1.64**	.79	**1.17**
3/8" (1.29 LF/fluid oz.)	Inst	LF	CA	.043	.30	1.16	---	1.46	**2.12**	1.06	**1.53**
1/2" (.729 LF/fluid oz.)	Inst	LF	CA	.048	.54	1.29	---	1.83	**2.62**	1.35	**1.90**
Latex, premium quality											
1/8" (11.6 LF/fluid oz.)	Inst	LF	CA	.026	.04	.70	---	.74	**1.11**	.52	**.77**
1/4" (2.91 LF/fluid oz.)	Inst	LF	CA	.036	.18	.97	---	1.15	**1.69**	.82	**1.20**
3/8" (1.29 LF/fluid oz.)	Inst	LF	CA	.043	.39	1.16	---	1.55	**2.23**	1.14	**1.63**
1/2" (.729 LF/fluid oz.)	Inst	LF	CA	.048	.71	1.29	---	2.00	**2.82**	1.49	**2.07**

Description	Oper	Unit	Crew Size	Man-Hours Per Unit	Avg Mat'l Unit Cost	Avg Labor Unit Cost	Avg Equip Unit Cost	Avg Total Unit Cost	Avg Price Incl O&P	Large Volume Avg Total Unit Cost	Large Volume Avg Price Incl O&P
Latex caulk, good quality											
1/8" (11.6 LF/fluid oz.)	Inst	LF	CA	.026	.04	.70	---	.74	**1.11**	.52	**.77**
1/4" (2.91 LF/fluid oz.)	Inst	LF	CA	.036	.14	.97	---	1.11	**1.64**	.79	**1.17**
3/8" (1.29 LF/fluid oz.)	Inst	LF	CA	.043	.30	1.16	---	1.46	**2.12**	1.06	**1.53**
1/2" (.729 LF/fluid oz.)	Inst	LF	CA	.048	.54	1.29	---	1.83	**2.62**	1.35	**1.90**
Oil base caulk, good quality											
1/8" (11.6 LF/fluid oz.)	Inst	LF	CA	.026	.04	.70	---	.74	**1.11**	.52	**.77**
1/4" (2.91 LF/fluid oz.)	Inst	LF	CA	.036	.15	.97	---	1.12	**1.66**	.80	**1.18**
3/8" (1.29 LF/fluid oz.)	Inst	LF	CA	.043	.34	1.16	---	1.50	**2.17**	1.10	**1.58**
1/2" (.729 LF/fluid oz.)	Inst	LF	CA	.048	.60	1.29	---	1.89	**2.69**	1.40	**1.96**
Oil base caulk, economy quality											
1/8" (11.6 LF/fluid oz.)	Inst	LF	CA	.026	.01	.70	---	.71	**1.08**	.50	**.75**
1/4" (2.91 LF/fluid oz.)	Inst	LF	CA	.036	.08	.97	---	1.05	**1.57**	.74	**1.11**
3/8" (1.29 LF/fluid oz.)	Inst	LF	CA	.043	.18	1.16	---	1.34	**1.98**	.96	**1.41**
1/2" (.729 LF/fluid oz.)	Inst	LF	CA	.048	.31	1.29	---	1.60	**2.34**	1.15	**1.66**
Silicone caulk, good quality											
1/8" (11.6 LF/fluid oz.)	Inst	LF	CA	.026	.05	.70	---	.75	**1.13**	.54	**.80**
1/4" (2.91 LF/fluid oz.)	Inst	LF	CA	.036	.19	.97	---	1.16	**1.70**	.83	**1.22**
3/8" (1.29 LF/fluid oz.)	Inst	LF	CA	.043	.44	1.16	---	1.60	**2.29**	1.18	**1.67**
1/2" (.729 LF/fluid oz.)	Inst	LF	CA	.048	.76	1.29	---	2.05	**2.88**	1.53	**2.12**
Silicone caulk, premium quality											
1/8" (11.6 LF/fluid oz.)	Inst	LF	CA	.026	.05	.70	---	.75	**1.13**	.54	**.80**
1/4" (2.91 LF/fluid oz.)	Inst	LF	CA	.036	.23	.97	---	1.20	**1.75**	.87	**1.26**
3/8" (1.29 LF/fluid oz.)	Inst	LF	CA	.043	.52	1.16	---	1.68	**2.39**	1.25	**1.76**
1/2" (.729 LF/fluid oz.)	Inst	LF	CA	.048	.91	1.29	---	2.20	**3.06**	1.66	**2.28**
Tub caulk, white siliconized											
1/8" (11.6 LF/fluid oz.)	Inst	LF	CA	.026	.01	.70	---	.71	**1.08**	.50	**.75**
1/4" (2.91 LF/fluid oz.)	Inst	LF	CA	.036	.08	.97	---	1.05	**1.57**	.74	**1.11**
3/8" (1.29 LF/fluid oz.)	Inst	LF	CA	.043	.18	1.16	---	1.34	**1.98**	.96	**1.41**
1/2" (.729 LF/fluid oz.)	Inst	LF	CA	.048	.31	1.29	---	1.60	**2.34**	1.15	**1.66**
Anti-algae and mildew resistant tub caulk, premium quality white or clear silicone											
1/8" (11.6 LF/fluid oz.)	Inst	LF	CA	.026	.05	.70	---	.75	**1.13**	.54	**.80**
1/4" (2.91 LF/fluid oz.)	Inst	LF	CA	.036	.23	.97	---	1.20	**1.75**	.87	**1.26**
3/8" (1.29 LF/fluid oz.)	Inst	LF	CA	.043	.53	1.16	---	1.69	**2.40**	1.26	**1.77**
1/2" (.729 LF/fluid oz.)	Inst	LF	CA	.048	.94	1.29	---	2.23	**3.10**	1.68	**2.30**
Elastomeric caulk, premium quality											
1/8" (11.6 LF/fluid oz.)	Inst	LF	CA	.026	.05	.70	---	.75	**1.13**	.54	**.80**
1/4" (2.91 LF/fluid oz.)	Inst	LF	CA	.036	.22	.97	---	1.19	**1.74**	.85	**1.24**
3/8" (1.29 LF/fluid oz.)	Inst	LF	CA	.043	.50	1.16	---	1.66	**2.36**	1.24	**1.75**
1/2" (.729 LF/fluid oz.)	Inst	LF	CA	.048	.90	1.29	---	2.19	**3.05**	1.65	**2.26**
Add for irregular surfaces such as vertical masonry or lap siding	Inst	%	CA	---	5.0	---	---	---	**---**	---	**---**
Caulking gun, heavy duty, professional type	Inst	EA	CA	---	40.80	---	---	40.80	**49.00**	34.50	**41.40**
Caulking gun, economy grade											
11-oz. cartridge	Inst	EA	CA	---	7.72	---	---	7.72	**9.26**	6.53	**7.84**
29-oz. cartridge	Inst	EA	CA	---	27.00	---	---	27.00	**32.40**	22.90	**27.40**

Ceramic tile

1. **Dimensions**. There are many sizes of ceramic tile. Only $4^1/_4$" x $4^1/_4$" and 1" x 1" will be discussed here.
   a. $4^1/_4$" x $4^1/_4$" tile is furnished both unmounted and back-mounted. Back-mounted tile are usually furnished in sheets of 12 tile.
   b. 1" x 1" mosaic tile is furnished face-mounted and back-mounted in sheets; normally, 2'-0" x 1'-0".
2. **Installation.** There are three methods:
   a. Conventional, which uses portland cement, sand and wet tile grout.
   b. Dry-set, which uses dry-set mix and dry tile grout mix.
   c. Organic adhesive, which uses adhesive and dry tile grout mix.

The conventional method is the most expensive and is used less frequently than the other methods.

3. **Estimating Technique**. For tile, determine the area and add 5% to 10% for waste. For cove, base or trim, determine the length in linear feet and add 5% to 10% for waste.

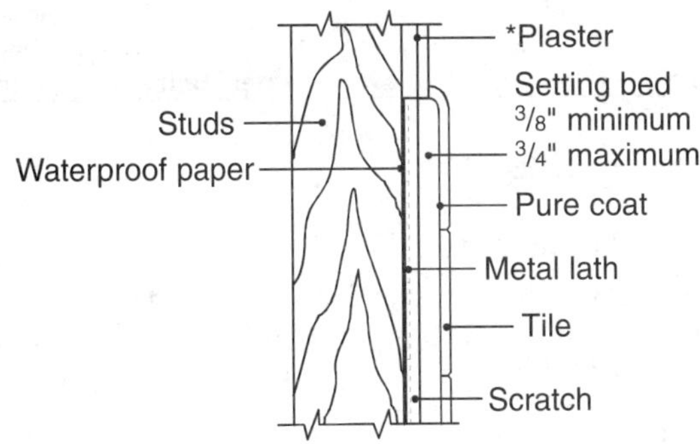

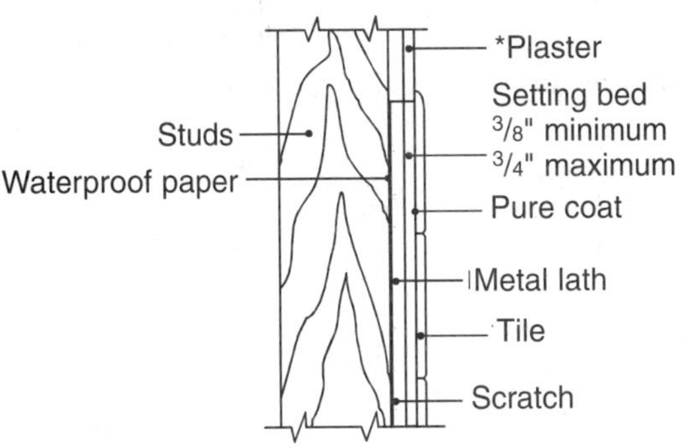

*This installation would be similar if wall finish above tile wainscot were of other material such as wallboard, plywood, etc.

Wood or steel construction with plaster above tile wainscot

Courtesy: ***Ceramic Tile Institute of America***
700 N. Virgil Ave., Ste 300, Los Angeles, CA 90029

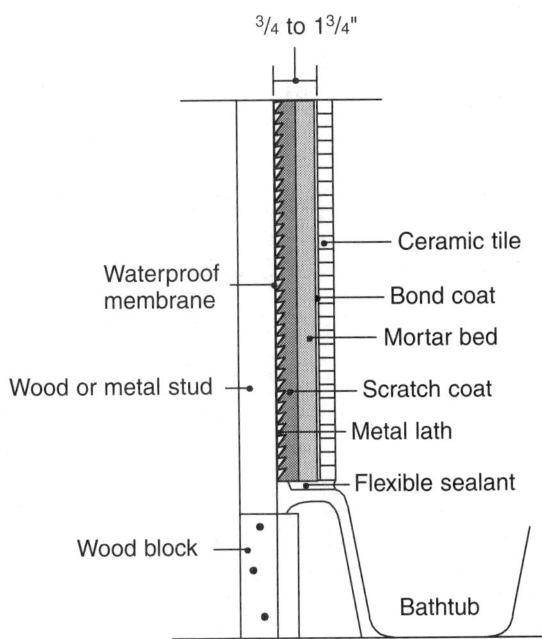

Cross section of bathtub wall using cement mortar

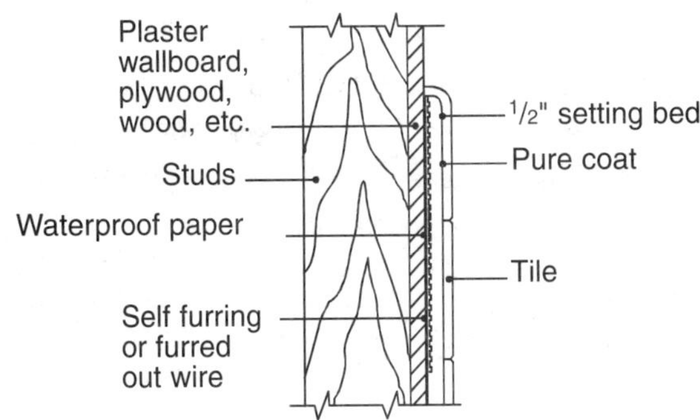

Wood or steel construction with solid covered backing

Description	Oper	Unit	Crew Size	Man-Hours Per Unit	Avg Mat'l Unit Cost	Avg Labor Unit Cost	Avg Equip Unit Cost	Avg Total Unit Cost	Avg Price Incl O&P	Large Volume Avg Total Unit Cost	Large Volume Avg Price Incl O&P
Ceramic tile											
Countertop/backsplash											
Adhesive set with backmounted tile											
1" x 1"	Inst	SF	TB	.381	6.24	8.77	---	15.01	**20.30**	10.56	**13.90**
4-1/4" x 4-1/4"	Inst	SF	TB	.333	4.98	7.67	---	12.65	**17.20**	8.84	**11.70**
Cove/base											
Adhesive set with unmounted tile											
4-1/4" x 4-1/4"	Inst	LF	TB	.333	4.79	7.67	---	12.46	**16.90**	8.68	**11.50**
6" x 4-1/4"	Inst	LF	TB	.281	4.68	6.47	---	11.15	**15.00**	7.84	**10.30**
Conventional mortar set with unmounted tile											
4-1/4" x 4-1/4"	Inst	LF	TB	.593	4.69	13.70	---	18.39	**25.80**	12.18	**16.80**
6" x 4-1/4"	Inst	LF	TB	.533	4.54	12.30	---	16.84	**23.50**	11.22	**15.40**
Dry-set mortar with unmounted tile											
4-1/4" x 4-1/4"	Inst	LF	TB	.410	4.72	9.44	---	14.16	**19.50**	9.67	**13.00**
6" x 4-1/4"	Inst	LF	TB	.356	4.57	8.20	---	12.77	**17.50**	8.79	**11.80**
Floors											
1-1/2" x 1-1/2"											
Adhesive or dry-set base	Demo	SF	LB	.048	---	1.05	---	1.05	**1.59**	.64	**.96**
Conventional mortar base	Demo	SF	LB	.056	---	1.23	---	1.23	**1.86**	.75	**1.13**
Adhesive set with backmounted tile											
1-1/2" x 1-1/2"	Inst	SF	TB	.222	6.24	5.11	---	11.35	**14.80**	8.35	**10.70**
4-1/4" x 4-1/4"	Inst	SF	TB	.205	4.98	4.72	---	9.70	**12.80**	7.06	**9.09**
Conventional mortar set with backmounted tile											
1-1/2" x 1-1/2"	Inst	SF	TB	.444	5.95	10.20	---	16.15	**22.10**	11.19	**15.00**
4-1/4" x 4-1/4"	Inst	SF	TB	.381	4.70	8.77	---	13.47	**18.50**	9.26	**12.50**
Dry-set mortar with backmounted tile											
1-1/2" x 1-1/2"	Inst	SF	TB	.281	6.01	6.47	---	12.48	**16.60**	8.97	**11.60**
4-1/4" x 4-1/4"	Inst	SF	TB	.254	4.76	5.85	---	10.61	**14.20**	7.54	**9.86**
Wainscot cap											
Adhesive set with unmounted tile											
2" x 6"	Inst	LF	TB	.222	2.44	5.11	---	7.55	**10.40**	5.13	**6.95**
Conventional mortar set with unmounted tile											
2" x 6"	Inst	LF	TB	.356	2.40	8.20	---	10.60	**15.00**	6.94	**9.64**
Dry-set mortar with unmounted tile											
2" x 6"	Inst	LF	TB	.267	2.40	6.15	---	8.55	**11.90**	5.71	**7.83**
Walls, 1-1/2" x 1-1/2" or 4-1/4" x 4-1/4"											
Adhesive or dry-set base	Demo	SF	LB	.056	---	1.23	---	1.23	**1.86**	.72	**1.09**
Conventional mortar base	Demo	SF	LB	.067	---	1.47	---	1.47	**2.22**	.88	**1.33**
Adhesive set with backmounted tile											
1-1/2" x 1-1/2"	Inst	SF	TB	.267	6.24	6.15	---	12.39	**16.30**	8.97	**11.60**
4-1/4" x 4-1/4"	Inst	SF	TB	.242	4.98	5.57	---	10.55	**14.00**	7.57	**9.84**
Conventional mortar set with backmounted tile											
1-1/2" x 1-1/2"	Inst	SF	TB	.533	5.95	12.30	---	18.25	**25.10**	12.41	**16.80**
4-1/4" x 4-1/4"	Inst	SF	TB	.444	4.70	10.20	---	14.90	**20.60**	10.14	**13.80**
Dry-set mortar with backmounted tile											
1-1/2" x 1-1/2"	Inst	SF	TB	.333	6.01	7.67	---	13.68	**18.30**	9.71	**12.70**
4-1/4" x 4-1/4"	Inst	SF	TB	.296	4.76	6.82	---	11.58	**15.60**	8.14	**10.80**

Closet door systems

Labor costs include hanging and fitting of doors and hardware

Bi-folding units

Includes hardware and pine fascia trim

Unfinished

Description	Oper	Unit	Crew Size	Man-Hours Per Unit	Avg Mat'l Unit Cost	Avg Labor Unit Cost	Avg Equip Unit Cost	Avg Total Unit Cost	Avg Price Incl O&P	Large Volume Avg Total Unit Cost	Large Volume Avg Price Incl O&P
Birch, flush face, 1-3/8" T, hollow core											
2'-0" x 6'-8", 2 doors	Inst	Set	2C	2.67	84.90	72.00	---	156.90	**207.00**	118.00	**152.00**
2'-6" x 6'-8", 2 doors	Inst	Set	2C	2.67	95.30	72.00	---	167.30	**219.00**	127.20	**162.00**
3'-0" x 6'-8", 2 doors	Inst	Set	2C	2.67	105.00	72.00	---	177.00	**230.00**	135.70	**172.00**
4'-0" x 6'-8", 4 doors	Inst	Set	2C	3.33	149.00	89.80	---	238.80	**308.00**	185.90	**233.00**
6'-0" x 6'-8", 4 doors	Inst	Set	2C	3.33	184.00	89.80	---	273.80	**348.00**	215.90	**269.00**
8'-0" x 6'-8", 4 doors	Inst	Set	2C	3.33	244.00	89.80	---	333.80	**417.00**	268.90	**330.00**
4'-0" x 8'-0", 4 doors	Inst	Set	2C	3.33	254.00	89.80	---	343.80	**428.00**	277.90	**339.00**
6'-0" x 8'-0", 4 doors	Inst	Set	2C	3.33	321.00	89.80	---	410.80	**506.00**	337.90	**408.00**
8'-0" x 8'-0", 4 doors	Inst	Set	2C	3.33	403.00	89.80	---	492.80	**600.00**	409.90	**491.00**
Lauan, flush face, 1-3/8" T, hollow core											
2'-0" x 6'-8", 2 doors	Inst	Set	2C	2.67	89.90	72.00	---	161.90	**213.00**	122.40	**157.00**
2'-6" x 6'-8", 2 doors	Inst	Set	2C	2.67	96.40	72.00	---	168.40	**220.00**	128.20	**163.00**
3'-0" x 6'-8", 2 doors	Inst	Set	2C	2.67	105.00	72.00	---	177.00	**231.00**	136.00	**172.00**
4'-0" x 6'-8", 4 doors	Inst	Set	2C	3.33	158.00	89.80	---	247.80	**319.00**	193.90	**243.00**
6'-0" x 6'-8", 4 doors	Inst	Set	2C	3.33	193.00	89.80	---	282.80	**359.00**	224.90	**278.00**
8'-0" x 6'-8", 4 doors	Inst	Set	2C	3.33	164.00	89.80	---	253.80	**325.00**	198.90	**249.00**
4'-0" x 8'-0", 4 doors	Inst	Set	2C	3.33	252.00	89.80	---	341.80	**426.00**	275.90	**337.00**
6'-0" x 8'-0", 4 doors	Inst	Set	2C	3.33	316.00	89.80	---	405.80	**499.00**	331.90	**402.00**
8'-0" x 8'-0", 4 doors	Inst	Set	2C	3.33	398.00	89.80	---	487.80	**594.00**	404.90	**486.00**
Ponderosa pine, colonial raised panel (2/door), 1-3/8" T, solid core											
2'-0" x 6'-8", 2 doors	Inst	Set	2C	2.67	92.80	72.00	---	164.80	**216.00**	125.00	**160.00**
4'-0" x 6'-8", 4 doors	Inst	Set	2C	3.33	173.00	89.80	---	262.80	**336.00**	206.90	**258.00**
6'-0" x 6'-8", 4 doors	Inst	Set	2C	3.33	199.00	89.80	---	288.80	**366.00**	229.90	**284.00**
8'-0" x 6'-8", 4 doors	Inst	Set	2C	3.33	275.00	89.80	---	364.80	**453.00**	296.90	**362.00**
Ponderosa pine, raised panel louver, 1-3/8" T, solid core											
2'-0" x 6'-8", 2 doors	Inst	Set	2C	2.67	95.70	72.00	---	167.70	**220.00**	127.60	**163.00**
4'-0" x 6'-8", 4 doors	Inst	Set	2C	3.33	179.00	89.80	---	268.80	**342.00**	211.90	**264.00**
6'-0" x 6'-8", 4 doors	Inst	Set	2C	3.33	205.00	89.80	---	294.80	**373.00**	234.90	**290.00**
Hardboard, flush face, 1-3/8" T, hollow core											
2'-0" x 6'-8", 2 doors	Inst	Set	2C	2.67	75.20	72.00	---	147.20	**196.00**	109.50	**142.00**
2'-6" x 6'-8", 2 doors	Inst	Set	2C	2.67	83.70	72.00	---	155.70	**206.00**	116.90	**150.00**
3'-0" x 6'-8", 2 doors	Inst	Set	2C	2.67	89.80	72.00	---	161.80	**213.00**	122.30	**157.00**
4'-0" x 6'-8", 4 doors	Inst	Set	2C	3.33	132.00	89.80	---	221.80	**289.00**	170.90	**216.00**
6'-0" x 6'-8", 4 doors	Inst	Set	2C	3.33	147.00	89.80	---	236.80	**305.00**	183.90	**231.00**
8'-0" x 6'-8", 4 doors	Inst	Set	2C	3.33	230.00	89.80	---	319.80	**401.00**	256.90	**315.00**
4'-0" x 8'-0", 4 doors	Inst	Set	2C	3.33	220.00	89.80	---	309.80	**389.00**	247.90	**305.00**
6'-0" x 8'-0", 4 doors	Inst	Set	2C	3.33	264.00	89.80	---	353.80	**440.00**	286.90	**349.00**
8'-0" x 8'-0", 4 doors	Inst	Set	2C	3.33	367.00	89.80	---	456.80	**559.00**	377.90	**455.00**
Sen (ash), flush face, 1-3/8" T, hollow core											
2'-0" x 6'-8", 2 doors	Inst	Set	2C	2.67	108.00	72.00	---	180.00	**233.00**	138.20	**175.00**
2'-6" x 6'-8", 2 doors	Inst	Set	2C	2.67	119.00	72.00	---	191.00	**246.00**	148.10	**186.00**
3'-0" x 6'-8", 2 doors	Inst	Set	2C	2.67	130.00	72.00	---	202.00	**259.00**	158.10	**198.00**
4'-0" x 6'-8", 4 doors	Inst	Set	2C	3.33	188.00	89.80	---	277.80	**353.00**	219.90	**273.00**

Description	Oper	Unit	Crew Size	Man-Hours Per Unit	Avg Mat'l Unit Cost	Avg Labor Unit Cost	Avg Equip Unit Cost	Avg Total Unit Cost	Avg Price Incl O&P	Large Volume Avg Total Unit Cost	Large Volume Avg Price Incl O&P
6'-0" x 6'-8", 4 doors	Inst	Set	2C	3.33	233.00	89.80	---	322.80	**405.00**	259.90	**319.00**
8'-0" x 6'-8", 4 doors	Inst	Set	2C	3.33	316.00	89.80	---	405.80	**500.00**	332.90	**403.00**
4'-0" x 8'-0", 4 doors	Inst	Set	2C	3.33	148.00	89.80	---	237.80	**306.00**	183.90	**232.00**
6'-0" x 8'-0", 4 doors	Inst	Set	2C	3.33	435.00	89.80	---	524.80	**637.00**	437.90	**523.00**
8'-0" x 8'-0", 4 doors	Inst	Set	2C	3.33	607.00	89.80	---	696.80	**834.00**	589.90	**698.00**
Prefinished											
Walnut tone, mar resistant finish											
Embossed (distressed wood appearance) lauan, flush face, 1-3/8" T, hollow core											
2'-0" x 6'-8", 2 doors	Inst	Set	2C	2.67	64.70	72.00	---	136.70	**184.00**	100.20	**131.00**
4'-0" x 6'-8", 4 doors	Inst	Set	2C	3.33	126.00	89.80	---	215.80	**282.00**	165.90	**210.00**
6'-0" x 6'-8", 4 doors	Inst	Set	2C	3.33	168.00	89.80	---	257.80	**330.00**	201.90	**252.00**
Lauan, flush face, 1-3/8" T, hollow core											
2'-0" x 6'-8", 2 doors	Inst	Set	2C	2.67	71.80	72.00	---	143.80	**192.00**	106.50	**138.00**
4'-0" x 6'-8", 4 doors	Inst	Set	2C	3.33	142.00	89.80	---	231.80	**299.00**	178.90	**226.00**
6'-0" x 6'-8", 4 doors	Inst	Set	2C	3.33	191.00	89.80	---	280.80	**356.00**	222.90	**276.00**
Ponderosa pine, full louver, 1-3/8" T, hollow core											
2'-0" x 6'-8", 2 doors	Inst	Set	2C	2.67	150.00	72.00	---	222.00	**282.00**	175.10	**217.00**
4'-0" x 6'-8", 4 doors	Inst	Set	2C	3.33	291.00	89.80	---	380.80	**471.00**	309.90	**377.00**
6'-0" x 6'-8", 4 doors	Inst	Set	2C	3.33	369.00	89.80	---	458.80	**561.00**	379.90	**457.00**
Ponderosa pine, raised louver, 1-3/8" T, hollow core											
2'-0" x 6'-8", 2 doors	Inst	Set	2C	2.67	157.00	72.00	---	229.00	**290.00**	182.10	**225.00**
4'-0" x 6'-8", 4 doors	Inst	Set	2C	3.33	308.00	89.80	---	397.80	**491.00**	325.90	**395.00**
6'-0" x 6'-8", 4 doors	Inst	Set	2C	3.33	369.00	89.80	---	458.80	**561.00**	379.90	**457.00**
Sliding or bypassing units											
Includes hardware, 4-5/8" jambs, header, and fascia											
Wood inserts; 1-3/8" T, hollow core											
Unfinished birch											
4'-0" x 6'-8", 2 doors	Inst	Set	2C	3.81	139.00	103.00	---	242.00	**315.00**	183.70	**234.00**
6'-0" x 6'-8", 2 doors	Inst	Set	2C	3.81	169.00	103.00	---	272.00	**350.00**	210.70	**265.00**
8'-0" x 6'-8", 2 doors	Inst	Set	2C	3.81	313.00	103.00	---	416.00	**516.00**	337.70	**412.00**
4'-0" x 8'-0", 2 doors	Inst	Set	2C	3.81	246.00	103.00	---	349.00	**438.00**	278.70	**343.00**
6'-0" x 8'-0", 2 doors	Inst	Set	2C	3.81	285.00	103.00	---	388.00	**484.00**	313.70	**383.00**
8'-0" x 8'-0", 2 doors	Inst	Set	2C	3.81	393.00	103.00	---	496.00	**608.00**	408.70	**493.00**
10'-0" x 6'-8", 3 doors	Inst	Set	2C	5.33	438.00	144.00	---	582.00	**722.00**	472.30	**576.00**
12'-0" x 6'-8", 3 doors	Inst	Set	2C	5.33	473.00	144.00	---	617.00	**763.00**	504.30	**612.00**
10'-0" x 8'-0", 3 doors	Inst	Set	2C	5.33	546.00	144.00	---	690.00	**846.00**	567.30	**685.00**
12'-0" x 8'-0", 3 doors	Inst	Set	2C	5.33	593.00	144.00	---	737.00	**900.00**	609.30	**733.00**
Unfinished hardboard											
4'-0" x 6'-8", 2 doors	Inst	Set	2C	3.81	113.00	103.00	---	216.00	**286.00**	161.20	**208.00**
6'-0" x 6'-8", 2 doors	Inst	Set	2C	3.81	140.00	103.00	---	243.00	**317.00**	184.70	**236.00**
8'-0" x 6'-8", 2 doors	Inst	Set	2C	3.81	244.00	103.00	---	347.00	**437.00**	276.70	**341.00**
4'-0" x 8'-0", 2 doors	Inst	Set	2C	3.81	169.00	103.00	---	272.00	**351.00**	210.70	**265.00**
6'-0" x 8'-0", 2 doors	Inst	Set	2C	3.81	197.00	103.00	---	300.00	**382.00**	235.70	**294.00**
8'-0" x 8'-0", 2 doors	Inst	Set	2C	3.81	261.00	103.00	---	364.00	**456.00**	291.70	**359.00**
10'-0" x 6'-8", 3 doors	Inst	Set	2C	5.33	319.00	144.00	---	463.00	**585.00**	368.30	**455.00**
12'-0" x 6'-8", 3 doors	Inst	Set	2C	5.33	370.00	144.00	---	514.00	**644.00**	412.30	**506.00**
10'-0" x 8'-0", 3 doors	Inst	Set	2C	5.33	421.00	144.00	---	565.00	**703.00**	458.30	**558.00**
12'-0" x 8'-0", 3 doors	Inst	Set	2C	5.33	396.00	144.00	---	540.00	**673.00**	435.30	**533.00**

Description	Oper	Unit	Crew Size	Man-Hours Per Unit	Avg Mat'l Unit Cost	Avg Labor Unit Cost	Avg Equip Unit Cost	Avg Total Unit Cost	Avg Price Incl O&P	Large Volume Avg Total Unit Cost	Large Volume Avg Price Incl O&P
Unfinished lauan											
4'-0" x 6'-8", 2 doors	Inst	Set	2C	3.81	123.00	103.00	---	226.00	**297.00**	169.70	**218.00**
6'-0" x 6'-8", 2 doors	Inst	Set	2C	3.81	157.00	103.00	---	260.00	**337.00**	199.70	**253.00**
8'-0" x 6'-8", 2 doors	Inst	Set	2C	3.81	246.00	103.00	---	349.00	**439.00**	278.70	**344.00**
4'-0" x 8'-0", 2 doors	Inst	Set	2C	3.81	234.00	103.00	---	337.00	**425.00**	268.70	**331.00**
6'-0" x 8'-0", 2 doors	Inst	Set	2C	3.81	279.00	103.00	---	382.00	**476.00**	307.70	**376.00**
8'-0" x 8'-0", 2 doors	Inst	Set	2C	3.81	370.00	103.00	---	473.00	**582.00**	388.70	**469.00**
10'-0" x 6'-8", 3 doors	Inst	Set	2C	5.33	334.00	144.00	---	478.00	**603.00**	381.30	**471.00**
12'-0" x 6'-8", 3 doors	Inst	Set	2C	5.33	373.00	144.00	---	517.00	**647.00**	415.30	**510.00**
10'-0" x 8'-0", 3 doors	Inst	Set	2C	5.33	522.00	144.00	---	666.00	**819.00**	547.30	**661.00**
12'-0" x 8'-0", 3 doors	Inst	Set	2C	5.33	559.00	144.00	---	703.00	**861.00**	579.30	**698.00**
Unfinished red oak											
4'-0" x 6'-8", 2 doors	Inst	Set	2C	3.81	159.00	103.00	---	262.00	**339.00**	201.70	**255.00**
6'-0" x 6'-8", 2 doors	Inst	Set	2C	3.81	211.00	103.00	---	314.00	**399.00**	247.70	**308.00**
8'-0" x 6'-8", 2 doors	Inst	Set	2C	3.81	354.00	103.00	---	457.00	**564.00**	374.70	**453.00**
4'-0" x 8'-0", 2 doors	Inst	Set	2C	3.81	317.00	103.00	---	420.00	**521.00**	341.70	**416.00**
6'-0" x 8'-0", 2 doors	Inst	Set	2C	3.81	351.00	103.00	---	454.00	**560.00**	371.70	**450.00**
8'-0" x 8'-0", 2 doors	Inst	Set	2C	3.81	504.00	103.00	---	607.00	**735.00**	505.70	**605.00**
10'-0" x 6'-8", 3 doors	Inst	Set	2C	5.33	501.00	144.00	---	645.00	**794.00**	528.30	**639.00**
12'-0" x 6'-8", 3 doors	Inst	Set	2C	5.33	535.00	144.00	---	679.00	**834.00**	558.30	**674.00**
10'-0" x 8'-0", 3 doors	Inst	Set	2C	5.33	745.00	144.00	---	889.00	**1080.00**	744.30	**887.00**
12'-0" x 8'-0", 3 doors	Inst	Set	2C	5.33	759.00	144.00	---	903.00	**1090.00**	756.30	**902.00**
Mirror bypass units											
Frameless unit with 1/2" beveled mirror edges											
4'-0" x 6'-8", 2 doors	Inst	Set	2C	4.44	240.00	120.00	---	360.00	**458.00**	284.00	**353.00**
6'-0" x 6'-8", 2 doors	Inst	Set	2C	4.44	311.00	120.00	---	431.00	**539.00**	346.00	**425.00**
8'-0" x 6'-8", 2 doors	Inst	Set	2C	4.44	375.00	120.00	---	495.00	**613.00**	403.00	**490.00**
4'-0" x 8'-0", 2 doors	Inst	Set	2C	4.44	272.00	120.00	---	392.00	**495.00**	312.00	**386.00**
6'-0" x 8'-0", 2 doors	Inst	Set	2C	4.44	352.00	120.00	---	472.00	**586.00**	382.00	**466.00**
8'-0" x 8'-0", 2 doors	Inst	Set	2C	4.44	436.00	120.00	---	556.00	**683.00**	457.00	**552.00**
10'-0" x 6'-8", 3 doors	Inst	Set	2C	5.33	530.00	144.00	---	674.00	**828.00**	553.30	**669.00**
12'-0" x 6'-8", 3 doors	Inst	Set	2C	5.33	577.00	144.00	---	721.00	**882.00**	595.30	**717.00**
10'-0" x 8'-0", 3 doors	Inst	Set	2C	5.33	608.00	144.00	---	752.00	**918.00**	623.30	**748.00**
12'-0" x 8'-0", 3 doors	Inst	Set	2C	5.33	666.00	144.00	---	810.00	**984.00**	673.30	**807.00**
Aluminum frame unit											
4'-0" x 6'-8", 2 doors	Inst	Set	2C	4.44	234.00	120.00	---	354.00	**451.00**	279.00	**347.00**
6'-0" x 6'-8", 2 doors	Inst	Set	2C	4.44	302.00	120.00	---	422.00	**530.00**	339.00	**416.00**
8'-0" x 6'-8", 2 doors	Inst	Set	2C	4.44	368.00	120.00	---	488.00	**605.00**	397.00	**483.00**
4'-0" x 8'-0", 2 doors	Inst	Set	2C	4.44	267.00	120.00	---	387.00	**489.00**	308.00	**381.00**
6'-0" x 8'-0", 2 doors	Inst	Set	2C	4.44	343.00	120.00	---	463.00	**577.00**	375.00	**458.00**
8'-0" x 8'-0", 2 doors	Inst	Set	2C	4.44	417.00	120.00	---	537.00	**661.00**	440.00	**533.00**
10'-0" x 6'-8", 3 doors	Inst	Set	2C	5.33	530.00	144.00	---	674.00	**828.00**	553.30	**669.00**
12'-0" x 6'-8", 3 doors	Inst	Set	2C	5.33	587.00	144.00	---	731.00	**894.00**	604.30	**727.00**
10'-0" x 8'-0", 3 doors	Inst	Set	2C	5.33	603.00	144.00	---	747.00	**912.00**	618.30	**743.00**
12'-0" x 8'-0", 3 doors	Inst	Set	2C	5.33	627.00	144.00	---	771.00	**939.00**	639.30	**767.00**
Golden oak frame unit											
4'-0" x 6'-8", 2 doors	Inst	Set	2C	4.44	311.00	120.00	---	431.00	**539.00**	346.00	**425.00**
6'-0" x 6'-8", 2 doors	Inst	Set	2C	4.44	392.00	120.00	---	512.00	**633.00**	418.00	**508.00**
8'-0" x 6'-8", 2 doors	Inst	Set	2C	4.44	475.00	120.00	---	595.00	**728.00**	491.00	**591.00**
4'-0" x 8'-0", 2 doors	Inst	Set	2C	4.44	353.00	120.00	---	473.00	**588.00**	383.00	**467.00**
6'-0" x 8'-0", 2 doors	Inst	Set	2C	4.44	446.00	120.00	---	566.00	**695.00**	466.00	**562.00**
8'-0" x 8'-0", 2 doors	Inst	Set	2C	4.44	535.00	120.00	---	655.00	**797.00**	544.00	**652.00**

Description	Oper	Unit	Crew Size	Man-Hours Per Unit	Avg Mat'l Unit Cost	Avg Labor Unit Cost	Avg Equip Unit Cost	Avg Total Unit Cost	Avg Price Incl O&P	Large Volume Avg Total Unit Cost	Large Volume Avg Price Incl O&P
10'-0" x 6'-8", 3 doors	Inst	Set	2C	5.33	675.00	144.00	---	819.00	**995.00**	682.30	**816.00**
12'-0" x 6'-8", 3 doors	Inst	Set	2C	5.33	738.00	144.00	---	882.00	**1070.00**	737.30	**879.00**
10'-0" x 8'-0", 3 doors	Inst	Set	2C	5.33	760.00	144.00	---	904.00	**1090.00**	757.30	**903.00**
12'-0" x 8'-0", 3 doors	Inst	Set	2C	5.33	834.00	144.00	---	978.00	**1180.00**	822.30	**977.00**
Steel frame unit											
4'-0" x 6'-8", 2 doors	Inst	Set	2C	4.44	169.00	120.00	---	289.00	**376.00**	221.00	**280.00**
6'-0" x 6'-8", 2 doors	Inst	Set	2C	4.44	221.00	120.00	---	341.00	**436.00**	267.00	**333.00**
8'-0" x 6'-8", 2 doors	Inst	Set	2C	4.44	270.00	120.00	---	390.00	**493.00**	310.00	**383.00**
4'-0" x 8'-0", 2 doors	Inst	Set	2C	4.44	185.00	120.00	---	305.00	**395.00**	236.00	**298.00**
6'-0" x 8'-0", 2 doors	Inst	Set	2C	4.44	251.00	120.00	---	371.00	**470.00**	293.00	**364.00**
8'-0" x 8'-0", 2 doors	Inst	Set	2C	4.44	313.00	120.00	---	433.00	**542.00**	348.00	**427.00**
10'-0" x 6'-8", 3 doors	Inst	Set	2C	5.33	396.00	144.00	---	540.00	**674.00**	435.30	**533.00**
12'-0" x 6'-8", 3 doors	Inst	Set	2C	5.33	442.00	144.00	---	586.00	**726.00**	476.30	**579.00**
10'-0" x 8'-0", 3 doors	Inst	Set	2C	5.33	450.00	144.00	---	594.00	**736.00**	483.30	**587.00**
12'-0" x 8'-0", 3 doors	Inst	Set	2C	5.33	505.00	144.00	---	649.00	**800.00**	532.30	**644.00**
Accordion doors											
Custom prefinished woodgrain print											
2'-0" x 6'-8"	Inst	Set	2C	2.67	266.00	72.00	---	338.00	**415.00**	278.10	**336.00**
3'-0" x 6'-8"	Inst	Set	2C	2.67	360.00	72.00	---	432.00	**524.00**	361.10	**431.00**
4'-0" x 6'-8"	Inst	Set	2C	2.67	465.00	72.00	---	537.00	**645.00**	454.10	**538.00**
5'-0" x 6'-8"	Inst	Set	2C	2.67	538.00	72.00	---	610.00	**728.00**	518.10	**612.00**
6'-0" x 6'-8"	Inst	Set	2C	2.67	657.00	72.00	---	729.00	**866.00**	623.10	**733.00**
7'-0" x 6'-8"	Inst	Set	2C	3.33	782.00	89.80	---	871.80	**1040.00**	743.90	**875.00**
8'-0" x 6'-8"	Inst	Set	2C	3.33	853.00	89.80	---	942.80	**1120.00**	805.90	**947.00**
9'-0" x 6'-8"	Inst	Set	2C	3.33	954.00	89.80	---	1043.80	**1230.00**	895.90	**1050.00**
10'-0" x 6'-8"	Inst	Set	2C	3.33	1100.00	89.80	---	1189.80	**1400.00**	1021.90	**1200.00**
For 8'-0"H, ADD	Inst	%	2C	---	12.0	---	---	---	**---**	---	**---**
Heritage prefinished real wood veneer											
2'-0" x 6'-8"	Inst	Set	2C	2.67	518.00	72.00	---	590.00	**705.00**	500.10	**591.00**
3'-0" x 6'-8"	Inst	Set	2C	2.67	727.00	72.00	---	799.00	**946.00**	685.10	**804.00**
4'-0" x 6'-8"	Inst	Set	2C	2.67	943.00	72.00	---	1015.00	**1190.00**	875.10	**1020.00**
5'-0" x 6'-8"	Inst	Set	2C	2.67	1090.00	72.00	---	1162.00	**1360.00**	1002.10	**1170.00**
6'-0" x 6'-8"	Inst	Set	2C	2.67	1300.00	72.00	---	1372.00	**1610.00**	1193.10	**1390.00**
7'-0" x 6'-8"	Inst	Set	2C	3.33	1520.00	89.80	---	1609.80	**1880.00**	1393.90	**1620.00**
8'-0" x 6'-8"	Inst	Set	2C	3.33	1660.00	89.80	---	1749.80	**2050.00**	1523.90	**1770.00**
9'-0" x 6'-8"	Inst	Set	2C	3.33	1880.00	89.80	---	1969.80	**2300.00**	1713.90	**1990.00**
10'-0" x 6'-8"	Inst	Set	2C	3.33	2090.00	89.80	---	2179.80	**2540.00**	1903.90	**2210.00**
For 8'-0"H, ADD	Inst	%	2C	---	20.0	---	---	---	**---**	---	**---**
Track and hardware only											
4'-0" x 6'-8", 2 doors	Inst	Set	---	---	19.60	---	---	19.60	**22.60**	17.30	**19.90**
6'-0" x 6'-8", 2 doors	Inst	Set	---	---	25.60	---	---	25.60	**29.40**	22.60	**26.00**
8'-0" x 6'-8", 2 doors	Inst	Set	---	---	33.00	---	---	33.00	**38.00**	29.10	**33.50**
4'-0" x 8'-0", 2 doors	Inst	Set	---	---	19.60	---	---	19.60	**22.60**	17.30	**19.90**
6'-0" x 8'-0", 2 doors	Inst	Set	---	---	25.60	---	---	25.60	**29.40**	22.60	**26.00**
8'-0" x 8'-0", 2 doors	Inst	Set	---	---	33.00	---	---	33.00	**38.00**	29.10	**33.50**
10'-0" x 6'-8", 3 doors	Inst	Set	---	---	45.20	---	---	45.20	**52.00**	39.90	**45.90**
12'-0" x 6'-8", 3 doors	Inst	Set	---	---	55.90	---	---	55.90	**64.30**	49.40	**56.80**
10'-0" x 8'-0", 3 doors	Inst	Set	---	---	45.20	---	---	45.20	**52.00**	39.90	**45.90**
12'-0" x 8'-0", 3 doors	Inst	Set	---	---	55.90	---	---	55.90	**64.30**	49.40	**56.80**

Description	Oper	Unit	Crew Size	Man-Hours Per Unit	Avg Mat'l Unit Cost	Avg Labor Unit Cost	Avg Equip Unit Cost	Avg Total Unit Cost	Avg Price Incl O&P	Large Volume Avg Total Unit Cost	Large Volume Avg Price Incl O&P

(Costs Based On Small Volume: Crew Size through Avg Price Incl O&P; Large Volume: last two columns)

Columns

See also Framing, page 103

Aluminum, extruded; self supporting

Designed as decorative, loadbearing elements for porches, entrances, colonnades, etc.; primed, knocked-down, and carton packed complete with cap and base

Description	Oper	Unit	Crew Size	Man-Hours Per Unit	Avg Mat'l Unit Cost	Avg Labor Unit Cost	Avg Equip Unit Cost	Avg Total Unit Cost	Avg Price Incl O&P	Large Volume Avg Total Unit Cost	Large Volume Avg Price Incl O&P
Column with standard cap and base											
8" dia. x 8' to 12' H	Inst	Ea	CS	6.67	134.00	169.00	89.40	392.40	**506.00**	271.60	**347.00**
10" dia. x 8' to 12' H	Inst	Ea	CS	6.67	196.00	169.00	89.40	454.40	**580.00**	325.60	**412.00**
10" dia. x 16' to 20' H	Inst	Ea	CS	8.89	264.00	225.00	119.00	608.00	**777.00**	436.50	**553.00**
12" dia. x 9' to 12' H	Inst	Ea	CS	6.67	422.00	169.00	89.40	680.40	**853.00**	523.60	**650.00**
12" dia. x 16' to 24' H	Inst	Ea	CS	8.89	487.00	225.00	119.00	831.00	**1040.00**	631.50	**787.00**
Column with Corinthian cap and decorative base											
8" dia. x 8' to 12' H	Inst	Ea	CS	6.67	253.00	169.00	89.40	511.40	**650.00**	376.60	**473.00**
10" dia. x 8' to 12' H	Inst	Ea	CS	6.67	327.00	169.00	89.40	585.40	**738.00**	439.60	**550.00**
10" dia. x 16' to 20' H	Inst	Ea	CS	8.89	401.00	225.00	119.00	745.00	**942.00**	557.50	**697.00**
12" dia. x 9' to 12' H	Inst	Ea	CS	6.67	644.00	169.00	89.40	902.40	**1120.00**	717.60	**883.00**
12" dia. x 16' to 24' H	Inst	Ea	CS	8.89	709.00	225.00	119.00	1053.00	**1310.00**	825.50	**1020.00**

Brick. See Masonry, page 161

Wood, treated, No. 1 common and better white pine, T&G construction

Designed as decorative, loadbearing elements for porches, entrances, colonnades, etc.; primed, knocked-down, and carton packed complete with cap and base

Description	Oper	Unit	Crew Size	Man-Hours Per Unit	Avg Mat'l Unit Cost	Avg Labor Unit Cost	Avg Equip Unit Cost	Avg Total Unit Cost	Avg Price Incl O&P	Large Volume Avg Total Unit Cost	Large Volume Avg Price Incl O&P
Plain column with standard cap and base											
8" dia. x 8' to 12' H	Inst	Ea	CS	8.00	375.00	202.00	107.00	684.00	**865.00**	513.30	**642.00**
10" dia. x 8' to 12' H	Inst	Ea	CS	8.00	452.00	202.00	107.00	761.00	**957.00**	580.30	**723.00**
12" dia. x 8' to 12' H	Inst	Ea	CS	8.89	518.00	225.00	119.00	862.00	**1080.00**	658.50	**819.00**
14" dia. x 12' to 16' H	Inst	Ea	CS	11.4	832.00	288.00	153.00	1273.00	**1590.00**	991.90	**1230.00**
16" dia. x 18' to 20' H	Inst	Ea	CS	12.5	1210.00	316.00	168.00	1694.00	**2100.00**	1351.00	**1660.00**
Plain column with Corinthian cap and decorative base											
8" dia. x 8' to 12' H	Inst	Ea	CS	8.00	433.00	202.00	107.00	742.00	**935.00**	564.30	**703.00**
10" dia. x 8' to 12' H	Inst	Ea	CS	8.00	565.00	202.00	107.00	874.00	**1090.00**	679.30	**842.00**
12" dia. x 8' to 12' H	Inst	Ea	CS	8.89	631.00	225.00	119.00	975.00	**1220.00**	757.50	**938.00**
14" dia. x 12' to 16' H	Inst	Ea	CS	11.4	931.00	288.00	153.00	1372.00	**1710.00**	1077.90	**1330.00**
16" dia. x 18' to 20' H	Inst	Ea	CS	12.5	1300.00	316.00	168.00	1784.00	**2200.00**	1421.00	**1750.00**

Concrete

Concrete Footings

1. **Dimensions.** 6" T, 8" T, 12" T x 12" W, 16" W, 20" W; 12" T x 24" W.
2. **Installation**
   a. Forms. 2" side forms equal in height to the thickness of the footing. 2" x 4" stakes 4'-0" oc, no less than the thickness of the footing. 2" x 4" bracing for stakes 8'-0" oc for 6" and 8" thick footings and 4'-0" oc for 12" thick footings. 1" x 2" or 1" x 3" spreaders 4'-0" oc

   b. Concrete. 1-2-4 mix is used in this section.

   c. Reinforcing steel. Various sizes, but usually only #3, #4, or #5 straight rods with end ties are used.
3. **Notes on Labor**
   a. Forming. Output based on a crew of two carpenters and one laborer.

   b. Grading, finish. Output based on what one laborer can do in one day.

   c. Reinforcing steel. Output based on what one laborer or one ironworker can do in one day.

   d. Concrete. Output based on a crew of two laborers and one carpenter.

   e. Forms, wrecking and cleaning. Output based on what one laborer can do in one day.
4. **Estimating Technique**. Determine the linear feet of footing.

Concrete Foundations

1. **Dimensions**. 8" T, 12" T x 4' H, 8' H, or 12' H.
2. **Installation**
   a. Forms. 4' x 8' panels made of $^{3}/_{4}$" form grade plywood backed with 2" x 4" studs and sills (studs approximately 16" oc), three sets and six sets of 2" x 4" wales for 4', 8' and 12' high walls. 2" x 4" wales for 4', 8' and 12' high walls. 2" x 4" diagonal braces (with stakes) 12'-0" oc one side. Snap ties spaced 22" oc, 20" oc and 17" oc along each wale for 4', 8' and 12' high walls. Paraffin oil coating for forms. Twelve uses are estimated for panels; twenty uses for wales and braces; snap ties are used only once.

   b. Concrete. 1-2-4 mix.

   c. Reinforcing steel. Sizes #3 to #7. Bars are straight except dowels which may on occasion be bent rods.
3. **Notes on Labor**
   a. Concrete, placing. Output based on a crew of one carpenter and five laborers.

   b. Forming. Output based on a crew of four carpenters and one laborer.

   c. Reinforcing steel rods. Output based on what two laborers or two ironworkers can do in one day.

   d. Wrecking and cleaning forms. Output based on what two laborers can do in one day.
4. **Estimating Technique**
   a. Determine linear feet of wall if wall is 8" or 12" x 4', 8' or 12', or determine square feet of wall. Then calculate and add the linear feet of rods.

Concrete Interior Floor Finishes

1. **Dimensions**. $3^{1}/_{2}$", 4", 5", 6" thick x various areas.
2. **Installation**
   a. Forms. A wood form may or may not be required. A foundation wall may serve as a form for both basement and first floor slabs. In this section, only 2" x 4" and 2" x 6" side forms with stakes 4'-0" oc are considered.

   b. Finish grading. Dirt or gravel.

   c. Screeds (wood strips placed in area where concrete is to be placed). The concrete when placed will be finished even with top of the screeds. Screeds must be pulled before concrete sets up and the voids filled with concrete. 2" x 2" and 2" x 4" screeds with 2" x 2" stakes 6'-0" oc will be covered in this section.

   d. Steel reinforcing. Items to be covered are: #3, #4, #5 rods; 6 x 6/10-10 and 6 x 6/6-6 welded wire mesh.

   e. Concrete. 1-2-4 mix.

3. **Notes on Labor**

   a. Forms and screeds. Output based on a crew of two masons and one laborer.

   b. Finish grading. Output based on what one laborer can do in one day.

   c. Reinforcing. Output based on what two laborers or two ironworkers can do in one day.

   d. Concrete, place and finish. Output based on three cement masons and five laborers as a crew.

   e. Wrecking and cleaning forms. Output based on what one laborer can do in one day.

4. **Estimating Technique**

   a. Finish grading, mesh, and concrete. Determine the area and add waste.

   b. Forms, screeds and rods. Determine the linear feet.

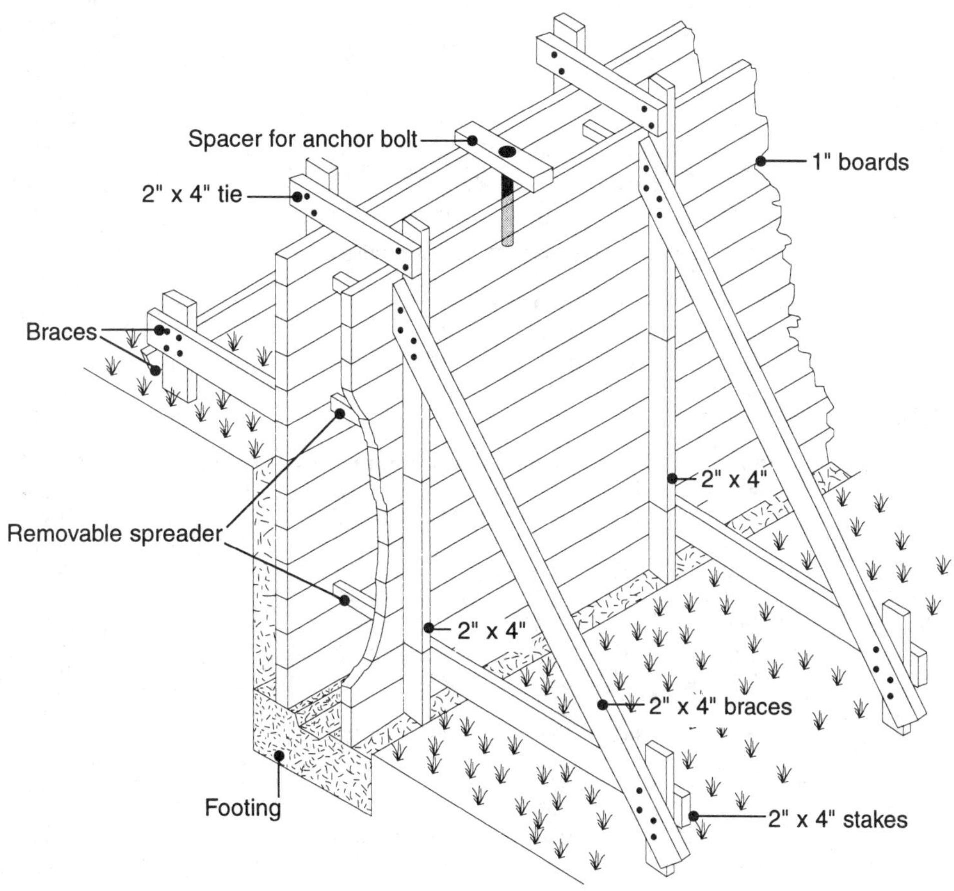

Description	Oper	Unit	Crew Size	Man-Hours Per Unit	Avg Mat'l Unit Cost	Avg Labor Unit Cost	Avg Equip Unit Cost	Avg Total Unit Cost	Avg Price Incl O&P	Large Volume Avg Total Unit Cost	Large Volume Avg Price Incl O&P
Concrete, cast in place											
Footings, with air tools; reinforced											
8" T x 12" W	Demo	LF	AD	.202	---	4.88	.78	5.66	**8.10**	4.79	**6.86**
8" T x 16" W	Demo	LF	AD	.214	---	5.17	.83	6.00	**8.58**	5.10	**7.29**
8" T x 20" W	Demo	LF	AD	.229	---	5.53	.88	6.41	**9.18**	5.44	**7.78**
12" T x 12" W	Demo	LF	AD	.224	---	5.41	.87	6.28	**8.99**	5.33	**7.62**
12" T x 16" W	Demo	LF	AD	.235	---	5.68	.91	6.59	**9.42**	5.60	**8.02**
12" T x 20" W	Demo	LF	AD	.247	---	5.97	.96	6.93	**9.91**	5.91	**8.45**
12" T x 24" W	Demo	LF	AD	.261	---	6.30	1.01	7.31	**10.50**	6.22	**8.90**
Footings, per LF poured footing											
Forming, 4 uses											
2" x 6"	Inst	LF	CS	.116	.78	2.93	---	3.71	**5.24**	3.35	**4.75**
2" x 8"	Inst	LF	CS	.130	.98	3.29	---	4.27	**5.98**	3.84	**5.42**
2" x 12"	Inst	LF	CS	.145	1.40	3.67	---	5.07	**6.97**	4.51	**6.26**
Grading, finish by hand											
6", 8", 12" T x 12" W	Inst	LF	1L	.026	---	.57	---	.57	**.86**	.50	**.76**
6", 8", 12" T x 16" W	Inst	LF	1L	.030	---	.66	---	.66	**.99**	.59	**.89**
6", 8", 12" T x 20" W	Inst	LF	1L	.034	---	.75	---	.75	**1.13**	.68	**1.03**
12" T x 24" W	Inst	LF	1L	.039	---	.86	---	.86	**1.29**	.77	**1.16**
Reinforcing steel in place, material costs include lap, waste, and tie wire											
Two (No. 3) 3/8" rods	Inst	LF	1L	.015	.44	.33	---	.77	**.94**	.67	**.82**
Two (No. 4) 1/2" rods	Inst	LF	1L	.016	.50	.35	---	.85	**1.03**	.72	**.87**
Two (No. 5) 5/8" rods	Inst	LF	1L	.016	.78	.35	---	1.13	**1.31**	.97	**1.14**
Three (No. 3) 3/8" rods	Inst	LF	1L	.019	.66	.42	---	1.08	**1.29**	.91	**1.10**
Three (No. 4) 1/2" rods	Inst	LF	1L	.020	.75	.44	---	1.19	**1.41**	1.02	**1.22**
Three (No. 5) 5/8" rods	Inst	LF	1L	.021	1.17	.46	---	1.63	**1.87**	1.38	**1.59**
Concrete, pour from truck into forms; using 3,000 PSI, 1-1/2" aggregate, 7.7 sack mix											
6" T x 12" W (54.0 LF/CY)	Inst	LF	LC	.043	1.71	1.02	---	2.73	**3.24**	2.32	**2.79**
6" T x 16" W (40.5 LF/CY)	Inst	LF	LC	.048	1.71	1.13	---	2.84	**3.42**	2.42	**2.93**
6" T x 20" W (32.3 LF/CY)	Inst	LF	LC	.049	2.57	1.16	---	3.73	**4.32**	3.14	**3.67**
8" T x 12" W (40.5 LF/CY)	Inst	LF	LC	.048	1.71	1.13	---	2.84	**3.42**	2.42	**2.93**
8" T x 16" W (30.4 LF/CY)	Inst	LF	LC	.052	2.57	1.23	---	3.80	**4.42**	3.21	**3.78**
8" T x 20" W (24.3 LF/CY)	Inst	LF	LC	.055	3.42	1.30	---	4.72	**5.38**	3.96	**4.55**
12" T x 12" W (27.0 LF/CY)	Inst	LF	LC	.050	3.42	1.18	---	4.60	**5.20**	3.86	**4.41**
12" T x 16" W (20.3 LF/CY)	Inst	LF	LC	.058	4.28	1.37	---	5.65	**6.35**	4.73	**5.35**
12" T x 20" W (16.2 LF/CY)	Inst	LF	LC	.063	5.13	1.49	---	6.62	**7.38**	5.55	**6.23**
12" T x 24" W (13.5 LF/CY)	Inst	LF	LC	.055	5.99	1.30	---	7.29	**7.95**	6.07	**6.66**
Forms, wreck, remove, and clean											
2" x 6"	Inst	LF	1L	.051	---	1.12	---	1.12	**1.69**	1.01	**1.52**
2" x 8"	Inst	LF	1L	.056	---	1.23	---	1.23	**1.86**	1.10	**1.66**
2" x 12"	Inst	LF	1L	.061	---	1.34	---	1.34	**2.02**	1.21	**1.82**

Description	Oper	Unit	Crew Size	Man-Hours Per Unit	Avg Mat'l Unit Cost	Avg Labor Unit Cost	Avg Equip Unit Cost	Avg Total Unit Cost	Avg Price Incl O&P	Large Volume Avg Total Unit Cost	Large Volume Avg Price Incl O&P
Foundations and retaining walls with air tools, per LF wall											
With reinforcing											
4'-0" H											
8" T	Demo	LF	AD	.353	---	8.52	1.36	9.88	**14.20**	8.41	**12.00**
12" T	Demo	LF	AD	.490	---	11.80	1.89	13.69	**19.60**	11.60	**16.60**
8'-0" H											
8" T	Demo	LF	AD	.381	---	9.20	1.47	10.67	**15.30**	9.07	**13.00**
12" T	Demo	LF	AD	.545	---	13.20	2.11	15.31	**21.90**	12.98	**18.50**
12"-0" H											
8" T	Demo	LF	AD	.414	---	10.00	1.60	11.60	**16.60**	9.88	**14.20**
12" T	Demo	LF	AD	.649	---	15.70	2.50	18.20	**26.00**	15.31	**21.90**
Without reinforcing											
4'-0" H											
8" T	Demo	LF	AD	.308	---	7.44	1.19	8.63	**12.40**	7.31	**10.50**
12" T	Demo	LF	AD	.414	---	10.00	1.60	11.60	**16.60**	9.88	**14.20**
8'-0" H											
8" T	Demo	LF	AD	.338	---	8.16	1.31	9.47	**13.60**	8.01	**11.50**
12" T	Demo	LF	AD	.471	---	11.40	1.82	13.22	**18.90**	11.20	**16.00**
12"-0" H											
8" T	Demo	LF	AD	.369	---	8.91	1.43	10.34	**14.80**	8.85	**12.70**
12" T	Demo	LF	AD	.545	---	13.20	2.11	15.31	**21.90**	12.98	**18.50**
Forming only											
Material price includes panel forms, wales, braces, snap ties, paraffin oil and nails											
8" or 12" T x 4'-0" H											
Make (@12 uses)	Inst	SF	CU	.063	1.21	1.64	---	2.85	**3.70**	2.50	**3.28**
Erect and coat	Inst	SF	CU	.106	.03	2.75	---	2.78	**4.21**	2.56	**3.89**
Wreck and clean	Inst	SF	LB	.048	---	1.05	---	1.05	**1.59**	.94	**1.43**
8" or 12" T x 8'-0" H											
Make (@12 uses)	Inst	SF	CU	.066	1.11	1.71	---	2.82	**3.71**	2.49	**3.32**
Erect and coat	Inst	SF	CU	.128	.03	3.32	---	3.35	**5.08**	3.08	**4.68**
Wreck and clean	Inst	SF	LB	.058	---	1.27	---	1.27	**1.92**	1.14	**1.72**
8" or 12" T x 12'-0" H											
Make (@12 uses)	Inst	SF	CU	.065	1.16	1.69	---	2.85	**3.72**	2.51	**3.32**
Erect and coat	Inst	SF	CU	.150	.03	3.89	---	3.92	**5.95**	3.60	**5.47**
Wreck and clean	Inst	SF	LB	.068	---	1.49	---	1.49	**2.25**	1.36	**2.05**
Reinforcing steel rods. 5% waste included; pricing based on LF of rod											
No. 3 (3/8" rod)	Inst	LF	LB	.013	.21	.29	---	.50	**.64**	.43	**.57**
No. 4 (1/2" rod)	Inst	LF	LB	.013	.24	.29	---	.53	**.67**	.46	**.60**
No. 5 (5/8" rod)	Inst	LF	LB	.014	.37	.31	---	.68	**.83**	.60	**.74**
No. 6 (3/4" rod)	Inst	LF	LB	.016	.54	.35	---	.89	**1.07**	.75	**.90**
No. 7 (7/8" rod)	Inst	LF	LB	.018	.73	.40	---	1.13	**1.33**	.95	**1.13**
Concrete, placed from trucks into forms, material cost includes 5% waste and assumes use of 3,000 PSI, 1-1/2" aggregate, 5.7 sack mix											
8" T x 4'-0" H (10.12 LF/CY)	Inst	LF	LG	.077	8.55	1.75	---	10.30	**11.20**	8.61	**9.42**
8" T x 4'-0" H (40.5 SF/CY)	Inst	SF	LG	.019	1.71	.43	---	2.14	**2.36**	1.79	**1.98**
8" T x 8'-0" H (5.06 LF/CY)	Inst	LF	LG	.154	17.10	3.51	---	20.61	**22.40**	17.17	**18.80**
8" T x 8'-0" H (40.5 SF/CY)	Inst	SF	LG	.019	1.71	.43	---	2.14	**2.36**	1.79	**1.98**
8" T x 12'-0" H (3.37 LF/CY)	Inst	LF	LG	.232	25.70	5.29	---	30.99	**33.60**	25.76	**28.20**
8" T x 12'-0" H (40.5 SF/CY)	Inst	SF	LG	.019	1.71	.43	---	2.14	**2.36**	1.79	**1.98**

Description	Oper	Unit	Crew Size	Man-Hours Per Unit	Avg Mat'l Unit Cost	Avg Labor Unit Cost	Avg Equip Unit Cost	Avg Total Unit Cost	Avg Price Incl O&P	Large Volume Avg Total Unit Cost	Large Volume Avg Price Incl O&P
12' T x 4'-0" H (6.75 LF/CY)	Inst	LF	LG	.115	12.80	2.62	---	15.42	**16.80**	12.87	**14.10**
12" T x 4'-0" H (27.0 LF/CY)	Inst	SF	LG	.029	3.42	.66	---	4.08	**4.42**	3.39	**3.69**
12" T x 8'-0" H (3.38 LF/CY)	Inst	LF	LG	.231	25.70	5.26	---	30.96	**33.60**	25.74	**28.20**
12" T x 8'-0" H (27.0 SF/CY)	Inst	SF	LG	.026	3.42	.59	---	4.01	**4.31**	3.39	**3.69**
12" T x 12'-0" H (2.25 LF/CY)	Inst	LF	LG	.312	37.60	7.11	---	44.71	**48.40**	37.91	**41.60**
12" T x 12'-0" H (27.0 SF/CY)	Inst	SF	LG	.026	3.42	.59	---	4.01	**4.31**	3.39	**3.69**
Slabs, with air tools											
With reinforcing (6 x 6/10 x 10)											
4" T	Demo	SF	AD	.074	---	1.79	.20	1.99	**2.88**	1.69	**2.45**
5" T	Demo	SF	AD	.085	---	2.05	.23	2.28	**3.31**	1.96	**2.84**
6" T	Demo	SF	AD	.097	---	2.34	.26	2.60	**3.77**	2.22	**3.23**
Without reinforcing											
4" T	Demo	SF	AD	.051	---	1.23	.14	1.37	**1.99**	1.18	**1.71**
5" T	Demo	SF	AD	.059	---	1.42	.16	1.58	**2.30**	1.37	**1.99**
6" T	Demo	SF	AD	.067	---	1.62	.18	1.80	**2.61**	1.53	**2.22**
Slabs and interior floor finishes											
Forming; 4 uses											
2" x 4" (4" T slab)	Inst	LF	DD	.045	.26	1.11	---	1.37	**1.93**	1.23	**1.74**
2" x 6" (5" T and 6" T slab)	Inst	LF	DD	.045	.34	1.11	---	1.45	**2.01**	1.29	**1.80**
Grading											
Dirt, cut and fill, +/- 1/10 ft.	Inst	SF	1L	.011	---	.24	---	.24	**.36**	.22	**.33**
Gravel, 3/4" to 1-1/2" stone	Inst	SF	1L	.012	.42	.26	---	.68	**.82**	.58	**.70**
Screeds; 3 uses											
2" x 2"	Inst	LF	DD	.031	.11	.77	---	.88	**1.26**	.81	**1.17**
2" x 4"	Inst	LF	DD	.034	.20	.84	---	1.04	**1.46**	.95	**1.35**
Reinforcing steel rods includes 5% waste, costs are per LF of rod or SF of mesh											
No. 3 (3/8" rod)	Inst	LF	LB	.012	.21	.26	---	.47	**.61**	.41	**.53**
No. 4 (1/2" rod)	Inst	LF	LB	.013	.24	.29	---	.53	**.67**	.44	**.56**
No. 5 (5/8" rod)	Inst	LF	LB	.013	.37	.29	---	.66	**.80**	.57	**.71**
6x6/10-10 @21 lbs/CSF	Inst	SF	LB	.003	.11	.07	---	.18	**.21**	.16	**.19**
6x6/6-6 @42 lbs/CSF	Inst	SF	LB	.004	.21	.09	---	.30	**.34**	.27	**.31**
Concrete, pour and finish (steel trowel), material cost includes 5% waste and assumes use of 2500 PSI, 1" aggregate, 5.5 sack mix											
3-1/2" T (92.57 SF/CY)	Inst	SF	DF	.025	.86	.59	---	1.45	**1.74**	1.24	**1.51**
4" T (81.00 SF/CY)	Inst	SF	DF	.026	.86	.61	---	1.47	**1.78**	1.26	**1.55**
5" T (64.80 SF/CY)	Inst	SF	DF	.027	1.71	.63	---	2.34	**2.66**	1.99	**2.28**
6" T (54.00 SF/CY)	Inst	SF	DF	.028	1.71	.66	---	2.37	**2.70**	2.01	**2.32**
Forms, wreck and clean											
2" x 4"	Inst	LF	1L	.025	---	.55	---	.55	**.83**	.48	**.73**
2" x 6"	Inst	LF	1L	.026	---	.57	---	.57	**.86**	.53	**.80**
Ready mix											
Material costs only, prices are typical for most cities, and assume a ten mile haul of more than 7.25 CY											
Using 1-1/2" aggregate											
2000 PSI, 4.8 sack mix	Inst	CY	---	---	75.70	---	---	75.70	**75.70**	62.10	**62.10**
2500 PSI, 5.2 sack mix	Inst	CY	---	---	78.20	---	---	78.20	**78.20**	64.10	**64.10**
3000 PSI, 5.7 sack mix	Inst	CY	---	---	81.40	---	---	81.40	**81.40**	66.70	**66.70**
3500 PSI, 6.3 sack mix	Inst	CY	---	---	83.10	---	---	83.10	**83.10**	68.10	**68.10**
4000 PSI, 6.9 sack mix	Inst	CY	---	---	88.80	---	---	88.80	**88.80**	72.80	**72.80**

Description	Oper	Unit	Crew Size	Man-Hours Per Unit	Avg Mat'l Unit Cost	Avg Labor Unit Cost	Avg Equip Unit Cost	Avg Total Unit Cost	Avg Price Incl O&P	Large Volume Avg Total Unit Cost	Large Volume Avg Price Incl O&P
Using 1" aggregate											
2000 PSI, 5.0 sack mix	Inst	CY	---	---	77.40	---	---	77.40	**77.40**	63.40	**63.40**
2500 PSI, 5.5 sack mix	Inst	CY	---	---	80.60	---	---	80.60	**80.60**	66.10	**66.10**
3000 PSI, 6.0 sack mix	Inst	CY	---	---	83.10	---	---	83.10	**83.10**	68.10	**68.10**
3500 PSI, 6.6 sack mix	Inst	CY	---	---	87.10	---	---	87.10	**87.10**	71.40	**71.40**
4000 PSI, 7.1 sack mix	Inst	CY	---	---	89.60	---	---	89.60	**89.60**	73.40	**73.40**
Using 3/8" aggregate											
2000 PSI, 6.0 sack mix	Inst	CY	---	---	83.10	---	---	83.10	**83.10**	68.10	**68.10**
2500 PSI, 6.5 sack mix	Inst	CY	---	---	86.30	---	---	86.30	**86.30**	70.70	**70.70**
3000 PSI, 7.2 sack mix	Inst	CY	---	---	90.40	---	---	90.40	**90.40**	74.10	**74.10**
3500 PSI, 7.9 sack mix	Inst	CY	---	---	94.50	---	---	94.50	**94.50**	77.40	**77.40**
4000 PSI, 8.5 sack mix	Inst	CY	---	---	100.00	---	---	100.00	**100.00**	82.10	**82.10**
Adjustments											
Delivery over 10 miles, ADD	Inst	Mi	---	---	.70	---	---	.70	**.70**	.61	**.61**
High early strength, ADD											
5.0 sack mix	Inst	CY	---	---	9.02	---	---	9.02	**9.02**	7.84	**7.84**
6.0 sack mix	Inst	CY	---	---	11.30	---	---	11.30	**11.30**	9.81	**9.81**
Lightweight aggregate, ADD											
Mix from truck to forms	Inst	CY	---	---	46.80	---	---	46.80	**46.80**	40.70	**40.70**
Pump mix	Inst	CY	---	---	49.40	---	---	49.40	**49.40**	42.90	**42.90**
For 1% Calcium chloride	Inst	Sack	---	---	.32	---	---	.32	**.32**	.28	**.28**
For 2% Calcium chloride	Inst	Sack	---	---	.46	---	---	.46	**.46**	.40	**.40**
Chemical compensated shrinkage (WRDA Admix)											
ADD	Inst	Sack	---	---	.26	---	---	.26	**.26**	.23	**.23**
Loads 7.25 CY or less, ADD											
7.24 CY	Inst	LS	---	---	5.20	---	---	5.20	**5.20**	4.52	**4.52**
6.0 CY	Inst	LS	---	---	18.20	---	---	18.20	**18.20**	15.80	**15.80**
5.0 CY	Inst	LS	---	---	28.60	---	---	28.60	**28.60**	24.90	**24.90**
4.0 CY	Inst	LS	---	---	39.00	---	---	39.00	**39.00**	33.90	**33.90**
3.0 CY	Inst	LS	---	---	49.40	---	---	49.40	**49.40**	42.90	**42.90**
2.0 CY	Inst	LS	---	---	59.80	---	---	59.80	**59.80**	52.00	**52.00**
1.0 CY	Inst	LS	---	---	70.20	---	---	70.20	**70.20**	61.00	**61.00**
Coloring of concrete, ADD											
Light or sand colors	Inst	CY	---	---	20.50	---	---	20.50	**20.50**	17.80	**17.80**
Medium or buff colors	Inst	CY	---	---	30.20	---	---	30.20	**30.20**	26.30	**26.30**
Dark colors	Inst	CY	---	---	45.50	---	---	45.50	**45.50**	39.60	**39.60**
Green	Inst	CY	---	---	50.70	---	---	50.70	**50.70**	44.10	**44.10**
Standby charge for time in excess of 5 Min/CY											
ADD	Inst	Min	---	---	1.30	---	---	1.30	**1.30**	1.13	**1.13**

Concrete block. See Masonry, page 164

Countertops

Formica

One piece tops; straight, "L", or "U" shapes; surfaced with laminated plastic cemented to particleboard base

Post formed countertop with raised front drip edge

Description	Oper	Unit	Crew Size	Man-Hours Per Unit	Avg Mat'l Unit Cost	Avg Labor Unit Cost	Avg Equip Unit Cost	Avg Total Unit Cost	Avg Price Incl O&P	Large Volume Avg Total Unit Cost	Large Volume Avg Price Incl O&P
25" W, 1-1/2" H front edge, 4" H coved backsplash											
Satin/suede patterns	Inst	LF	2C	.808	20.90	21.80	---	42.70	**57.10**	31.80	**41.40**
Solid patterns	Inst	LF	2C	.808	23.20	21.80	---	45.00	**59.80**	33.90	**43.80**
Specialty finish patterns	Inst	LF	2C	.808	27.80	21.80	---	49.60	**65.10**	38.10	**48.60**
Wood tone patterns	Inst	LF	2C	.808	29.00	21.80	---	50.80	**66.50**	39.10	**49.80**

Description	Oper	Unit	Crew Size	Man-Hours Per Unit	Avg Mat'l Unit Cost	Avg Labor Unit Cost	Avg Equip Unit Cost	Avg Total Unit Cost	Avg Price Incl O&P	Large Volume Avg Total Unit Cost	Large Volume Avg Price Incl O&P
			Costs Based On Small Volume							Large Volume	
Double roll top, 1-1/2" H front and back edges, no backsplash											
25" W											
Satin/suede patterns	Inst	LF	2C	.762	26.10	20.50	---	46.60	**61.20**	35.70	**45.60**
Solid patterns	Inst	LF	2C	.762	29.00	20.50	---	49.50	**64.60**	38.30	**48.60**
Specialty finish patterns	Inst	LF	2C	.762	34.80	20.50	---	55.30	**71.30**	43.50	**54.60**
Wood tone patterns	Inst	LF	2C	.762	36.30	20.50	---	56.80	**72.90**	44.80	**56.10**
36" W											
Satin/suede patterns	Inst	LF	2C	.833	40.70	22.50	---	63.20	**81.00**	50.00	**62.50**
Solid patterns	Inst	LF	2C	.833	45.20	22.50	---	67.70	**86.20**	54.10	**67.10**
Specialty finish patterns	Inst	LF	2C	.833	54.30	22.50	---	76.80	**96.60**	62.20	**76.50**
Wood tone patterns	Inst	LF	2C	.833	56.60	22.50	---	79.10	**99.20**	64.20	**78.80**
Post formed countertop with square edge veneer front											
25" W, 1-1/2" H front edge, 4" H coved backsplash											
Satin/suede patterns	Inst	LF	2C	.808	31.30	21.80	---	53.10	**69.10**	41.20	**52.20**
Solid patterns	Inst	LF	2C	.808	34.80	21.80	---	56.60	**73.10**	44.30	**55.80**
Specialty finish patterns	Inst	LF	2C	.808	41.80	21.80	---	63.60	**81.10**	50.50	**62.90**
Wood tone patterns	Inst	LF	2C	.808	43.50	21.80	---	65.30	**83.10**	52.10	**64.70**
Self-edge countertop with square edge veneer front											
25" W, 1-1/2" H front edge, 4" H coved backsplash (@90 degree angle to deck)											
Satin/suede patterns	Inst	LF	2C	.808	64.70	21.80	---	86.50	**108.00**	71.10	**86.60**
Solid patterns	Inst	LF	2C	.808	71.90	21.80	---	93.70	**116.00**	77.60	**94.00**
Specialty finish patterns	Inst	LF	2C	.808	86.30	21.80	---	108.10	**132.00**	90.50	**109.00**
Wood tone patterns	Inst	LF	2C	.808	89.90	21.80	---	111.70	**137.00**	93.70	**113.00**
Material cost adjustments, ADD											
Diagonal corner cut											
Standard	Inst	Ea	---	---	20.90	---	---	20.90	**20.90**	18.70	**18.70**
With plateau shelf	Inst	Ea	---	---	98.60	---	---	98.60	**98.60**	88.40	**88.40**
Radius corner cut											
3" or 6"	Inst	Ea	---	---	13.90	---	---	13.90	**13.90**	12.50	**12.50**
12"	Inst	Ea	---	---	13.90	---	---	13.90	**13.90**	12.50	**12.50**
Quarter radius end	Inst	Ea	---	---	13.90	---	---	13.90	**13.90**	12.50	**12.50**
Half radius end	Inst	Ea	---	---	25.50	---	---	25.50	**25.50**	22.90	**22.90**
Miter corner, shop assembled	Inst	Ea	---	---	21.00	---	---	21.00	**21.00**	18.80	**18.80**
Splicing any top or leg 12' or longer, shop assembled	Inst	Ea	---	---	24.40	---	---	24.40	**24.40**	21.80	**21.80**
Endsplash with finished sides & edges	Inst	Ea	---	---	18.60	---	---	18.60	**18.60**	16.60	**16.60**
Sink or range cutout	Inst	Ea	---	---	8.12	---	---	8.12	**8.12**	7.28	**7.28**

Ceramic tile

Description	Oper	Unit	Crew Size	Man-Hours Per Unit	Avg Mat'l Unit Cost	Avg Labor Unit Cost	Avg Equip Unit Cost	Avg Total Unit Cost	Avg Price Incl O&P	Large Volume Avg Total Unit Cost	Large Volume Avg Price Incl O&P
Countertop/backsplash											
Adhesive set with backmounted tile											
1" x 1"	Inst	SF	TB	.381	6.24	8.77	---	15.01	**20.30**	10.56	**15.50**
4-1/4" x 4-1/4"	Inst	SF	TB	.333	4.98	7.67	---	12.65	**17.20**	8.84	**13.10**
Cove/base											
Adhesive set with unmounted tile											
4-1/4" x 4-1/4"	Inst	LF	TB	.333	4.79	7.67	---	12.46	**16.90**	8.68	**11.50**
6" x 4-1/4"	Inst	LF	TB	.281	4.68	6.47	---	11.15	**15.00**	7.84	**10.30**

Description	Oper	Unit	Crew Size	Man-Hours Per Unit	Avg Mat'l Unit Cost	Avg Labor Unit Cost	Avg Equip Unit Cost	Avg Total Unit Cost	Avg Price Incl O&P	Large Volume Avg Total Unit Cost	Large Volume Avg Price Incl O&P
Conventional/mortar set with unmounted tile											
4-1/4" x 4-1/4"	Inst	LF	TB	.593	4.69	13.70	---	18.39	**25.80**	12.18	**16.80**
6" x 4-1/4"	Inst	LF	TB	.533	4.54	12.30	---	16.84	**23.50**	11.22	**15.40**
Dry-set mortar with unmounted tile											
4-1/4" x 4-1/4"	Inst	LF	TB	.410	4.72	9.44	---	14.16	**19.50**	9.67	**13.00**
6" x 4-1/4"	Inst	LF	TB	.356	4.57	8.20	---	12.77	**17.50**	8.79	**11.80**

Wood

Butcher block construction throughout top; custom, straight, "L", or "U" shapes

Description	Oper	Unit	Crew Size	Man-Hours Per Unit	Avg Mat'l Unit Cost	Avg Labor Unit Cost	Avg Equip Unit Cost	Avg Total Unit Cost	Avg Price Incl O&P	Large Volume Avg Total Unit Cost	Large Volume Avg Price Incl O&P
Self-edge top; 26" W, with 4" H backsplash, 1-1/2" H front and back edges	Inst	LF	2C	.746	67.20	20.10	---	87.30	**108.00**	72.30	**88.00**
Material cost adjustments, ADD											
Miter corner	Inst	Ea	---	---	16.80	---	---	16.80	**16.80**	14.80	**14.80**
Sink or surface saver cutout	Inst	Ea	---	---	50.40	---	---	50.40	**50.40**	44.40	**44.40**
45 degree plateau corner	Inst	Ea	---	---	252.00	---	---	252.00	**252.00**	222.00	**222.00**
End splash	Inst	Ea	---	---	58.80	---	---	58.80	**58.80**	51.80	**51.80**

Cupolas

Description	Oper	Unit	Crew Size	Man-Hours Per Unit	Avg Mat'l Unit Cost	Avg Labor Unit Cost	Avg Equip Unit Cost	Avg Total Unit Cost	Avg Price Incl O&P	Large Volume Avg Total Unit Cost	Large Volume Avg Price Incl O&P
Natural finish redwood, aluminum roof											
24" x 24", 25" H	Inst	Ea	CA	1.90	226.00	51.20	---	277.20	**361.00**	234.70	**302.00**
30" x 30", 30" H	Inst	Ea	CA	2.22	346.00	59.90	---	405.90	**524.00**	347.90	**444.00**
35" x 35", 33" H	Inst	Ea	CA	2.67	402.00	72.00	---	474.00	**612.00**	405.10	**518.00**
Natural finish redwood, copper roof											
22" x 22", 33" H	Inst	Ea	CA	1.90	226.00	51.20	---	277.20	**361.00**	234.70	**302.00**
25" x 25", 39" H	Inst	Ea	CA	2.22	346.00	59.90	---	405.90	**524.00**	347.90	**444.00**
35" x 35", 33" H	Inst	Ea	CA	2.67	402.00	72.00	---	474.00	**612.00**	405.10	**518.00**
Natural finish redwood, octagonal shape copper roof											
31" W, 37" H	Inst	Ea	CA	2.67	515.00	72.00	---	587.00	**754.00**	507.10	**646.00**
35" W, 43" H	Inst	Ea	CA	3.33	655.00	89.80	---	744.80	**955.00**	643.90	**819.00**
Weathervanes for above, aluminum											
18" H, black finish	Inst	Ea	---	---	56.90	---	---	56.90	**56.90**	51.20	**51.20**
24" H, black finish	Inst	Ea	---	---	72.60	---	---	72.60	**72.60**	65.40	**65.40**
36" H, black and gold finish	Inst	Ea	---	---	113.00	---	---	113.00	**113.00**	101.00	**101.00**

Demolition

Average Manhours per Unit is based on what the designated crew can do in one day. In this section, the crew might be a laborer, a carpenter, a floor layer, etc. Who does the wrecking depends on the quantity of wrecking to be done.

A contractor might use laborers exclusively on a large volume job, but use a carpenter or a carpenter and a laborer on a small volume job.

The choice of tools and equipment greatly affects the quantity of work that a worker can accomplish in a day. A person can remove more brick with a compressor and pneumatic tool than with a sledgehammer or pry bar. In this section, the phrase "by hand" includes the use of hand tools, i.e., sledgehammers, wrecking bars, claw hammers, etc. When the Average Manhours per Unit is based on the use of equipment (not hand tools), a description of the equipment is provided.

The Average Manhours per Unit is not based on the use of "heavy" equipment such as bulldozers, cranes with wrecking balls, etc. The Average Manhours per Unit does include the labor involved in hauling wrecked material or debris to a dumpster located at the site. Average rental costs for dumpsters are: $350.00 for 40 CY (20' x 8' x 8' H); $275.00 for 30 CY (20' x 8' x 6' H). Rental period includes delivery and pickup when full.

Description	Oper	Unit	Crew Size	Man-Hours Per Unit	Avg Mat'l Unit Cost	Avg Labor Unit Cost	Avg Equip Unit Cost	Avg Total Unit Cost	Avg Price Incl O&P	Large Volume Avg Total Unit Cost	Large Volume Avg Price Incl O&P
Demolition											
Wreck and remove to dumpster											
Concrete											
Footings, with air tools; reinforced											
8" T x 12" W (.67 CF/LF)	Demo	LF	AB	.205	---	4.84	1.19	6.03	**8.45**	3.91	**5.48**
8" T x 16" W (.89 CF/LF)	Demo	LF	AB	.246	---	5.81	1.43	7.24	**10.10**	4.71	**6.59**
8" T x 20" W (1.11 CF/LF)	Demo	LF	AB	.274	---	6.47	1.58	8.05	**11.30**	5.23	**7.33**
12" T x 12" W (1.00 CF/LF)	Demo	LF	AB	.352	---	8.31	2.04	10.35	**14.50**	6.72	**9.43**
12" T x 16" W (1.33 CF/LF)	Demo	LF	AB	.410	---	9.68	2.38	12.06	**16.90**	7.84	**11.00**
12" T x 20" W (1.67 CF/LF)	Demo	LF	AB	.448	---	10.60	2.59	13.19	**18.50**	8.55	**12.00**
12" T x 24" W (2.00 CF/LF)	Demo	LF	AB	.492	---	11.60	2.85	14.45	**20.30**	9.40	**13.20**
Foundations and retaining walls, with air tools, per LF wall											
With reinforcing											
4'-0" H											
8" T (2.67 CF/LF)	Demo	SF	AB	.410	---	9.68	2.38	12.06	**16.90**	7.84	**11.00**
12" T (4.00 CF/LF)	Demo	SF	AB	.547	---	12.90	3.17	16.07	**22.50**	10.46	**14.70**
8'-0" H											
8" T (5.33 CF/LF)	Demo	SF	AB	.448	---	10.60	2.59	13.19	**18.50**	8.55	**12.00**
12" T (8.00 CF/LF)	Demo	SF	AB	.615	---	14.50	3.56	18.06	**25.30**	11.76	**16.50**
12'-0" H											
8" T (8.00 CF/LF)	Demo	SF	AB	.492	---	11.60	2.85	14.45	**20.30**	9.40	**13.20**
12" T (12.00 CF/LF)	Demo	SF	AB	.703	---	16.60	4.07	20.67	**29.00**	13.45	**18.80**
Without reinforcing											
4'-0" H											
8" T (2.67 CF/LF)	Demo	SF	AB	.308	---	7.27	1.78	9.05	**12.70**	5.88	**8.24**
12" T (4.00 CF/LF)	Demo	SF	AB	.410	---	9.68	2.38	12.06	**16.90**	7.84	**11.00**
8'-0" H											
8" T (5.33 CF/LF)	Demo	SF	AB	.352	---	8.31	2.04	10.35	**14.50**	6.72	**9.43**
12" T (8.00 CF/LF)	Demo	SF	AB	.492	---	11.60	2.85	14.45	**20.30**	9.40	**13.20**
12'-0" H											
8" T (8.00 CF/LF)	Demo	SF	AB	.410	---	9.68	2.38	12.06	**16.90**	7.84	**11.00**
12" T (12.00 CF/LF)	Demo	SF	AB	.615	---	14.50	3.56	18.06	**25.30**	11.76	**16.50**
Slabs with air tools											
With reinforcing											
4" T	Demo	SF	AB	.058	---	1.37	.34	1.71	**2.39**	1.12	**1.57**
5" T	Demo	SF	AB	.067	---	1.58	.39	1.97	**2.76**	1.26	**1.77**
6" T	Demo	SF	AB	.070	---	1.65	.41	2.06	**2.89**	1.35	**1.89**
Without reinforcing											
4" T	Demo	SF	AB	.049	---	1.16	.29	1.45	**2.02**	.95	**1.32**
5" T	Demo	SF	AB	.058	---	1.37	.34	1.71	**2.39**	1.12	**1.57**
6" T	Demo	SF	AB	.067	---	1.58	.39	1.97	**2.76**	1.26	**1.77**
Masonry											
Brick											
Chimneys											
4" T wall	Demo	VLF	LB	.985	---	21.60	---	21.60	**32.60**	14.10	**21.20**
8" T wall	Demo	VLF	LB	2.46	---	54.00	---	54.00	**81.50**	35.10	**53.00**

Description	Oper	Unit	Crew Size	Man-Hours Per Unit	Avg Mat'l Unit Cost	Avg Labor Unit Cost	Avg Equip Unit Cost	Avg Total Unit Cost	Avg Price Incl O&P	Large Volume Avg Total Unit Cost	Large Volume Avg Price Incl O&P
Columns, 12" x 12" o.d.	Demo	VLF	LB	.492	---	10.80	---	10.80	**16.30**	7.02	**10.60**
Veneer, 4" T, with air tools	Demo	SF	AB	.091	---	2.15	.53	2.68	**3.75**	1.73	**2.43**
Walls, with air tools											
8" T wall	Demo	SF	AB	.182	---	4.30	1.06	5.36	**7.50**	3.50	**4.90**
12" T wall	Demo	SF	AB	.259	---	6.11	1.50	7.61	**10.70**	4.94	**6.93**
Concrete block; lightweight (haydite), standard or heavyweight											
Foundations and retaining walls; no excavation included											
Without reinforcing or with only lateral reinforcing											
With air tools											
8" W x 8" H x 16" L	Demo	SF	AB	.088	---	2.08	.51	2.59	**3.63**	1.68	**2.35**
12" W x 8" H x 16" L	Demo	SF	AB	.103	---	2.43	.59	3.02	**4.24**	1.97	**2.76**
Without air tools											
8" W x 8" H x 16" L	Demo	SF	LB	.109	---	2.39	---	2.39	**3.61**	1.56	**2.35**
12" W x 8" H x 16" L	Demo	SF	LB	.130	---	2.85	---	2.85	**4.31**	1.84	**2.78**
With vertical reinforcing in every other core (2 cores per block) with cores filled											
With air tools											
8" W x 8" H x 16" L	Demo	SF	AB	.145	---	3.42	.84	4.26	**5.97**	2.77	**3.88**
12" W x 8" H x 16" L	Demo	SF	AB	.170	---	4.01	.98	4.99	**7.00**	3.24	**4.53**
Exterior walls (above grade) and partitions, no shoring included											
Without reinforcing or with only lateral reinforcing											
With air tools											
8" W x 8" H x 16" L	Demo	SF	AB	.070	---	1.65	.41	2.06	**2.89**	1.35	**1.89**
12" W x 8" H x 16" L	Demo	SF	AB	.082	---	1.94	.48	2.42	**3.38**	1.56	**2.19**
Without air tools											
8" W x 8" H x 16" L	Demo	SF	LB	.088	---	1.93	---	1.93	**2.92**	1.25	**1.89**
12" W x 8" H x 16" L	Demo	SF	LB	.103	---	2.26	---	2.26	**3.41**	1.47	**2.22**
Fences											
Without reinforcing or with only lateral reinforcing											
With air tools											
6" W x 4" H x 16" L	Demo	SF	AB	.063	---	1.49	.37	1.86	**2.60**	1.21	**1.69**
6" W x 6" H x 16" L	Demo	SF	AB	.067	---	1.58	.39	1.97	**2.76**	1.26	**1.77**
8" W x 8" H x 16" L	Demo	SF	AB	.070	---	1.65	.41	2.06	**2.89**	1.35	**1.89**
12" W x 8" H x 16" L	Demo	SF	AB	.082	---	1.94	.48	2.42	**3.38**	1.56	**2.19**
Without air tools											
6" W x 4" H x 16" L	Demo	SF	LB	.079	---	1.73	---	1.73	**2.62**	1.14	**1.72**
6" W x 6" H x 16" L	Demo	SF	LB	.083	---	1.82	---	1.82	**2.75**	1.19	**1.79**
8" W x 8" H x 16" L	Demo	SF	LB	.088	---	1.93	---	1.93	**2.92**	1.25	**1.89**
12" W x 8" H x 16" L	Demo	SF	LB	.103	---	2.26	---	2.26	**3.41**	1.47	**2.22**
Quarry tile, 6" or 9" squares											
Floors											
Conventional mortar set	Demo	SF	LB	.055	---	1.21	---	1.21	**1.82**	.79	**1.19**
Dry-set mortar	Demo	SF	LB	.048	---	1.05	---	1.05	**1.59**	.68	**1.03**

Description	Oper	Unit	Crew Size	Man-Hours Per Unit	Avg Mat'l Unit Cost	Avg Labor Unit Cost	Avg Equip Unit Cost	Avg Total Unit Cost	Avg Price Incl O&P	Large Volume Avg Total Unit Cost	Large Volume Avg Price Incl O&P
Rough carpentry (framing)											
Dimension lumber											
Beams, set on steel columns											
Built-up from 2" lumber											
4" T x 10" W - 10' L (2 pcs)	Demo	LF	LB	.029	---	.64	---	.64	**.96**	.42	**.63**
4" T x 12" W - 12' L (2 pcs)	Demo	LF	LB	.024	---	.53	---	.53	**.80**	.35	**.53**
6" T x 10" W - 10' L (3 pcs)	Demo	LF	LB	.029	---	.64	---	.64	**.96**	.42	**.63**
6" T x 12" W - 12' L (3 pcs)	Demo	LF	LB	.024	---	.53	---	.53	**.80**	.35	**.53**
Single member (solid lumber)											
3" T x 12" W - 12' L	Demo	LF	LB	.024	---	.53	---	.53	**.80**	.35	**.53**
4" T x 12" W - 12' L	Demo	LF	LB	.024	---	.53	---	.53	**.80**	.35	**.53**
Bracing, diagonal, notched-in, studs oc											
1" x 6" - 10'	Demo	LF	LB	.041	---	.90	---	.90	**1.36**	.57	**.86**
Bridging, "X" type, 1" x 3", (8", 10", 12" T) 16" oc	Demo	LF	LB	.070	---	1.54	---	1.54	**2.32**	1.01	**1.52**
Columns or posts											
4" x 4" -8' L	Demo	LF	LB	.032	---	.70	---	.70	**1.06**	.46	**.70**
6" x 6" -8' L	Demo	LF	LB	.032	---	.70	---	.70	**1.06**	.46	**.70**
6" x 8" -8' L	Demo	LF	LB	.038	---	.83	---	.83	**1.26**	.55	**.83**
8" x 8" -8' L	Demo	LF	LB	.040	---	.88	---	.88	**1.33**	.57	**.86**
Fascia											
1" x 4" - 12" L	Demo	LF	LB	.019	---	.42	---	.42	**.63**	.26	**.40**
Firestops or stiffeners											
2" x 4" - 16"	Demo	LF	LB	.052	---	1.14	---	1.14	**1.72**	.75	**1.13**
2" x 6" - 16"	Demo	LF	LB	.052	---	1.14	---	1.14	**1.72**	.75	**1.13**
Furring strips, 1" x 4" - 8' L											
Walls; strips 12" oc											
Studs 16" oc	Demo	SF	LB	.020	---	.44	---	.44	**.66**	.29	**.43**
Studs 24" oc	Demo	SF	LB	.018	---	.40	---	.40	**.60**	.26	**.40**
Masonry (concrete blocks)	Demo	SF	LB	.022	---	.48	---	.48	**.73**	.33	**.50**
Concrete	Demo	SF	LB	.038	---	.83	---	.83	**1.26**	.55	**.83**
Ceiling; joists 16" oc											
Strips 12" oc	Demo	SF	LB	.029	---	.64	---	.64	**.96**	.42	**.63**
Strips 16" oc	Demo	SF	LB	.023	---	.50	---	.50	**.76**	.33	**.50**
Headers or lintels, over openings											
Built-up or single member											
4" T x 6" W - 4' L	Demo	LF	LB	.044	---	.97	---	.97	**1.46**	.64	**.96**
4" T x 8" W - 8' L	Demo	LF	LB	.034	---	.75	---	.75	**1.13**	.48	**.73**
4" T x 10" W - 10' L	Demo	LF	LB	.029	---	.64	---	.64	**.96**	.42	**.63**
4" T x 12" W - 12' L	Demo	LF	LB	.024	---	.53	---	.53	**.80**	.35	**.53**
4" T x 14" W - 14' L	Demo	LF	LB	.024	---	.53	---	.53	**.80**	.35	**.53**
Joists											
Ceiling											
2" x 4" - 8' L	Demo	LF	LB	.020	---	.44	---	.44	**.66**	.29	**.43**
2" x 4" - 10' L	Demo	LF	LB	.018	---	.40	---	.40	**.60**	.24	**.36**

Description	Oper	Unit	Crew Size	Man-Hours Per Unit	Avg Mat'l Unit Cost	Avg Labor Unit Cost	Avg Equip Unit Cost	Avg Total Unit Cost	Avg Price Incl O&P	Large Volume Avg Total Unit Cost	Large Volume Avg Price Incl O&P
2" x 8" - 12' L	Demo	LF	LB	.017	---	.37	---	.37	**.56**	.24	**.36**
2" x 10" - 14' L	Demo	LF	LB	.016	---	.35	---	.35	**.53**	.22	**.33**
2" x 12" - 16' L	Demo	LF	LB	.016	---	.35	---	.35	**.53**	.22	**.33**
Floor; seated on sill plate											
2" x 8" - 12' L	Demo	LF	LB	.015	---	.33	---	.33	**.50**	.22	**.33**
2" x 10" - 14' L	Demo	LF	LB	.014	---	.31	---	.31	**.46**	.20	**.30**
2" x 12" - 16' L	Demo	LF	LB	.014	---	.31	---	.31	**.46**	.20	**.30**
Ledgers											
Nailed, 2" x 6" - 12' L	Demo	LF	LB	.035	---	.77	---	.77	**1.16**	.50	**.76**
Bolted, 3" x 8" - 12' L	Demo	LF	LB	.025	---	.55	---	.55	**.83**	.35	**.53**
Plates; 2" x 4" or 2" x 6"											
Double top nailed	Demo	LF	LB	.030	---	.66	---	.66	**.99**	.44	**.66**
Sill, nailed	Demo	LF	LB	.018	---	.40	---	.40	**.60**	.26	**.40**
Sill or bottom, bolted	Demo	LF	LB	.036	---	.79	---	.79	**1.19**	.50	**.76**
Rafters											
Common											
2" x 4" - 14' L	Demo	LF	LB	.014	---	.31	---	.31	**.46**	.20	**.30**
2" x 6" - 14' L	Demo	LF	LB	.015	---	.33	---	.33	**.50**	.22	**.33**
2" x 8" - 14' L	Demo	LF	LB	.017	---	.37	---	.37	**.56**	.24	**.36**
2" x 10" - 14' L	Demo	LF	LB	.021	---	.46	---	.46	**.70**	.29	**.43**
Hip and/or valley											
2" x 4" - 16' L	Demo	LF	LB	.012	---	.26	---	.26	**.40**	.18	**.27**
2" x 6" - 16' L	Demo	LF	LB	.013	---	.29	---	.29	**.43**	.18	**.27**
2" x 8" - 16' L	Demo	LF	LB	.015	---	.33	---	.33	**.50**	.22	**.33**
2" x 10" - 16' L	Demo	LF	LB	.018	---	.40	---	.40	**.60**	.26	**.40**
Jack											
2" x 4" - 6' L	Demo	LF	LB	.031	---	.68	---	.68	**1.03**	.44	**.66**
2" x 6" - 6' L	Demo	LF	LB	.033	---	.72	---	.72	**1.09**	.48	**.73**
2" x 8" - 6' L	Demo	LF	LB	.038	---	.83	---	.83	**1.26**	.53	**.80**
2" x 10" - 6' L	Demo	LF	LB	.046	---	1.01	---	1.01	**1.52**	.66	**.99**
Roof decking, solid T&G											
2" x 6" - 12' L	Demo	LF	LB	.038	---	.83	---	.83	**1.26**	.55	**.83**
2" x 8" - 12' L	Demo	LF	LB	.027	---	.59	---	.59	**.89**	.40	**.60**
Studs											
2" x 4" - 8' L	Demo	LF	LB	.018	---	.40	---	.40	**.60**	.26	**.40**
2" x 6" - 8' L	Demo	LF	LB	.018	---	.40	---	.40	**.60**	.26	**.40**
Stud partitions, studs 16" oc with bottom plate and double top plates and firestops; per LF partition											
2" x 4" - 8' L	Demo	LF	LB	.246	---	5.40	---	5.40	**8.15**	3.51	**5.30**
2" x 6" - 8' L	Demo	LF	LB	.246	---	5.40	---	5.40	**8.15**	3.51	**5.30**
Boards											
Sheathing, regular or diagonal											
1" x 8"											
Roof	Demo	SF	LB	.018	---	.40	---	.40	**.60**	.26	**.40**
Sidewall	Demo	SF	LB	.015	---	.33	---	.33	**.50**	.22	**.33**
Subflooring, regular or diagonal											
1" x 8" - 16' L	Demo	SF	LB	.016	---	.35	---	.35	**.53**	.22	**.33**
1" x 10" - 16' L	Demo	SF	LB	.013	---	.29	---	.29	**.43**	.18	**.27**

Description	Oper	Unit	Crew Size	Man-Hours Per Unit	Avg Mat'l Unit Cost	Avg Labor Unit Cost	Avg Equip Unit Cost	Avg Total Unit Cost	Avg Price Incl O&P	Large Volume Avg Total Unit Cost	Large Volume Avg Price Incl O&P
Plywood											
Sheathing											
Roof											
1/2" T, CDX	Demo	SF	LB	.012	---	.26	---	.26	**.40**	.18	**.27**
5/8" T, CDX	Demo	SF	LB	.013	---	.29	---	.29	**.43**	.18	**.27**
Wall											
3/8" or 1/2" T, CDX	Demo	SF	LB	.010	---	.22	---	.22	**.33**	.13	**.20**
5/8" T, CDX	Demo	SF	LB	.010	---	.22	---	.22	**.33**	.15	**.23**
Subflooring											
5/8", 3/4" T, CDX	Demo	SF	LB	.011	---	.24	---	.24	**.36**	.15	**.23**
1-1/8" T, 2-4-1, T&G long edges	Demo	SF	LB	.015	---	.33	---	.33	**.50**	.22	**.33**
Trusses, "W" pattern with gin pole, 24' to 30' spans											
3 - in - 12 slope	Demo	SF	LB	.947	---	20.80	---	20.80	**31.40**	13.50	**20.40**
5 - in - 12 slope	Demo	SF	LB	.947	---	20.80	---	20.80	**31.40**	13.50	**20.40**
Finish carpentry											
Bath accessories, screw on type	Demo	Ea	LB	.197	---	4.32	---	4.32	**6.53**	2.81	**4.24**
Cabinets											
Kitchen, to 3' x 4', wood; base, wall, or peninsula	Demo	Ea	LB	.985	---	21.60	---	21.60	**32.60**	14.10	**21.20**
Medicine, metal	Demo	Ea	LB	.821	---	18.00	---	18.00	**27.20**	11.70	**17.70**
Vanity, cabinet and sink top											
Disconnect plumbing and remove to dumpster	Demo	Ea	LB	1.54	---	33.80	---	33.80	**51.00**	22.00	**33.10**
Remove old unit, replace with new unit, reconnect plumbing	Demo	Ea	SB	3.52	---	92.30	---	92.30	**137.00**	60.00	**89.40**
Hardwood flooring (over wood subfloors)											
Block, set in mastic	Demo	SF	LB	.021	---	.46	---	.46	**.70**	.29	**.43**
Strip, nailed	Demo	SF	LB	.027	---	.59	---	.59	**.89**	.40	**.60**
Marlite panels, 4' x 8', adhesive set	Demo	SF	LB	.029	---	.64	---	.64	**.96**	.42	**.63**
Molding and trim											
At base (floor)	Demo	LF	LB	.023	---	.50	---	.50	**.76**	.33	**.50**
At ceiling	Demo	LF	LB	.021	---	.46	---	.46	**.70**	.29	**.43**
On walls or cabinets	Demo	LF	LB	.015	---	.33	---	.33	**.50**	.22	**.33**
Paneling											
Plywood, prefinished	Demo	SF	LB	.013	---	.29	---	.29	**.43**	.20	**.30**
Wood	Demo	SF	LB	.015	---	.33	---	.33	**.50**	.22	**.33**
Weather protection											
Insulation											
Batt/roll, with wall or ceiling finish already removed											
Joists, 16" or 24" oc	Demo	SF	LB	.008	---	.18	---	.18	**.27**	.11	**.17**
Rafters, 16" or 24" oc	Demo	SF	LB	.010	---	.22	---	.22	**.33**	.13	**.20**
Studs, 16" or 24" oc	Demo	SF	LB	.007	---	.15	---	.15	**.23**	.11	**.17**

Description	Oper	Unit	Crew Size	Man-Hours Per Unit	Avg Mat'l Unit Cost	Avg Labor Unit Cost	Avg Equip Unit Cost	Avg Total Unit Cost	Avg Price Incl O&P	Large Volume Avg Total Unit Cost	Large Volume Avg Price Incl O&P
Loose, with ceiling finish already removed											
Joists, 16" or 24" oc											
4" T	Demo	SF	LB	.006	---	.13	---	.13	**.20**	.09	**.13**
6" T	Demo	SF	LB	.011	---	.24	---	.24	**.36**	.15	**.23**
Rigid											
Roofs											
1/2" T	Demo	SQ	LB	1.45	---	31.80	---	31.80	**48.10**	20.70	**31.20**
1" T	Demo	SQ	LB	1.64	---	36.00	---	36.00	**54.40**	23.50	**35.50**
Walls, 1/2" T	Demo	SF	LB	.012	---	.26	---	.26	**.40**	.15	**.23**
Sheet metal											
Gutter and downspouts											
Aluminum	Demo	LF	LB	.029	---	.64	---	.64	**.96**	.42	**.63**
Galvanized	Demo	LF	LB	.038	---	.83	---	.83	**1.26**	.55	**.83**
Roofing and siding											
Aluminum											
Roofing, nailed to wood											
Corrugated (2-1/2"), 26" W with 3-3/4" side lap and 6" end lap	Demo	SQ	LB	2.46	---	54.00	---	54.00	**81.50**	35.10	**53.00**
Siding, nailed to wood											
Clapboard (i.e., lap drop)											
8" exposure	Demo	SF	LB	.024	---	.53	---	.53	**.80**	.35	**.53**
10" exposure	Demo	SF	LB	.019	---	.42	---	.42	**.63**	.29	**.43**
Corrugated (2-1/2"), 26" W with 2-1/2" side lap and 4" end lap	Demo	SF	LB	.021	---	.46	---	.46	**.70**	.29	**.43**
Panels, 4' x 8'	Demo	SF	LB	.010	---	.22	---	.22	**.33**	.15	**.23**
Shingle, 24" L with 12" exposure	Demo	SF	LB	.017	---	.37	---	.37	**.56**	.24	**.36**
Asphalt shingle roofing											
240 lb./SQ., strip, 3 tab 5" exposure	Demo	SQ	LB	1.54	---	33.80	---	33.80	**51.00**	22.00	**33.10**
Built-up/hot roofing (to wood deck)											
3 ply											
With gravel	Demo	SQ	LB	2.24	---	49.20	---	49.20	**74.20**	31.80	**48.10**
Without gravel	Demo	SQ	LB	1.76	---	38.60	---	38.60	**58.30**	25.00	**37.80**
5 ply											
With gravel	Demo	SQ	LB	2.46	---	54.00	---	54.00	**81.50**	35.10	**53.00**
Without gravel	Demo	SQ	LB	1.89	---	41.50	---	41.50	**62.60**	27.00	**40.80**
Clay tile roofing											
2 piece interlocking	Demo	SQ	LB	2.46	---	54.00	---	54.00	**81.50**	35.10	**53.00**
1 piece	Demo	SQ	LB	2.24	---	49.20	---	49.20	**74.20**	31.80	**48.10**
Hardboard siding											
Lap, 1/2" T x 12" W x 16' L, with 11" exposure	Demo	SF	LB	.018	---	.40	---	.40	**.60**	.26	**.40**
Panels, 7/16" T x 4' W x 8' H	Demo	SF	LB	.010	---	.22	---	.22	**.33**	.13	**.20**

Description	Oper	Unit	Crew Size	Man-Hours Per Unit	Avg Mat'l Unit Cost	Avg Labor Unit Cost	Avg Equip Unit Cost	Avg Total Unit Cost	Avg Price Incl O&P	Avg Total Unit Cost	Avg Price Incl O&P
			Costs Based On Small Volume							Large Volume	
Mineral surfaced roll roofing											
Single coverage 90 lb/SQ roll with 6" end lap and 2" headlap	Demo	SQ	LB	.769	---	16.90	---	16.90	**25.50**	11.00	**16.60**
Double coverage selvage roll, with 6" end lap and 17" exposure	Demo	SQ	LB	1.12	---	24.60	---	24.60	**37.10**	16.00	**24.10**
Wood											
Roofing											
Shakes											
24" L with 10" exposure											
1/2" to 3/4" T	Demo	SQ	LB	.985	---	21.60	---	21.60	**32.60**	14.10	**21.20**
3/4" to 5/4" T	Demo	SQ	LB	1.07	---	23.50	---	23.50	**35.50**	15.30	**23.10**
Shingles											
16" L with 5" exposure	Demo	SQ	LB	2.05	---	45.00	---	45.00	**67.90**	29.20	**44.10**
18" L with 5-1/2" exposure	Demo	SQ	LB	1.89	---	41.50	---	41.50	**62.60**	27.00	**40.80**
24" L with 7-1/2" exposure	Demo	SQ	LB	1.37	---	30.10	---	30.10	**45.40**	19.50	**29.50**
Siding											
Bevel											
1/2" x 8" with 6-3/4" exposure	Demo	SF	LB	.027	---	.59	---	.59	**.89**	.40	**.60**
5/8" x 10" with 8-3/4" exposure	Demo	SF	LB	.021	---	.46	---	.46	**.70**	.31	**.46**
3/4" x 12" with 10-3/4" exposure	Demo	SF	LB	.017	---	.37	---	.37	**.56**	.24	**.36**
Drop (horizontal), 1/4" T&G											
1" x 8" with 7" exposure	Demo	SF	LB	.026	---	.57	---	.57	**.86**	.37	**.56**
1" x 10" with 9" exposure	Demo	SF	LB	.020	---	.44	---	.44	**.66**	.29	**.43**
Board (1" x 12") and batten (1" x 2") @ 12" oc											
Horizontal	Demo	SF	LB	.019	---	.42	---	.42	**.63**	.29	**.43**
Vertical											
Standard	Demo	SF	LB	.024	---	.53	---	.53	**.80**	.35	**.53**
Reverse	Demo	SF	LB	.023	---	.50	---	.50	**.76**	.33	**.50**
Board on board (1" x 12" with 1-1/2" overlap), vertical	Demo	SF	LB	.026	---	.57	---	.57	**.86**	.37	**.56**
Plywood (1/2" T) with battens (1" x 2")											
16" oc battens	Demo	SF	LB	.011	---	.24	---	.24	**.36**	.15	**.23**
24" oc battens	Demo	SF	LB	.010	---	.22	---	.22	**.33**	.15	**.23**
Shakes											
24" L with 11-1/2" exposure											
1/2" to 3/4" T	Demo	SF	LB	.015	---	.33	---	.33	**.50**	.22	**.33**
3/4" to 5/4" T	Demo	SF	LB	.017	---	.37	---	.37	**.56**	.24	**.36**
Shingles											
16" L with 7-1/2" exposure	Demo	SF	LB	.025	---	.55	---	.55	**.83**	.35	**.53**
18" L with 8-1/2" exposure	Demo	SF	LB	.022	---	.48	---	.48	**.73**	.31	**.46**
24" L with 11-1/2" exposure	Demo	SF	LB	.016	---	.35	---	.35	**.53**	.22	**.33**
Doors, windows and glazing											
Doors with related trim and frame											
Closet, with track											
Folding, 4 doors	Demo	Set	LB	2.05	---	45.00	---	45.00	**67.90**	29.20	**44.10**
Sliding, 2 or 3 doors	Demo	Set	LB	2.05	---	45.00	---	45.00	**67.90**	29.20	**44.10**

Description	Oper	Unit	Crew Size	Man-Hours Per Unit	Avg Mat'l Unit Cost	Avg Labor Unit Cost	Avg Equip Unit Cost	Avg Total Unit Cost	Avg Price Incl O&P	Large Volume Avg Total Unit Cost	Large Volume Avg Price Incl O&P
Entry, 3' x 7'	Demo	Ea	LB	1.76	---	38.60	---	38.60	**58.30**	25.00	**37.80**
Fire, 3' x 7'	Demo	Ea	LB	1.76	---	38.60	---	38.60	**58.30**	25.00	**37.80**
Garage											
Wood, aluminum, or hardboard											
Single	Demo	Ea	LB	3.08	---	67.60	---	67.60	**102.00**	43.90	**66.30**
Double	Demo	Ea	LB	4.10	---	90.00	---	90.00	**136.00**	58.60	**88.50**
Steel											
Single	Demo	Ea	LB	3.52	---	77.30	---	77.30	**117.00**	50.30	**75.90**
Double	Demo	Ea	LB	4.92	---	108.00	---	108.00	**163.00**	70.20	**106.00**
Glass sliding, with track											
2 lites wide	Demo	Set	LB	3.08	---	67.60	---	67.60	**102.00**	43.90	**66.30**
3 lites wide	Demo	Set	LB	4.10	---	90.00	---	90.00	**136.00**	58.60	**88.50**
4 lites wide	Demo	Set	LB	6.15	---	135.00	---	135.00	**204.00**	87.80	**133.00**
Interior, 3' x 7'	Demo	Ea	LB	1.54	---	33.80	---	33.80	**51.00**	22.00	**33.10**
Screen, 3' x 7'	Demo	Ea	LB	1.23	---	27.00	---	27.00	**40.80**	17.60	**26.50**
Storm combination, 3' x 7'	Demo	Ea	LB	1.23	---	27.00	---	27.00	**40.80**	17.60	**26.50**
Windows, with related trim and frame											
To 12 SF											
Aluminum	Demo	Ea	LB	1.17	---	25.70	---	25.70	**38.80**	16.70	**25.30**
Wood	Demo	Ea	LB	1.54	---	33.80	---	33.80	**51.00**	22.00	**33.10**
13 SF to 50 SF											
Aluminum	Demo	Ea	LB	1.89	---	41.50	---	41.50	**62.60**	27.00	**40.80**
Wood	Demo	Ea	LB	2.46	---	54.00	---	54.00	**81.50**	35.10	**53.00**
Glazing, clean sash and remove old putty or rubber											
3/32" T float, putty or rubber											
8" x 12" (0.667 SF)	Demo	SF	GA	.492	---	12.40	---	12.40	**18.50**	8.04	**12.10**
12" x 16" (1.333 SF)	Demo	SF	GA	.274	---	6.88	---	6.88	**10.30**	4.47	**6.71**
14" x 20" (1.944 SF)	Demo	SF	GA	.224	---	5.62	---	5.62	**8.44**	3.64	**5.46**
16" x 24" (2.667 SF)	Demo	SF	GA	.176	---	4.42	---	4.42	**6.63**	2.86	**4.29**
24" x 26" (4.333 SF)	Demo	SF	GA	.130	---	3.26	---	3.26	**4.90**	2.11	**3.16**
36" x 24" (6.000 SF)	Demo	SF	GA	.098	---	2.46	---	2.46	**3.69**	1.61	**2.41**
1/8" T float, putty, steel sash											
12" x 16" (1.333 SF)	Demo	SF	GA	.274	---	6.88	---	6.88	**10.30**	4.47	**6.71**
16" x 20" (2.222 SF)	Demo	SF	GA	.189	---	4.75	---	4.75	**7.12**	3.09	**4.63**
16" x 24" (2.667 SF)	Demo	SF	GA	.176	---	4.42	---	4.42	**6.63**	2.86	**4.29**
24" x 26" (4.333 SF)	Demo	SF	GA	.130	---	3.26	---	3.26	**4.90**	2.11	**3.16**
28" x 32" (6.222 SF)	Demo	SF	GA	.095	---	2.39	---	2.39	**3.58**	1.56	**2.34**
36" x 36" (9.000 SF)	Demo	SF	GA	.077	---	1.93	---	1.93	**2.90**	1.26	**1.88**
36" x 48" (12.000 SF)	Demo	SF	GA	.065	---	1.63	---	1.63	**2.45**	1.05	**1.58**
1/4" T float											
Wood sash with putty											
72" x 48" (24.0 SF)	Demo	SF	GA	.067	---	1.68	---	1.68	**2.52**	1.08	**1.62**
Aluminum sash with aluminum channel and rigid neoprene rubber											
48" x 96" (32.0 SF)	Demo	SF	GA	.070	---	1.76	---	1.76	**2.64**	1.16	**1.73**
96" x 96" (64.0 SF)	Demo	SF	GA	.068	---	1.71	---	1.71	**2.56**	1.10	**1.66**
1" T insulating glass; with 2 pieces 1/4" float and 1/2" air space											
To 6.0 SF	Demo	SF	GA	.246	---	6.18	---	6.18	**9.27**	4.02	**6.03**
6.1 to 12.0 SF	Demo	SF	GA	.112	---	2.81	---	2.81	**4.22**	1.83	**2.75**
12.1 to 18.0 SF	Demo	SF	GA	.085	---	2.13	---	2.13	**3.20**	1.38	**2.07**
18.1 to 24.0 SF	Demo	SF	GA	.082	---	2.06	---	2.06	**3.09**	1.33	**2.00**

Description	Oper	Unit	Crew Size	Man-Hours Per Unit	Avg Mat'l Unit Cost	Avg Labor Unit Cost	Avg Equip Unit Cost	Avg Total Unit Cost	Avg Price Incl O&P	Avg Total Unit Cost	Avg Price Incl O&P
			Costs Based On Small Volume							Large Volume	
Aluminum sliding door glass with aluminum channel and rigid neoprene rubber											
34" x 76" (17.944 SF)											
5/8" T insulating glass with 2 pieces 5/32" T tempered with 1-1/4" air space	Demo	SF	GA	.072	---	1.81	---	1.81	**2.71**	1.18	**1.77**
5/32" T tempered	Demo	SF	GA	.063	---	1.58	---	1.58	**2.37**	1.03	**1.54**
46" x 76" (24.278 SF)											
5/8" T insulating glass with 2 pieces 5/32" T tempered with 1-1/4" air space	Demo	SF	GA	.057	---	1.43	---	1.43	**2.15**	.93	**1.39**
5/32" T tempered	Demo	SF	GA	.050	---	1.26	---	1.26	**1.88**	.83	**1.24**
Finishes											
Exterior and interior, with hand tools; no insulation removal included											
Plaster and stucco; remove to studs or sheathing											
Lath (wood or metal) and plaster, walls and ceiling											
2 coats	Demo	SY	LB	.189	---	4.15	---	4.15	**6.26**	2.70	**4.08**
3 coats	Demo	SY	LB	.205	---	4.50	---	4.50	**6.79**	2.92	**4.41**
Stucco and metal netting											
2 coats	Demo	SY	LB	.259	---	5.69	---	5.69	**8.58**	3.69	**5.57**
3 coats	Demo	SY	LB	.308	---	6.76	---	6.76	**10.20**	4.39	**6.63**
Wallboard, gypsum (drywall), walls and ceilings	Demo	SF	LB	.016	---	.35	---	.35	**.53**	.22	**.33**
Ceramic, metal, plastic tile											
Floors, 1" x 1"											
Adhesive or dry-set base	Demo	SF	LB	.045	---	.99	---	.99	**1.49**	.64	**.96**
Conventional mortar base	Demo	SF	LB	.052	---	1.14	---	1.14	**1.72**	.75	**1.13**
Walls, 1" x 1" or 4-1/4" x 4-1/4"											
Adhesive or dry-set base	Demo	SF	LB	.051	---	1.12	---	1.12	**1.69**	.72	**1.09**
Conventional mortar base	Demo	SF	LB	.062	---	1.36	---	1.36	**2.05**	.88	**1.33**
Acoustical or insulating ceiling tile											
Adhesive set, tile only	Demo	SF	LB	.019	---	.42	---	.42	**.63**	.26	**.40**
Stapled, tile only	Demo	SF	LB	.021	---	.46	---	.46	**.70**	.31	**.46**
Stapled, tile and furring strips	Demo	SF	LB	.016	---	.35	---	.35	**.53**	.22	**.33**
Suspended ceiling system; panels and grid system	Demo	SF	LB	.014	---	.31	---	.31	**.46**	.20	**.30**
Resilient flooring, adhesive set											
Sheet products	Demo	SY	LB	.154	---	3.38	---	3.38	**5.10**	2.20	**3.31**
Tile products	Demo	SF	LB	.016	---	.35	---	.35	**.53**	.24	**.36**
Wallpaper											
Average output is expressed in rolls (36.0 SF/single roll)											
Single layer of paper from plaster with steaming equipment	Demo	Roll	1L	.615	---	13.50	---	13.50	**20.40**	8.78	**13.30**
Several layers of paper from plaster with steaming equipment	Demo	Roll	1L	1.03	---	22.60	---	22.60	**34.10**	14.60	**22.10**

Description	Oper	Unit	Crew Size	Man-Hours Per Unit	Avg Mat'l Unit Cost	Avg Labor Unit Cost	Avg Equip Unit Cost	Avg Total Unit Cost	Avg Price Incl O&P	Large Volume Avg Total Unit Cost	Large Volume Avg Price Incl O&P
Vinyls (with non-woven, woven, or synthetic fiber backings) from plaster with steaming equipment											
	Demo	Roll	1L	.410	---	9.00	---	9.00	**13.60**	5.86	**8.85**
Single layer of paper from drywall with steaming equipment											
	Demo	Roll	1L	.615	---	13.50	---	13.50	**20.40**	8.78	**13.30**
Several layers of paper from drywall with steaming equipment											
	Demo	Roll	1L	1.03	---	22.60	---	22.60	**34.10**	14.60	**22.10**
Vinyls (with synthetic fiber backing) from drywall with steaming equipment											
	Demo	Roll	1L	.410	---	9.00	---	9.00	**13.60**	5.86	**8.85**
Vinyls (with other backings), from drywall with steaming equipment											
	Demo	Roll	1L	.615	---	13.50	---	13.50	**20.40**	8.78	**13.30**

Dishwashers

High quality units. Labor cost includes rough-in

Built-in front loading

Description	Oper	Unit	Crew Size	Man-Hours Per Unit	Avg Mat'l Unit Cost	Avg Labor Unit Cost	Avg Equip Unit Cost	Avg Total Unit Cost	Avg Price Incl O&P	Large Volume Avg Total Unit Cost	Large Volume Avg Price Incl O&P
Six cycles	Inst	Ea	SC	10.3	759.00	305.00	---	1064.00	**1330.00**	874.00	**1070.00**
Five cycles	Inst	Ea	SC	10.3	699.00	305.00	---	1004.00	**1260.00**	820.00	**1010.00**
Four cycles	Inst	Ea	SC	10.3	633.00	305.00	---	938.00	**1180.00**	762.00	**943.00**
Three cycles	Inst	Ea	SC	10.3	446.00	305.00	---	751.00	**967.00**	595.00	**751.00**
Two cycles	Inst	Ea	SC	10.3	264.00	305.00	---	569.00	**759.00**	433.00	**565.00**
Adjustments											
Remove and reset dishwasher only	Reset	Ea	SA	2.05	---	62.40	---	62.40	**93.10**	40.50	**60.40**
Front and side panel kits											
White or prime finish	Inst	Ea	---	---	37.40	---	---	37.40	**37.40**	33.30	**33.30**
Regular color finish	Inst	Ea	---	---	37.40	---	---	37.40	**37.40**	33.30	**33.30**
Brushed chrome finish	Inst	Ea	---	---	44.00	---	---	44.00	**44.00**	39.20	**39.20**
Stainless steel finish	Inst	Ea	---	---	60.50	---	---	60.50	**60.50**	53.90	**53.90**
Black-glass acrylic finish with trim kit	Inst	Ea	---	---	99.00	---	---	99.00	**99.00**	88.20	**88.20**
Stainless steel trim kit	Inst	Ea	---	---	44.00	---	---	44.00	**44.00**	39.20	**39.20**

Portable front loading

Convertible, front loading, portable dishwashers; hardwood top, front & side panels

Description	Oper	Unit	Crew Size	Man-Hours Per Unit	Avg Mat'l Unit Cost	Avg Labor Unit Cost	Avg Equip Unit Cost	Avg Total Unit Cost	Avg Price Incl O&P	Large Volume Avg Total Unit Cost	Large Volume Avg Price Incl O&P
Six cycles											
White	Inst	Ea	SC	6.15	814.00	182.00	---	996.00	**1210.00**	844.00	**1010.00**
Colors	Inst	Ea	SC	6.15	842.00	182.00	---	1024.00	**1240.00**	869.00	**1040.00**
Four cycles											
White	Inst	Ea	SC	6.15	754.00	182.00	---	936.00	**1140.00**	790.00	**949.00**
Colors	Inst	Ea	SC	6.15	781.00	182.00	---	963.00	**1170.00**	815.00	**977.00**
Three cycles											
White	Inst	Ea	SC	6.15	710.00	182.00	---	892.00	**1090.00**	751.00	**904.00**
Colors	Inst	Ea	SC	6.15	743.00	182.00	---	925.00	**1130.00**	781.00	**937.00**
Front loading portables											
Three cycles with hardwood top											
White	Inst	Ea	SC	6.15	633.00	182.00	---	815.00	**999.00**	683.00	**825.00**
Colors	Inst	Ea	SC	6.15	671.00	182.00	---	853.00	**1040.00**	717.00	**864.00**
Two cycles with porcelain top											
White	Inst	Ea	SC	6.15	572.00	182.00	---	754.00	**930.00**	629.00	**763.00**
Colors	Inst	Ea	SC	6.15	638.00	182.00	---	820.00	**1010.00**	687.00	**830.00**

Description	Oper	Unit	Crew Size	Man-Hours Per Unit	Avg Mat'l Unit Cost	Avg Labor Unit Cost	Avg Equip Unit Cost	Avg Total Unit Cost	Avg Price Incl O&P	Avg Total Unit Cost	Avg Price Incl O&P
			Costs Based On Small Volume							Large Volume	
Dishwasher - sink combination, includes good quality fittings, 48" cabinet, faucets and water supply kit											
Six cycles											
White	Inst	Ea	SC	16.3	1200.00	483.00	---	1683.00	**2110.00**	1387.00	**1710.00**
Colors	Inst	Ea	SC	16.3	1240.00	483.00	---	1723.00	**2150.00**	1427.00	**1750.00**
Three cycles											
White	Inst	Ea	SC	16.3	1110.00	483.00	---	1593.00	**2000.00**	1307.00	**1610.00**
Remove and reset combination dishwasher - sink only	Reset	Ea	SA	3.42	---	104.00	---	104.00	**155.00**	67.60	**101.00**
Material adjustments											
For standard quality, DEDUCT	Inst	%		---	-20.0	---	---	---	**---**	---	**---**
For economy quality, DEDUCT	Inst	%		---	-30.0	---	---	---	**---**	---	**---**

Door frames

Exterior wood door frames

Description	Oper	Unit	Crew Size	Man-Hours Per Unit	Avg Mat'l Unit Cost	Avg Labor Unit Cost	Avg Equip Unit Cost	Avg Total Unit Cost	Avg Price Incl O&P	Avg Total Unit Cost	Avg Price Incl O&P
Exterior frame with exterior trim											
5/4 x 4-9/16" deep											
Pine	Inst	LF	2C	.076	3.31	2.05	---	5.36	**6.92**	4.26	**5.36**
Oak	Inst	LF	2C	.082	5.65	2.21	---	7.86	**9.86**	6.48	**7.94**
Walnut	Inst	LF	2C	.089	7.07	2.40	---	9.47	**11.80**	7.88	**9.59**
5/4 x 5-3/16" deep											
Pine	Inst	LF	2C	.076	3.59	2.05	---	5.64	**7.24**	4.52	**5.66**
Oak	Inst	LF	2C	.082	6.10	2.21	---	8.31	**10.40**	6.88	**8.40**
Walnut	Inst	LF	2C	.089	9.18	2.40	---	11.58	**14.20**	9.80	**11.80**
5/4 x 6-9/16" deep											
Pine	Inst	LF	2C	.076	4.20	2.05	---	6.25	**7.94**	5.07	**6.29**
Oak	Inst	LF	2C	.082	7.18	2.21	---	9.39	**11.60**	7.87	**9.54**
Walnut	Inst	LF	2C	.089	10.80	2.40	---	13.20	**16.00**	11.26	**13.50**
Exterior sills											
8/4 x 8" deep											
No horns	Inst	LF	2C	.333	9.41	8.98	---	18.39	**24.50**	13.97	**18.10**
2" horns	Inst	LF	2C	.333	10.10	8.98	---	19.08	**25.30**	14.59	**18.80**
3" horns	Inst	LF	2C	.333	10.90	8.98	---	19.88	**26.20**	15.37	**19.70**
8/4 x 10" deep											
No horns	Inst	LF	2C	.444	12.20	12.00	---	24.20	**32.20**	18.30	**23.70**
2" horns	Inst	LF	2C	.444	13.40	12.00	---	25.40	**33.60**	19.40	**25.00**
3" horns	Inst	LF	2C	.444	14.70	12.00	---	26.70	**35.00**	20.60	**26.30**
Exterior, colonial frame and trim											
3' opening, in swing	Inst	Ea	2C	1.33	473.00	35.90	---	508.90	**599.00**	453.60	**529.00**
5' 4" opening, in/out swing	Inst	Ea	2C	1.78	764.00	48.00	---	812.00	**951.00**	725.90	**845.00**
6' opening, in/out swing	Inst	Ea	2C	2.67	946.00	72.00	---	1018.00	**1200.00**	906.10	**1060.00**

Interior wood door frames

Description	Oper	Unit	Crew Size	Man-Hours Per Unit	Avg Mat'l Unit Cost	Avg Labor Unit Cost	Avg Equip Unit Cost	Avg Total Unit Cost	Avg Price Incl O&P	Avg Total Unit Cost	Avg Price Incl O&P
Interior frame											
11/16" x 3-5/8" deep											
Pine	Inst	LF	2C	.076	3.34	2.05	---	5.39	**6.96**	4.29	**5.39**
Oak	Inst	LF	2C	.082	4.10	2.21	---	6.31	**8.08**	5.06	**6.31**
Walnut	Inst	LF	2C	.089	6.16	2.40	---	8.56	**10.70**	7.05	**8.63**

Description	Oper	Unit	Crew Size	Man-Hours Per Unit	Avg Mat'l Unit Cost	Avg Labor Unit Cost	Avg Equip Unit Cost	Avg Total Unit Cost	Avg Price Incl O&P	Avg Total Unit Cost (Large Volume)	Avg Price Incl O&P (Large Volume)
			Costs Based On Small Volume							Large Volume	
11/16" x 4-9/16" deep											
Pine	Inst	LF	2C	.076	3.44	2.05	---	5.49	**7.07**	4.38	**5.50**
Oak	Inst	LF	2C	.082	4.10	2.21	---	6.31	**8.08**	5.06	**6.31**
Walnut	Inst	LF	2C	.089	6.16	2.40	---	8.56	**10.70**	7.05	**8.63**
11/16" x 5-3/16" deep											
Pine	Inst	LF	2C	.076	3.90	2.05	---	5.95	**7.60**	4.80	**5.98**
Oak	Inst	LF	2C	.082	4.22	2.21	---	6.43	**8.21**	5.17	**6.44**
Walnut	Inst	LF	2C	.089	6.38	2.40	---	8.78	**11.00**	7.25	**8.86**
Pocket door frame											
2'-0", to 3'-0" x 6'-8"	Inst	Ea	2C	2.29	85.50	61.70	---	147.20	**192.00**	113.90	**144.00**
3'-6" x 6'-8"	Inst	Ea	2C	2.29	157.00	61.70	---	218.70	**275.00**	179.90	**220.00**
4'-0" x 6'-8"	Inst	Ea	2C	2.29	160.00	61.70	---	221.70	**277.00**	181.90	**222.00**
Threshold, oak											
5/8" x 3-5/8" deep	Inst	LF	2C	.167	2.23	4.50	---	6.73	**9.41**	4.74	**6.44**
5/8" x 4-5/8" deep	Inst	LF	2C	.178	2.91	4.80	---	7.71	**10.60**	5.53	**7.43**
5/8" x 5-5/8" deep	Inst	LF	2C	.190	3.47	5.12	---	8.59	**11.80**	6.23	**8.31**
Door hardware											
Locksets											
Outside locks											
Knobs (pin tumbler)											
Excellent quality	Inst	Ea	CA	1.00	107.00	27.00	---	134.00	**164.00**	111.10	**134.00**
Good quality	Inst	Ea	CA	.800	67.70	21.60	---	89.30	**111.00**	72.30	**88.10**
Average quality	Inst	Ea	CA	.800	44.00	21.60	---	65.60	**83.40**	51.70	**64.40**
Handlesets											
Excellent quality	Inst	Ea	CA	1.00	171.00	27.00	---	198.00	**238.00**	167.00	**198.00**
Good quality	Inst	Ea	CA	.800	85.50	21.60	---	107.10	**131.00**	87.80	**106.00**
Average quality	Inst	Ea	CA	.800	44.00	21.60	---	65.60	**83.40**	51.70	**64.40**
Bath or bedroom locks											
Excellent quality	Inst	Ea	CA	.800	51.30	21.60	---	72.90	**91.80**	59.00	**73.10**
Good quality	Inst	Ea	CA	.615	34.20	16.60	---	50.80	**64.50**	40.50	**50.60**
Average quality	Inst	Ea	CA	.615	25.00	16.60	---	41.60	**53.90**	32.50	**41.30**
Passage latches											
Excellent quality	Inst	Ea	CA	.615	68.40	16.60	---	85.00	**104.00**	70.20	**84.70**
Good quality	Inst	Ea	CA	.471	33.90	12.70	---	46.60	**58.20**	37.70	**46.40**
Average quality	Inst	Ea	CA	.471	21.40	12.70	---	34.10	**43.90**	26.90	**34.00**
Dead locks (double key)											
Excellent quality	Inst	Ea	CA	.800	68.40	21.60	---	90.00	**111.00**	73.80	**90.20**
Good quality	Inst	Ea	CA	.615	42.00	16.60	---	58.60	**73.50**	47.30	**58.30**
Average quality	Inst	Ea	CA	.615	23.00	16.60	---	39.60	**51.60**	30.80	**39.30**
Kickplates, 8" x 30"											
Aluminum	Inst	Ea	CA	.471	25.10	12.70	---	37.80	**48.10**	30.10	**37.70**
Brass or bronze	Inst	Ea	CA	.471	39.90	12.70	---	52.60	**65.20**	43.00	**52.50**
Thresholds											
Aluminum, pre-notched, draft-proof, standard											
3/4" high											
32" long	Inst	Ea	CA	.381	7.01	10.30	---	17.31	**23.70**	12.83	**17.30**
36" long	Inst	Ea	CA	.381	7.81	10.30	---	18.11	**24.60**	13.52	**18.00**

Description	Oper	Unit	Crew Size	Man-Hours Per Unit	Avg Mat'l Unit Cost	Avg Labor Unit Cost	Avg Equip Unit Cost	Avg Total Unit Cost	Avg Price Incl O&P	Avg Total Unit Cost	Avg Price Incl O&P
			Costs Based On Small Volume							Large Volume	
42" long	Inst	Ea	CA	.381	9.18	10.30	---	19.48	**26.20**	14.71	**19.40**
48" long	Inst	Ea	CA	.381	10.40	10.30	---	20.70	**27.60**	15.80	**20.70**
60" long	Inst	Ea	CA	.381	13.10	10.30	---	23.40	**30.60**	18.04	**23.30**
72" long	Inst	Ea	CA	.381	15.70	10.30	---	26.00	**33.70**	20.34	**25.90**
For high-rug type, 1-1/8"											
ADD	Inst	%	CA	---	25.0	---	---	---	**---**	---	**---**
Wood, oak threshold with vinyl weather seal											
5/8" x 3-1/2"											
33" long	Inst	Ea	CA	.381	5.81	10.30	---	16.11	**22.30**	11.79	**16.10**
37" long	Inst	Ea	CA	.381	6.33	10.30	---	16.63	**22.90**	12.23	**16.60**
43" long	Inst	Ea	CA	.381	6.95	10.30	---	17.25	**23.60**	12.78	**17.20**
49" long	Inst	Ea	CA	.381	7.70	10.30	---	18.00	**24.50**	13.42	**17.90**
61" long	Inst	Ea	CA	.381	8.66	10.30	---	18.96	**25.60**	14.26	**18.90**
73" long	Inst	Ea	CA	.381	10.60	10.30	---	20.90	**27.80**	15.95	**20.80**
3/4" x 3-1/2"											
33" long	Inst	Ea	CA	.381	6.73	10.30	---	17.03	**23.40**	12.58	**17.00**
37" long	Inst	Ea	CA	.381	8.66	10.30	---	18.96	**25.60**	14.26	**18.90**

Doors

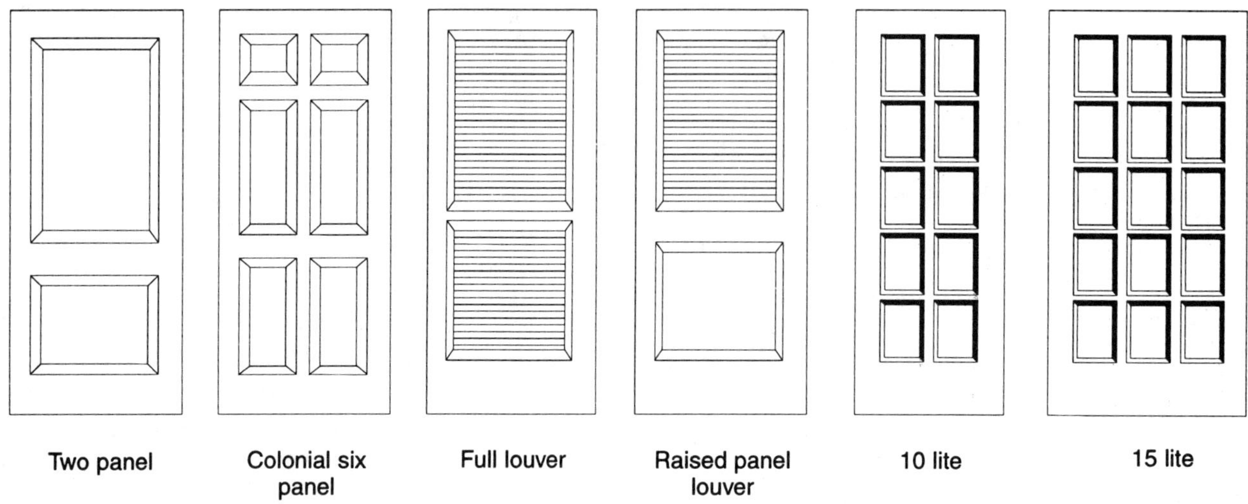

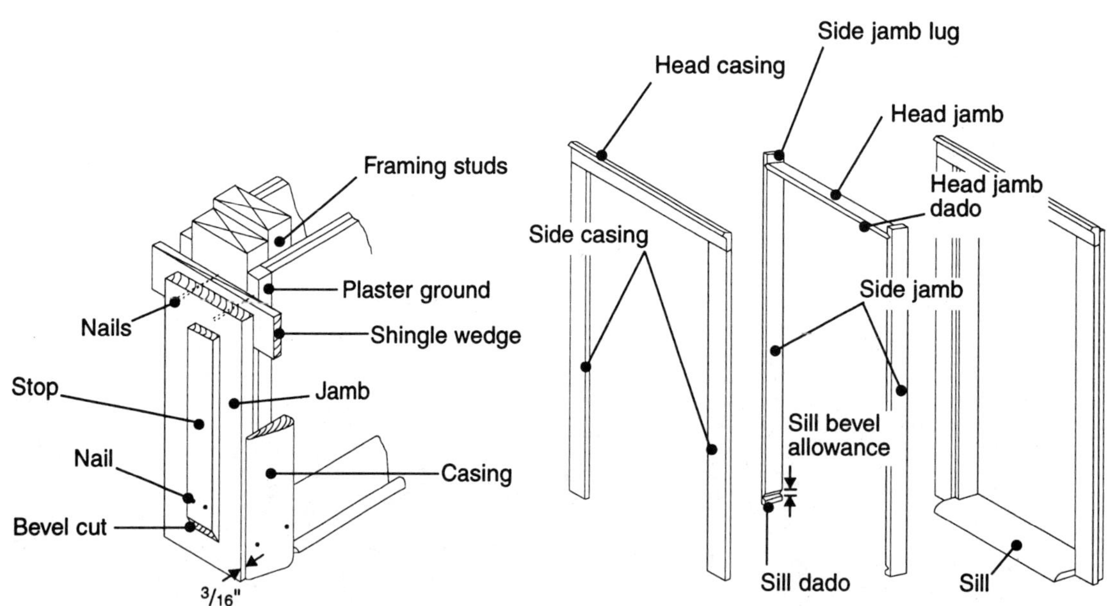

Parts of Exterior Door Frame

Assembled Package Door Units (Interior)

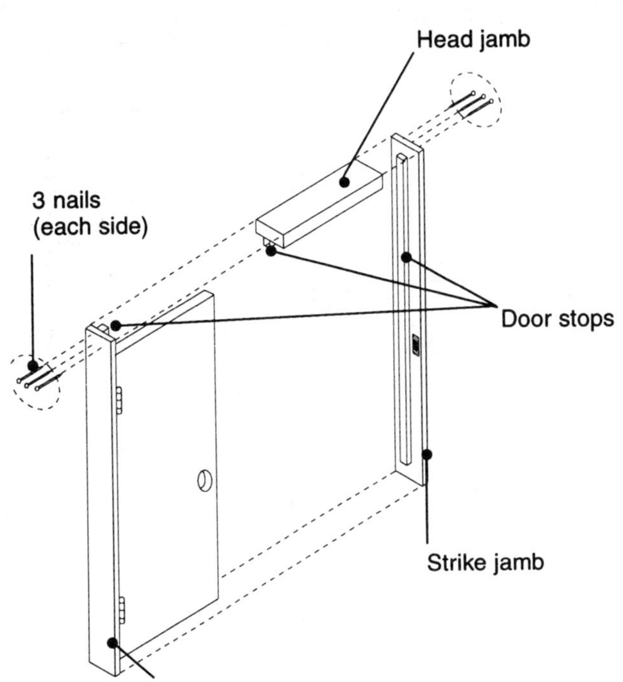

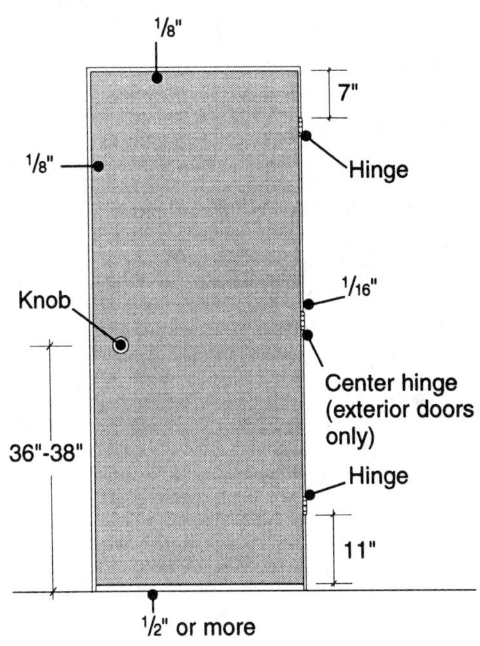

Unit Includes:

Jamb: 4 9/16" finger joint pine w/ stops applied.

Butts: 1 pair 3 1/2" x 3 1/2" applied to door and jamb.

Door: 3 degree bevel one side 3/16" under std. width - net 80"
H center bored - 2 1/8" bore
2 3/8" backset - 1" edge bore

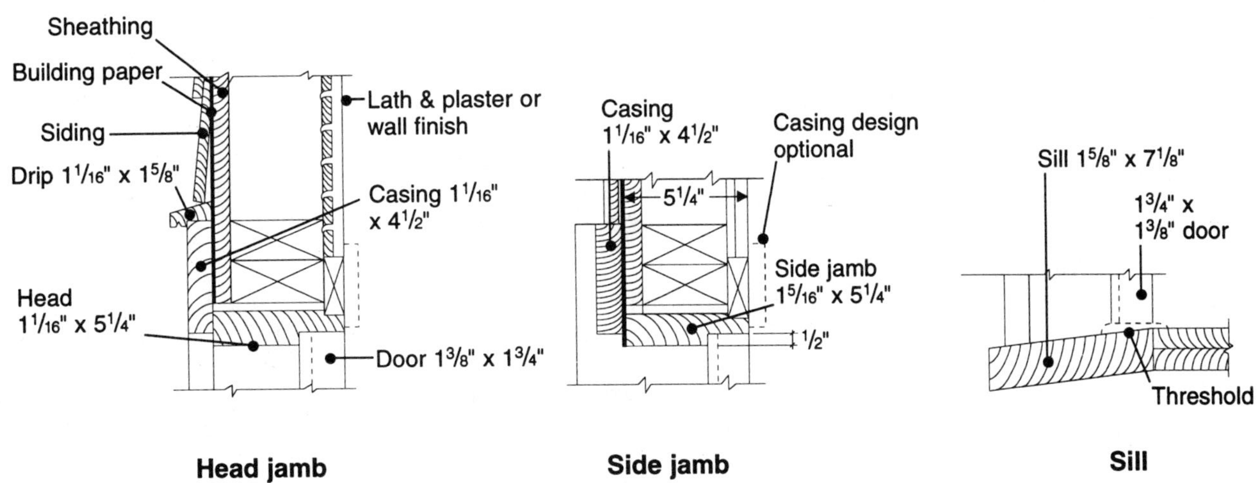

			Costs Based On Small Volume						Large Volume	
Description	Oper	Unit	Crew Size	Man-Hours Per Unit	Avg Mat'l Unit Cost	Avg Labor Unit Cost	Avg Equip Unit Cost	Avg Total Unit Cost	Avg Price Incl O&P	Avg Total Unit Cost	Avg Price Incl O&P
Doors											
Entrance doors											
Decorative glass doors and matching sidelights											
Lafayette collection, 1-3/8" panels											
Astoria style											
2'-6", 2'-8", 3'-0" W x 6'-8" H doors											
Fir & pine	Inst	Ea	2C	2.05	926.00	55.30	---	981.30	**1150.00**	750.90	**877.00**
Oak	Inst	Ea	2C	2.05	1210.00	55.30	---	1265.30	**1480.00**	972.90	**1130.00**
1'-0", 1'-2" x 6'-8-1/2" sidelights											
Fir & pine	Inst	Ea	2C	1.54	410.00	41.50	---	451.50	**535.00**	344.00	**405.00**
Oak	Inst	Ea	2C	1.54	521.00	41.50	---	562.50	**662.00**	429.00	**503.00**
Bourbon Royale style											
2'-6", 2'-8", 3'-0" W x 6'-8" H doors											
Fir & pine	Inst	Ea	2C	2.05	910.00	55.30	---	965.30	**1130.00**	737.90	**862.00**
Oak	Inst	Ea	2C	2.05	1200.00	55.30	---	1255.30	**1460.00**	962.90	**1120.00**
1'-0", 1'-2" x 6'-8-1/2" sidelights											
Fir & pine	Inst	Ea	2C	1.54	431.00	41.50	---	472.50	**559.00**	360.00	**424.00**
Oak	Inst	Ea	2C	1.54	535.00	41.50	---	576.50	**678.00**	440.00	**515.00**
Jubilee style											
2'-6", 2'-8", 3'-0" W x 6'-8" H doors											
Fir & pine	Inst	Ea	2C	2.05	958.00	55.30	---	1013.30	**1190.00**	774.90	**904.00**
Oak	Inst	Ea	2C	2.05	1250.00	55.30	---	1305.30	**1520.00**	998.90	**1160.00**
1'-0", 1'-2" x 6'-8-1/2" sidelights											
Fir & pine	Inst	Ea	2C	1.54	402.00	41.50	---	443.50	**525.00**	337.00	**398.00**
Oak	Inst	Ea	2C	1.54	611.00	41.50	---	652.50	**766.00**	498.00	**583.00**
Lexington style											
2'-6", 2'-8", 3'-0" W x 6'-8" H doors											
Fir & pine	Inst	Ea	2C	2.05	955.00	55.30	---	1010.30	**1180.00**	772.90	**902.00**
Oak	Inst	Ea	2C	2.05	1250.00	55.30	---	1305.30	**1520.00**	996.90	**1160.00**
1'-0", 1'-2" x 6'-8-1/2" sidelights											
Fir & pine	Inst	Ea	2C	1.54	429.00	41.50	---	470.50	**556.00**	358.00	**421.00**
Oak	Inst	Ea	2C	1.54	532.00	41.50	---	573.50	**675.00**	437.00	**513.00**
Marquis style											
2'-6", 2'-8", 3'-0" W x 6'-8" H doors											
Fir & pine	Inst	Ea	2C	2.05	998.00	55.30	---	1053.30	**1230.00**	805.90	**940.00**
Oak	Inst	Ea	2C	2.05	1290.00	55.30	---	1345.30	**1570.00**	1031.90	**1200.00**
1'-0", 1'-2" x 6'-8-1/2" sidelights											
Fir & pine	Inst	Ea	2C	1.54	451.00	41.50	---	492.50	**582.00**	375.00	**441.00**
Oak	Inst	Ea	2C	1.54	502.00	41.50	---	543.50	**641.00**	415.00	**487.00**
Monaco style											
2'-6", 2'-8", 3'-0" W x 6'-8" H doors											
Fir & pine	Inst	Ea	2C	2.05	1170.00	55.30	---	1225.30	**1430.00**	940.90	**1090.00**
Oak	Inst	Ea	2C	2.05	1450.00	55.30	---	1505.30	**1760.00**	1155.90	**1350.00**
1'-0", 1'-2" x 6'-8-1/2" sidelights											
Fir & pine	Inst	Ea	2C	1.54	537.00	41.50	---	578.50	**681.00**	442.00	**518.00**
Oak	Inst	Ea	2C	1.54	640.00	41.50	---	681.50	**799.00**	521.00	**609.00**

Description	Oper	Unit	Crew Size	Man-Hours Per Unit	Avg Mat'l Unit Cost	Avg Labor Unit Cost	Avg Equip Unit Cost	Avg Total Unit Cost	Avg Price Incl O&P	Large Volume Avg Total Unit Cost	Large Volume Avg Price Incl O&P
Windsor style											
2'-6", 2'-8", 3'-0" W x 6'-8" H doors											
Fir & pine	Inst	Ea	2C	2.05	987.00	55.30	---	1042.30	**1220.00**	797.90	**931.00**
Oak	Inst	Ea	2C	2.05	1280.00	55.30	---	1335.30	**1550.00**	1019.90	**1190.00**
1'-0", 1'-2" x 6'-8-1/2" sidelights											
Fir & pine	Inst	Ea	2C	1.54	423.00	41.50	---	464.50	**550.00**	354.00	**417.00**
Oak	Inst	Ea	2C	1.54	529.00	41.50	---	570.50	**671.00**	435.00	**510.00**
Stile and rail raised panels											
No lites											
2 raised panels											
2'-6" to 3'-0" x 6'-8" x 1-3/4" T	Inst	Ea	2C	2.05	358.00	55.30	---	413.30	**496.00**	311.90	**372.00**
3'-6" x 6'-8" x 1-3/4" T	Inst	Ea	2C	2.22	358.00	59.90	---	417.90	**503.00**	315.10	**377.00**
4 raised panels											
2'-6" to 3'-0" x 6'-8" x 1-3/4" T	Inst	Ea	2C	2.05	326.00	55.30	---	381.30	**459.00**	286.90	**344.00**
3'-6" x 6'-8" x 1-3/4" T	Inst	Ea	2C	2.22	478.00	59.90	---	537.90	**641.00**	408.10	**484.00**
6 raised panels											
2'-0" to 2'-8" x 6'-8" x 1-3/4" T	Inst	Ea	2C	2.05	315.00	55.30	---	370.30	**446.00**	278.90	**334.00**
3'-0" x 6'-8" x 1-3/4" T	Inst	Ea	2C	2.22	324.00	59.90	---	383.90	**463.00**	289.10	**347.00**
3'-6" x 6'-8" x 1-3/4" T	Inst	Ea	2C	2.46	474.00	66.30	---	540.30	**646.00**	409.10	**486.00**
8 raised panels											
2'-6" to 2'-8" x 6'-8" x 1-3/4" T	Inst	Ea	2C	2.05	317.00	55.30	---	372.30	**449.00**	280.90	**336.00**
3'-0" x 6'-8" x 1-3/4" T	Inst	Ea	2C	2.22	324.00	59.90	---	383.90	**463.00**	289.10	**347.00**
3'-6" x 6'-8" x 1-3/4" T	Inst	Ea	2C	2.46	474.00	66.30	---	540.30	**646.00**	409.10	**486.00**
2'-6" to 2'-8" x 7'-0" x 1-3/4" T	Inst	Ea	2C	2.05	341.00	55.30	---	396.30	**477.00**	298.90	**357.00**
3'-0" x 7'-0" x 1-3/4" T	Inst	Ea	2C	2.22	367.00	59.90	---	426.90	**513.00**	322.10	**385.00**
3'-6" x 7'-0" x 1-3/4" T	Inst	Ea	2C	2.46	498.00	66.30	---	564.30	**673.00**	427.10	**507.00**
10 raised panels											
2'-6" to 2'-8" x 8'-0" x 1-3/4" T	Inst	Ea	2C	2.22	414.00	59.90	---	473.90	**567.00**	359.10	**427.00**
3'-0" x 8'-0" x 1-3/4" T	Inst	Ea	2C	2.46	434.00	66.30	---	500.30	**600.00**	378.10	**451.00**
3'-6" x 8'-0" x 1-3/4" T	Inst	Ea	2C	2.71	549.00	73.10	---	622.10	**743.00**	472.00	**560.00**
Lites											
1 raised panel, 1 lite											
2'-6" to 3'-0" x 6'-8" x 1-3/4" T											
Open w/stops	Inst	Ea	2C	2.05	313.00	55.30	---	368.30	**444.00**	276.90	**332.00**
Tempered clear	Inst	Ea	2C	2.05	326.00	55.30	---	381.30	**459.00**	286.90	**344.00**
Acrylic or tempered amber	Inst	Ea	2C	2.05	391.00	55.30	---	446.30	**534.00**	337.90	**402.00**
1 raised panel, 6 lites											
2'-6" to 3'-0" x 6'-8" x 1-3/4" T											
Open w/stops	Inst	Ea	2C	2.05	337.00	55.30	---	392.30	**471.00**	295.90	**353.00**
Tempered clear	Inst	Ea	2C	2.05	361.00	55.30	---	416.30	**499.00**	313.90	**375.00**
Acrylic or tempered amber	Inst	Ea	2C	2.05	408.00	55.30	---	463.30	**554.00**	350.90	**417.00**
1 raised panel, 9 lites											
2'-6" to 3'-0" x 6'-8" x 1-3/4" T											
Open w/stops	Inst	Ea	2C	2.05	343.00	55.30	---	398.30	**479.00**	300.90	**359.00**
Tempered clear	Inst	Ea	2C	2.05	369.00	55.30	---	424.30	**508.00**	320.90	**382.00**
Acrylic or tempered amber	Inst	Ea	2C	2.05	447.00	55.30	---	502.30	**598.00**	380.90	**451.00**
1 raised panel, 12 diamond lites											
2'-6" to 3'-0" x 6'-8" x 1-3/4" T											
Open w/stops	Inst	Ea	2C	2.05	351.00	55.30	---	406.30	**488.00**	306.90	**366.00**
Tempered clear	Inst	Ea	2C	2.05	447.00	55.30	---	502.30	**598.00**	380.90	**451.00**
Acrylic or tempered amber	Inst	Ea	2C	2.05	558.00	55.30	---	613.30	**725.00**	465.90	**549.00**

Description	Oper	Unit	Crew Size	Man-Hours Per Unit	Avg Mat'l Unit Cost	Avg Labor Unit Cost	Avg Equip Unit Cost	Avg Total Unit Cost	Avg Price Incl O&P	Large Volume Avg Total Unit Cost	Large Volume Avg Price Incl O&P
Dutch doors, country style, fir, stile and rail											
Two raised panels in lower door											
1 lite in upper door											
2'-6" to 3'-0" x 6'-8" x 1-3/4" T											
Empty	Inst	Ea	2C	2.05	431.00	55.30	---	486.30	**579.00**	367.90	**437.00**
Tempered clear	Inst	Ea	2C	2.05	447.00	55.30	---	502.30	**598.00**	380.90	**451.00**
Tempered amber	Inst	Ea	2C	2.05	558.00	55.30	---	613.30	**725.00**	465.90	**549.00**
4 lites in upper door											
2'-6" to 3'-0" x 6'-8" x 1-3/4" T											
Empty	Inst	Ea	2C	2.05	441.00	55.30	---	496.30	**591.00**	375.90	**446.00**
Tempered clear	Inst	Ea	2C	2.05	484.00	55.30	---	539.30	**641.00**	408.90	**484.00**
Tempered amber	Inst	Ea	2C	2.05	508.00	55.30	---	563.30	**668.00**	427.90	**505.00**
9 lites in upper door											
2'-6" to 3'-0" x 6'-8" x 1-3/4" T											
Empty	Inst	Ea	2C	2.05	436.00	55.30	---	491.30	**586.00**	372.90	**442.00**
Tempered clear	Inst	Ea	2C	2.05	453.00	55.30	---	508.30	**605.00**	384.90	**456.00**
Tempered amber	Inst	Ea	2C	2.05	562.00	55.30	---	617.30	**730.00**	469.90	**553.00**
12 diamond lites in upper door											
2'-6" to 3'-0" x 6'-8" x 1-3/4" T											
Empty	Inst	Ea	2C	2.05	479.00	55.30	---	534.30	**635.00**	405.90	**480.00**
Tempered clear	Inst	Ea	2C	2.05	515.00	55.30	---	570.30	**676.00**	433.90	**512.00**
Tempered amber	Inst	Ea	2C	2.05	637.00	55.30	---	692.30	**816.00**	526.90	**620.00**
Four diamond shaped raised panels in lower door											
1 lite in upper door											
2'-6" to 3'-0" x 6'-8" x 1-3/4" T											
Empty	Inst	Ea	2C	2.05	434.00	55.30	---	489.30	**583.00**	370.90	**440.00**
Tempered clear	Inst	Ea	2C	2.05	451.00	55.30	---	506.30	**602.00**	383.90	**455.00**
Tempered amber	Inst	Ea	2C	2.05	561.00	55.30	---	616.30	**729.00**	468.90	**553.00**
6 lites in upper door											
2'-6" to 3'-0" x 6'-8" x 1-3/4" T											
Empty	Inst	Ea	2C	2.05	469.00	55.30	---	524.30	**624.00**	397.90	**471.00**
Tempered clear	Inst	Ea	2C	2.05	488.00	55.30	---	543.30	**645.00**	411.90	**487.00**
Tempered amber	Inst	Ea	2C	2.05	604.00	55.30	---	659.30	**778.00**	501.90	**590.00**
9 lites in upper door											
2'-6" to 3'-0" x 6'-8" x 1-3/4" T											
Empty	Inst	Ea	2C	2.05	440.00	55.30	---	495.30	**590.00**	374.90	**445.00**
Tempered clear	Inst	Ea	2C	2.05	456.00	55.30	---	511.30	**609.00**	387.90	**459.00**
Tempered amber	Inst	Ea	2C	2.05	566.00	55.30	---	621.30	**735.00**	472.90	**557.00**
12 diamond lites in upper door											
2'-6" to 3'-0" x 6'-8" x 1-3/4" T											
Empty	Inst	Ea	2C	2.05	483.00	55.30	---	538.30	**639.00**	408.90	**483.00**
Tempered clear	Inst	Ea	2C	2.05	519.00	55.30	---	574.30	**681.00**	435.90	**515.00**
Tempered amber	Inst	Ea	2C	2.05	640.00	55.30	---	695.30	**820.00**	529.90	**623.00**
French doors											
Douglas fir or hemlock 4-1/2" stiles and top rail, 9-1/2" bottom rail, tempered glass with stops											
1 lite											
1-3/8" T and 1-3/4" T											
2'-0" x 6'-8"	Inst	Ea	2C	1.76	221.00	47.50	---	268.50	**326.00**	200.70	**243.00**
2'-4", 2'-6", 2'-8" x 6'-8"	Inst	Ea	2C	2.05	224.00	55.30	---	279.30	**342.00**	208.90	**254.00**
3'-0" x 6'-8"	Inst	Ea	2C	2.46	230.00	66.30	---	296.30	**365.00**	221.10	**270.00**

Description	Oper	Unit	Crew Size	Man-Hours Per Unit	Avg Mat'l Unit Cost	Avg Labor Unit Cost	Avg Equip Unit Cost	Avg Total Unit Cost	Avg Price Incl O&P	Large Volume Avg Total Unit Cost	Large Volume Avg Price Incl O&P
			Costs Based On Small Volume							Large Volume	
5 lites, 5 high											
1-3/8" T and 1-3/4" T											
2'-0" x 6'-8"	Inst	Ea	2C	1.76	221.00	47.50	---	268.50	**326.00**	200.70	**243.00**
2'-4", 2'-6", 2'-8" x 6'-8"	Inst	Ea	2C	2.05	229.00	55.30	---	284.30	**347.00**	212.90	**258.00**
3'-0" x 6'-8"	Inst	Ea	2C	2.46	235.00	66.30	---	301.30	**371.00**	224.10	**274.00**
10 lites, 5 high											
1-3/8" T and 1-3/4" T											
2'-0" x 6'-8"	Inst	Ea	2C	1.76	224.00	47.50	---	271.50	**330.00**	203.70	**246.00**
2'-4", 2'-6", 2'-8" x 6'-8"	Inst	Ea	2C	2.05	228.00	55.30	---	283.30	**346.00**	211.90	**257.00**
3'-0" x 6'-8"	Inst	Ea	2C	2.46	234.00	66.30	---	300.30	**370.00**	223.10	**273.00**
15 lites, 5 high											
1-3/8" T and 1-3/4" T											
2'-6", 2'-8" x 6'-8"	Inst	Ea	2C	2.05	262.00	55.30	---	317.30	**386.00**	237.90	**287.00**
3'-0" x 6'-8"	Inst	Ea	2C	2.46	269.00	66.30	---	335.30	**410.00**	250.10	**304.00**
French sidelites											
Douglas fir or hemlock 4-1/2" stiles and top rail, 9-1/2" bottom rail, tempered glass with stops											
1 lite											
1-3/4" T											
1'-0", 1'-2" x 6'-8"	Inst	Ea	2C	1.54	161.00	41.50	---	202.50	**248.00**	151.00	**184.00**
1'-4", 1'-6" x 6'-8"	Inst	Ea	2C	1.54	170.00	41.50	---	211.50	**259.00**	158.00	**192.00**
5 lites, 5 high											
1-3/4" T											
1'-0", 1'-2" x 6'-8"	Inst	Ea	2C	1.54	167.00	41.50	---	208.50	**256.00**	156.00	**190.00**
1'-4", 1'-6" x 6'-8"	Inst	Ea	2C	1.54	182.00	41.50	---	223.50	**273.00**	168.00	**203.00**
Garden doors											
1 door, "X" unit											
2'-6" x 6'-8"	Inst	LS	2C	2.05	284.00	55.30	---	339.30	**411.00**	254.90	**307.00**
3'-0" x 6'-8"	Inst	LS	2C	2.46	305.00	66.30	---	371.30	**451.00**	278.10	**336.00**
2 doors, "XO/OX" unit											
5'-0" x 6'-8"	Inst	LS	2C	3.72	532.00	100.00	---	632.00	**764.00**	475.20	**571.00**
6'-0" x 6'-8"	Inst	LS	2C	4.44	570.00	120.00	---	690.00	**838.00**	518.50	**625.00**
3 doors, "XOX" unit											
7'-6" x 6'-8"	Inst	LS	2C	5.52	759.00	149.00	---	908.00	**1100.00**	684.10	**823.00**
9'-0" x 6'-8"	Inst	LS	2C	6.67	822.00	180.00	---	1002.00	**1220.00**	750.00	**906.00**
Sliding doors											
Glass sliding doors, with 5-1/2" anodized aluminum frame, trim, weatherstripping, tempered 3/16" T clear single glazed glass, with screen											
6'-8" H											
5' W, 2 lites, 1 sliding	Inst	Set	2C	6.15	347.00	166.00	---	513.00	**651.00**	376.00	**472.00**
6' W, 2 lites, 1 sliding	Inst	Set	2C	6.15	383.00	166.00	---	549.00	**692.00**	403.00	**504.00**
7' W, 2 lites, 1 sliding	Inst	Set	2C	6.15	419.00	166.00	---	585.00	**733.00**	431.00	**535.00**
8' W, 2 lites, 1 sliding	Inst	Set	2C	6.15	454.00	166.00	---	620.00	**775.00**	459.00	**567.00**
10' W, 2 lites, 1 sliding	Inst	Set	2C	6.15	603.00	166.00	---	769.00	**945.00**	573.00	**699.00**
9' W, 3 lites, 1 sliding	Inst	Set	2C	8.00	520.00	216.00	---	736.00	**926.00**	545.00	**680.00**
12' W, 3 lites, 1 sliding	Inst	Set	2C	8.00	625.00	216.00	---	841.00	**1050.00**	626.00	**773.00**
15' W, 3 lites, 1 sliding	Inst	Set	2C	8.00	845.00	216.00	---	1061.00	**1300.00**	796.00	**969.00**

Description	Oper	Unit	Crew Size	Man-Hours Per Unit	Avg Mat'l Unit Cost	Avg Labor Unit Cost	Avg Equip Unit Cost	Avg Total Unit Cost	Avg Price Incl O&P	Large Volume Avg Total Unit Cost	Large Volume Avg Price Incl O&P
8'-0" H											
5' W, 2 lites, 1 sliding	Inst	Set	2C	6.15	412.00	166.00	---	578.00	**726.00**	426.00	**530.00**
6' W, 2 lites, 1 sliding	Inst	Set	2C	6.15	443.00	166.00	---	609.00	**761.00**	450.00	**557.00**
7' W, 2 lites, 1 sliding	Inst	Set	2C	6.15	523.00	166.00	---	689.00	**854.00**	512.00	**629.00**
8' W, 2 lites, 1 sliding	Inst	Set	2C	6.15	605.00	166.00	---	771.00	**948.00**	575.00	**701.00**
10' W, 2 lites, 1 sliding	Inst	Set	2C	6.15	698.00	166.00	---	864.00	**1060.00**	647.00	**784.00**
9' W, 3 lites, 1 sliding	Inst	Set	2C	8.00	609.00	216.00	---	825.00	**1030.00**	614.00	**759.00**
12' W, 3 lites, 1 sliding	Inst	Set	2C	8.00	845.00	216.00	---	1061.00	**1300.00**	796.00	**969.00**
15' W, 3 lites, 1 sliding	Inst	Set	2C	8.00	983.00	216.00	---	1199.00	**1460.00**	902.00	**1090.00**
French sliding door units											
Douglas fir doors, solid brass hardware, screen, frames, and exterior molding included											
1 lite, 2 panels, single glazed											
1-3/4" T											
5'-0" x 6'-8"	Inst	Ea	2C	8.00	2640.00	216.00	---	2856.00	**3360.00**	2174.00	**2560.00**
6'-0" x 6'-8"	Inst	Ea	2C	8.00	2810.00	216.00	---	3026.00	**3560.00**	2314.00	**2710.00**
7'-0" x 6'-8"	Inst	Ea	2C	8.00	3040.00	216.00	---	3256.00	**3820.00**	2494.00	**2920.00**
8'-0" x 6'-8"	Inst	Ea	2C	8.00	3200.00	216.00	---	3416.00	**4010.00**	2614.00	**3060.00**
5'-0" x 7'-0"	Inst	Ea	2C	8.00	2700.00	216.00	---	2916.00	**3430.00**	2224.00	**2610.00**
6'-0" x 7'-0"	Inst	Ea	2C	8.00	2840.00	216.00	---	3056.00	**3590.00**	2334.00	**2740.00**
7'-0" x 7'-0"	Inst	Ea	2C	8.00	3130.00	216.00	---	3346.00	**3930.00**	2554.00	**2990.00**
8'-0" x 7'-0"	Inst	Ea	2C	8.00	3320.00	216.00	---	3536.00	**4150.00**	2704.00	**3170.00**
5'-0" x 8'-0"	Inst	Ea	2C	8.00	3050.00	216.00	---	3266.00	**3840.00**	2504.00	**2930.00**
6'-0" x 8'-0"	Inst	Ea	2C	8.00	3270.00	216.00	---	3486.00	**4090.00**	2674.00	**3120.00**
7'-0" x 8'-0"	Inst	Ea	2C	8.00	3520.00	216.00	---	3736.00	**4370.00**	2854.00	**3340.00**
8'-0" x 8'-0"	Inst	Ea	2C	8.00	3740.00	216.00	---	3956.00	**4630.00**	3034.00	**3540.00**
10 lites, 2 panels, single glazed											
1-3/4" T											
5'-0" x 6'-8"	Inst	Ea	2C	8.00	2840.00	216.00	---	3056.00	**3590.00**	2334.00	**2740.00**
6'-0" x 6'-8"	Inst	Ea	2C	8.00	2970.00	216.00	---	3186.00	**3750.00**	2434.00	**2860.00**
5'-0" x 7'-0"	Inst	Ea	2C	8.00	2900.00	216.00	---	3116.00	**3660.00**	2384.00	**2790.00**
6'-0" x 7'-0"	Inst	Ea	2C	8.00	3050.00	216.00	---	3266.00	**3840.00**	2504.00	**2930.00**
5'-0" x 8'-0"	Inst	Ea	2C	8.00	3230.00	216.00	---	3446.00	**4040.00**	2634.00	**3080.00**
6'-0" x 8'-0"	Inst	Ea	2C	8.00	3390.00	216.00	---	3606.00	**4230.00**	2764.00	**3230.00**
15 lites, 2 panels, single glazed											
1-3/4" T											
5'-0" x 6'-8"	Inst	Ea	2C	8.00	2990.00	216.00	---	3206.00	**3770.00**	2454.00	**2880.00**
6'-0" x 6'-8"	Inst	Ea	2C	8.00	3130.00	216.00	---	3346.00	**3930.00**	2564.00	**3000.00**
7'-0" x 6'-8"	Inst	Ea	2C	8.00	3330.00	216.00	---	3546.00	**4160.00**	2714.00	**3180.00**
8'-0" x 6'-8"	Inst	Ea	2C	8.00	3480.00	216.00	---	3696.00	**4330.00**	2834.00	**3310.00**
5'-0" x 7'-0"	Inst	Ea	2C	8.00	3050.00	216.00	---	3266.00	**3840.00**	2504.00	**2930.00**
6'-0" x 7'-0"	Inst	Ea	2C	8.00	3210.00	216.00	---	3426.00	**4020.00**	2624.00	**3070.00**
7'-0" x 7'-0"	Inst	Ea	2C	8.00	3420.00	216.00	---	3636.00	**4260.00**	2784.00	**3250.00**
8'-0" x 7'-0"	Inst	Ea	2C	8.00	3590.00	216.00	---	3806.00	**4450.00**	2914.00	**3400.00**
5'-0" x 8'-0"	Inst	Ea	2C	8.00	3380.00	216.00	---	3596.00	**4220.00**	2754.00	**3220.00**
6'-0" x 8'-0"	Inst	Ea	2C	8.00	3550.00	216.00	---	3766.00	**4410.00**	2884.00	**3370.00**
7'-0" x 8'-0"	Inst	Ea	2C	8.00	3810.00	216.00	---	4026.00	**4710.00**	3084.00	**3600.00**
8'-0" x 8'-0"	Inst	Ea	2C	8.00	3950.00	216.00	---	4166.00	**4880.00**	3194.00	**3730.00**

Description	Oper	Unit	Crew Size	Man-Hours Per Unit	Avg Mat'l Unit Cost	Avg Labor Unit Cost	Avg Equip Unit Cost	Avg Total Unit Cost	Avg Price Incl O&P	Large Volume Avg Total Unit Cost	Large Volume Avg Price Incl O&P
			Costs Based On Small Volume							Large Volume	
Fire doors, natural birch											
One hour rating											
2'-6" x 6'-8" x 1-3/4" T	Inst	Ea	2C	1.88	276.00	50.70	---	326.70	**394.00**	246.20	**295.00**
3'-6" x 6'-8" x 1-3/4" T	Inst	Ea	2C	2.22	442.00	59.90	---	501.90	**600.00**	380.10	**452.00**
2'-6" x 7'-0" x 1-3/4" T	Inst	Ea	2C	1.88	298.00	50.70	---	348.70	**420.00**	263.20	**315.00**
3'-6" x 7'-0" x 1-3/4" T	Inst	Ea	2C	2.22	442.00	59.90	---	501.90	**599.00**	380.10	**451.00**
1.5 hour rating											
2'-6" x 6'-8" x 1-3/4" T	Inst	Ea	2C	1.88	357.00	50.70	---	407.70	**487.00**	308.20	**367.00**
3'-6" x 6'-8" x 1-3/4" T	Inst	Ea	2C	2.22	500.00	59.90	---	559.90	**665.00**	425.10	**503.00**
2'-6" x 7'-0" x 1-3/4" T	Inst	Ea	2C	1.88	384.00	50.70	---	434.70	**518.00**	329.20	**391.00**
3'-6" x 7'-0" x 1-3/4" T	Inst	Ea	2C	2.22	513.00	59.90	---	572.90	**681.00**	435.10	**515.00**
Interior doors											
Passage doors, flush face, includes hinge hardware											
Hollow core prehung door units											
Hardboard											
2'-6" x 6'-8" x 1-3/8" T	Inst	Ea	2C	1.63	35.70	43.90	---	79.60	**108.00**	56.40	**75.50**
2'-8" x 6'-8" x 1-3/8" T	Inst	Ea	2C	1.76	37.90	47.50	---	85.40	**116.00**	60.00	**80.40**
3'-0" x 6'-8" x 1-3/8" T	Inst	Ea	2C	1.76	40.00	47.50	---	87.50	**118.00**	61.60	**82.20**
2'-6" x 7'-0" x 1-3/8" T	Inst	Ea	2C	1.63	45.40	43.90	---	89.30	**119.00**	63.90	**84.10**
2'-8" x 7'-0" x 1-3/8" T	Inst	Ea	2C	1.76	47.60	47.50	---	95.10	**127.00**	67.40	**89.00**
3'-0" x 7'-0" x 1-3/8" T	Inst	Ea	2C	1.76	49.80	47.50	---	97.30	**129.00**	69.10	**90.90**
Lauan											
2'-6" x 6'-8" x 1-3/8" T	Inst	Ea	2C	1.63	47.20	43.90	---	91.10	**121.00**	65.40	**85.80**
2'-8" x 6'-8" x 1-3/8" T	Inst	Ea	2C	1.76	49.00	47.50	---	96.50	**129.00**	68.50	**90.20**
3'-0" x 6'-8" x 1-3/8" T	Inst	Ea	2C	1.76	52.60	47.50	---	100.10	**133.00**	71.30	**93.40**
2'-6" x 7'-0" x 1-3/8" T	Inst	Ea	2C	1.63	55.90	43.90	---	99.80	**131.00**	72.10	**93.50**
2'-8" x 7'-0" x 1-3/8" T	Inst	Ea	2C	1.76	58.00	47.50	---	105.50	**139.00**	75.40	**98.20**
3'-0" x 7'-0" x 1-3/8" T	Inst	Ea	2C	1.76	61.60	47.50	---	109.10	**143.00**	78.20	**101.00**
Birch											
2'-6" x 6'-8" x 1-3/8" T	Inst	Ea	2C	1.63	57.10	43.90	---	101.00	**132.00**	73.00	**94.50**
2'-8" x 6'-8" x 1-3/8" T	Inst	Ea	2C	1.76	59.10	47.50	---	106.60	**140.00**	76.30	**99.10**
3'-0" x 6'-8" x 1-3/8" T	Inst	Ea	2C	1.76	62.50	47.50	---	110.00	**144.00**	78.90	**102.00**
2'-6" x 7'-0" x 1-3/8" T	Inst	Ea	2C	1.63	71.00	43.90	---	114.90	**148.00**	83.70	**107.00**
2'-8" x 7'-0" x 1-3/8" T	Inst	Ea	2C	1.76	73.70	47.50	---	121.20	**157.00**	87.60	**112.00**
3'-0" x 7'-0" x 1-3/8" T	Inst	Ea	2C	1.76	77.70	47.50	---	125.20	**162.00**	90.70	**116.00**
Solid core prehung door units											
Hardboard											
2'-6" x 6'-8" x 1-3/8" T	Inst	Ea	2C	1.76	72.70	47.50	---	120.20	**156.00**	86.80	**111.00**
2'-8" x 6'-8" x 1-3/8" T	Inst	Ea	2C	1.88	75.40	50.70	---	126.10	**164.00**	91.40	**117.00**
3'-0" x 6'-8" x 1-3/8" T	Inst	Ea	2C	1.88	77.20	50.70	---	127.90	**166.00**	92.80	**119.00**
2'-6" x 7'-0" x 1-3/8" T	Inst	Ea	2C	1.76	80.50	47.50	---	128.00	**165.00**	92.80	**118.00**
2'-8" x 7'-0" x 1-3/8" T	Inst	Ea	2C	1.88	82.90	50.70	---	133.60	**172.00**	97.10	**124.00**
3'-0" x 7'-0" x 1-3/8" T	Inst	Ea	2C	1.88	82.90	50.70	---	133.60	**172.00**	97.10	**124.00**
Lauan											
2'-6" x 6'-8" x 1-3/8" T	Inst	Ea	2C	1.76	83.10	47.50	---	130.60	**168.00**	94.80	**120.00**
2'-8" x 6'-8" x 1-3/8" T	Inst	Ea	2C	1.88	85.80	50.70	---	136.50	**176.00**	99.40	**127.00**
3'-0" x 6'-8" x 1-3/8" T	Inst	Ea	2C	1.88	89.20	50.70	---	139.90	**180.00**	102.00	**130.00**
2'-6" x 7'-0" x 1-3/8" T	Inst	Ea	2C	1.76	91.10	47.50	---	138.60	**177.00**	101.00	**128.00**
2'-8" x 7'-0" x 1-3/8" T	Inst	Ea	2C	1.88	93.20	50.70	---	143.90	**184.00**	105.10	**133.00**
3'-0" x 7'-0" x 1-3/8" T	Inst	Ea	2C	1.88	96.70	50.70	---	147.40	**188.00**	107.90	**136.00**

Description	Oper	Unit	Crew Size	Man-Hours Per Unit	Avg Mat'l Unit Cost	Avg Labor Unit Cost	Avg Equip Unit Cost	Avg Total Unit Cost	Avg Price Incl O&P	Large Volume Avg Total Unit Cost	Large Volume Avg Price Incl O&P
Birch											
2'-6" x 6'-8" x 1-3/8" T	Inst	Ea	2C	1.76	90.80	47.50	---	138.30	**177.00**	100.80	**127.00**
2'-8" x 6'-8" x 1-3/8" T	Inst	Ea	2C	1.88	92.60	50.70	---	143.30	**184.00**	104.70	**133.00**
3'-0" x 6'-8" x 1-3/8" T	Inst	Ea	2C	1.88	96.40	50.70	---	147.10	**188.00**	107.60	**136.00**
2'-6" x 7'-0" x 1-3/8" T	Inst	Ea	2C	1.76	98.70	47.50	---	146.20	**186.00**	106.90	**134.00**
2'-8" x 7'-0" x 1-3/8" T	Inst	Ea	2C	1.88	101.00	50.70	---	151.70	**193.00**	110.90	**140.00**
3'-0" x 7'-0" x 1-3/8" T	Inst	Ea	2C	1.88	105.00	50.70	---	155.70	**197.00**	114.00	**143.00**
Prehung package door units											
Hollow core, flush face door units											
Hardboard											
2'-6" x 6'-8" x 1-3/8" T	Inst	Ea	2C	2.05	111.00	55.30	---	166.30	**212.00**	121.90	**153.00**
2'-8" x 6'-8" x 1-3/8" T	Inst	Ea	2C	2.05	114.00	55.30	---	169.30	**215.00**	123.70	**155.00**
3'-0" x 6'-8" x 1-3/8" T	Inst	Ea	2C	2.05	116.00	55.30	---	171.30	**217.00**	125.30	**157.00**
Lauan											
2'-6" x 6'-8" x 1-3/8" T	Inst	Ea	2C	2.05	123.00	55.30	---	178.30	**226.00**	130.90	**164.00**
2'-8" x 6'-8" x 1-3/8" T	Inst	Ea	2C	2.05	125.00	55.30	---	180.30	**228.00**	132.30	**165.00**
3'-0" x 6'-8" x 1-3/8" T	Inst	Ea	2C	2.05	128.00	55.30	---	183.30	**232.00**	135.00	**168.00**
Birch											
2'-6" x 6'-8" x 1-3/8" T	Inst	Ea	2C	2.05	133.00	55.30	---	188.30	**237.00**	138.90	**172.00**
2'-8" x 6'-8" x 1-3/8" T	Inst	Ea	2C	2.05	135.00	55.30	---	190.30	**239.00**	139.90	**174.00**
3'-0" x 6'-8" x 1-3/8" T	Inst	Ea	2C	2.05	138.00	55.30	---	193.30	**243.00**	142.90	**177.00**
Ash											
2'-6" x 6'-8" x 1-3/8" T	Inst	Ea	2C	2.05	150.00	55.30	---	205.30	**256.00**	150.90	**187.00**
2'-8" x 6'-8" x 1-3/8" T	Inst	Ea	2C	2.05	157.00	55.30	---	212.30	**264.00**	156.90	**194.00**
3'-0" x 6'-8" x 1-3/8" T	Inst	Ea	2C	2.05	161.00	55.30	---	216.30	**269.00**	159.90	**197.00**
Solid core, flush face door units											
Hardboard											
2'-6" x 6'-8" x 1-3/8" T	Inst	Ea	2C	2.22	150.00	59.90	---	209.90	**264.00**	155.10	**193.00**
2'-8" x 6'-8" x 1-3/8" T	Inst	Ea	2C	2.22	153.00	59.90	---	212.90	**267.00**	157.10	**195.00**
3'-0" x 6'-8" x 1-3/8" T	Inst	Ea	2C	2.22	155.00	59.90	---	214.90	**269.00**	158.10	**197.00**
Lauan											
2'-6" x 6'-8" x 1-3/8" T	Inst	Ea	2C	2.22	160.00	59.90	---	219.90	**276.00**	163.10	**202.00**
2'-8" x 6'-8" x 1-3/8" T	Inst	Ea	2C	2.22	163.00	59.90	---	222.90	**279.00**	165.10	**204.00**
3'-0" x 6'-8" x 1-3/8" T	Inst	Ea	2C	2.22	167.00	59.90	---	226.90	**283.00**	168.10	**207.00**
Birch											
2'-6" x 6'-8" x 1-3/8" T	Inst	Ea	2C	2.22	168.00	59.90	---	227.90	**284.00**	169.10	**209.00**
2'-8" x 6'-8" x 1-3/8" T	Inst	Ea	2C	2.22	170.00	59.90	---	229.90	**287.00**	170.10	**210.00**
3'-0" x 6'-8" x 1-3/8" T	Inst	Ea	2C	2.22	174.00	59.90	---	233.90	**291.00**	173.10	**214.00**
Ash											
2'-6" x 6'-8" x 1-3/8" T	Inst	Ea	2C	2.22	182.00	59.90	---	241.90	**300.00**	180.10	**221.00**
2'-8" x 6'-8" x 1-3/8" T	Inst	Ea	2C	2.22	186.00	59.90	---	245.90	**305.00**	182.10	**224.00**
3'-0" x 6'-8" x 1-3/8" T	Inst	Ea	2C	2.22	192.00	59.90	---	251.90	**312.00**	187.10	**230.00**
Adjustments											
For mitered casing, both sides, supplied but not applied											
9/16" x 1-1/2" finger joint, ADD	Inst	Ea	---	---	23.00	---	---	23.00	**26.50**	25.00	**28.80**
9/16" x 2-1/4" solid pine, ADD	Inst	Ea	---	---	32.20	---	---	32.20	**37.00**	24.90	**28.60**
9/16" x 2-1/2" solid pine, ADD	Inst	Ea	---	---	30.40	---	---	30.40	**34.90**	23.40	**26.90**
For solid fir jamb, 4-9/16", ADD	Inst	Ea	---	---	29.60	---	---	29.60	**34.10**	22.90	**26.30**

Description	Oper	Unit	Crew Size	Man-Hours Per Unit	Avg Mat'l Unit Cost	Avg Labor Unit Cost	Avg Equip Unit Cost	Avg Total Unit Cost	Avg Price Incl O&P	Large Volume Avg Total Unit Cost	Large Volume Avg Price Incl O&P
Screen doors											
Aluminum frame with fiberglass wire screen, and plain grille, includes hardware, hinges, closer and latch											
3'-0" x 7'-0"	Inst	Ea	2C	1.54	55.20	41.50	---	96.70	**127.00**	69.60	**90.00**
Wood screen doors, pine, aluminum wire screen, 1-1/8" T											
2'-6" x 6'-9"	Inst	Ea	2C	1.54	92.00	41.50	---	133.50	**169.00**	98.00	**123.00**
2'-8" x 6'-9"	Inst	Ea	2C	1.54	92.00	41.50	---	133.50	**169.00**	98.00	**123.00**
3'-0" x 6'-9"	Inst	Ea	2C	1.54	101.00	41.50	---	142.50	**179.00**	105.10	**131.00**
Half screen											
2'-6" x 6'-9"	Inst	Ea	2C	1.54	115.00	41.50	---	156.50	**195.00**	115.80	**143.00**
2'-8" x 6'-9"	Inst	Ea	2C	1.54	115.00	41.50	---	156.50	**195.00**	115.80	**143.00**
3'-0" x 6'-9"	Inst	Ea	2C	1.54	138.00	41.50	---	179.50	**222.00**	134.00	**163.00**
Steel doors											
24 gauge steel faces, fully primed, reversible, prices are for a standard 2-3/8" diameter bore											
Decorative doors											
Dual glazed, tempered clear											
2'-6", 2'-8", 3'-0" x 6'-8" x 1-3/4" T	Inst	Ea	2C	2.22	218.00	59.90	---	277.90	**342.00**	207.10	**253.00**
Dual glazed, tempered clear, two bevel											
2'-6", 2'-8", 3'-0" x 6'-8" x 1-3/4" T	Inst	Ea	2C	2.22	256.00	59.90	---	315.90	**385.00**	236.10	**286.00**
Embossed raised panel doors											
6 raised panels											
2'-6", 2'-8", 3'-0" x 6'-8" x 1-3/4" T	Inst	Ea	2C	2.22	123.00	59.90	---	182.90	**232.00**	133.90	**168.00**
2'-8" x 8'-0" x 1-3/4" T	Inst	Ea	2C	2.46	207.00	66.30	---	273.30	**339.00**	203.10	**249.00**
3'-0" x 8'-0" x 1-3/4" T	Inst	Ea	2C	2.46	226.00	66.30	---	292.30	**361.00**	218.10	**266.00**
6 raised panels, dual glazed, tempered clear											
2'-6", 2'-8", 3'-0" x 6'-8" x 1-3/4" T	Inst	Ea	2C	2.22	191.00	59.90	---	250.90	**311.00**	187.10	**229.00**
8 raised panels											
2'-8", 3'-0" x 6'-8" x 1-3/4" T	Inst	Ea	2C	2.22	125.00	59.90	---	184.90	**235.00**	135.70	**170.00**
Embossed raised panels and dual glazed tempered clear glass											
2 raised panels with 1 lite											
2'-6", 2'-8", 3'-0" x 6'-8" x 1-3/4" T	Inst	Ea	2C	2.22	207.00	59.90	---	266.90	**329.00**	198.10	**243.00**
2 raised panels with 9 lites											
2'-6", 2'-8", 3'-0" x 6'-8" x 1-3/4" T	Inst	Ea	2C	2.22	210.00	59.90	---	269.90	**333.00**	201.10	**246.00**
2 raised panels with 12 diamond lites											
2'-6", 2'-8", 3'-0" x 6'-8" x 1-3/4" T	Inst	Ea	2C	2.22	218.00	59.90	---	277.90	**342.00**	207.10	**253.00**
4 diamond raised panels with 1 lite											
2'-6", 2'-8", 3'-0" x 6'-8" x 1-3/4" T	Inst	Ea	2C	2.22	207.00	59.90	---	266.90	**329.00**	198.10	**243.00**
4 diamond raised panels with 9 lites											
2'-6", 2'-8", 3'-0" x 6'-8" x 1-3/4" T	Inst	Ea	2C	2.22	210.00	59.90	---	269.90	**333.00**	201.10	**246.00**
4 diamond raised panels with 12 diamond lites											
2'-6", 2'-8", 3'-0" x 6'-8" x 1-3/4" T	Inst	Ea	2C	2.22	218.00	59.90	---	277.90	**342.00**	207.10	**253.00**

Description	Oper	Unit	Crew Size	Man-Hours Per Unit	Avg Mat'l Unit Cost	Avg Labor Unit Cost	Avg Equip Unit Cost	Avg Total Unit Cost	Avg Price Incl O&P	Large Volume Avg Total Unit Cost	Large Volume Avg Price Incl O&P
Flush doors											
2'-0" to 3'-0" x 6'-8" x 1-3/4" T	Inst	Ea	2C	2.22	118.00	59.90	---	177.90	**226.00**	130.00	**164.00**
2'-0" to 3'-0" x 8'-0" x 1-3/4" T	Inst	Ea	2C	2.05	213.00	55.30	---	268.30	**329.00**	200.90	**244.00**
Dual glazed, tempered clear											
2'-6", 2'-8", 3'-0" x 6'-8" x 1-3/4" T	Inst	Ea	2C	2.22	180.00	59.90	---	239.90	**298.00**	178.10	**219.00**
1-1/2 hour fire label											
2'-0" to 3'-0" x 6'-8" x 1-3/4" T	Inst	Ea	2C	2.22	129.00	59.90	---	188.90	**239.00**	138.50	**174.00**
2'-0" to 3'-0" x 7'-0" x 1-3/4" T	Inst	Ea	2C	2.05	156.00	55.30	---	211.30	**264.00**	156.90	**193.00**
French doors, flush face											
1 lite											
2'-6", 2'-8", 3'-0" x 6'-8" x 1-3/4" T	Inst	Ea	2C	2.22	235.00	59.90	---	294.90	**361.00**	220.10	**268.00**
2'-6", 2'-8", 3'-0" x 8,-0" x 1-3/4" T	Inst	Ea	2C	2.46	386.00	66.30	---	452.30	**545.00**	341.10	**409.00**
10 lite											
2'-6", 2'-8", 3'-0" x 6'-8" x 1-3/4" T	Inst	Ea	2C	2.22	248.00	59.90	---	307.90	**377.00**	231.10	**280.00**
2'-6", 2'-8", 3'-0" x 8,-0" x 1-3/4" T	Inst	Ea	2C	2.46	409.00	66.30	---	475.30	**572.00**	359.10	**429.00**
15 lite											
2'-6", 2'-8", 3'-0" x 6'-8" x 1-3/4" T	Inst	Ea	2C	2.22	253.00	59.90	---	312.90	**382.00**	234.10	**284.00**
2'-6", 2'-8", 3'-0" x 8,-0" x 1-3/4" T	Inst	Ea	2C	2.46	419.00	66.30	---	485.30	**582.00**	366.10	**437.00**
French sidelites											
Embossed raised panel											
3 lite											
1'-0", 1'-2" x 6'-8"	Inst	Ea	2C	1.54	129.00	41.50	---	170.50	**211.00**	126.40	**155.00**
Flush face											
1'-0", 1'-2" x 6'-8"	Inst	Ea	2C	1.54	63.50	41.50	---	105.00	**136.00**	76.00	**97.30**
Dual glazed, tempered clear											
1'-0", 1'-2" x 6'-8"	Inst	Ea	2C	1.54	133.00	41.50	---	174.50	**217.00**	130.00	**159.00**
Prehung package door units											
Flush steel door											
2'-6" x 6'-8" x 1-3/8" T	Inst	Ea	2C	2.05	219.00	55.30	---	274.30	**336.00**	204.90	**249.00**

Drywall (Sheetrock or wallboard)

The types of drywall are Standard, Fire Resistant, Water Resistant, and Fire and Water Resistant.

1. **Dimensions.** $^1/_4$", $^3/_8$" x 6' to 12'; $^1/_2$", $^5/_8$" x 6' to 14'.
2. **Installation**
   a. One ply. Sheets are nailed to studs and joists using $1^1/_4$" or $1^3/_8$" x .101 type 500 nails. Nails are usually spaced 8" oc on walls and 7" oc on ceilings. The joints between sheets are filled, taped and sanded. Tape and compound are available in kits containing 250 LF of tape and 18 lbs of compound or 75 LF of tape and 5 lbs of compound.
   b. Two plies. Initial ply is nailed to studs or joists. Second ply is laminated to first with taping compounds as the adhesive. Only the joints in the second ply are filled, taped and finished.
3. **Estimating Technique.** Determine area, deduct openings and add approximately 10% for waste.

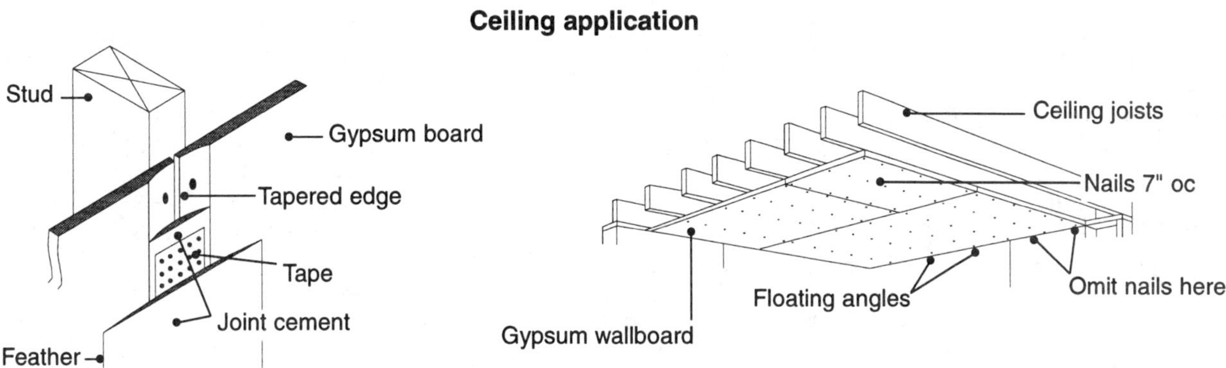

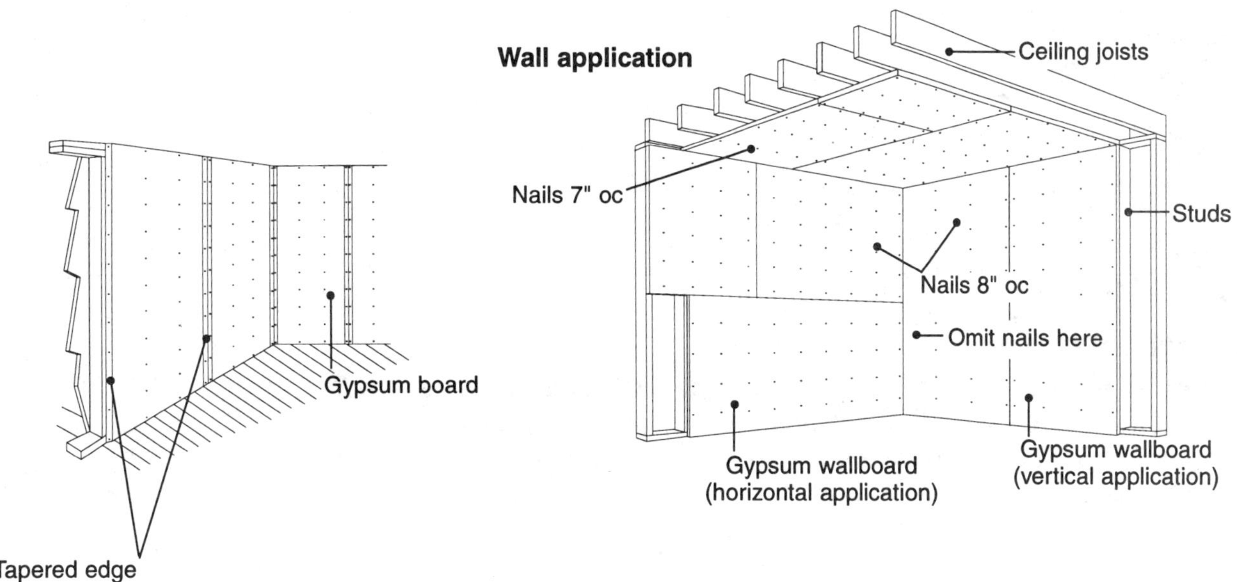

Description	Oper	Unit	Crew Size	Man-Hours Per Unit	Avg Mat'l Unit Cost	Avg Labor Unit Cost	Avg Equip Unit Cost	Avg Total Unit Cost	Avg Price Incl O&P	Large Volume Avg Total Unit Cost	Large Volume Avg Price Incl O&P
Drywall											
Drywall or Sheetrock, walls and ceilings											
	Demo	SF	1L	.023	---	.50	---	.50	**.76**	.22	**.33**
Drywall or Sheetrock (gypsum plasterboard)											
Includes tape, finish and nails, 5% waste included; applied to wood studs or joists											
Walls only, 4' x 6', 8', 9', 10', 12'											
3/8" T, standard	Inst	SF	CA	.024	.23	.65	---	.88	**1.24**	.77	**1.10**
1/2" T, standard	Inst	SF	CA	.024	.23	.65	---	.88	**1.24**	.78	**1.11**
5/8" T, standard	Inst	SF	CA	.025	.28	.67	---	.95	**1.33**	.84	**1.18**
Laminated 3/8" T sheets, std.	Inst	SF	CA	.033	.43	.89	---	1.32	**1.83**	1.12	**1.56**
Ceilings only, 4' x 6', 8', 9', 10', 12'											
3/8" T, standard	Inst	SF	CA	.029	.23	.78	---	1.01	**1.44**	.88	**1.26**
1/2" T, standard	Inst	SF	CA	.029	.23	.78	---	1.01	**1.44**	.89	**1.27**
5/8" T, standard	Inst	SF	CA	.031	.28	.84	---	1.12	**1.58**	.97	**1.39**
Laminated 3/8" T sheets, std.	Inst	SF	CA	.040	.43	1.08	---	1.51	**2.11**	1.31	**1.85**
Average for ceilings and walls											
3/8" T, standard	Inst	SF	CA	.025	.23	.67	---	.90	**1.28**	.80	**1.14**
1/2" T, standard	Inst	SF	CA	.025	.23	.67	---	.90	**1.28**	.81	**1.15**
5/8" T, standard	Inst	SF	CA	.026	.28	.70	---	.98	**1.37**	.87	**1.23**
Laminated 3/8" T sheets, std.	Inst	SF	CA	.034	.43	.92	---	1.35	**1.87**	1.15	**1.60**
Tape and bed joints on repaired sheetrock											
Walls	Inst	SF	CA	.024	.06	.65	---	.71	**1.05**	.64	**.96**
Ceilings	Inst	SF	CA	.028	.06	.75	---	.81	**1.21**	.72	**1.08**
Thin coat plaster	Inst	SF	CA	.016	.12	.43	---	.55	**.79**	.49	**.69**
Material and Labor adjustments											
No tape and finish, DEDUCT	Inst	SF	CA	-.013	-.06	-.35	---	-.41	**---**	-.35	**---**
For ceilings 9' H											
Ceiling only, ADD	Inst	SF	CA	.009	---	.24	---	.24	**.37**	.22	**.33**
Walls & ceilings (average), ADD	Inst	SF	CA	.004	---	.11	---	.11	**.16**	.08	**.12**
Material adjustments											
For fire resistant (1/2" and 5/8" T) ADD	Inst	SF	CA	---	.03	---	---	.03	**.03**	.03	**.03**
For water resistant (1/2" and 5/8" T) ADD	Inst	SF	CA	---	.09	---	---	.09	**.10**	.09	**.10**
For fire and water resistant (5/8" T only) ADD	Inst	SF	CA	---	.10	---	---	.10	**.11**	.10	**.11**
Electrical											
General work											
Residential service, single phase system. Prices given on a cost per each basis for a unit price system which includes a weathercap, service entrance cable, meter socket, entrance disconnect switch, ground rod with clamp, ground cable, EMT, and panelboard.											
Weathercap											
100 AMP service	Inst	Ea	EA	1.43	6.18	41.20	---	47.38	**68.10**	34.31	**49.00**
200 AMP service	Inst	Ea	EA	2.29	18.20	66.00	---	84.20	**119.00**	62.30	**87.00**

Description	Oper	Unit	Crew Size	Man-Hours Per Unit	Avg Mat'l Unit Cost	Avg Labor Unit Cost	Avg Equip Unit Cost	Avg Total Unit Cost	Avg Price Incl O&P	Large Volume Avg Total Unit Cost	Large Volume Avg Price Incl O&P
Service entrance cable (typical allowance is 20 LF)											
100 AMP service	Inst	LF	EA	.176	.14	5.08	---	5.22	**7.67**	3.68	**5.40**
200 AMP service	Inst	LF	EA	.254	.31	7.33	---	7.64	**11.20**	5.41	**7.92**
Meter socket											
100 AMP service	Inst	Ea	EA	5.71	34.10	165.00	---	199.10	**283.00**	145.50	**206.00**
200 AMP service	Inst	Ea	EA	8.89	53.30	256.00	---	309.30	**441.00**	232.60	**328.00**
Entrance disconnect switch											
100 AMP service	Inst	Ea	EA	8.89	169.00	256.00	---	425.00	**574.00**	336.00	**447.00**
200 AMP service	Inst	Ea	EA	13.3	338.00	384.00	---	722.00	**956.00**	573.00	**748.00**
Ground rod, with clamp											
100 AMP service	Inst	Ea	EA	3.81	18.20	110.00	---	128.20	**184.00**	93.20	**133.00**
200 AMP service	Inst	Ea	EA	3.81	35.10	110.00	---	145.10	**203.00**	108.30	**150.00**
Ground cable (typical allowance is 10 LF)											
100 AMP service	Inst	LF	EA	.114	.17	3.29	---	3.46	**5.06**	2.46	**3.59**
200 AMP service	Inst	LF	EA	.143	.27	4.12	---	4.39	**6.41**	3.12	**4.54**
3/4" EMT (typical allowance is 10 LF)											
200 AMP service	Inst	LF	EA	.152	.68	4.38	---	5.06	**7.27**	3.69	**5.26**
Panelboard											
100 AMP service - 12 circuit	Inst	Ea	EA	16.0	150.00	461.00	---	611.00	**855.00**	442.00	**610.00**
200 AMP service - 24 circuit	Inst	Ea	EA	26.7	358.00	770.00	---	1128.00	**1550.00**	896.00	**1220.00**
Adjustments for other than normal working situations											
Cut and patch, ADD	Inst	%	EA	---	50.0	---	---	---	**---**	---	**---**
Dust protection, ADD	Inst	%	EA	---	40.0	---	---	---	**---**	---	**---**
Protect existing work, ADD	Inst	%	EA	---	50.0	---	---	---	**---**	---	**---**

Wiring per outlet or switch; wall or ceiling

Description	Oper	Unit	Crew Size	Man-Hours Per Unit	Avg Mat'l Unit Cost	Avg Labor Unit Cost	Avg Equip Unit Cost	Avg Total Unit Cost	Avg Price Incl O&P	Large Volume Avg Total Unit Cost	Large Volume Avg Price Incl O&P
Romex, non-metallic sheathed cable, 600 volt, copper with ground wire	Inst	Ea	EA	.714	9.10	20.60	---	29.70	**40.90**	22.52	**30.70**
BX, flexible armored cable, 600 volt, copper	Inst	Ea	EA	.952	15.60	27.50	---	43.10	**58.60**	33.10	**44.50**
EMT with wire, electric metallic thinwall, 1/2"	Inst	Ea	EA	1.90	19.50	54.80	---	74.30	**104.00**	55.80	**76.80**
Rigid with wire, 1/2"	Inst	Ea	EA	2.86	26.00	82.50	---	108.50	**152.00**	80.90	**112.00**
Wiring, connection, and installation in closed wall structure ADD	Inst	Ea	EA	1.43	---	41.20	---	41.20	**61.00**	28.80	**42.70**

Lighting. See Lighting fixtures, page 156

Special systems

Burglary detection systems

Description	Oper	Unit	Crew Size	Man-Hours Per Unit	Avg Mat'l Unit Cost	Avg Labor Unit Cost	Avg Equip Unit Cost	Avg Total Unit Cost	Avg Price Incl O&P	Large Volume Avg Total Unit Cost	Large Volume Avg Price Incl O&P
Alarm bell	Inst	Ea	EA	2.86	78.00	82.50	---	160.50	**212.00**	127.30	**165.00**
Burglar alarm, battery operated											
Mechanical trigger	Inst	Ea	EA	2.86	280.00	82.50	---	362.50	**443.00**	306.70	**372.00**
Electrical trigger	Inst	Ea	EA	2.86	338.00	82.50	---	420.50	**511.00**	359.70	**432.00**

Description	Oper	Unit	Crew Size	Man-Hours Per Unit	Avg Mat'l Unit Cost	Avg Labor Unit Cost	Avg Equip Unit Cost	Avg Total Unit Cost	Avg Price Incl O&P	Large Volume Avg Total Unit Cost	Large Volume Avg Price Incl O&P
Adjustments, ADD											
Outside key control	Inst	Ea	---	---	78.00	---	---	78.00	**89.70**	69.60	**80.00**
Remote signaling circuitry	Inst	Ea	---	---	126.00	---	---	126.00	**145.00**	113.00	**129.00**
Card reader											
Standard	Inst	Ea	EA	4.44	943.00	128.00	---	1071.00	**1270.00**	933.30	**1100.00**
Multi-code	Inst	Ea	EA	4.44	1220.00	128.00	---	1348.00	**1590.00**	1172.30	**1380.00**
Detectors											
Motion											
Infrared photoelectric	Inst	Ea	EA	5.71	195.00	165.00	---	360.00	**468.00**	289.00	**371.00**
Passive infrared	Inst	Ea	EA	5.71	260.00	165.00	---	425.00	**543.00**	347.00	**438.00**
Ultrasonic, 12 volt	Inst	Ea	EA	5.71	234.00	165.00	---	399.00	**513.00**	324.00	**411.00**
Microwave											
10' to 200'	Inst	Ea	EA	5.71	676.00	165.00	---	841.00	**1020.00**	718.00	**864.00**
10' to 350'	Inst	Ea	EA	5.71	1950.00	165.00	---	2115.00	**2490.00**	1855.00	**2170.00**
Door switches											
Hinge switch	Inst	Ea	EA	2.29	59.80	66.00	---	125.80	**167.00**	99.50	**130.00**
Magnetic switch	Inst	Ea	EA	2.29	71.50	66.00	---	137.50	**180.00**	109.90	**142.00**
Exit control locks											
Horn alarm	Inst	Ea	EA	2.86	351.00	82.50	---	433.50	**526.00**	370.70	**446.00**
Flashing light alarm	Inst	Ea	EA	2.86	397.00	82.50	---	479.50	**578.00**	411.70	**492.00**
Glass break alarm switch	Inst	Ea	EA	1.43	48.10	41.20	---	89.30	**116.00**	71.70	**92.00**
Indicating panels											
1 channel	Inst	Ea	EA	4.44	371.00	128.00	---	499.00	**616.00**	423.30	**517.00**
10 channel	Inst	Ea	EA	7.27	1280.00	210.00	---	1490.00	**1780.00**	1294.00	**1540.00**
20 channel	Inst	Ea	EA	11.4	2470.00	329.00	---	2799.00	**3330.00**	2431.00	**2880.00**
40 channel	Inst	Ea	EA	20.0	4550.00	577.00	---	5127.00	**6090.00**	4521.00	**5350.00**
Police connect panel	Inst	Ea	EA	2.86	247.00	82.50	---	329.50	**406.00**	277.70	**339.00**
Siren	Inst	Ea	EA	2.86	148.00	82.50	---	230.50	**292.00**	189.70	**237.00**
Switchmats											
30" x 5'	Inst	Ea	EA	2.29	85.80	66.00	---	151.80	**196.00**	122.70	**156.00**
30" x 25'	Inst	Ea	EA	2.86	205.00	82.50	---	287.50	**358.00**	240.70	**296.00**
Telephone dialer	Inst	Ea	EA	2.29	390.00	66.00	---	456.00	**546.00**	394.10	**468.00**
Doorbell systems											
Includes transformer, button, bell											
Door chimes, 2 notes	Inst	Ea	EA	1.14	78.00	32.90	---	110.90	**138.00**	92.70	**114.00**
Tube type chimes	Inst	Ea	EA	1.43	195.00	41.20	---	236.20	**285.00**	202.80	**243.00**
Transformer and button only	Inst	Ea	EA	.952	45.50	27.50	---	73.00	**93.00**	59.80	**75.20**
Push button only	Inst	Ea	EA	.714	19.50	20.60	---	40.10	**52.90**	31.80	**41.40**
Fire alarm systems											
Battery and rack	Inst	Ea	EA	3.81	780.00	110.00	---	890.00	**1060.00**	773.00	**914.00**
Automatic charger	Inst	Ea	EA	1.90	501.00	54.80	---	555.80	**657.00**	485.40	**570.00**
Detector											
Fixed temperature	Inst	Ea	EA	1.90	31.20	54.80	---	86.00	**117.00**	66.20	**88.80**
Rate of rise	Inst	Ea	EA	1.90	39.00	54.80	---	93.80	**126.00**	73.20	**96.80**

Description	Oper	Unit	Crew Size	Man-Hours Per Unit	Avg Mat'l Unit Cost	Avg Labor Unit Cost	Avg Equip Unit Cost	Avg Total Unit Cost	Avg Price Incl O&P	Large Volume Avg Total Unit Cost	Large Volume Avg Price Incl O&P
			Costs Based On Small Volume							Large Volume	
Door holder											
Electro-magnetic	Inst	Ea	EA	2.86	87.10	82.50	---	169.60	**222.00**	135.40	**175.00**
Combination holder/closer	Inst	Ea	EA	3.81	488.00	110.00	---	598.00	**723.00**	512.00	**614.00**
Fire drill switch	Inst	Ea	EA	1.90	97.50	54.80	---	152.30	**193.00**	125.40	**157.00**
Glass break alarm switch	Inst	Ea	EA	1.43	55.90	41.20	---	97.10	**125.00**	78.70	**100.00**
Signal bell	Inst	Ea	EA	1.90	55.90	54.80	---	110.70	**145.00**	88.30	**114.00**
Smoke detector											
Ceiling type	Inst	Ea	EA	1.90	72.80	54.80	---	127.60	**165.00**	103.40	**131.00**
Duct type	Inst	Ea	EA	3.81	286.00	110.00	---	396.00	**492.00**	332.00	**407.00**
Intercom systems											
Master stations, with digital AM/FM receiver, up to 20 remote stations, & telephone interface											
Solid state with antenna, with 200' wire, 4 speakers	Inst	Set	EA	5.71	1560.00	165.00	---	1725.00	**2040.00**	1505.00	**1770.00**
Solid state with antenna, with 1000' wire, 8 speakers	Inst	Set	EA	11.4	2080.00	329.00	---	2409.00	**2880.00**	2091.00	**2480.00**
Room to door intercoms											
Master station, door station, with transformer, wire	Inst	Set	EA	2.86	390.00	82.50	---	472.50	**571.00**	405.70	**486.00**
Second door station, ADD	Inst	Ea	EA	1.43	52.00	41.20	---	93.20	**121.00**	75.20	**96.00**
Installation, wiring, and transformer, wire ADD	Inst	LS	EA	2.29	---	66.00	---	66.00	**97.70**	46.10	**68.30**
Telephone, phone-jack wiring											
Pre-wiring, per outlet or jack	Inst	Ea	EA	.952	26.00	27.50	---	53.50	**70.50**	42.40	**55.20**
Wiring, connection, and installation in closed wall structure ADD	Inst	Ea	EA	1.43	---	41.20	---	41.20	**61.00**	28.80	**42.70**
Television antenna											
Television antenna outlet, with 300 OHM	Inst	Ea	EA	.952	15.60	27.50	---	43.10	**58.60**	33.10	**44.50**
Wiring, connection, and installation in closed wall structure ADD	Inst	Ea	EA	1.43	---	41.20	---	41.20	**61.00**	28.80	**42.70**
Thermostat wiring											
For heating units located on the first floor											
Thermostat on first floor	Inst	Ea	EA	1.43	26.00	41.20	---	67.20	**90.90**	52.00	**69.40**
Wiring, connection, and installation in closed wall structure ADD	Inst	Ea	EA	1.43	---	41.20	---	41.20	**61.00**	28.80	**42.70**
Thermostat on second floor	Inst	Ea	EA	2.29	32.50	66.00	---	98.50	**135.00**	75.10	**102.00**
Wiring, connection, and installation in closed wall structure ADD	Inst	Ea	EA	1.90	---	54.80	---	54.80	**81.10**	38.40	**56.80**

Description	Oper	Unit	Crew Size	Man-Hours Per Unit	Avg Mat'l Unit Cost	Avg Labor Unit Cost	Avg Equip Unit Cost	Avg Total Unit Cost	Avg Price Incl O&P	Large Volume Avg Total Unit Cost	Large Volume Avg Price Incl O&P
Entrances											
Single and double door entrances	Demo	Ea	LB	.889	---	19.50	---	19.50	**29.50**	14.10	**21.20**
Colonial design											
White pine includes frames, pediments, and pilasters											
Plain carved archway											
Single door units											
3'-0" W x 6'-8" H	Inst	Ea	2C	3.08	360.00	83.00	---	443.00	**487.00**	373.90	**402.00**
Double door units											
Two - 2'-6" W x 6'-8" H	Inst	Ea	2C	4.10	583.00	111.00	---	694.00	**751.00**	590.00	**627.00**
Two - 2'-8" W x 6'-8" H	Inst	Ea	2C	4.10	623.00	111.00	---	734.00	**791.00**	625.00	**663.00**
Two - 3'-0" W x 6'-8" H	Inst	Ea	2C	4.10	663.00	111.00	---	774.00	**831.00**	661.00	**698.00**
Decorative carved archway											
Single door units											
3'-0" W x 6'-8" H	Inst	Ea	2C	3.08	315.00	83.00	---	398.00	**441.00**	333.90	**362.00**
Double door units											
Two - 2'-6" W x 6'-8" H	Inst	Ea	2C	4.10	469.00	111.00	---	580.00	**637.00**	489.00	**526.00**
Two - 2'-8" W x 6'-8" H	Inst	Ea	2C	4.10	509.00	111.00	---	620.00	**677.00**	524.00	**562.00**
Two - 3 -0" W x 6'-8" H	Inst	Ea	2C	4.10	549.00	111.00	---	660.00	**717.00**	560.00	**597.00**
Excavation											
Digging out or trenching	Demo	CY	LB	.056	---	1.23	---	1.23	**1.86**	.75	**1.13**
Pits, medium earth, piled											
With front end loader, track mounted, 1-1/2 CY capacity; 55 CY per hour	Demo	CY	VB	.061	---	1.62	.88	2.50	**3.31**	1.48	**1.96**
By hand											
To 4'-0" D	Demo	CY	LB	1.78	---	39.10	---	39.10	**59.00**	23.50	**35.50**
4'-0" to 6'-0" D	Demo	CY	LB	2.67	---	58.60	---	58.60	**88.50**	35.10	**53.00**
6'-0" to 8'-0" D	Demo	CY	LB	4.00	---	87.80	---	87.80	**133.00**	58.60	**88.50**
Continuous footing or trench, medium earth, piled											
With tractor backhoe, 3/8 CY capacity (48 HP); 15 CY per hour	Demo	CY	VB	.222	---	5.89	3.21	9.10	**12.00**	5.46	**7.22**
By hand, to 4'-0" D	Demo	CY	LB	1.78	---	39.10	---	39.10	**59.00**	23.50	**35.50**
Backfilling, by hand, medium soil											
Without compaction	Demo	CY	LB	.941	---	20.70	---	20.70	**31.20**	12.50	**18.90**
With hand compaction											
6" layers	Demo	CY	LB	1.60	---	35.10	---	35.10	**53.00**	20.70	**31.20**
12" layers	Demo	CY	LB	1.23	---	27.00	---	27.00	**40.80**	16.00	**24.10**
With vibrating plate compaction											
6" layers	Demo	CY	AB	1.33	---	31.40	1.86	33.26	**48.90**	20.02	**29.40**
12" layers	Demo	CY	AB	1.14	---	26.90	1.60	28.50	**42.00**	16.63	**24.50**

Facebrick. See Masonry, page 161

Description	Oper	Unit	Crew Size	Man-Hours Per Unit	Avg Mat'l Unit Cost	Avg Labor Unit Cost	Avg Equip Unit Cost	Avg Total Unit Cost	Avg Price Incl O&P	Large Volume Avg Total Unit Cost	Large Volume Avg Price Incl O&P
Fences											
Basketweave											
Redwood, preassembled, 8' L panels, includes 4" x 4" line posts, horizontal or vertical weave											
5' H	Inst	LF	CS	.133	10.20	3.36	---	13.56	**16.90**	11.07	**13.70**
6' H	Inst	LF	CS	.133	11.90	3.36	---	15.26	**18.80**	12.47	**15.30**
Adjustments											
Corner or end posts, 8' H	Inst	Ea	CA	.800	19.80	21.60	---	41.40	**55.60**	33.20	**44.30**
3-1/2' W x 5' H gate, with hardware	Inst	Ea	CA	1.00	86.00	27.00	---	113.00	**140.00**	93.40	**115.00**
Board, per LF complete fence system											
6' H boards nailed to wood frame, on 1 side only; 8' L redwood 4" x 4"(milled) posts, set 2' D in concrete filled holes @ 6' oc; frame members 2" x 4" (milled) as 2 rails between posts per 6' L fence section; costs are per LF of fence											
Douglas fir frame members with redwood posts											
Milled boards, "dog-eared" one end											
Cedar											
1" x 6" - 6' H	Inst	LF	CS	.533	32.30	13.50	---	45.80	**57.70**	37.10	**46.40**
1" x 8" - 6' H	Inst	LF	CS	.453	35.60	11.50	---	47.10	**58.30**	38.37	**47.40**
1" x 10" - 6' H	Inst	LF	CS	.400	36.90	10.10	---	47.00	**57.80**	38.39	**46.90**
Douglas fir											
1" x 6" - 6' H	Inst	LF	CS	.533	31.00	13.50	---	44.50	**56.10**	36.00	**45.10**
1" x 8" - 6' H	Inst	LF	CS	.453	33.20	11.50	---	44.70	**55.60**	36.37	**45.10**
1" x 10" - 6' H	Inst	LF	CS	.400	34.70	10.10	---	44.80	**55.30**	36.49	**44.80**
Redwood											
1" x 6" - 6' H	Inst	LF	CS	.533	32.30	13.50	---	45.80	**57.70**	37.10	**46.40**
1" x 8" - 6' H	Inst	LF	CS	.453	35.60	11.50	---	47.10	**58.30**	38.37	**47.40**
1" x 10" - 6' H	Inst	LF	CS	.400	40.50	10.10	---	50.60	**61.90**	41.39	**50.40**
Rough boards, both ends squared											
Cedar											
1" x 6" - 6' H	Inst	LF	CS	.490	31.80	12.40	---	44.20	**55.50**	35.93	**44.80**
1" x 8" - 6' H	Inst	LF	CS	.429	34.90	10.90	---	45.80	**56.60**	37.19	**45.80**
1" x 10" - 6' H	Inst	LF	CS	.375	35.80	9.48	---	45.28	**55.60**	37.03	**45.20**
Douglas fir											
1" x 6" - 6' H	Inst	LF	CS	.490	30.60	12.40	---	43.00	**54.00**	34.83	**43.60**
1" x 8" - 6' H	Inst	LF	CS	.429	32.70	10.90	---	43.60	**54.10**	35.39	**43.70**
1" x 10" - 6' H	Inst	LF	CS	.375	34.20	9.48	---	43.68	**53.70**	35.63	**43.60**
Redwood											
1" x 6" - 6' H	Inst	LF	CS	.490	31.80	12.40	---	44.20	**55.50**	35.93	**44.80**
1" x 8" - 6' H	Inst	LF	CS	.429	34.90	10.90	---	45.80	**56.60**	37.19	**45.80**
1" x 10" - 6' H	Inst	LF	CS	.375	39.50	9.48	---	48.98	**59.80**	40.13	**48.80**

Description	Oper	Unit	Crew Size	Man-Hours Per Unit	Avg Mat'l Unit Cost	Avg Labor Unit Cost	Avg Equip Unit Cost	Avg Total Unit Cost	Avg Price Incl O&P	Large Volume Avg Total Unit Cost	Large Volume Avg Price Incl O&P
Redwood frame members with redwood posts											
Milled boards, "dog-eared" one end											
Cedar											
1" x 6" - 6' H	Inst	LF	CS	.533	32.40	13.50	---	45.90	**57.70**	37.10	**46.40**
1" x 8" - 6' H	Inst	LF	CS	.453	35.60	11.50	---	47.10	**58.40**	38.37	**47.40**
1" x 10" - 6' H	Inst	LF	CS	.400	36.90	10.10	---	47.00	**57.80**	38.39	**47.00**
Douglas fir											
1" x 6" - 6' H	Inst	LF	CS	.533	31.00	13.50	---	44.50	**56.20**	36.00	**45.20**
1" x 8" - 6' H	Inst	LF	CS	.453	33.30	11.50	---	44.80	**55.70**	36.47	**45.10**
1" x 10" - 6' H	Inst	LF	CS	.400	34.70	10.10	---	44.80	**55.30**	36.59	**44.90**
Redwood											
1" x 6" - 6' H	Inst	LF	CS	.533	32.40	13.50	---	45.90	**57.70**	37.10	**46.40**
1" x 8" - 6' H	Inst	LF	CS	.453	35.60	11.50	---	47.10	**58.40**	38.37	**47.40**
1" x 10" - 6' H	Inst	LF	CS	.400	40.50	10.10	---	50.60	**62.00**	41.39	**50.40**
Rough boards, both ends squared											
Cedar											
1" x 6" - 6' H	Inst	LF	CS	.490	31.90	12.40	---	44.30	**55.50**	35.93	**44.80**
1" x 8" - 6' H	Inst	LF	CS	.429	34.90	10.90	---	45.80	**56.60**	37.19	**45.80**
1" x 10" - 6' H	Inst	LF	CS	.375	35.80	9.48	---	45.28	**55.60**	37.03	**45.20**
Douglas fir											
1" x 6" - 6' H	Inst	LF	CS	.490	30.60	12.40	---	43.00	**54.00**	34.93	**43.60**
1" x 8" - 6' H	Inst	LF	CS	.429	32.60	10.90	---	43.50	**54.00**	35.29	**43.60**
1" x 10" - 6' H	Inst	LF	CS	.375	34.20	9.48	---	43.68	**53.70**	35.63	**43.70**
Redwood											
1" x 6" - 6' H	Inst	LF	CS	.490	31.90	12.40	---	44.30	**55.50**	35.93	**44.80**
1" x 8" - 6' H	Inst	LF	CS	.429	34.90	10.90	---	45.80	**56.60**	37.19	**45.80**
1" x 10" - 6' H	Inst	LF	CS	.375	39.60	9.48	---	49.08	**59.90**	40.13	**48.80**
Chain link											
9 gauge galvanized steel, includes top rail (1-5/8" o.d.), line posts (2" o.d.) @10' oc and sleeves											
36" H	Inst	LF	HB	.145	5.30	3.83	---	9.13	**11.90**	7.31	**9.45**
42" H	Inst	LF	HB	.152	5.78	4.02	---	9.80	**12.70**	7.84	**10.10**
48" H	Inst	LF	HB	.160	6.27	4.23	---	10.50	**13.60**	8.40	**10.80**
60" H	Inst	LF	HB	.178	7.23	4.70	---	11.93	**15.40**	9.56	**12.30**
72" H	Inst	LF	HB	.200	8.36	5.29	---	13.65	**17.60**	10.94	**14.00**
Adjustments											
11-1/2" gauge galvanized steel fabric DEDUCT	Inst	%		---	-27.0	---	---	---	**---**	---	**---**
12 gauge galvanized steel fabric DEDUCT	Inst	%		---	-29.0	---	---	---	**---**	---	**---**
9 gauge green vinyl-coated fabric DEDUCT	Inst	%		---	-4.0	---	---	---	**---**	---	**---**
11 gauge green vinyl-coated fabric DEDUCT	Inst	%		---	-3.8	---	---	---	**---**	---	**---**

Description	Oper	Unit	Crew Size	Man-Hours Per Unit	Avg Mat'l Unit Cost	Avg Labor Unit Cost	Avg Equip Unit Cost	Avg Total Unit Cost	Avg Price Incl O&P	Large Volume Avg Total Unit Cost	Large Volume Avg Price Incl O&P
Filler strips, ADD											
Aluminum, baked on enamel finish											
Diagonal, 1-7/8" W											
48" H	Inst	LF	LB	.213	3.59	4.68	---	8.27	**11.20**	6.51	**8.75**
60" H	Inst	LF	LB	.213	4.35	4.68	---	9.03	**12.10**	7.14	**9.48**
72" H	Inst	LF	LB	.213	5.09	4.68	---	9.77	**12.90**	7.76	**10.20**
Vertical, 1-1/4" W											
48" H	Inst	LF	LB	.213	3.59	4.68	---	8.27	**11.20**	6.51	**8.75**
60" H	Inst	LF	LB	.213	4.22	4.68	---	8.90	**11.90**	7.03	**9.35**
72" H	Inst	LF	LB	.213	4.97	4.68	---	9.65	**12.80**	7.66	**10.10**
Wood, redwood stain											
Vertical, 1-1/4" W											
48" H	Inst	LF	LB	.213	3.59	4.68	---	8.27	**11.20**	6.51	**8.75**
60" H	Inst	LF	LB	.213	4.35	4.68	---	9.03	**12.10**	7.14	**9.48**
72" H	Inst	LF	LB	.213	5.09	4.68	---	9.77	**12.90**	7.76	**10.20**
Corner posts (2-1/2" o.d.), installed, heavyweight											
36" H	Inst	Ea	HA	.381	20.70	10.90	---	31.60	**40.30**	25.50	**32.30**
42" H	Inst	Ea	HA	.400	23.50	11.50	---	35.00	**44.30**	28.09	**35.30**
48" H	Inst	Ea	HA	.421	24.80	12.10	---	36.90	**46.80**	29.88	**37.70**
60" H	Inst	Ea	HA	.471	29.70	13.50	---	43.20	**54.50**	35.20	**44.30**
72" H	Inst	Ea	HA	.533	34.50	15.30	---	49.80	**62.80**	40.30	**50.40**
End or gate posts (2-1/2" o.d.), installed, heavyweight											
36" H	Inst	Ea	HA	.381	15.90	10.90	---	26.80	**34.80**	21.50	**27.60**
42" H	Inst	Ea	HA	.400	17.30	11.50	---	28.80	**37.20**	22.89	**29.40**
48" H	Inst	Ea	HA	.421	19.30	12.10	---	31.40	**40.50**	25.28	**32.40**
60" H	Inst	Ea	HA	.471	22.80	13.50	---	36.30	**46.60**	29.40	**37.60**
72" H	Inst	Ea	HA	.533	26.20	15.30	---	41.50	**53.20**	33.40	**42.50**
Gates, square corner frame, 9 gauge wire, installed											
3' W walkway gates											
36" H	Inst	Ea	HA	.800	58.00	22.90	---	80.90	**101.00**	66.00	**82.30**
42" H	Inst	Ea	HA	.800	60.70	22.90	---	83.60	**104.00**	68.30	**84.90**
48" H	Inst	Ea	HA	.889	62.10	25.50	---	87.60	**110.00**	70.90	**88.50**
60" H	Inst	Ea	HA	.889	74.50	25.50	---	100.00	**124.00**	81.30	**100.00**
72" H	Inst	Ea	HA	1.00	86.90	28.70	---	115.60	**143.00**	93.50	**115.00**
12' W double driveway gates											
36" H	Inst	Ea	HA	1.60	153.00	45.90	---	198.90	**245.00**	166.10	**205.00**
42" H	Inst	Ea	HA	1.60	161.00	45.90	---	206.90	**255.00**	173.10	**213.00**
48" H	Inst	Ea	HA	2.00	167.00	57.40	---	224.40	**279.00**	184.90	**230.00**
60" H	Inst	Ea	HA	2.00	200.00	57.40	---	257.40	**317.00**	212.90	**261.00**
72" H	Inst	Ea	HA	2.67	221.00	76.60	---	297.60	**370.00**	241.40	**299.00**

Split rail

Red cedar, 10' L sectional spans

Description	Oper	Unit	Crew Size	Man-Hours Per Unit	Avg Mat'l Unit Cost	Avg Labor Unit Cost	Avg Equip Unit Cost	Avg Total Unit Cost	Avg Price Incl O&P	Large Volume Avg Total Unit Cost	Large Volume Avg Price Incl O&P
Rails only	Inst	Ea	---	---	9.38	---	---	9.38	**9.38**	7.83	**7.83**
Bored 2 rail posts											
5'-6" line or end posts	Inst	Ea	CA	.800	11.00	21.60	---	32.60	**45.50**	25.82	**35.80**
5'-6" corner posts	Inst	Ea	CA	.800	12.40	21.60	---	34.00	**47.10**	27.00	**37.10**
Bored 3 rail posts											
6'-6" line or end posts	Inst	Ea	CA	.800	13.80	21.60	---	35.40	**48.70**	28.10	**38.50**
6'-6" corner posts	Inst	Ea	CA	.800	15.20	21.60	---	36.80	**50.20**	29.30	**39.80**

Description	Oper	Unit	Crew Size	Man-Hours Per Unit	Avg Mat'l Unit Cost	Avg Labor Unit Cost	Avg Equip Unit Cost	Avg Total Unit Cost	Avg Price Incl O&P	Large Volume Avg Total Unit Cost	Large Volume Avg Price Incl O&P
Complete fence estimate (does not include gates)											
2 rail, 36" H, 5'-6" post	Inst	Ea	CJ	.056	3.12	1.37	---	4.49	**5.67**	3.63	**4.55**
3 rail, 48" H, 6'-6" post	Inst	Ea	CJ	.065	4.06	1.59	---	5.65	**7.09**	4.56	**5.68**
Gate											
2 rails											
3-1/2' W	Inst	Ea	CA	.800	55.20	21.60	---	76.80	**96.30**	62.70	**78.20**
5' W	Inst	Ea	CA	1.00	71.80	27.00	---	98.80	**124.00**	81.50	**102.00**
3 rails											
3-1/2' W	Inst	Ea	CA	.800	69.00	21.60	---	90.60	**112.00**	74.20	**91.40**
5' W	Inst	Ea	CA	1.00	80.00	27.00	---	107.00	**133.00**	88.40	**110.00**

Fiberglass panels

Description	Oper	Unit	Crew Size	Man-Hours Per Unit	Avg Mat'l Unit Cost	Avg Labor Unit Cost	Avg Equip Unit Cost	Avg Total Unit Cost	Avg Price Incl O&P	Large Volume Avg Total Unit Cost	Large Volume Avg Price Incl O&P
Corrugated; 8', 10', or 12' L panels; 2-1/2" W x 1/2" D corrugation											
Nailed on wood frame											
4 oz., 0.03" T, 26" W	Inst	SF	CA	.053	1.39	1.43	---	2.82	**3.77**	2.35	**3.10**
Fire retardant	Inst	SF	CA	.053	1.91	1.43	---	3.34	**4.37**	2.84	**3.66**
5 oz., 0.037" T, 26" W	Inst	SF	CA	.053	1.71	1.43	---	3.14	**4.14**	2.65	**3.44**
Fire retardant	Inst	SF	CA	.053	2.35	1.43	---	3.78	**4.87**	3.24	**4.12**
6 oz., 0.045" T, 26" W	Inst	SF	CA	.053	2.02	1.43	---	3.45	**4.49**	2.94	**3.78**
Fire retardant	Inst	SF	CA	.053	2.79	1.43	---	4.22	**5.38**	3.65	**4.59**
8 oz., 0.06" T, 26" W	Inst	SF	CA	.053	2.34	1.43	---	3.77	**4.86**	3.23	**4.11**
Fire retardant	Inst	SF	CA	.053	3.23	1.43	---	4.66	**5.89**	4.05	**5.05**
Flat panels; 8', 10', or 12' L panels; clear, green, or white											
5 oz., 48" W	Inst	SF	CA	.053	.92	1.43	---	2.35	**3.23**	1.93	**2.62**
Fire retardant	Inst	SF	CA	.053	.92	1.43	---	2.35	**3.23**	1.93	**2.62**
6 oz., 48" W	Inst	SF	CA	.053	1.82	1.43	---	3.25	**4.26**	2.75	**3.56**
Fire retardant	Inst	SF	CA	.053	3.14	1.43	---	4.57	**5.78**	3.96	**4.95**
8 oz., 48" W	Inst	SF	CA	.053	2.34	1.43	---	3.77	**4.86**	3.23	**4.11**
Fire retardant	Inst	SF	CA	.053	4.05	1.43	---	5.48	**6.83**	4.80	**5.92**
Solar block; 8', 10', 12' L panels; 2-1/2" W x 1/2" D corrugation; nailed on wood frame											
5 oz., 26" W	Inst	SF	CA	.053	1.05	1.43	---	2.48	**3.38**	2.04	**2.74**
Accessories											
Wood corrugated											
2-1/2" W x 1-1/2" D x 6' L	Inst	Ea	---	---	2.75	---	---	2.75	**2.75**	2.53	**2.53**
2-1/2" W x 1-1/2" D x 8' L	Inst	Ea	---	---	3.34	---	---	3.34	**3.34**	3.07	**3.07**
2-1/2" W x 3/4" D x 6' L	Inst	Ea	---	---	1.27	---	---	1.27	**1.27**	1.16	**1.16**
2-1/2" W x 3/4" D x 8' L	Inst	Ea	---	---	1.63	---	---	1.63	**1.63**	1.49	**1.49**
Rubber corrugated											
1" x 3"	Inst	Ea	---	---	1.27	---	---	1.27	**1.27**	1.16	**1.16**
Polyfoam corrugated											
1" x 3"	Inst	Ea	---	---	.89	---	---	.89	**.89**	.82	**.82**
Vertical crown molding											
Wood											
1-1/2" x 6' L	Inst	Ea	---	---	2.30	---	---	2.30	**2.30**	2.11	**2.11**
1-1/2" x 8' L	Inst	Ea	---	---	3.04	---	---	3.04	**3.04**	2.79	**2.79**
Polyfoam, 1" x 1" x 3' L	Inst	Ea	---	---	1.11	---	---	1.11	**1.11**	1.02	**1.02**
Rubber, 1" x 1" x 3' L	Inst	Ea	---	---	2.10	---	---	2.10	**2.10**	1.93	**1.93**

Fireplaces

Woodburning, prefabricated. No masonry support required, installs directly on floor. Ceramic backed firebox with black vitreous enamel side panels. No finish plastering or brick hearthwork included. Fire screen, 9" (i.d.) factory-built insulated chimneys with flue, lining, damper, and flashing with rain cap included. Chimney height from floor to where chimney exits through roof.

Description	Oper	Unit	Crew Size	Man-Hours Per Unit	Avg Mat'l Unit Cost	Avg Labor Unit Cost	Avg Equip Unit Cost	Avg Total Unit Cost	Avg Price Incl O&P	Avg Total Unit Cost	Avg Price Incl O&P
			Costs Based On Small Volume							Large Volume	
36" W fireplace unit with:											
Up to 9'-0" chimney height	Inst	LS	CJ	18.8	865.00	460.00	---	1325.00	**1690.00**	1066.00	**1340.00**
9'-3" to 12'-2" chimney	Inst	LS	CJ	20.5	910.00	501.00	---	1411.00	**1810.00**	1130.00	**1420.00**
12'-3" to 15'-1" chimney	Inst	LS	CJ	22.2	953.00	543.00	---	1496.00	**1920.00**	1198.00	**1510.00**
15'-2" to 18'-0" chimney	Inst	LS	CJ	24.6	999.00	602.00	---	1601.00	**2060.00**	1275.00	**1610.00**
18'-1" to 20'-11" chimney	Inst	LS	CJ	27.1	1040.00	663.00	---	1703.00	**2210.00**	1358.00	**1720.00**
21'-3" to 23'-10" chimney	Inst	LS	CJ	30.8	1090.00	753.00	---	1843.00	**2400.00**	1451.00	**1850.00**
23'-11" to 24'-9" chimney	Inst	LS	CJ	34.8	1130.00	851.00	---	1981.00	**2600.00**	1560.00	**2000.00**
42" W fireplace unit with:											
Up to 9'-0" chimney height	Inst	LS	CJ	18.8	978.00	460.00	---	1438.00	**1820.00**	1166.00	**1450.00**
9'-3" to 12'-2" chimney	Inst	LS	CJ	20.5	1030.00	501.00	---	1531.00	**1950.00**	1236.00	**1540.00**
12'-3" to 15'-1" chimney	Inst	LS	CJ	22.2	1080.00	543.00	---	1623.00	**2070.00**	1311.00	**1640.00**
15'-2" to 18'-0" chimney	Inst	LS	CJ	24.6	1130.00	602.00	---	1732.00	**2220.00**	1391.00	**1750.00**
18'-1" to 20'-11" chimney	Inst	LS	CJ	27.1	1180.00	663.00	---	1843.00	**2370.00**	1485.00	**1860.00**
21'-3" to 23'-10" chimney	Inst	LS	CJ	30.8	1230.00	753.00	---	1983.00	**2560.00**	1579.00	**2000.00**
23'-11" to 24'-9" chimney	Inst	LS	CJ	34.8	1290.00	851.00	---	2141.00	**2770.00**	1700.00	**2160.00**
Accessories											
Log lighter with gas valve (straight or angle pattern)	Inst	Ea	SA	2.05	29.00	62.40	---	91.40	**126.00**	66.20	**89.90**
Log lighter, less gas valve (straight, angle, tee pattern)	Inst	Ea	SA	1.03	10.90	31.40	---	42.30	**59.30**	29.93	**41.40**
Gas valve for log lighter	Inst	Ea	SA	1.03	19.40	31.40	---	50.80	**69.00**	37.40	**50.00**
Spare parts											
Gas valve key	Inst	Ea	---	---	1.33	---	---	1.33	**1.33**	1.18	**1.18**
Stem extender	Inst	Ea	---	---	2.42	---	---	2.42	**2.42**	2.14	**2.14**
Extra long	Inst	Ea	---	---	3.33	---	---	3.33	**3.33**	2.94	**2.94**
Valve floor plate	Inst	Ea	---	---	2.78	---	---	2.78	**2.78**	2.46	**2.46**
Lighter burner tube (12" to 17")	Inst	Ea	---	---	5.57	---	---	5.57	**5.57**	4.92	**4.92**

Fireplace mantels. See Mantels, fireplace, page 158

Flashing. See Sheet metal, page 212

Floor finishes. See individual items.

Floor joists. See Framing, page 110

Description	Oper	Unit	Crew Size	Man-Hours Per Unit	Avg Mat'l Unit Cost	Avg Labor Unit Cost	Avg Equip Unit Cost	Avg Total Unit Cost	Avg Price Incl O&P	Large Volume Avg Total Unit Cost	Large Volume Avg Price Incl O&P
Food centers											
Includes wiring, connection and installation in exposed drainboard only											
Built-in models, 1/4 hp, 4-1/4" x 6-3/4" x 10" rough cut, 110 volts, 6 speed											
	Inst	Ea	EA	3.81	500.00	110.00	---	610.00	**738.00**	510.00	**611.00**
Options											
Blender	Inst	Ea	---	---	53.70	---	---	53.70	**53.70**	46.50	**46.50**
Citrus fruit juicer	Inst	Ea	---	---	26.30	---	---	26.30	**26.30**	22.80	**22.80**
Food processor	Inst	Ea	---	---	272.00	---	---	272.00	**272.00**	235.00	**235.00**
Ice crusher	Inst	Ea	---	---	85.30	---	---	85.30	**85.30**	73.80	**73.80**
Knife sharpener	Inst	Ea	---	---	56.00	---	---	56.00	**56.00**	48.40	**48.40**
Meat grinder, shredder/slicer with power post	Inst	Ea	---	---	326.00	---	---	326.00	**326.00**	282.00	**282.00**
Mixer	Inst	Ea	---	---	153.00	---	---	153.00	**153.00**	132.00	**132.00**

Footings. See Concrete, page 60

Formica. See Countertops, page 63

Forming. See Concrete, page 61

Foundations. See Concrete, page 61

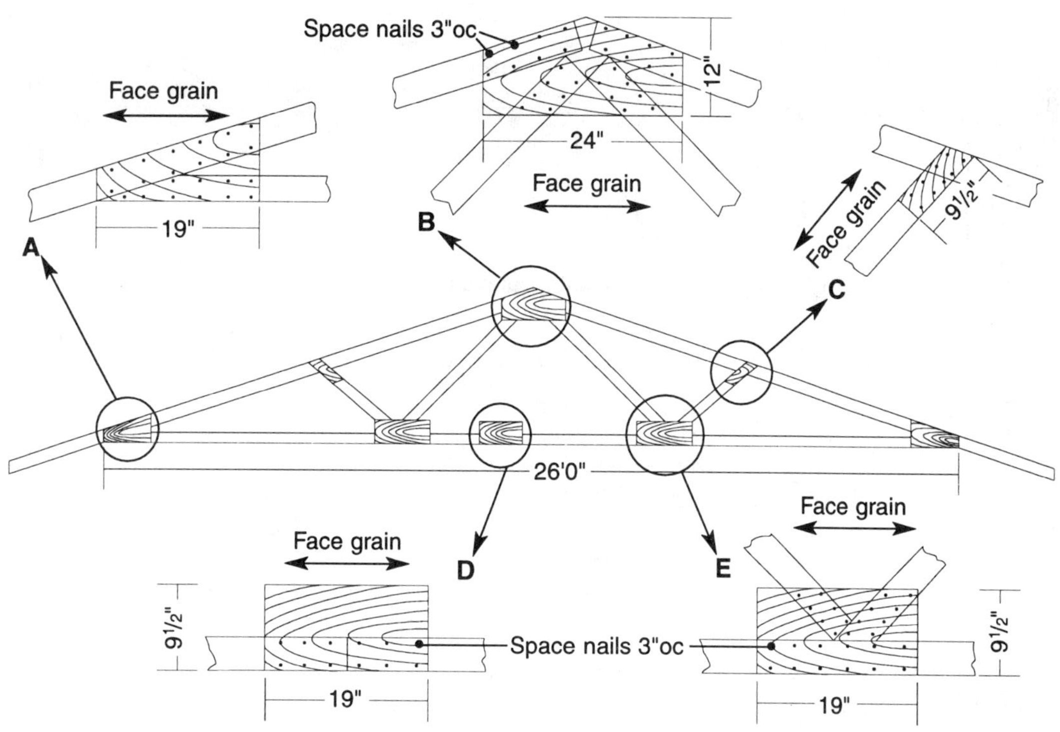

Construction of a 26 foot W truss:

- **A** Bevel-heel gusset
- **B** Peak gusset
- **C** Upper chord intermediate gusset
- **D** Splice of lower chord
- **E** Lower chord intermediate gusset

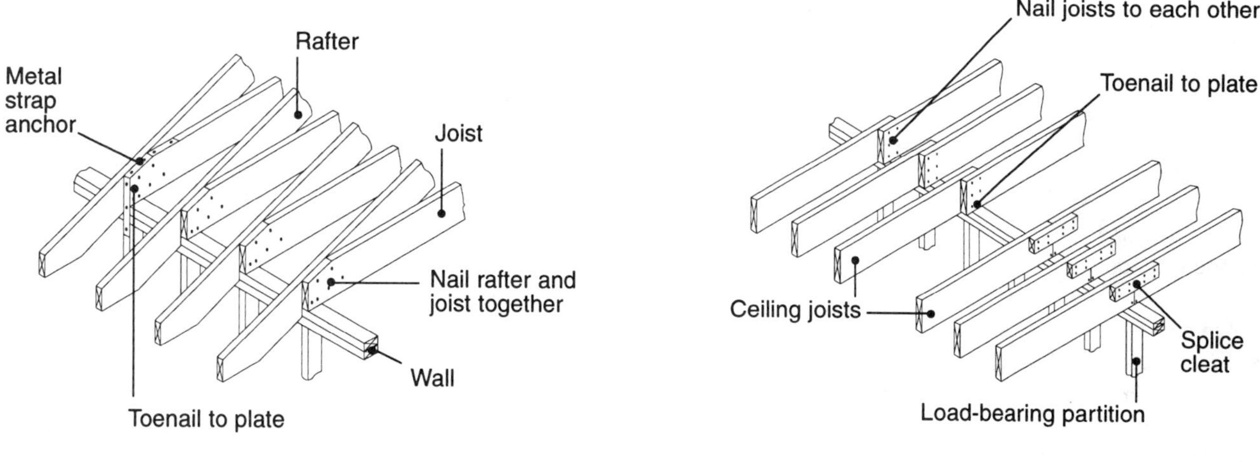

Metal joist hanger

Wood hanger

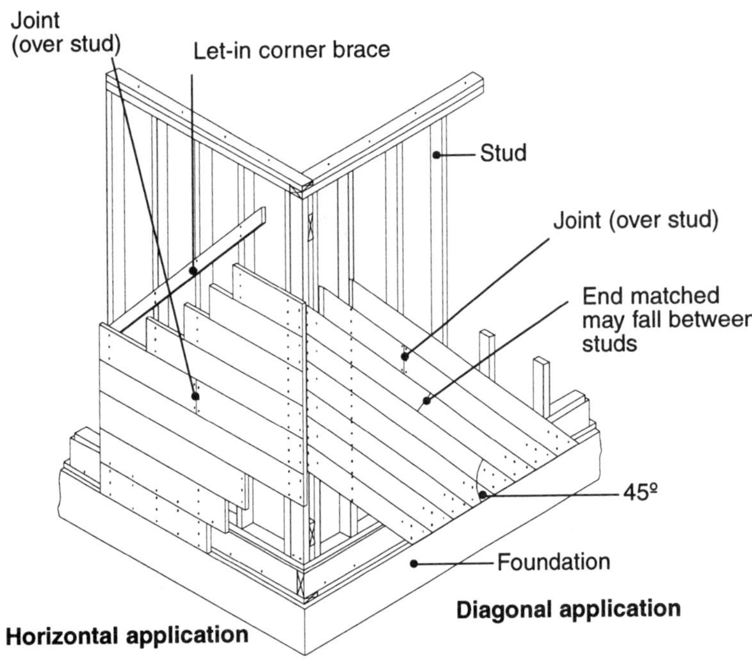

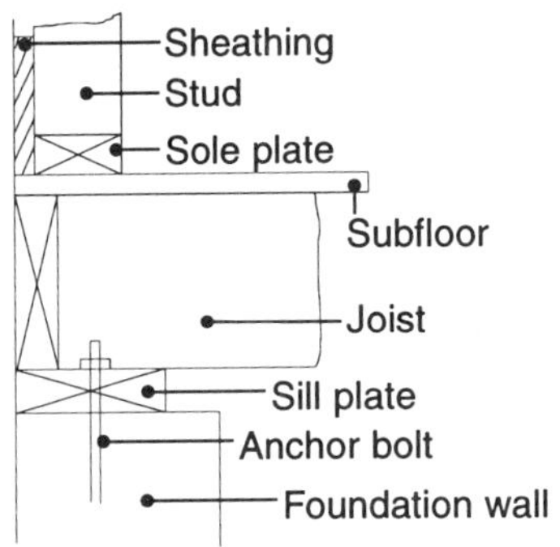

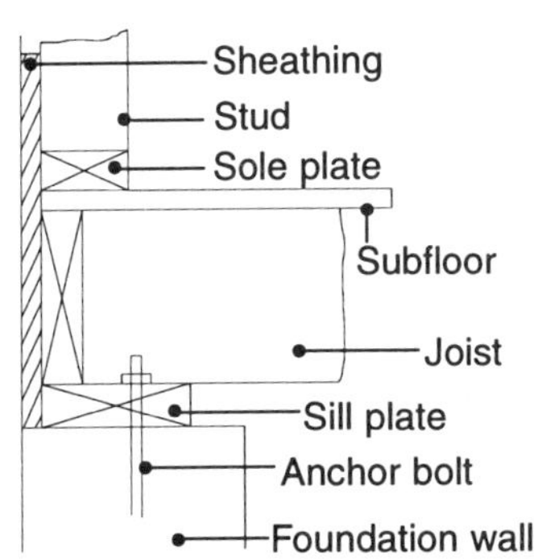

Application of wood sheathing:
A Horizontal and diagonal
B Started at subfloor
C Started at foundation wall

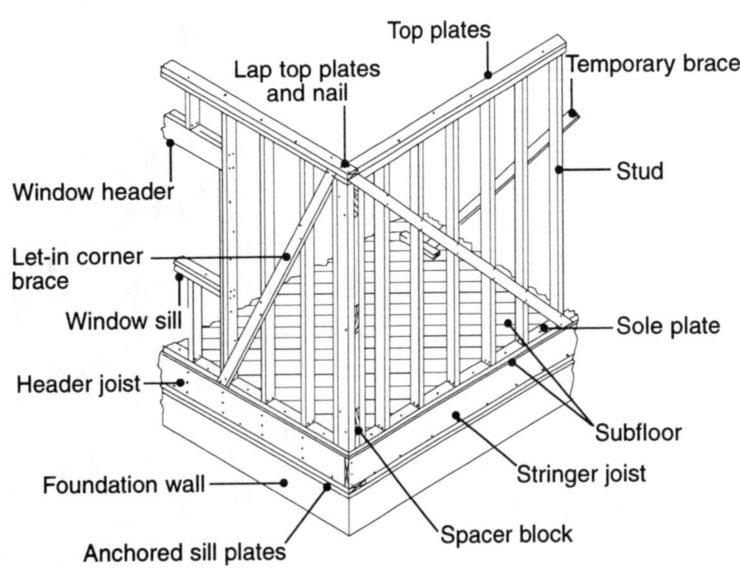

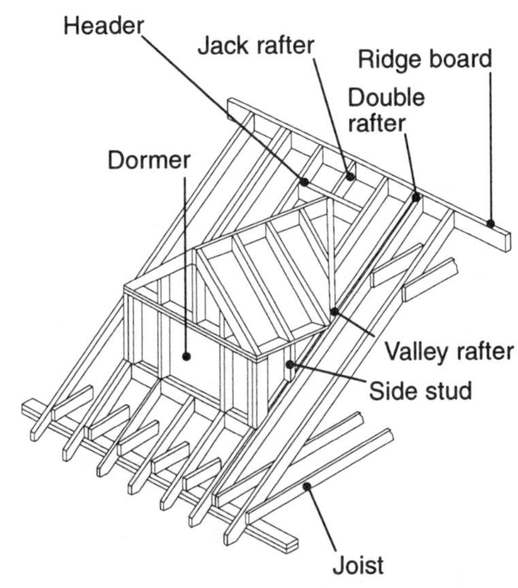

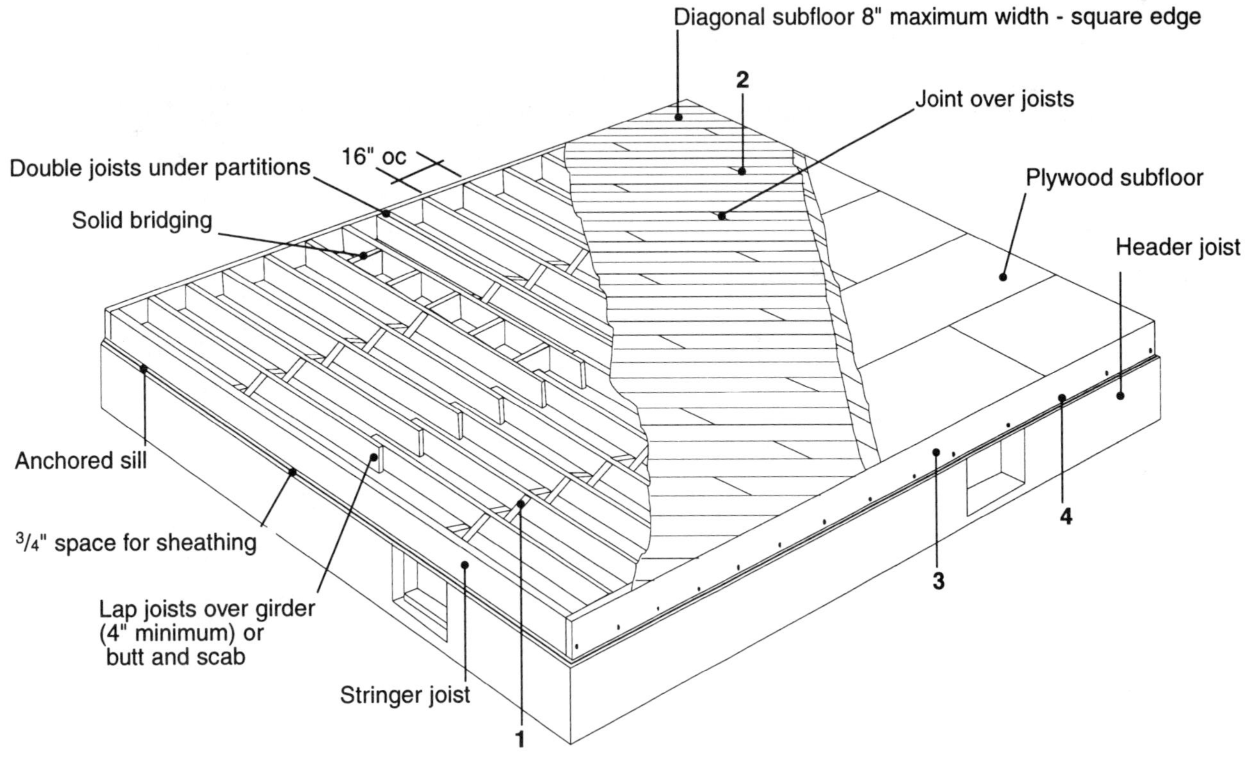

Floor framing:

1. Nailing bridge to joists
2. Nailing board subfloor to joists
3. Nailing header to joists
4. Toenailing header to sill

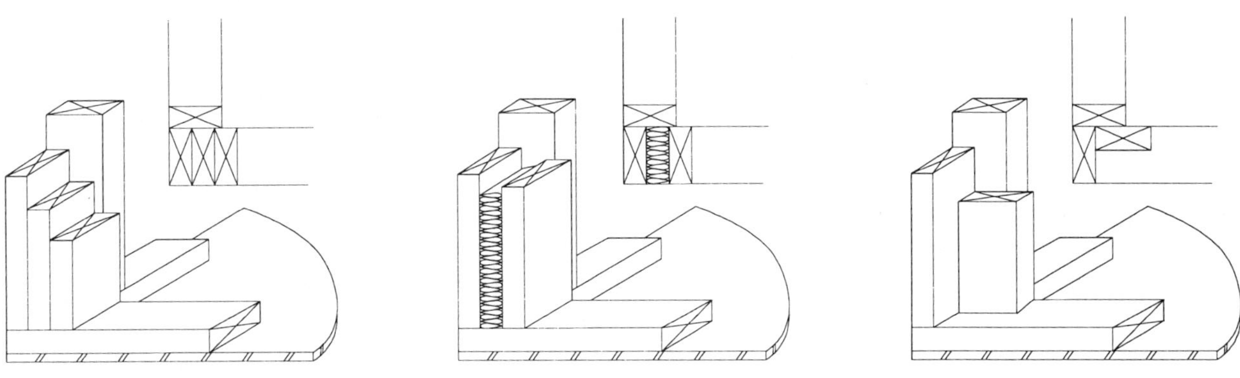

Stud arrangements at exterior corners

Description	Oper	Unit	Crew Size	Man-Hours Per Unit	Avg Mat'l Unit Cost	Avg Labor Unit Cost	Avg Equip Unit Cost	Avg Total Unit Cost	Avg Price Incl O&P	Avg Total Unit Cost	Avg Price Incl O&P
				Costs Based On Small Volume						Large Volume	
Framing, rough carpentry											
Dimension lumber											
Beams; set on steel columns, not wood columns											
Built-up (2 pieces)											
4" x 6" - 10'	Demo	LF	LB	.028	---	.61	---	.61	**.93**	.53	**.80**
4" x 6" - 10'	Inst	LF	2C	.056	1.90	1.51	.16	3.57	**4.64**	2.97	**3.86**
4" x 8" - 10'	Demo	LF	LB	.032	---	.70	---	.70	**1.06**	.59	**.89**
4" x 8" - 10'	Inst	LF	2C	.054	2.69	1.46	.16	4.31	**5.47**	3.57	**4.55**
4" x 10" - 10'	Demo	LF	LB	.035	---	.77	---	.77	**1.16**	.66	**.99**
4" x 10" - 10'	Inst	LF	2C	.058	3.44	1.56	.17	5.17	**6.50**	4.27	**5.38**
4" x 12" - 12'	Demo	LF	LB	.038	---	.83	---	.83	**1.26**	.70	**1.06**
4" x 12" - 12'	Inst	LF	2C	.062	4.25	1.67	.18	6.10	**7.61**	5.07	**6.34**
6" x 8" - 10'	Demo	LF	LD	.051	---	1.25	---	1.25	**1.88**	1.05	**1.59**
6" x 8" - 10'	Inst	LF	CW	.086	5.26	2.10	.13	7.49	**9.38**	6.25	**7.84**
6" x 10" - 10'	Demo	LF	LD	.056	---	1.37	---	1.37	**2.07**	1.15	**1.74**
6" x 10" - 10'	Inst	LF	CW	.093	6.56	2.27	.14	8.97	**11.10**	7.44	**9.25**
6" x 12" - 12'	Demo	LF	LD	.061	---	1.49	---	1.49	**2.25**	1.27	**1.92**
6" x 12" - 12'	Inst	LF	CW	.099	7.86	2.42	.14	10.42	**12.90**	8.63	**10.70**
Built-up (3 pieces)											
6" x 8" - 10'	Demo	LF	LD	.051	---	1.25	---	1.25	**1.88**	1.08	**1.62**
6" x 8" - 10'	Inst	LF	CW	.089	4.06	2.18	.13	6.37	**8.11**	5.27	**6.73**
6" x 10" - 10'	Demo	LF	LD	.056	---	1.37	---	1.37	**2.07**	1.17	**1.77**
6" x 10" - 10'	Inst	LF	CW	.095	5.21	2.32	.14	7.67	**9.66**	6.34	**8.01**
6" x 12" - 12'	Demo	LF	LD	.062	---	1.52	---	1.52	**2.29**	1.30	**1.96**
6" x 12" - 12'	Inst	LF	CW	.102	6.42	2.49	.15	9.06	**11.30**	7.53	**9.42**
9" x 10" - 12'	Demo	LF	LD	.070	---	1.71	---	1.71	**2.59**	1.44	**2.18**
9" x 10" - 12'	Inst	LF	CW	.111	9.88	2.72	.16	12.76	**15.70**	10.58	**13.00**
9" x 12" - 12'	Demo	LF	LD	.076	---	1.86	---	1.86	**2.81**	1.59	**2.40**
9" x 12" - 12'	Inst	LF	CW	.119	11.80	2.91	.17	14.88	**18.20**	12.36	**15.10**
Single member											
2" x 6"	Demo	LF	LB	.020	---	.44	---	.44	**.66**	.37	**.56**
2" x 6"	Inst	LF	2C	.036	1.06	.97	.11	2.14	**2.80**	1.80	**2.36**
2" x 8"	Demo	LF	LB	.022	---	.48	---	.48	**.73**	.40	**.60**
2" x 8"	Inst	LF	2C	.038	1.45	1.02	.11	2.58	**3.33**	2.18	**2.82**
2" x 10"	Demo	LF	LB	.023	---	.50	---	.50	**.76**	.44	**.66**
2" x 10"	Inst	LF	2C	.041	1.84	1.11	.12	3.07	**3.92**	2.52	**3.22**
2" x 12"	Demo	LF	LB	.025	---	.55	---	.55	**.83**	.46	**.70**
2" x 12"	Inst	LF	2C	.043	2.24	1.16	.12	3.52	**4.46**	2.92	**3.70**
3" x 6"	Demo	LF	LB	.023	---	.50	---	.50	**.76**	.42	**.63**
3" x 6"	Inst	LF	2C	.040	2.06	1.08	.12	3.26	**4.13**	2.71	**3.44**
3" x 8"	Demo	LF	LB	.025	---	.55	---	.55	**.83**	.46	**.70**
3" x 8"	Inst	LF	2C	.043	2.74	1.16	.12	4.02	**5.03**	3.33	**4.17**
3" x 10"	Demo	LF	LB	.028	---	.61	---	.61	**.93**	.53	**.80**
3" x 10"	Inst	LF	2C	.046	3.39	1.24	.13	4.76	**5.91**	3.95	**4.92**

			Costs Based On Small Volume							Large Volume	
Description	Oper	Unit	Crew Size	Man-Hours Per Unit	Avg Mat'l Unit Cost	Avg Labor Unit Cost	Avg Equip Unit Cost	Avg Total Unit Cost	Avg Price Incl O&P	Avg Total Unit Cost	Avg Price Incl O&P
3" x 12"	Demo	LF	LB	.030	---	.66	---	.66	**.99**	.57	**.86**
3" x 12"	Inst	LF	2C	.049	4.04	1.32	.14	5.50	**6.79**	4.57	**5.66**
3" x 14"	Demo	LF	LB	.033	---	.72	---	.72	**1.09**	.61	**.93**
3" x 14"	Inst	LF	2C	.052	4.88	1.40	.15	6.43	**7.89**	5.35	**6.59**
4" x 6"	Demo	LF	LB	.025	---	.55	---	.55	**.83**	.46	**.70**
4" x 6"	Inst	LF	2C	.043	2.48	1.16	.13	3.77	**4.74**	3.14	**3.96**
4" x 8"	Demo	LF	LB	.029	---	.64	---	.64	**.96**	.55	**.83**
4" x 8"	Inst	LF	2C	.048	3.32	1.29	.14	4.75	**5.93**	3.96	**4.94**
4" x 10"	Demo	LF	LB	.032	---	.70	---	.70	**1.06**	.59	**.89**
4" x 10"	Inst	LF	2C	.052	4.10	1.40	.15	5.65	**7.00**	4.70	**5.82**
4" x 12"	Demo	LF	LB	.036	---	.79	---	.79	**1.19**	.68	**1.03**
4" x 12"	Inst	LF	2C	.057	4.90	1.54	.16	6.60	**8.13**	5.47	**6.75**
4" x 14"	Demo	LF	LH	.044	---	.97	---	.97	**1.46**	.81	**1.23**
4" x 14"	Inst	LF	CS	.070	5.95	1.77	.14	7.86	**9.67**	6.51	**8.02**
4" x 16"	Demo	LF	LH	.048	---	1.05	---	1.05	**1.59**	.88	**1.33**
4" x 16"	Inst	LF	CS	.075	7.06	1.90	.14	9.10	**11.10**	7.52	**9.22**
6" x 8"	Demo	LF	LJ	.048	---	1.05	---	1.05	**1.59**	.90	**1.36**
6" x 8"	Inst	LF	CW	.081	6.39	1.98	.12	8.49	**10.50**	7.01	**8.67**
6" x 10"	Demo	LF	LJ	.054	---	1.19	---	1.19	**1.79**	1.01	**1.52**
6" x 10"	Inst	LF	CW	.088	8.12	2.15	.13	10.40	**12.70**	8.60	**10.60**
6" x 12"	Demo	LF	LJ	.060	---	1.32	---	1.32	**1.99**	1.12	**1.69**
6" x 12"	Inst	LF	CW	.095	10.10	2.32	.14	12.56	**15.30**	10.40	**12.70**
6" x 14"	Demo	LF	LJ	.066	---	1.45	---	1.45	**2.19**	1.23	**1.86**
6" x 14"	Inst	LF	CW	.103	11.80	2.52	.15	14.47	**17.50**	11.95	**14.50**
6" x 16"	Demo	LF	LJ	.072	---	1.58	---	1.58	**2.39**	1.34	**2.02**
6" x 16"	Inst	LF	CW	.110	14.00	2.69	.16	16.85	**20.30**	13.94	**16.80**
8" x 8"	Demo	LF	LJ	.056	---	1.23	---	1.23	**1.86**	1.05	**1.59**
8" x 8"	Inst	LF	CW	.091	9.23	2.23	.13	11.59	**14.10**	9.56	**11.70**
8" x 10"	Demo	LF	LJ	.068	---	1.49	---	1.49	**2.25**	1.27	**1.92**
8" x 10"	Inst	LF	CW	.104	11.70	2.54	.15	14.39	**17.50**	11.94	**14.50**
8" x 12"	Demo	LF	LJ	.073	---	1.60	---	1.60	**2.42**	1.36	**2.05**
8" x 12"	Inst	LF	CW	.110	14.70	2.69	.16	17.55	**21.10**	14.44	**17.40**
8" x 14"	Demo	LF	LJ	.080	---	1.76	---	1.76	**2.65**	1.49	**2.25**
8" x 14"	Inst	LF	CW	.119	17.60	2.91	.17	20.68	**24.80**	17.02	**20.40**
8" x 16"	Demo	LF	LJ	.089	---	1.95	---	1.95	**2.95**	1.67	**2.52**
8" x 16"	Inst	LF	CW	.129	20.50	3.16	.19	23.85	**28.50**	19.65	**23.60**
Blocking, horizontal, for studs											
2" x 4" - 12"	Demo	LF	1L	.013	---	.29	---	.29	**.43**	.24	**.36**
2" x 4" - 12"	Inst	LF	CA	.042	.62	1.13	.25	2.00	**2.68**	1.69	**2.27**
2" x 4" - 16"	Demo	LF	1L	.012	---	.26	---	.26	**.40**	.24	**.36**
2" x 4" - 16"	Inst	LF	CA	.034	.62	.92	.20	1.74	**2.31**	1.46	**1.94**
2" x 4" - 24"	Demo	LF	1L	.012	---	.26	---	.26	**.40**	.22	**.33**
2" x 4" - 24"	Inst	LF	CA	.027	.62	.73	.16	1.51	**1.98**	1.26	**1.66**

			Costs Based On Small Volume							Large Volume	
Description	Oper	Unit	Crew Size	Man-Hours Per Unit	Avg Mat'l Unit Cost	Avg Labor Unit Cost	Avg Equip Unit Cost	Avg Total Unit Cost	Avg Price Incl O&P	Avg Total Unit Cost	Avg Price Incl O&P
2" x 6" - 12"	Demo	LF	1L	.015	---	.33	---	.33	**.50**	.29	**.43**
2" x 6" - 12"	Inst	LF	CA	.045	.93	1.21	.27	2.41	**3.18**	2.04	**2.70**
2" x 6" - 16"	Demo	LF	1L	.015	---	.33	---	.33	**.50**	.26	**.40**
2" x 6" - 16"	Inst	LF	CA	.037	.93	1.00	.22	2.15	**2.81**	1.81	**2.38**
2" x 6" - 24"	Demo	LF	1L	.014	---	.31	---	.31	**.46**	.26	**.40**
2" x 6" - 24"	Inst	LF	CA	.029	.93	.78	.17	1.88	**2.43**	1.57	**2.04**
2" x 8" - 12"	Demo	LF	1L	.018	---	.40	---	.40	**.60**	.33	**.50**
2" x 8" - 12"	Inst	LF	CA	.049	1.32	1.32	.28	2.92	**3.81**	2.43	**3.16**
2" x 8" - 16"	Demo	LF	1L	.017	---	.37	---	.37	**.56**	.31	**.46**
2" x 8" - 16"	Inst	LF	CA	.040	1.32	1.08	.23	2.63	**3.39**	2.20	**2.84**
2" x 8" - 24"	Demo	LF	1L	.016	---	.35	---	.35	**.53**	.31	**.46**
2" x 8" - 24"	Inst	LF	CA	.031	1.32	.84	.18	2.34	**2.97**	1.97	**2.51**
Bracing, diagonal let-ins											
Studs, 12" oc											
1" x 6"	Demo	Set	1L	.013	---	.29	---	.29	**.43**	.24	**.36**
1" x 6"	Inst	Set	CA	.066	.47	1.78	.38	2.63	**3.63**	2.23	**3.07**
Studs, 16" oc											
1" x 6"	Demo	Set	1L	.012	---	.26	---	.26	**.40**	.22	**.33**
1" x 6"	Inst	Set	CA	.051	.47	1.37	.30	2.14	**2.93**	1.80	**2.46**
Studs, 24" oc											
1" x 6"	Demo	Set	1L	.011	---	.24	---	.24	**.36**	.22	**.33**
1" x 6"	Inst	Set	CA	.039	.47	1.05	.23	1.75	**2.37**	1.47	**1.99**
Bridging, "X" type											
For joists 12" oc											
1" x 3"	Demo	Set	1L	.017	---	.37	---	.37	**.56**	.33	**.50**
1" x 3"	Inst	Set	CA	.067	.72	1.81	.39	2.92	**3.96**	2.46	**3.34**
2" x 2"	Demo	Set	1L	.018	---	.40	---	.40	**.60**	.33	**.50**
2" x 2"	Inst	Set	CA	.068	.99	1.83	.40	3.22	**4.33**	2.71	**3.65**
For joists 16" oc											
1" x 3"	Demo	Set	1L	.018	---	.40	---	.40	**.60**	.33	**.50**
1" x 3"	Inst	Set	CA	.069	.77	1.86	.40	3.03	**4.11**	2.53	**3.44**
2" x 2"	Demo	Set	1L	.018	---	.40	---	.40	**.60**	.35	**.53**
2" x 2"	Inst	Set	CA	.069	1.05	1.86	.40	3.31	**4.44**	2.79	**3.75**
For joists 24" oc											
1" x 3"	Demo	Set	1L	.019	---	.42	---	.42	**.63**	.35	**.53**
1" x 3"	Inst	Set	CA	.019	.85	.51	.11	1.47	**1.87**	1.22	**1.55**
2" x 2"	Demo	Set	1L	.070	---	1.54	---	1.54	**2.32**	1.32	**1.99**
2" x 2"	Inst	Set	CA	.071	1.17	1.91	.41	3.49	**4.67**	2.92	**3.90**
Bridging, solid, between joists											
2" x 6" - 12"	Demo	Set	1L	.016	---	.35	---	.35	**.53**	.31	**.46**
2" x 6" - 12"	Inst	Set	CA	.047	.93	1.27	.28	2.48	**3.28**	2.07	**2.74**
2" x 8" - 12"	Demo	Set	1L	.018	---	.40	---	.40	**.60**	.33	**.50**
2" x 8" - 12"	Inst	Set	CA	.051	1.32	1.37	.30	2.99	**3.91**	2.49	**3.25**

Description	Oper	Unit	Crew Size	Man-Hours Per Unit	Avg Mat'l Unit Cost	Avg Labor Unit Cost	Avg Equip Unit Cost	Avg Total Unit Cost	Avg Price Incl O&P	Avg Total Unit Cost	Avg Price Incl O&P
				Costs Based On Small Volume						Large Volume	
2" x 10" - 12"	Demo	Set	1L	.021	---	.46	---	.46	**.70**	.40	**.60**
2" x 10" - 12"	Inst	Set	CA	.054	1.70	1.46	.32	3.48	**4.49**	2.90	**3.75**
2" x 12" - 12"	Demo	Set	1L	.023	---	.50	---	.50	**.76**	.42	**.63**
2" x 12" - 12"	Inst	Set	CA	.058	2.10	1.56	.34	4.00	**5.13**	3.33	**4.28**
2" x 6" - 16"	Demo	Set	1L	.015	---	.33	---	.33	**.50**	.29	**.43**
2" x 6" - 16"	Inst	Set	CA	.039	.93	1.05	.23	2.21	**2.90**	1.84	**2.42**
2" x 8" - 16"	Demo	Set	1L	.017	---	.37	---	.37	**.56**	.33	**.50**
2" x 8" - 16"	Inst	Set	CA	.042	1.32	1.13	.24	2.69	**3.48**	2.23	**2.89**
2" x 10" - 16"	Demo	Set	1L	.020	---	.44	---	.44	**.66**	.37	**.56**
2" x 10" - 16"	Inst	Set	CA	.045	1.70	1.21	.26	3.17	**4.06**	2.63	**3.38**
2" x 12" - 16"	Demo	Set	1L	.022	---	.48	---	.48	**.73**	.40	**.60**
2" x 12" - 16"	Inst	Set	CA	.048	2.10	1.29	.28	3.67	**4.66**	3.05	**3.87**
2" x 6" - 24"	Demo	Set	1L	.014	---	.31	---	.31	**.46**	.26	**.40**
2" x 6" - 24"	Inst	Set	CA	.030	.93	.81	.18	1.92	**2.48**	1.61	**2.09**
2" x 8" - 24"	Demo	Set	1L	.016	---	.35	---	.35	**.53**	.31	**.46**
2" x 8" - 24"	Inst	Set	CA	.033	1.32	.89	.19	2.40	**3.06**	1.99	**2.55**
2" x 10" - 24"	Demo	Set	1L	.019	---	.42	---	.42	**.63**	.35	**.53**
2" x 10" - 24"	Inst	Set	CA	.035	1.70	.94	.20	2.84	**3.59**	2.37	**3.00**
2" x 12" - 24"	Demo	Set	1L	.021	---	.46	---	.46	**.70**	.40	**.60**
2" x 12" - 24"	Inst	Set	CA	.037	2.10	1.00	.22	3.32	**4.15**	2.78	**3.49**
Columns or posts, without base or cap, hardware, or chamfer corners											
4" x 4" - 8'	Demo	LF	LB	.032	---	.70	---	.70	**1.06**	.59	**.89**
4" x 4" - 8'	Inst	LF	2C	.060	1.59	1.62	.17	3.38	**4.46**	2.83	**3.75**
4" x 6" - 8'	Demo	LF	LB	.036	---	.79	---	.79	**1.19**	.66	**.99**
4" x 6" - 8'	Inst	LF	2C	.064	2.38	1.73	.19	4.30	**5.55**	3.57	**4.62**
4" x 8" - 8'	Demo	LF	LB	.039	---	.86	---	.86	**1.29**	.72	**1.09**
4" x 8" - 8'	Inst	LF	2C	.069	3.22	1.86	.20	5.28	**6.73**	4.38	**5.59**
6" x 6" - 8'	Demo	LF	LB	.044	---	.97	---	.97	**1.46**	.81	**1.23**
6" x 6" - 8'	Inst	LF	2C	.076	4.73	2.05	.22	7.00	**8.77**	5.80	**7.29**
6" x 8" - 8'	Demo	LF	LB	.049	---	1.08	---	1.08	**1.62**	.92	**1.39**
6" x 8" - 8'	Inst	LF	2C	.082	6.29	2.21	.24	8.74	**10.80**	7.23	**8.98**
6" x 10" - 8'	Demo	LF	LB	.055	---	1.21	---	1.21	**1.82**	1.01	**1.52**
6" x 10" - 8'	Inst	LF	2C	.089	8.02	2.40	.26	10.68	**13.10**	8.82	**10.90**
8" x 8" - 8'	Demo	LF	LB	.059	---	1.30	---	1.30	**1.96**	1.10	**1.66**
8" x 8" - 8'	Inst	LF	2C	.096	9.13	2.59	.28	12.00	**14.70**	9.93	**12.20**
8" x 10" - 8'	Demo	LF	LB	.070	---	1.54	---	1.54	**2.32**	1.30	**1.96**
8" x 10" - 8'	Inst	LF	2C	.108	11.60	2.91	.32	14.83	**18.10**	12.32	**15.10**
Fascia											
1" x 4" - 12'	Demo	LF	LB	.014	---	.31	---	.31	**.46**	.26	**.40**
1" x 4" - 12'	Inst	LF	2C	.049	.30	1.32	.14	1.76	**2.49**	1.48	**2.09**
Furring strips on ceilings											
1" x 3" on wood	Demo	LF	LB	.030	---	.66	---	.66	**.99**	.57	**.86**
1" x 3" on wood	Inst	LF	2C	.033	.22	.89	.10	1.21	**1.71**	1.01	**1.43**

Description	Oper	Unit	Crew Size	Man-Hours Per Unit	Avg Mat'l Unit Cost	Avg Labor Unit Cost	Avg Equip Unit Cost	Avg Total Unit Cost	Avg Price Incl O&P	Large Volume Avg Total Unit Cost	Large Volume Avg Price Incl O&P
Furring strips on walls											
1" x 3" on wood	Demo	LF	LB	.023	---	.50	---	.50	**.76**	.44	**.66**
1" x 3" on wood	Inst	LF	2C	.025	.22	.67	.07	.96	**1.35**	.83	**1.17**
1" x 3" on masonry	Demo	LF	LB	.030	---	.66	---	.66	**.99**	.57	**.86**
1" x 3" on masonry	Inst	LF	2C	.033	.22	.89	.10	1.21	**1.71**	1.01	**1.43**
1" x 3" on concrete	Demo	LF	LB	.041	---	.90	---	.90	**1.36**	.77	**1.16**
1" x 3" on concrete	Inst	LF	2C	.045	.22	1.21	.13	1.56	**2.23**	1.31	**1.87**
Headers or lintels, over openings											
4 feet wide											
4" x 6"	Demo	LF	1L	.028	---	.61	---	.61	**.93**	.53	**.80**
4" x 6"	Inst	LF	CA	.048	2.36	1.29	.28	3.93	**4.96**	3.26	**4.11**
4" x 8"	Demo	LF	1L	.031	---	.68	---	.68	**1.03**	.57	**.86**
4" x 8"	Inst	LF	CA	.051	3.20	1.37	.30	4.87	**6.07**	4.05	**5.05**
4" x 12"	Demo	LF	1L	.038	---	.83	---	.83	**1.26**	.70	**1.06**
4" x 12"	Inst	LF	CA	.058	4.78	1.56	.34	6.68	**8.21**	5.54	**6.82**
4" x 14"	Demo	LF	1L	.040	---	.88	---	.88	**1.33**	.75	**1.13**
4" x 14"	Inst	LF	CA	.060	5.83	1.62	.35	7.80	**9.51**	6.47	**7.91**
6 feet wide											
4" x 12"	Demo	LF	LB	.032	---	.70	---	.70	**1.06**	.59	**.89**
4" x 12"	Inst	LF	2C	.048	4.78	1.29	.14	6.21	**7.60**	5.17	**6.33**
8 feet wide											
4" x 12"	Demo	LF	LB	.029	---	.64	---	.64	**.96**	.53	**.80**
4" x 12"	Inst	LF	2C	.042	4.78	1.13	.12	6.03	**7.34**	4.98	**6.07**
10 feet wide											
4" x 12"	Demo	LF	LB	.025	---	.55	---	.55	**.83**	.48	**.73**
4" x 12"	Inst	LF	2C	.035	4.78	.94	.10	5.82	**7.03**	4.84	**5.85**
4" x 14"	Demo	LF	LB	.030	---	.66	---	.66	**.99**	.57	**.86**
4" x 14"	Inst	LF	2C	.041	5.83	1.11	.12	7.06	**8.50**	5.84	**7.05**
12 feet wide											
4" x 14"	Demo	LF	LB	.028	---	.61	---	.61	**.93**	.53	**.80**
4" x 14"	Inst	LF	2C	.038	5.83	1.02	.11	6.96	**8.37**	5.79	**6.97**
4" x 16"	Demo	LF	LB	.033	---	.72	---	.72	**1.09**	.61	**.93**
4" x 16"	Inst	LF	2C	.044	6.95	1.19	.13	8.27	**9.93**	6.82	**8.19**
14 feet wide											
4" x 16"	Demo	LF	LB	.032	---	.70	---	.70	**1.06**	.59	**.89**
4" x 16"	Inst	LF	2C	.042	6.95	1.13	.12	8.20	**9.83**	6.75	**8.10**
16 feet wide											
4" x 16"	Demo	LF	LB	.031	---	.68	---	.68	**1.03**	.57	**.86**
4" x 16"	Inst	LF	2C	.040	6.95	1.08	.12	8.15	**9.75**	6.73	**8.06**
18 feet wide											
4" x 16"	Demo	LF	LB	.030	---	.66	---	.66	**.99**	.55	**.83**
4" x 16"	Inst	LF	2C	.038	6.95	1.02	.11	8.08	**9.66**	6.66	**7.97**
Joists, ceiling/floor, per LF of stick											
2" x 4" -6'	Demo	LF	LB	.018	---	.40	---	.40	**.60**	.35	**.53**
2" x 4" -6'	Inst	LF	2C	.022	.62	.59	.07	1.28	**1.68**	1.08	**1.43**

			Costs Based On Small Volume							Large Volume	
Description	**Oper**	**Unit**	**Crew Size**	**Man-Hours Per Unit**	**Avg Mat'l Unit Cost**	**Avg Labor Unit Cost**	**Avg Equip Unit Cost**	**Avg Total Unit Cost**	**Avg Price Incl O&P**	**Avg Total Unit Cost**	**Avg Price Incl O&P**
2" x 4" -8'	Demo	LF	LB	.016	---	.35	---	.35	**.53**	.29	**.43**
2" x 4" -8'	Inst	LF	2C	.019	.62	.51	.05	1.18	**1.54**	.99	**1.29**
2" x 4" -10'	Demo	LF	LB	.014	---	.31	---	.31	**.46**	.26	**.40**
2" x 4" -10'	Inst	LF	2C	.017	.62	.46	.05	1.13	**1.46**	.93	**1.20**
2" x 4" -12'	Demo	LF	LB	.013	---	.29	---	.29	**.43**	.24	**.36**
2" x 4" -12'	Inst	LF	2C	.015	.62	.40	.04	1.06	**1.37**	.90	**1.16**
2" x 6" - 8'	Demo	LF	LB	.018	---	.40	---	.40	**.60**	.33	**.50**
2" x 6" - 8'	Inst	LF	2C	.021	.93	.57	.06	1.56	**1.99**	1.30	**1.66**
2" x 6" - 10'	Demo	LF	LB	.016	---	.35	---	.35	**.53**	.29	**.43**
2" x 6" - 10'	Inst	LF	2C	.019	.93	.51	.05	1.49	**1.90**	1.24	**1.58**
2" x 6" - 12'	Demo	LF	LB	.014	---	.31	---	.31	**.46**	.26	**.40**
2" x 6" - 12'	Inst	LF	2C	.017	.93	.46	.05	1.44	**1.82**	1.20	**1.53**
2" x 6" - 14'	Demo	LF	LB	.013	---	.29	---	.29	**.43**	.24	**.36**
2" x 6" - 14'	Inst	LF	2C	.016	.93	.43	.05	1.41	**1.78**	1.15	**1.45**
2" x 8" - 10'	Demo	LF	LB	.018	---	.40	---	.40	**.60**	.33	**.50**
2" x 8" - 10'	Inst	LF	2C	.021	1.32	.57	.06	1.95	**2.44**	1.62	**2.03**
2" x 8" - 12'	Demo	LF	LB	.016	---	.35	---	.35	**.53**	.31	**.46**
2" x 8" - 12'	Inst	LF	2C	.019	1.32	.51	.06	1.89	**2.36**	1.56	**1.95**
2" x 8" - 14'	Demo	LF	LB	.015	---	.33	---	.33	**.50**	.29	**.43**
2" x 8" - 14'	Inst	LF	2C	.018	1.32	.49	.05	1.86	**2.31**	1.52	**1.90**
2" x 8" - 16'	Demo	LF	LB	.014	---	.31	---	.31	**.46**	.26	**.40**
2" x 8" - 16'	Inst	LF	2C	.017	1.32	.46	.05	1.83	**2.26**	1.50	**1.86**
2" x 10" - 12'	Demo	LF	LB	.018	---	.40	---	.40	**.60**	.33	**.50**
2" x 10" - 12'	Inst	LF	2C	.021	1.70	.57	.06	2.33	**2.88**	1.93	**2.39**
2" x 10" - 14'	Demo	LF	LB	.017	---	.37	---	.37	**.56**	.31	**.46**
2" x 10" - 14'	Inst	LF	2C	.020	1.70	.54	.06	2.30	**2.83**	1.90	**2.35**
2" x 10" - 16'	Demo	LF	LB	.016	---	.35	---	.35	**.53**	.31	**.46**
2" x 10" - 16'	Inst	LF	2C	.019	1.70	.51	.05	2.26	**2.78**	1.87	**2.30**
2" x 10" - 18'	Demo	LF	LB	.015	---	.33	---	.33	**.50**	.29	**.43**
2" x 10" - 18'	Inst	LF	2C	.018	1.70	.49	.05	2.24	**2.74**	1.83	**2.25**
2" x 12" - 14'	Demo	LF	LB	.019	---	.42	---	.42	**.63**	.35	**.53**
2" x 12" - 14'	Inst	LF	2C	.022	2.10	.59	.06	2.75	**3.38**	2.27	**2.78**
2" x 12" - 16'	Demo	LF	LB	.018	---	.40	---	.40	**.60**	.33	**.50**
2" x 12" - 16'	Inst	LF	2C	.020	2.10	.54	.06	2.70	**3.29**	2.24	**2.74**
2" x 12" - 18'	Demo	LF	LB	.017	---	.37	---	.37	**.56**	.31	**.46**
2" x 12" - 18'	Inst	LF	2C	.019	2.10	.51	.06	2.67	**3.25**	2.24	**2.74**
2" x 12" - 20'	Demo	LF	LB	.016	---	.35	---	.35	**.53**	.31	**.46**
2" x 12" - 20'	Inst	LF	2C	.019	2.10	.51	.05	2.66	**3.24**	2.21	**2.70**
3" x 8" - 12'	Demo	LF	LB	.020	---	.44	---	.44	**.66**	.37	**.56**
3" x 8" - 12'	Inst	LF	2C	.023	2.61	.62	.07	3.30	**4.01**	2.75	**3.35**
3" x 8" - 14'	Demo	LF	LB	.019	---	.42	---	.42	**.63**	.35	**.53**
3" x 8" - 14'	Inst	LF	2C	.021	2.61	.57	.06	3.24	**3.92**	2.69	**3.26**
3" x 8" - 16'	Demo	LF	LB	.018	---	.40	---	.40	**.60**	.33	**.50**
3" x 8" - 16'	Inst	LF	2C	.020	2.61	.54	.06	3.21	**3.88**	2.66	**3.22**
3" x 8" - 18'	Demo	LF	LB	.017	---	.37	---	.37	**.56**	.31	**.46**
3" x 8" - 18'	Inst	LF	2C	.019	2.61	.51	.06	3.18	**3.84**	2.63	**3.18**

Description	Oper	Unit	Crew Size	Man-Hours Per Unit	Avg Mat'l Unit Cost	Avg Labor Unit Cost	Avg Equip Unit Cost	Avg Total Unit Cost	Avg Price Incl O&P	Large Volume Avg Total Unit Cost	Large Volume Avg Price Incl O&P
3" x 10" - 16'	Demo	LF	LB	.021	---	.46	---	.46	**.70**	.37	**.56**
3" x 10" - 16'	Inst	LF	2C	.023	3.26	.62	.07	3.95	**4.76**	3.28	**3.96**
3" x 10" - 18'	Demo	LF	LB	.020	---	.44	---	.44	**.66**	.37	**.56**
3" x 10" - 18'	Inst	LF	2C	.022	3.26	.59	.06	3.91	**4.71**	3.24	**3.91**
3" x 10" - 20'	Demo	LF	LB	.019	---	.42	---	.42	**.63**	.35	**.53**
3" x 10" - 20'	Inst	LF	2C	.021	3.26	.57	.06	3.89	**4.67**	3.22	**3.87**
3" x 12" - 16'	Demo	LF	LB	.023	---	.50	---	.50	**.76**	.42	**.63**
3" x 12" - 16'	Inst	LF	2C	.026	3.26	.70	.08	4.04	**4.89**	3.33	**4.04**
3" x 12" - 18'	Demo	LF	LB	.022	---	.48	---	.48	**.73**	.40	**.60**
3" x 12" - 18'	Inst	LF	2C	.024	3.91	.65	.07	4.63	**5.55**	3.81	**4.57**
3" x 12" - 20'	Demo	LF	LB	.021	---	.46	---	.46	**.70**	.40	**.60**
3" x 12" - 20'	Inst	LF	2C	.023	3.91	.62	.07	4.60	**5.51**	3.81	**4.57**
3" x 12" - 22'	Demo	LF	LB	.020	---	.44	---	.44	**.66**	.37	**.56**
3" x 12" - 22'	Inst	LF	2C	.023	3.91	.62	.07	4.60	**5.51**	3.78	**4.53**
Joists, ceiling/floor, per SF of area											
2" x 4" - 6', 12" oc	Demo	SF	LB	.019	---	.42	---	.42	**.63**	.35	**.53**
2" x 4" - 6', 12" oc	Inst	SF	2C	.023	.64	.62	.07	1.33	**1.75**	1.09	**1.44**
2" x 4" - 6', 16" oc	Demo	SF	LB	.016	---	.35	---	.35	**.53**	.29	**.43**
2" x 4" - 6', 16" oc	Inst	SF	2C	.019	.48	.51	.06	1.05	**1.39**	.88	**1.17**
2" x 4" - 6', 24" oc	Demo	SF	LB	.012	---	.26	---	.26	**.40**	.22	**.33**
2" x 4" - 6', 24" oc	Inst	SF	2C	.015	.33	.40	.04	.77	**1.03**	.63	**.84**
2" x 6" - 8', 12" oc	Demo	SF	LB	.018	---	.40	---	.40	**.60**	.33	**.50**
2" x 6" - 8', 12" oc	Inst	SF	2C	.022	.95	.59	.06	1.60	**2.05**	1.32	**1.68**
2" x 6" - 8', 16" oc	Demo	SF	LB	.015	---	.33	---	.33	**.50**	.29	**.43**
2" x 6" - 8', 16" oc	Inst	SF	2C	.018	.73	.49	.05	1.27	**1.63**	1.05	**1.35**
2" x 6" - 8', 24" oc	Demo	SF	LB	.012	---	.26	---	.26	**.40**	.22	**.33**
2" x 6" - 8', 24" oc	Inst	SF	2C	.014	.50	.38	.04	.92	**1.19**	.76	**.99**
2" x 8" - 10', 12" oc	Demo	SF	LB	.018	---	.40	---	.40	**.60**	.33	**.50**
2" x 8" - 10', 12" oc	Inst	SF	2C	.021	1.36	.57	.06	1.99	**2.48**	1.65	**2.06**
2" x 8" - 10', 16" oc	Demo	SF	LB	.015	---	.33	---	.33	**.50**	.29	**.43**
2" x 8" - 10', 16" oc	Inst	SF	2C	.018	1.02	.49	.05	1.56	**1.96**	1.28	**1.62**
2" x 8" - 10', 24" oc	Demo	SF	LB	.012	---	.26	---	.26	**.40**	.22	**.33**
2" x 8" - 10', 24" oc	Inst	SF	2C	.014	.70	.38	.04	1.12	**1.42**	.92	**1.18**
2" x 10" - 12', 12" oc	Demo	SF	LB	.019	---	.42	---	.42	**.63**	.35	**.53**
2" x 10" - 12', 12" oc	Inst	SF	2C	.022	1.75	.59	.06	2.40	**2.97**	1.96	**2.42**
2" x 10" - 12', 16" oc	Demo	SF	LB	.016	---	.35	---	.35	**.53**	.29	**.43**
2" x 10" - 12', 16" oc	Inst	SF	2C	.018	1.32	.49	.05	1.86	**2.31**	1.53	**1.91**
2" x 10" - 12', 24" oc	Demo	SF	LB	.012	---	.26	---	.26	**.40**	.22	**.33**
2" x 10" - 12', 24" oc	Inst	SF	2C	.014	.91	.38	.04	1.33	**1.66**	1.09	**1.37**
2" x 12" - 14', 12" oc	Demo	SF	LB	.019	---	.42	---	.42	**.63**	.35	**.53**
2" x 12" - 14', 12" oc	Inst	SF	2C	.022	2.16	.59	.06	2.81	**3.45**	2.34	**2.87**

Description	Oper	Unit	Crew Size	Man-Hours Per Unit	Avg Mat'l Unit Cost	Avg Labor Unit Cost	Avg Equip Unit Cost	Avg Total Unit Cost	Avg Price Incl O&P	Large Volume Avg Total Unit Cost	Large Volume Avg Price Incl O&P
			Costs Based On Small Volume							Large Volume	
2" x 12" - 14', 16" oc	Demo	SF	LB	.016	---	.35	---	.35	**.53**	.31	**.46**
2" x 12" - 14', 16" oc	Inst	SF	2C	.019	1.63	.51	.05	2.19	**2.70**	1.82	**2.25**
2" x 12" - 14', 24" oc	Demo	SF	LB	.012	---	.26	---	.26	**.40**	.24	**.36**
2" x 12" - 14', 24" oc	Inst	SF	2C	.014	1.11	.38	.04	1.53	**1.89**	1.27	**1.58**
3" x 8" - 14', 12" oc	Demo	SF	LB	.019	---	.42	---	.42	**.63**	.35	**.53**
3" x 8" - 14', 12" oc	Inst	SF	2C	.022	2.68	.59	.06	3.33	**4.04**	2.76	**3.36**
3" x 8" - 14', 16" oc	Demo	SF	LB	.016	---	.35	---	.35	**.53**	.31	**.46**
3" x 8" - 14', 16" oc	Inst	SF	2C	.019	2.03	.51	.05	2.59	**3.16**	2.15	**2.63**
3" x 8" - 14', 24" oc	Demo	SF	LB	.012	---	.26	---	.26	**.40**	.22	**.33**
3" x 8" - 14', 24" oc	Inst	SF	2C	.014	1.38	.38	.04	1.80	**2.20**	1.48	**1.82**
3" x 10" - 16', 12" oc	Demo	SF	LB	.021	---	.46	---	.46	**.70**	.40	**.60**
3" x 10" - 16', 12" oc	Inst	SF	2C	.024	3.34	.65	.07	4.06	**4.89**	3.34	**4.03**
3" x 10" - 16', 16" oc	Demo	SF	LB	.018	---	.40	---	.40	**.60**	.33	**.50**
3" x 10" - 16', 16" oc	Inst	SF	2C	.020	2.53	.54	.06	3.13	**3.79**	2.59	**3.14**
3" x 10" - 16', 24" oc	Demo	SF	LB	.013	---	.29	---	.29	**.43**	.24	**.36**
3" x 10" - 16', 24" oc	Inst	SF	2C	.015	1.71	.40	.04	2.15	**2.62**	1.80	**2.19**
3" x 12" - 18', 12" oc	Demo	SF	LB	.022	---	.48	---	.48	**.73**	.42	**.63**
3" x 12" - 18', 12" oc	Inst	SF	2C	.025	4.01	.67	.07	4.75	**5.71**	3.93	**4.72**
3" x 12" - 18', 16" oc	Demo	SF	LB	.019	---	.42	---	.42	**.63**	.35	**.53**
3" x 12" - 18', 16" oc	Inst	SF	2C	.021	3.04	.57	.06	3.67	**4.42**	3.04	**3.66**
3" x 12" - 18', 24" oc	Demo	SF	LB	.014	---	.31	---	.31	**.46**	.26	**.40**
3" x 12" - 18', 24" oc	Inst	SF	2C	.016	2.07	.43	.05	2.55	**3.09**	2.09	**2.53**
Ledgers											
Nailed											
2" x 4" - 12'	Demo	LF	LB	.015	---	.33	---	.33	**.50**	.29	**.43**
2" x 4" - 12'	Inst	LF	2C	.022	.69	.59	.06	1.34	**1.76**	1.13	**1.48**
2" x 6" - 12'	Demo	LF	LB	.017	---	.37	---	.37	**.56**	.33	**.50**
2" x 6" - 12'	Inst	LF	2C	.024	1.00	.65	.07	1.72	**2.20**	1.45	**1.86**
2" x 8" - 12'	Demo	LF	LB	.019	---	.42	---	.42	**.63**	.35	**.53**
2" x 8" - 12'	Inst	LF	2C	.027	1.39	.73	.08	2.20	**2.78**	1.83	**2.32**
Bolted											
Labor/material costs are for ledgers with pre-embedded bolts. Labor costs include securing ledgers.											
3" x 6" - 12'	Demo	LF	LB	.026	---	.57	---	.57	**.86**	.48	**.73**
3" x 6" - 12'	Inst	LF	2C	.037	1.91	1.00	.11	3.02	**3.82**	2.50	**3.17**
3" x 8" - 12'	Demo	LF	LB	.029	---	.64	---	.64	**.96**	.53	**.80**
3" x 8" - 12'	Inst	LF	2C	.040	2.60	1.08	.12	3.80	**4.75**	3.15	**3.94**
3" x 10" - 12'	Demo	LF	LB	.032	---	.70	---	.70	**1.06**	.59	**.89**
3" x 10" - 12'	Inst	LF	2C	.043	3.25	1.16	.13	4.54	**5.63**	3.78	**4.70**
3" x 12" - 12'	Demo	LF	LB	.034	---	.75	---	.75	**1.13**	.64	**.96**
3" x 12" - 12'	Inst	LF	2C	.046	3.89	1.24	.13	5.26	**6.49**	4.36	**5.39**

Description	Oper	Unit	Crew Size	Man-Hours Per Unit	Avg Mat'l Unit Cost	Avg Labor Unit Cost	Avg Equip Unit Cost	Avg Total Unit Cost	Avg Price Incl O&P	Large Volume Avg Total Unit Cost	Large Volume Avg Price Incl O&P
Patio framing											
Wood deck: 4" x 4" rough sawn beams (4'-0" oc) leveled 1/16" to 1/8" and nailed to pre-set concrete piers with woodblock on top; 2" T x 4" W decking (S4S) with 1/2" spacing, double-nailed beam junctures.											
Fir	Demo	SF	LB	.050	---	1.10	---	1.10	**1.66**	.92	**1.39**
Fir	Inst	SF	CN	.071	5.68	1.84	.17	7.69	**9.50**	6.43	**7.96**
Redwood	Demo	SF	LB	.050	---	1.10	---	1.10	**1.66**	.92	**1.39**
Redwood	Inst	SF	CN	.071	6.11	1.84	.17	8.12	**10.00**	6.79	**8.38**
Wood awning: 4" x 4" columns (10'-0" oc) nailed to wood; 2" x 6" beams nailed horizontally to either side of columns; 2" x 6" ledger nailed to wall studs; 2" x 6" joists (4'-0" oc) nailed to ledger and toe-nailed on top of beams; 2" x 2" (4" oc) nailed to joists for sunscreen.											
Redwood, rough sawn	Demo	SF	LB	.065	---	1.43	---	1.43	**2.15**	1.21	**1.82**
Redwood, rough sawn	Inst	SF	CS	.113	1.73	2.86	.22	4.81	**6.55**	4.04	**5.51**
Plates; joined with studs before setting											
Double top, nailed											
2" x 4" - 8'	Demo	LF	LB	.023	---	.50	---	.50	**.76**	.44	**.66**
2" x 4" - 8'	Inst	LF	2C	.046	1.15	1.24	.13	2.52	**3.34**	2.10	**2.79**
2" x 6" - 8'	Demo	LF	LB	.023	---	.50	---	.50	**.76**	.44	**.66**
2" x 6" - 8'	Inst	LF	2C	.046	1.87	1.24	.13	3.24	**4.17**	2.70	**3.48**
Single bottom, nailed											
2" x 4" - 8'	Demo	LF	LB	.014	---	.31	---	.31	**.46**	.26	**.40**
2" x 4" - 8'	Inst	LF	2C	.023	.63	.62	.07	1.32	**1.74**	1.12	**1.48**
2" x 6" - 8'	Demo	LF	LB	.014	---	.31	---	.31	**.46**	.26	**.40**
2" x 6" - 8'	Inst	LF	2C	.023	.93	.62	.07	1.62	**2.08**	1.36	**1.75**
Sill or bottom, bolted. Labor/material costs are for plates with pre-embedded bolts, labor cost includes securing plates.											
2" x 4" - 8'	Demo	LF	LB	.027	---	.59	---	.59	**.89**	.50	**.76**
2" x 4" - 8'	Inst	LF	2C	.034	.58	.92	.10	1.60	**2.16**	1.33	**1.81**
2" x 6" - 8'	Demo	LF	LB	.027	---	.59	---	.59	**.89**	.50	**.76**
2" x 6" - 8'	Inst	LF	2C	.034	.87	.92	.10	1.89	**2.49**	1.57	**2.08**
Rafters, per LF of stick											
Common, gable or hip, to 1/3 pitch											
2" x 4" - 8' Avg.	Demo	LF	LB	.025	---	.55	---	.55	**.83**	.48	**.73**
2" x 4" - 8' Avg.	Inst	LF	2C	.028	.62	.75	.08	1.45	**1.94**	1.20	**1.60**
2" x 4" - 10' Avg.	Demo	LF	LB	.022	---	.48	---	.48	**.73**	.42	**.63**
2" x 4" - 10' Avg.	Inst	LF	2C	.024	.62	.65	.07	1.34	**1.77**	1.11	**1.47**
2" x 4" - 12' Avg.	Demo	LF	LB	.020	---	.44	---	.44	**.66**	.37	**.56**
2" x 4" - 12' Avg.	Inst	LF	2C	.021	.62	.57	.06	1.25	**1.63**	1.05	**1.37**
2" x 4" - 14' Avg.	Demo	LF	LB	.018	---	.40	---	.40	**.60**	.33	**.50**
2" x 4" - 14' Avg.	Inst	LF	2C	.019	.62	.51	.06	1.19	**1.55**	.99	**1.29**
2" x 4" - 16' Avg.	Demo	LF	LB	.017	---	.37	---	.37	**.56**	.31	**.46**
2" x 4" - 16' Avg.	Inst	LF	2C	.018	.62	.49	.05	1.16	**1.50**	.95	**1.24**
2" x 6" - 10' Avg.	Demo	LF	LB	.025	---	.55	---	.55	**.83**	.46	**.70**
2" x 6" - 10' Avg.	Inst	LF	2C	.027	.93	.73	.08	1.74	**2.26**	1.45	**1.89**

Description	Oper	Unit	Crew Size	Man-Hours Per Unit	Avg Mat'l Unit Cost	Avg Labor Unit Cost	Avg Equip Unit Cost	Avg Total Unit Cost	Avg Price Incl O&P	Large Volume Avg Total Unit Cost	Large Volume Avg Price Incl O&P
2" x 6" - 12' Avg.	Demo	LF	LB	.022	---	.48	---	.48	**.73**	.42	**.63**
2" x 6" - 12' Avg.	Inst	LF	2C	.024	.93	.65	.07	1.65	**2.12**	1.39	**1.79**
2" x 6" - 14' Avg.	Demo	LF	LB	.020	---	.44	---	.44	**.66**	.37	**.56**
2" x 6" - 14' Avg.	Inst	LF	2C	.022	.93	.59	.06	1.58	**2.03**	1.32	**1.70**
2" x 6" - 16' Avg.	Demo	LF	LB	.019	---	.42	---	.42	**.63**	.35	**.53**
2" x 6" - 16' Avg.	Inst	LF	2C	.020	.93	.54	.06	1.53	**1.95**	1.27	**1.62**
2" x 6" - 18' Avg.	Demo	LF	LB	.017	---	.37	---	.37	**.56**	.33	**.50**
2" x 6" - 18' Avg.	Inst	LF	2C	.019	.93	.51	.05	1.49	**1.90**	1.24	**1.58**
2" x 8" - 12' Avg.	Demo	LF	LB	.026	---	.57	---	.57	**.86**	.48	**.73**
2" x 8" - 12' Avg.	Inst	LF	2C	.028	1.32	.75	.08	2.15	**2.75**	1.77	**2.25**
2" x 8" - 14' Avg.	Demo	LF	LB	.023	---	.50	---	.50	**.76**	.44	**.66**
2" x 8" - 14' Avg.	Inst	LF	2C	.025	1.32	.67	.07	2.06	**2.61**	1.71	**2.16**
2" x 8" - 16' Avg.	Demo	LF	LB	.022	---	.48	---	.48	**.73**	.42	**.63**
2" x 8" - 16' Avg.	Inst	LF	2C	.023	1.32	.62	.07	2.01	**2.53**	1.68	**2.12**
2" x 8" - 18' Avg.	Demo	LF	LB	.021	---	.46	---	.46	**.70**	.37	**.56**
2" x 8" - 18' Avg.	Inst	LF	2C	.022	1.32	.59	.06	1.97	**2.48**	1.64	**2.07**
2" x 8" - 20' Avg.	Demo	LF	LB	.019	---	.42	---	.42	**.63**	.37	**.56**
2" x 8" - 20' Avg.	Inst	LF	2C	.021	1.32	.57	.06	1.95	**2.44**	1.62	**2.03**
Common, gable or hip, to 3/8-1/2 pitch											
2" x 4" - 8' Avg.	Demo	LF	LB	.030	---	.66	---	.66	**.99**	.55	**.83**
2" x 4" - 8' Avg.	Inst	LF	2C	.032	.62	.86	.09	1.57	**2.11**	1.34	**1.81**
2" x 4" - 10' Avg.	Demo	LF	LB	.025	---	.55	---	.55	**.83**	.48	**.73**
2" x 4" - 10' Avg.	Inst	LF	2C	.028	.62	.75	.08	1.45	**1.94**	1.23	**1.64**
2" x 4" - 12' Avg.	Demo	LF	LB	.023	---	.50	---	.50	**.76**	.42	**.63**
2" x 4" - 12' Avg.	Inst	LF	2C	.024	.62	.65	.07	1.34	**1.77**	1.14	**1.51**
2" x 4" - 14' Avg.	Demo	LF	LB	.020	---	.44	---	.44	**.66**	.37	**.56**
2" x 4" - 14' Avg.	Inst	LF	2C	.022	.62	.59	.06	1.27	**1.67**	1.07	**1.42**
2" x 4" - 16' Avg.	Demo	LF	LB	.019	---	.42	---	.42	**.63**	.35	**.53**
2" x 4" - 16' Avg.	Inst	LF	2C	.020	.62	.54	.06	1.22	**1.59**	1.02	**1.33**
2" x 6" - 10' Avg.	Demo	LF	LB	.029	---	.64	---	.64	**.96**	.53	**.80**
2" x 6" - 10' Avg.	Inst	LF	2C	.031	.93	.84	.09	1.86	**2.43**	1.54	**2.02**
2" x 6" - 12' Avg.	Demo	LF	LB	.025	---	.55	---	.55	**.83**	.48	**.73**
2" x 6" - 12' Avg.	Inst	LF	2C	.027	.93	.73	.08	1.74	**2.26**	1.45	**1.89**
2" x 6" - 14' Avg.	Demo	LF	LB	.023	---	.50	---	.50	**.76**	.42	**.63**
2" x 6" - 14' Avg.	Inst	LF	2C	.025	.93	.67	.07	1.67	**2.16**	1.39	**1.79**
2" x 6" - 16' Avg.	Demo	LF	LB	.021	---	.46	---	.46	**.70**	.40	**.60**
2" x 6" - 16' Avg.	Inst	LF	2C	.023	.93	.62	.07	1.62	**2.08**	1.33	**1.71**
2" x 6" - 18' Avg.	Demo	LF	LB	.019	---	.42	---	.42	**.63**	.35	**.53**
2" x 6" - 18' Avg.	Inst	LF	2C	.021	.93	.57	.06	1.56	**1.99**	1.30	**1.66**
2" x 8" - 12' Avg.	Demo	LF	LB	.029	---	.64	---	.64	**.96**	.53	**.80**
2" x 8" - 12' Avg.	Inst	LF	2C	.031	1.32	.84	.09	2.25	**2.88**	1.86	**2.39**

			Costs Based On Small Volume							Large Volume	
Description	Oper	Unit	Crew Size	Man-Hours Per Unit	Avg Mat'l Unit Cost	Avg Labor Unit Cost	Avg Equip Unit Cost	Avg Total Unit Cost	Avg Price Incl O&P	Avg Total Unit Cost	Avg Price Incl O&P
2" x 8" - 14' Avg.	Demo	LF	LB	.026	---	.57	---	.57	**.86**	.48	**.73**
2" x 8" - 14' Avg.	Inst	LF	2C	.028	1.32	.75	.08	2.15	**2.75**	1.80	**2.30**
2" x 8" - 16' Avg.	Demo	LF	LB	.024	---	.53	---	.53	**.80**	.44	**.66**
2" x 8" - 16' Avg.	Inst	LF	2C	.026	1.32	.70	.08	2.10	**2.66**	1.73	**2.20**
2" x 8" - 18' Avg.	Demo	LF	LB	.023	---	.50	---	.50	**.76**	.42	**.63**
2" x 8" - 18' Avg.	Inst	LF	2C	.024	1.32	.65	.07	2.04	**2.57**	1.71	**2.16**
2" x 8" - 20' Avg.	Demo	LF	LB	.021	---	.46	---	.46	**.70**	.40	**.60**
2" x 8" - 20' Avg.	Inst	LF	2C	.023	1.32	.62	.07	2.01	**2.53**	1.65	**2.08**
Common, cut-up roofs, to 1/3 pitch											
2" x 4" - 8' Avg.	Demo	LF	LB	.032	---	.70	---	.70	**1.06**	.59	**.89**
2" x 4" - 8' Avg.	Inst	LF	2C	.035	.63	.94	.10	1.67	**2.26**	1.41	**1.91**
2" x 4" - 10' Avg.	Demo	LF	LB	.027	---	.59	---	.59	**.89**	.50	**.76**
2" x 4" - 10' Avg.	Inst	LF	2C	.030	.63	.81	.09	1.53	**2.04**	1.25	**1.68**
2" x 4" - 12' Avg.	Demo	LF	LB	.024	---	.53	---	.53	**.80**	.44	**.66**
2" x 4" - 12' Avg.	Inst	LF	2C	.026	.63	.70	.08	1.41	**1.87**	1.16	**1.55**
2" x 4" - 14' Avg.	Demo	LF	LB	.022	---	.48	---	.48	**.73**	.40	**.60**
2" x 4" - 14' Avg.	Inst	LF	2C	.024	.63	.65	.07	1.35	**1.78**	1.11	**1.47**
2" x 4" - 16' Avg.	Demo	LF	LB	.020	---	.44	---	.44	**.66**	.37	**.56**
2" x 4" - 16' Avg.	Inst	LF	2C	.022	.63	.59	.06	1.28	**1.69**	1.05	**1.37**
2" x 6" - 10' Avg.	Demo	LF	LB	.031	---	.68	---	.68	**1.03**	.57	**.86**
2" x 6" - 10' Avg.	Inst	LF	2C	.033	.94	.89	.10	1.93	**2.53**	1.60	**2.11**
2" x 6" - 12' Avg.	Demo	LF	LB	.027	---	.59	---	.59	**.89**	.50	**.76**
2" x 6" - 12' Avg.	Inst	LF	2C	.029	.94	.78	.08	1.80	**2.35**	.74	**1.09**
2" x 6" - 14' Avg.	Demo	LF	LB	.024	---	.53	---	.53	**.80**	.44	**.66**
2" x 6" - 14' Avg.	Inst	LF	2C	.026	.94	.70	.08	1.72	**2.23**	.65	**.96**
2" x 6" - 16' Avg.	Demo	LF	LB	.022	---	.48	---	.48	**.73**	.42	**.63**
2" x 6" - 16' Avg.	Inst	LF	2C	.024	.94	.65	.07	1.66	**2.13**	1.37	**1.77**
2" x 6" - 18' Avg.	Demo	LF	LB	.020	---	.44	---	.44	**.66**	.37	**.56**
2" x 6" - 18' Avg.	Inst	LF	2C	.022	.94	.59	.06	1.59	**2.04**	1.33	**1.71**
2" x 8" - 12' Avg.	Demo	LF	LB	.030	---	.66	---	.66	**.99**	.57	**.86**
2" x 8" - 12' Avg.	Inst	LF	2C	.032	1.34	.86	.09	2.29	**2.94**	1.93	**2.49**
2" x 8" - 14' Avg.	Demo	LF	LB	.027	---	.59	---	.59	**.89**	.50	**.76**
2" x 8" - 14' Avg.	Inst	LF	2C	.029	1.34	.78	.09	2.21	**2.82**	1.84	**2.36**
2" x 8" - 16' Avg.	Demo	LF	LB	.025	---	.55	---	.55	**.83**	.46	**.70**
2" x 8" - 16' Avg.	Inst	LF	2C	.027	1.34	.73	.08	2.15	**2.73**	1.79	**2.28**
2" x 8" - 18' Avg.	Demo	LF	LB	.023	---	.50	---	.50	**.76**	.44	**.66**
2" x 8" - 18' Avg.	Inst	LF	2C	.025	1.34	.67	.07	2.08	**2.64**	1.73	**2.19**
2" x 8" - 20' Avg.	Demo	LF	LB	.022	---	.48	---	.48	**.73**	.42	**.63**
2" x 8" - 20' Avg.	Inst	LF	2C	.024	1.34	.65	.07	2.06	**2.59**	1.70	**2.14**
Common, cut-up roofs, to 3/8-1/2 pitch											
2" x 4" - 8' Avg.	Demo	LF	LB	.036	---	.79	---	.79	**1.19**	.68	**1.03**
2" x 4" - 8' Avg.	Inst	LF	2C	.040	.63	1.08	.12	1.83	**2.48**	1.53	**2.08**
2" x 4" - 10' Avg.	Demo	LF	LB	.031	---	.68	---	.68	**1.03**	.57	**.86**
2" x 4" - 10' Avg.	Inst	LF	2C	.033	.63	.89	.10	1.62	**2.18**	1.34	**1.81**
2" x 4" - 12' Avg.	Demo	LF	LB	.027	---	.59	---	.59	**.89**	.50	**.76**
2" x 4" - 12' Avg.	Inst	LF	2C	.029	.63	.78	.09	1.50	**2.00**	1.25	**1.68**

Description	Oper	Unit	Crew Size	Man-Hours Per Unit	Avg Mat'l Unit Cost	Avg Labor Unit Cost	Avg Equip Unit Cost	Avg Total Unit Cost	Avg Price Incl O&P	Avg Total Unit Cost	Avg Price Incl O&P
			Costs Based On Small Volume							Large Volume	
2" x 4" - 14' Avg.	Demo	LF	LB	.024	---	.53	---	.53	**.80**	.46	**.70**
2" x 4" - 14' Avg.	Inst	LF	2C	.026	.63	.70	.08	1.41	**1.87**	1.17	**1.56**
2" x 4" - 16' Avg.	Demo	LF	LB	.022	---	.48	---	.48	**.73**	.42	**.63**
2" x 4" - 16' Avg.	Inst	LF	2C	.024	.63	.65	.07	1.35	**1.78**	1.14	**1.51**
2" x 6" - 10' Avg.	Demo	LF	LB	.034	---	.75	---	.75	**1.13**	.64	**.96**
2" x 6" - 10' Avg.	Inst	LF	2C	.037	.94	1.00	.11	2.05	**2.71**	1.70	**2.25**
2" x 6" - 12' Avg.	Demo	LF	LB	.030	---	.66	---	.66	**.99**	.55	**.83**
2" x 6" - 12' Avg.	Inst	LF	2C	.032	.94	.86	.09	1.89	**2.48**	1.58	**2.07**
2" x 6" - 14' Avg.	Demo	LF	LB	.027	---	.59	---	.59	**.89**	.50	**.76**
2" x 6" - 14' Avg.	Inst	LF	2C	.029	.94	.78	.08	1.80	**2.35**	1.51	**1.98**
2" x 6" - 16' Avg.	Demo	LF	LB	.024	---	.53	---	.53	**.80**	.46	**.70**
2" x 6" - 16' Avg.	Inst	LF	2C	.026	.94	.70	.08	1.72	**2.23**	1.43	**1.86**
2" x 6" - 18' Avg.	Demo	LF	LB	.022	---	.48	---	.48	**.73**	.42	**.63**
2" x 6" - 18' Avg.	Inst	LF	2C	.024	.94	.65	.07	1.66	**2.13**	1.40	**1.81**
2" x 8" - 12' Avg.	Demo	LF	LB	.033	---	.72	---	.72	**1.09**	.61	**.93**
2" x 8" - 12' Avg.	Inst	LF	2C	.036	1.34	.97	.10	2.41	**3.12**	2.00	**2.58**
2" x 8" - 14' Avg.	Demo	LF	LB	.030	---	.66	---	.66	**.99**	.55	**.83**
2" x 8" - 14' Avg.	Inst	LF	2C	.032	1.34	.86	.09	2.29	**2.94**	1.91	**2.45**
2" x 8" - 16' Avg.	Demo	LF	LB	.027	---	.59	---	.59	**.89**	.50	**.76**
2" x 8" - 16' Avg.	Inst	LF	2C	.030	1.34	.81	.09	2.24	**2.86**	1.84	**2.36**
2" x 8" - 18' Avg.	Demo	LF	LB	.025	---	.55	---	.55	**.83**	.48	**.73**
2" x 8" - 18' Avg.	Inst	LF	2C	.027	1.34	.73	.08	2.15	**2.73**	1.79	**2.28**
2" x 8" - 20' Avg.	Demo	LF	LB	.024	---	.53	---	.53	**.80**	.44	**.66**
2" x 8" - 20' Avg.	Inst	LF	2C	.026	1.34	.70	.08	2.12	**2.69**	1.75	**2.23**
Rafters, per SF of area											
Gable or hip, to 1/3 pitch											
2" x 4" - 12" oc	Demo	SF	LB	.022	---	.48	---	.48	**.73**	.42	**.63**
2" x 4" - 12" oc	Inst	SF	2C	.024	.69	.65	.07	1.41	**1.85**	1.20	**1.58**
2" x 4" - 16" oc	Demo	SF	LB	.019	---	.42	---	.42	**.63**	.35	**.53**
2" x 4" - 16" oc	Inst	SF	2C	.020	.54	.54	.06	1.14	**1.50**	.95	**1.25**
2" x 4" - 24" oc	Demo	SF	LB	.014	---	.31	---	.31	**.46**	.26	**.40**
2" x 4" - 24" oc	Inst	SF	2C	.016	.39	.43	.05	.87	**1.15**	.71	**.94**
2" x 6" - 12" oc	Demo	SF	LB	.023	---	.50	---	.50	**.76**	.42	**.63**
2" x 6" - 12" oc	Inst	SF	2C	.025	1.01	.67	.07	1.75	**2.26**	1.46	**1.88**
2" x 6" - 16" oc	Demo	SF	LB	.019	---	.42	---	.42	**.63**	.35	**.53**
2" x 6" - 16" oc	Inst	SF	2C	.021	.78	.57	.06	1.41	**1.82**	1.18	**1.52**
2" x 6" - 24" oc	Demo	SF	LB	.015	---	.33	---	.33	**.50**	.26	**.40**
2" x 6" - 24" oc	Inst	SF	2C	.016	.56	.43	.05	1.04	**1.35**	.85	**1.10**
2" x 8" - 12" oc	Demo	SF	LB	.024	---	.53	---	.53	**.80**	.44	**.66**
2" x 8" - 12" oc	Inst	SF	2C	.026	1.41	.70	.08	2.19	**2.77**	1.81	**2.30**
2" x 8" - 16" oc	Demo	SF	LB	.020	---	.44	---	.44	**.66**	.37	**.56**
2" x 8" - 16" oc	Inst	SF	2C	.022	1.08	.59	.06	1.73	**2.20**	1.43	**1.81**

Description	Oper	Unit	Crew Size	Man-Hours Per Unit	Avg Mat'l Unit Cost	Avg Labor Unit Cost	Avg Equip Unit Cost	Avg Total Unit Cost	Avg Price Incl O&P	Large Volume Avg Total Unit Cost	Large Volume Avg Price Incl O&P
2" x 8" - 24" oc	Demo	SF	LB	.015	---	.33	---	.33	**.50**	.29	**.43**
2" x 8" - 24" oc	Inst	SF	2C	.017	.76	.46	.05	1.27	**1.62**	1.04	**1.33**
Gable or hip, to 3/8-1/2 pitch											
2" x 4" - 12" oc	Demo	SF	LB	.026	---	.57	---	.57	**.86**	.48	**.73**
2" x 4" - 12" oc	Inst	SF	2C	.028	.69	.75	.08	1.52	**2.02**	1.29	**1.71**
2" x 4" - 16" oc	Demo	SF	LB	.022	---	.48	---	.48	**.73**	.42	**.63**
2" x 4" - 16" oc	Inst	SF	2C	.024	.54	.65	.07	1.26	**1.67**	1.04	**1.39**
2" x 4" - 24" oc	Demo	SF	LB	.017	---	.37	---	.37	**.56**	.31	**.46**
2" x 4" - 24" oc	Inst	SF	2C	.018	.39	.49	.05	.93	**1.24**	.77	**1.03**
2" x 6" - 12" oc	Demo	SF	LB	.026	---	.57	---	.57	**.86**	.48	**.73**
2" x 6" - 12" oc	Inst	SF	2C	.028	1.01	.75	.08	1.84	**2.39**	1.55	**2.01**
2" x 6" - 16" oc	Demo	SF	LB	.022	---	.48	---	.48	**.73**	.42	**.63**
2" x 6" - 16" oc	Inst	SF	2C	.024	.78	.65	.07	1.50	**1.95**	1.24	**1.62**
2" x 6" - 24" oc	Demo	SF	LB	.017	---	.37	---	.37	**.56**	.31	**.46**
2" x 6" - 24" oc	Inst	SF	2C	.018	.56	.49	.05	1.10	**1.43**	.90	**1.18**
2" x 8" - 12" oc	Demo	SF	LB	.027	---	.59	---	.59	**.89**	.50	**.76**
2" x 8" - 12" oc	Inst	SF	2C	.029	1.41	.78	.08	2.27	**2.89**	1.88	**2.39**
2" x 8" - 16" oc	Demo	SF	LB	.022	---	.48	---	.48	**.73**	.42	**.63**
2" x 8" - 16" oc	Inst	SF	2C	.024	1.08	.65	.07	1.80	**2.30**	1.49	**1.90**
2" x 8" - 24" oc	Demo	SF	LB	.014	---	.31	---	.31	**.46**	.26	**.40**
2" x 8" - 24" oc	Inst	SF	2C	.015	.76	.40	.04	1.20	**1.53**	1.01	**1.29**
Cut-up roofs, to 1/3 pitch											
2" x 4" - 12" oc	Demo	SF	LB	.028	---	.61	---	.61	**.93**	.53	**.80**
2" x 4" - 12" oc	Inst	SF	2C	.030	.71	.81	.09	1.61	**2.14**	1.36	**1.81**
2" x 4" - 16" oc	Demo	SF	LB	.023	---	.50	---	.50	**.76**	.44	**.66**
2" x 4" - 16" oc	Inst	SF	2C	.026	.55	.70	.07	1.32	**1.77**	1.10	**1.48**
2" x 4" - 24" oc	Demo	SF	LB	.018	---	.40	---	.40	**.60**	.33	**.50**
2" x 4" - 24" oc	Inst	SF	2C	.019	.39	.51	.06	.96	**1.29**	.83	**1.11**
2" x 6" - 12" oc	Demo	SF	LB	.027	---	.59	---	.59	**.89**	.50	**.76**
2" x 6" - 12" oc	Inst	SF	2C	.030	1.03	.81	.09	1.93	**2.50**	1.58	**2.06**
2" x 6" - 16" oc	Demo	SF	LB	.023	---	.50	---	.50	**.76**	.44	**.66**
2" x 6" - 16" oc	Inst	SF	2C	.025	.80	.67	.07	1.54	**2.01**	1.28	**1.67**
2" x 6" - 24" oc	Demo	SF	LB	.018	---	.40	---	.40	**.60**	.33	**.50**
2" x 6" - 24" oc	Inst	SF	2C	.019	.57	.51	.06	1.14	**1.49**	.94	**1.23**
2" x 8" - 12" oc	Demo	SF	LB	.028	---	.61	---	.61	**.93**	.53	**.80**
2" x 8" - 12" oc	Inst	SF	2C	.030	1.44	.81	.09	2.34	**2.98**	1.95	**2.49**
2" x 8" - 16" oc	Demo	SF	LB	.023	---	.50	---	.50	**.76**	.44	**.66**
2" x 8" - 16" oc	Inst	SF	2C	.025	1.10	.67	.07	1.84	**2.36**	1.55	**2.00**
2" x 8" - 24" oc	Demo	SF	LB	.018	---	.40	---	.40	**.60**	.33	**.50**
2" x 8" - 24" oc	Inst	SF	2C	.019	.77	.51	.06	1.34	**1.72**	1.11	**1.43**

Description	Oper	Unit	Crew Size	Man-Hours Per Unit	Avg Mat'l Unit Cost	Avg Labor Unit Cost	Avg Equip Unit Cost	Avg Total Unit Cost	Avg Price Incl O&P	Large Volume Avg Total Unit Cost	Large Volume Avg Price Incl O&P
Cut-up roofs, to 3/8-1/2 pitch											
2" x 4" - 12" oc	Demo	SF	LB	.032	---	.70	---	.70	**1.06**	.59	**.89**
2" x 4" - 12" oc	Inst	SF	2C	.034	.71	.92	.10	1.73	**2.31**	1.45	**1.95**
2" x 4" - 16" oc	Demo	SF	LB	.026	---	.57	---	.57	**.86**	.48	**.73**
2" x 4" - 16" oc	Inst	SF	2C	.029	.55	.78	.08	1.41	**1.90**	1.19	**1.61**
2" x 4" - 24" oc	Demo	SF	LB	.020	---	.44	---	.44	**.66**	.37	**.56**
2" x 4" - 24" oc	Inst	SF	2C	.022	.39	.59	.06	1.04	**1.41**	.88	**1.20**
2" x 6" - 12" oc	Demo	SF	LB	.030	---	.66	---	.66	**.99**	.57	**.86**
2" x 6" - 12" oc	Inst	SF	2C	.033	1.03	.89	.10	2.02	**2.64**	1.67	**2.19**
2" x 6" - 16" oc	Demo	SF	LB	.026	---	.57	---	.57	**.86**	.48	**.73**
2" x 6" - 16" oc	Inst	SF	2C	.028	.80	.75	.08	1.63	**2.15**	1.37	**1.80**
2" x 6" - 24" oc	Demo	SF	LB	.020	---	.44	---	.44	**.66**	.37	**.56**
2" x 6" - 24" oc	Inst	SF	2C	.021	.57	.57	.06	1.20	**1.58**	1.00	**1.32**
2" x 8" - 12" oc	Demo	SF	LB	.031	---	.68	---	.68	**1.03**	.57	**.86**
2" x 8" - 12" oc	Inst	SF	2C	.033	1.44	.89	.10	2.43	**3.11**	2.01	**2.58**
2" x 8" - 16" oc	Demo	SF	LB	.026	---	.57	---	.57	**.86**	.48	**.73**
2" x 8" - 16" oc	Inst	SF	2C	.028	1.10	.75	.08	1.93	**2.49**	1.62	**2.09**
2" x 8" - 24" oc	Demo	SF	LB	.020	---	.44	---	.44	**.66**	.37	**.56**
2" x 8" - 24" oc	Inst	SF	2C	.021	.77	.57	.06	1.40	**1.81**	1.17	**1.51**
Roof decking, solid, T&G, dry for plank-and-beam construction											
2" x 6" - 12'	Demo	SF	LB	.029	---	.64	---	.64	**.96**	.55	**.83**
2" x 6" - 12'	Inst	SF	2C	.039	.15	1.05	.11	1.31	**1.88**	1.11	**1.59**
2" x 8" - 12'	Demo	SF	LB	.021	---	.46	---	.46	**.70**	.40	**.60**
2" x 8" - 12'	Inst	SF	2C	.028	.10	.75	.08	.93	**1.34**	.80	**1.15**
Studs/plates, per LF of stick											
Walls or partitions											
2" x 4" - 8' Avg.	Demo	LF	LB	.018	---	.40	---	.40	**.60**	.35	**.53**
2" x 4" - 8' Avg.	Inst	LF	2C	.022	.59	.59	.06	1.24	**1.64**	1.05	**1.39**
2" x 4" - 10' Avg.	Demo	LF	LB	.016	---	.35	---	.35	**.53**	.31	**.46**
2" x 4" - 10' Avg.	Inst	LF	2C	.019	.59	.51	.06	1.16	**1.52**	.99	**1.30**
2" x 4" - 12' Avg.	Demo	LF	LB	.014	---	.31	---	.31	**.46**	.26	**.40**
2" x 4" - 12' Avg.	Inst	LF	2C	.017	.59	.46	.05	1.10	**1.43**	.92	**1.21**
2" x 6" - 8' Avg.	Demo	LF	LB	.021	---	.46	---	.46	**.70**	.40	**.60**
2" x 6" - 8' Avg.	Inst	LF	2C	.025	.89	.67	.07	1.63	**2.12**	1.36	**1.76**
2" x 6" - 10' Avg.	Demo	LF	LB	.018	---	.40	---	.40	**.60**	.35	**.53**
2" x 6" - 10' Avg.	Inst	LF	2C	.022	.89	.59	.06	1.54	**1.99**	1.29	**1.67**
2" x 6" - 12' Avg.	Demo	LF	LB	.016	---	.35	---	.35	**.53**	.31	**.46**
2" x 6" - 12' Avg.	Inst	LF	2C	.020	.89	.54	.06	1.49	**1.90**	1.24	**1.59**

Description	Oper	Unit	Crew Size	Man-Hours Per Unit	Avg Mat'l Unit Cost	Avg Labor Unit Cost	Avg Equip Unit Cost	Avg Total Unit Cost	Avg Price Incl O&P	Large Volume Avg Total Unit Cost	Large Volume Avg Price Incl O&P
2" x 8" - 8' Avg.	Demo	LF	LB	.023	---	.50	---	.50	**.76**	.44	**.66**
2" x 8" - 8' Avg.	Inst	LF	2C	.028	1.26	.75	.08	2.09	**2.68**	1.75	**2.24**
2" x 8" - 10' Avg.	Demo	LF	LB	.021	---	.46	---	.46	**.70**	.37	**.56**
2" x 8" - 10' Avg.	Inst	LF	2C	.024	1.26	.65	.07	1.98	**2.50**	1.66	**2.11**
2" x 8" - 12' Avg.	Demo	LF	LB	.019	---	.42	---	.42	**.63**	.35	**.53**
2" x 8" - 12' Avg.	Inst	LF	2C	.022	1.26	.59	.06	1.91	**2.41**	1.59	**2.01**
Gable ends											
2" x 4" - 3' Avg.	Demo	LF	LB	.029	---	.64	---	.64	**.96**	.53	**.80**
2" x 4" - 3' Avg.	Inst	LF	2C	.055	.62	1.48	.16	2.26	**3.13**	1.92	**2.65**
2" x 4" - 4' Avg.	Demo	LF	LB	.023	---	.50	---	.50	**.76**	.44	**.66**
2" x 4" - 4' Avg.	Inst	LF	2C	.044	.62	1.19	.13	1.94	**2.65**	1.62	**2.21**
2" x 4" - 5' Avg.	Demo	LF	LB	.020	---	.44	---	.44	**.66**	.37	**.56**
2" x 4" - 5' Avg.	Inst	LF	2C	.037	.62	1.00	.11	1.73	**2.34**	1.44	**1.95**
2" x 4" - 6' Avg.	Demo	LF	LB	.017	---	.37	---	.37	**.56**	.33	**.50**
2" x 4" - 6' Avg.	Inst	LF	2C	.032	.62	.86	.09	1.57	**2.11**	1.34	**1.81**
2" x 6" - 3' Avg.	Demo	LF	LB	.030	---	.66	---	.66	**.99**	.55	**.83**
2" x 6" - 3' Avg.	Inst	LF	2C	.056	.93	1.51	.16	2.60	**3.52**	2.17	**2.94**
2" x 6" - 4' Avg.	Demo	LF	LB	.024	---	.53	---	.53	**.80**	.46	**.70**
2" x 6" - 4' Avg.	Inst	LF	2C	.045	.93	1.21	.13	2.27	**3.04**	1.89	**2.54**
2" x 6" - 5' Avg.	Demo	LF	LB	.021	---	.46	---	.46	**.70**	.40	**.60**
2" x 6" - 5' Avg.	Inst	LF	2C	.038	.93	1.02	.11	2.06	**2.74**	1.71	**2.28**
2" x 6" - 6' Avg.	Demo	LF	LB	.019	---	.42	---	.42	**.63**	.35	**.53**
2" x 6" - 6' Avg.	Inst	LF	2C	.034	.93	.92	.10	1.95	**2.56**	1.62	**2.14**
2" x 8" - 3' Avg.	Demo	LF	LB	.031	---	.68	---	.68	**1.03**	.57	**.86**
2" x 8" - 3' Avg.	Inst	LF	2C	.057	1.32	1.54	.17	3.03	**4.02**	2.51	**3.35**
2" x 8" - 4' Avg.	Demo	LF	LB	.025	---	.55	---	.55	**.83**	.48	**.73**
2" x 8" - 4' Avg.	Inst	LF	2C	.046	1.32	1.24	.13	2.69	**3.53**	2.24	**2.95**
2" x 8" - 5' Avg.	Demo	LF	LB	.022	---	.48	---	.48	**.73**	.42	**.63**
2" x 8" - 5' Avg.	Inst	LF	2C	.039	1.32	1.05	.11	2.48	**3.23**	2.07	**2.69**
2" x 8" - 6' Avg.	Demo	LF	LB	.020	---	.44	---	.44	**.66**	.37	**.56**
2" x 8" - 6' Avg.	Inst	LF	2C	.035	1.32	.94	.10	2.36	**3.05**	1.98	**2.56**
Studs/plates, per SF of area											
Walls or partitions											
2" x 4" - 8', 12" oc	Demo	SF	LB	.031	---	.68	---	.68	**1.03**	.59	**.89**
2" x 4" - 8', 12" oc	Inst	SF	2C	.038	1.03	1.02	.11	2.16	**2.85**	1.79	**2.37**
2" x 4" - 8', 16" oc	Demo	SF	LB	.028	---	.61	---	.61	**.93**	.53	**.80**
2" x 4" - 8', 16" oc	Inst	SF	2C	.034	.84	.92	.10	1.86	**2.46**	1.54	**2.05**
2" x 4" - 8', 24" oc	Demo	SF	LB	.024	---	.53	---	.53	**.80**	.44	**.66**
2" x 4" - 8', 24" oc	Inst	SF	2C	.029	.64	.78	.08	1.50	**2.00**	1.26	**1.69**
2" x 4" - 10', 12" oc	Demo	SF	LB	.026	---	.57	---	.57	**.86**	.48	**.73**
2" x 4" - 10', 12" oc	Inst	SF	2C	.032	.98	.86	.09	1.93	**2.53**	1.61	**2.11**
2" x 4" - 10', 16" oc	Demo	SF	LB	.023	---	.50	---	.50	**.76**	.44	**.66**
2" x 4" - 10', 16" oc	Inst	SF	2C	.028	.79	.75	.08	1.62	**2.14**	1.37	**1.80**
2" x 4" - 10', 24" oc	Demo	SF	LB	.019	---	.42	---	.42	**.63**	.35	**.53**
2" x 4" - 10', 24" oc	Inst	SF	2C	.023	.60	.62	.07	1.29	**1.70**	1.09	**1.44**

Description	Oper	Unit	Crew Size	Man-Hours Per Unit	Avg Mat'l Unit Cost	Avg Labor Unit Cost	Avg Equip Unit Cost	Avg Total Unit Cost	Avg Price Incl O&P	Avg Total Unit Cost	Avg Price Incl O&P
			Costs Based On Small Volume							Large Volume	
2" x 4" - 12', 12" oc	Demo	SF	LB	.023	---	.50	---	.50	**.76**	.42	**.63**
2" x 4" - 12', 12" oc	Inst	SF	2C	.028	.96	.75	.08	1.79	**2.33**	1.47	**1.91**
2" x 4" - 12', 16" oc	Demo	SF	LB	.020	---	.44	---	.44	**.66**	.37	**.56**
2" x 4" - 12', 16" oc	Inst	SF	2C	.024	.76	.65	.07	1.48	**1.93**	1.25	**1.63**
2" x 4" - 12', 24" oc	Demo	SF	LB	.016	---	.35	---	.35	**.53**	.31	**.46**
2" x 4" - 12', 24" oc	Inst	SF	2C	.020	.57	.54	.06	1.17	**1.54**	.97	**1.28**
2" x 6" - 8', 12" oc	Demo	SF	LB	.036	---	.79	---	.79	**1.19**	.66	**.99**
2" x 6" - 8', 12" oc	Inst	SF	2C	.043	1.53	1.16	.13	2.82	**3.65**	2.34	**3.03**
2" x 6" - 8', 16" oc	Demo	SF	LB	.032	---	.70	---	.70	**1.06**	.59	**.89**
2" x 6" - 8', 16" oc	Inst	SF	2C	.038	1.25	1.02	.11	2.38	**3.10**	2.01	**2.63**
2" x 6" - 8', 24" oc	Demo	SF	LB	.027	---	.59	---	.59	**.89**	.50	**.76**
2" x 6" - 8', 24" oc	Inst	SF	2C	.033	.95	.89	.09	1.93	**2.53**	1.61	**2.12**
2" x 6" - 10', 12" oc	Demo	SF	LB	.030	---	.66	---	.66	**.99**	.55	**.83**
2" x 6" - 10', 12" oc	Inst	SF	2C	.036	1.46	.97	.10	2.53	**3.25**	2.10	**2.70**
2" x 6" - 10', 16" oc	Demo	SF	LB	.026	---	.57	---	.57	**.86**	.48	**.73**
2" x 6" - 10', 16" oc	Inst	SF	2C	.032	1.18	.86	.09	2.13	**2.76**	1.77	**2.29**
2" x 6" - 10', 24" oc	Demo	SF	LB	.022	---	.48	---	.48	**.73**	.42	**.63**
2" x 6" - 10', 24" oc	Inst	SF	2C	.026	.89	.70	.08	1.67	**2.17**	1.42	**1.85**
2" x 6" - 12', 12" oc	Demo	SF	LB	.026	---	.57	---	.57	**.86**	.48	**.73**
2" x 6" - 12', 12" oc	Inst	SF	2C	.031	1.42	.84	.09	2.35	**2.99**	1.97	**2.52**
2" x 6" - 12', 16" oc	Demo	SF	LB	.023	---	.50	---	.50	**.76**	.42	**.63**
2" x 6" - 12', 16" oc	Inst	SF	2C	.027	1.13	.73	.08	1.94	**2.49**	1.62	**2.08**
2" x 6" - 12', 24" oc	Demo	SF	LB	.019	---	.42	---	.42	**.63**	.35	**.53**
2" x 6" - 12', 24" oc	Inst	SF	2C	.023	.85	.62	.07	1.54	**1.99**	1.26	**1.63**
2" x 8" - 8', 12" oc	Demo	SF	LB	.040	---	.88	---	.88	**1.33**	.75	**1.13**
2" x 8" - 8', 12" oc	Inst	SF	2C	.048	2.17	1.29	.14	3.60	**4.60**	3.01	**3.85**
2" x 8" - 8', 16" oc	Demo	SF	LB	.036	---	.79	---	.79	**1.19**	.66	**.99**
2" x 8" - 8', 16" oc	Inst	SF	2C	.043	1.75	1.16	.12	3.03	**3.89**	2.52	**3.24**
2" x 8" - 8', 24" oc	Demo	SF	LB	.030	---	.66	---	.66	**.99**	.57	**.86**
2" x 8" - 8', 24" oc	Inst	SF	2C	.036	1.34	.97	.11	2.42	**3.13**	2.03	**2.63**
2" x 8" - 10', 12" oc	Demo	SF	LB	.034	---	.75	---	.75	**1.13**	.64	**.96**
2" x 8" - 10', 12" oc	Inst	SF	2C	.040	2.07	1.08	.12	3.27	**4.14**	2.72	**3.45**
2" x 8" - 10', 16" oc	Demo	SF	LB	.030	---	.66	---	.66	**.99**	.55	**.83**
2" x 8" - 10', 16" oc	Inst	SF	2C	.035	1.66	.94	.10	2.70	**3.44**	2.26	**2.88**
2" x 8" - 10', 24" oc	Demo	SF	LB	.025	---	.55	---	.55	**.83**	.46	**.70**
2" x 8" - 10', 24" oc	Inst	SF	2C	.030	1.25	.81	.09	2.15	**2.76**	1.76	**2.27**
2" x 8" - 12', 12" oc	Demo	SF	LB	.030	---	.66	---	.66	**.99**	.55	**.83**
2" x 8" - 12', 12" oc	Inst	SF	2C	.035	2.02	.94	.10	3.06	**3.86**	2.55	**3.22**

				Costs Based On Small Volume						Large Volume	
Description	Oper	Unit	Crew Size	Man-Hours Per Unit	Avg Mat'l Unit Cost	Avg Labor Unit Cost	Avg Equip Unit Cost	Avg Total Unit Cost	Avg Price Incl O&P	Avg Total Unit Cost	Avg Price Incl O&P
2" x 8" - 12', 16" oc	Demo	SF	LB	.026	---	.57	---	.57	**.86**	.48	**.73**
2" x 8" - 12', 16" oc	Inst	SF	2C	.031	1.60	.84	.09	2.53	**3.20**	2.10	**2.66**
2" x 8" - 12', 24" oc	Demo	SF	LB	.021	---	.46	---	.46	**.70**	.40	**.60**
2" x 8" - 12', 24" oc	Inst	SF	2C	.025	1.18	.67	.07	1.92	**2.45**	1.60	**2.04**
Gable ends											
2" x 4" - 3' Avg., 12" oc	Demo	SF	LB	.087	---	1.91	---	1.91	**2.88**	1.62	**2.45**
2" x 4" - 3' Avg., 12" oc	Inst	SF	2C	.167	1.88	4.50	.49	6.87	**9.50**	5.78	**8.00**
2" x 4" - 3' Avg., 16" oc	Demo	SF	LB	.082	---	1.80	---	1.80	**2.72**	1.51	**2.29**
2" x 4" - 3' Avg., 16" oc	Inst	SF	2C	.155	1.77	4.18	.45	6.40	**8.84**	5.39	**7.46**
2" x 4" - 3' Avg., 24" oc	Demo	SF	LB	.074	---	1.62	---	1.62	**2.45**	1.38	**2.09**
2" x 4" - 3' Avg., 24" oc	Inst	SF	2C	.143	1.62	3.86	.42	5.90	**8.14**	4.94	**6.84**
2" x 4" - 4' Avg., 12" oc	Demo	SF	LB	.058	---	1.27	---	1.27	**1.92**	1.10	**1.66**
2" x 4" - 4' Avg., 12" oc	Inst	SF	2C	.110	1.57	2.97	.32	4.86	**6.63**	4.08	**5.59**
2" x 4" - 4' Avg., 16" oc	Demo	SF	LB	.054	---	1.19	---	1.19	**1.79**	1.01	**1.52**
2" x 4" - 4' Avg., 16" oc	Inst	SF	2C	.103	1.46	2.78	.30	4.54	**6.20**	3.79	**5.18**
2" x 4" - 4' Avg., 24" oc	Demo	SF	LB	.049	---	1.08	---	1.08	**1.62**	.92	**1.39**
2" x 4" - 4' Avg., 24" oc	Inst	SF	2C	.093	1.32	2.51	.27	4.10	**5.60**	3.43	**4.70**
2" x 4" - 5' Avg., 12" oc	Demo	SF	LB	.046	---	1.01	---	1.01	**1.52**	.86	**1.29**
2" x 4" - 5' Avg., 12" oc	Inst	SF	2C	.086	1.44	2.32	.25	4.01	**5.43**	3.36	**4.56**
2" x 4" - 5' Avg., 16" oc	Demo	SF	LB	.041	---	.90	---	.90	**1.36**	.77	**1.16**
2" x 4" - 5' Avg., 16" oc	Inst	SF	2C	.078	1.32	2.10	.23	3.65	**4.94**	3.04	**4.13**
2" x 4" - 5' Avg., 24" oc	Demo	SF	LB	.037	---	.81	---	.81	**1.23**	.68	**1.03**
2" x 4" - 5' Avg., 24" oc	Inst	SF	2C	.069	1.17	1.86	.20	3.23	**4.37**	2.72	**3.69**
2" x 4" - 6' Avg., 12" oc	Demo	SF	LB	.038	---	.83	---	.83	**1.26**	.70	**1.06**
2" x 4" - 6' Avg., 12" oc	Inst	SF	2C	.070	1.34	1.89	.20	3.43	**4.61**	2.86	**3.85**
2" x 4" - 6' Avg., 16" oc	Demo	SF	LB	.033	---	.72	---	.72	**1.09**	.61	**.93**
2" x 4" - 6' Avg., 16" oc	Inst	SF	2C	.062	1.19	1.67	.18	3.04	**4.09**	2.55	**3.44**
2" x 4" - 6' Avg., 24" oc	Demo	SF	LB	.029	---	.64	---	.64	**.96**	.55	**.83**
2" x 4" - 6' Avg., 24" oc	Inst	SF	2C	.054	1.04	1.46	.16	2.66	**3.57**	2.22	**2.99**
2" x 6" - 3' Avg., 12" oc	Demo	SF	LB	.090	---	1.98	---	1.98	**2.98**	1.69	**2.55**
2" x 6" - 3' Avg., 12" oc	Inst	SF	2C	.170	2.84	4.58	.50	7.92	**10.70**	6.65	**9.03**
2" x 6" - 3' Avg., 16" oc	Demo	SF	LB	.085	---	1.87	---	1.87	**2.82**	1.58	**2.39**
2" x 6" - 3' Avg., 16" oc	Inst	SF	2C	.160	2.66	4.31	.47	7.44	**10.10**	6.25	**8.48**
2" x 6" - 3' Avg., 24" oc	Demo	SF	LB	.077	---	1.69	---	1.69	**2.55**	1.45	**2.19**
2" x 6" - 3' Avg., 24" oc	Inst	SF	2C	.145	2.44	3.91	.42	6.77	**9.17**	5.70	**7.74**
2" x 6" - 4' Avg., 12" oc	Demo	SF	LB	.061	---	1.34	---	1.34	**2.02**	1.14	**1.72**
2" x 6" - 4' Avg., 12" oc	Inst	SF	2C	.113	2.36	3.05	.33	5.74	**7.67**	4.81	**6.45**
2" x 6" - 4' Avg., 16" oc	Demo	SF	LB	.057	---	1.25	---	1.25	**1.89**	1.05	**1.59**
2" x 6" - 4' Avg., 16" oc	Inst	SF	2C	.105	2.19	2.83	.31	5.33	**7.13**	4.45	**5.97**

			Costs Based On Small Volume							Large Volume	
Description	Oper	Unit	Crew Size	Man-Hours Per Unit	Avg Mat'l Unit Cost	Avg Labor Unit Cost	Avg Equip Unit Cost	Avg Total Unit Cost	Avg Price Incl O&P	Avg Total Unit Cost	Avg Price Incl O&P
2" x 6" - 4' Avg., 24" oc	Demo	SF	LB	.051	---	1.12	---	1.12	**1.69**	.97	**1.46**
2" x 6" - 4' Avg., 24" oc	Inst	SF	2C	.096	1.98	2.59	.28	4.85	**6.49**	4.04	**5.42**
2" x 6" - 5' Avg., 12" oc	Demo	SF	LB	.048	---	1.05	---	1.05	**1.59**	.90	**1.36**
2" x 6" - 5' Avg., 12" oc	Inst	SF	2C	.089	2.16	2.40	.26	4.82	**6.39**	4.01	**5.33**
2" x 6" - 5' Avg., 16" oc	Demo	SF	LB	.044	---	.97	---	.97	**1.46**	.81	**1.23**
2" x 6" - 5' Avg., 16" oc	Inst	SF	2C	.080	1.98	2.16	.23	4.37	**5.79**	3.65	**4.85**
2" x 6" - 5' Avg., 24" oc	Demo	SF	LB	.039	---	.86	---	.86	**1.29**	.72	**1.09**
2" x 6" - 5' Avg., 24" oc	Inst	SF	2C	.071	1.75	1.91	.21	3.87	**5.13**	3.23	**4.28**
2" x 6" - 6' Avg., 12" oc	Demo	SF	LB	.040	---	.88	---	.88	**1.33**	.75	**1.13**
2" x 6" - 6' Avg., 12" oc	Inst	SF	2C	.073	2.02	1.97	.21	4.20	**5.52**	3.50	**4.62**
2" x 6" - 6' Avg., 16" oc	Demo	SF	LB	.036	---	.79	---	.79	**1.19**	.66	**.99**
2" x 6" - 6' Avg., 16" oc	Inst	SF	2C	.065	1.79	1.75	.19	3.73	**4.91**	3.10	**4.09**
2" x 6" - 6' Avg., 24" oc	Demo	SF	LB	.031	---	.68	---	.68	**1.03**	.57	**.86**
2" x 6" - 6' Avg., 24" oc	Inst	SF	2C	.056	1.56	1.51	.16	3.23	**4.25**	2.71	**3.58**
2" x 8" - 3' Avg., 12" oc	Demo	SF	LB	.094	---	2.06	---	2.06	**3.12**	1.76	**2.65**
2" x 8" - 3' Avg., 12" oc	Inst	SF	2C	.174	4.03	4.69	.51	9.23	**12.30**	7.72	**10.30**
2" x 8" - 3' Avg., 16" oc	Demo	SF	LB	.088	---	1.93	---	1.93	**2.92**	1.65	**2.49**
2" x 8" - 3' Avg., 16" oc	Inst	SF	2C	.162	3.78	4.37	.47	8.62	**11.50**	7.22	**9.62**
2" x 8" - 3' Avg., 24" oc	Demo	SF	LB	.080	---	1.76	---	1.76	**2.65**	1.49	**2.25**
2" x 8" - 3' Avg., 24" oc	Inst	SF	2C	.148	3.46	3.99	.43	7.88	**10.50**	6.61	**8.80**
2" x 8" - 4' Avg., 12" oc	Demo	SF	LB	.064	---	1.40	---	1.40	**2.12**	1.21	**1.82**
2" x 8" - 4' Avg., 12" oc	Inst	SF	2C	.117	3.35	3.15	.34	6.84	**8.99**	5.71	**7.51**
2" x 8" - 4' Avg., 16" oc	Demo	SF	LB	.059	---	1.30	---	1.30	**1.96**	1.12	**1.69**
2" x 8" - 4' Avg., 16" oc	Inst	SF	2C	.108	3.11	2.91	.32	6.34	**8.32**	5.30	**6.97**
2" x 8" - 4' Avg., 24" oc	Demo	SF	LB	.054	---	1.19	---	1.19	**1.79**	1.01	**1.52**
2" x 8" - 4' Avg., 24" oc	Inst	SF	2C	.098	2.80	2.64	.29	5.73	**7.53**	4.78	**6.29**
2" x 8" - 5' Avg., 12" oc	Demo	SF	LB	.051	---	1.12	---	1.12	**1.69**	.94	**1.43**
2" x 8" - 5' Avg., 12" oc	Inst	SF	2C	.091	3.07	2.45	.27	5.79	**7.53**	4.85	**6.32**
2" x 8" - 5' Avg., 16" oc	Demo	SF	LB	.047	---	1.03	---	1.03	**1.56**	.88	**1.33**
2" x 8" - 5' Avg., 16" oc	Inst	SF	2C	.083	2.75	2.24	.24	5.23	**6.80**	4.35	**5.67**
2" x 8" - 5' Avg., 24" oc	Demo	SF	LB	.041	---	.90	---	.90	**1.36**	.77	**1.16**
2" x 8" - 5' Avg., 24" oc	Inst	SF	2C	.073	2.42	1.97	.21	4.60	**5.98**	3.84	**5.01**
2" x 8" - 6' Avg., 12" oc	Demo	SF	LB	.043	---	.94	---	.94	**1.43**	.79	**1.19**
2" x 8" - 6' Avg., 12" oc	Inst	SF	2C	.075	2.85	2.02	.22	5.09	**6.57**	4.26	**5.50**
2" x 8" - 6' Avg., 16" oc	Demo	SF	LB	.038	---	.83	---	.83	**1.26**	.70	**1.06**
2" x 8" - 6' Avg., 16" oc	Inst	SF	2C	.067	2.53	1.81	.19	4.53	**5.85**	3.79	**4.90**
2" x 8" - 6' Avg., 24" oc	Demo	SF	LB	.033	---	.72	---	.72	**1.09**	.61	**.93**
2" x 8" - 6' Avg., 24" oc	Inst	SF	2C	.058	2.21	1.56	.17	3.94	**5.09**	3.27	**4.23**

			Costs Based On Small Volume						Large Volume	
Description	Oper	Unit	Crew Size	Man-Hours Per Unit	Avg Mat'l Unit Cost	Avg Labor Unit Cost	Avg Equip Unit Cost	Avg Total Unit Cost	Avg Price Incl O&P	Avg Total Unit Cost	Avg Price Incl O&P
Studs/plates, per LF of wall or partition											
Walls or partitions											
2" x 4" - 8', 12" oc	Demo	LF	LB	.250	---	5.49	---	5.49	**8.29**	4.68	**7.06**
2" x 4" - 8', 12" oc	Inst	LF	2C	.302	7.90	8.14	.88	16.92	**22.30**	14.16	**18.70**
2" x 4" - 10', 12" oc	Demo	LF	LB	.262	---	5.75	---	5.75	**8.68**	4.87	**7.36**
2" x 4" - 10', 12" oc	Inst	LF	2C	.320	9.43	8.63	.93	18.99	**24.90**	15.80	**20.80**
2" x 4" - 12', 12" oc	Demo	LF	LB	.271	---	5.95	---	5.95	**8.98**	5.09	**7.69**
2" x 4" - 12', 12" oc	Inst	LF	2C	.333	11.00	8.98	.97	20.95	**27.30**	17.38	**22.70**
2" x 4" - 8', 16" oc	Demo	LF	LB	.222	---	4.87	---	4.87	**7.36**	4.13	**6.23**
2" x 4" - 8', 16" oc	Inst	LF	2C	.267	6.38	7.20	.78	14.36	**19.10**	12.05	**16.10**
2" x 4" - 10', 16" oc	Demo	LF	LB	.232	---	5.09	---	5.09	**7.69**	4.35	**6.56**
2" x 4" - 10', 16" oc	Inst	LF	2C	.281	7.53	7.58	.82	15.93	**21.00**	13.30	**17.60**
2" x 4" - 12', 16" oc	Demo	LF	LB	.242	---	5.31	---	5.31	**8.02**	4.50	**6.79**
2" x 4" - 12', 16" oc	Inst	LF	2C	.291	8.62	7.85	.85	17.32	**22.70**	14.39	**18.90**
2" x 4" - 8', 24" oc	Demo	LF	LB	.190	---	4.17	---	4.17	**6.30**	3.56	**5.37**
2" x 4" - 8', 24" oc	Inst	LF	2C	.232	4.80	6.25	.68	11.73	**15.70**	9.84	**13.20**
2" x 4" - 10', 24" oc	Demo	LF	LB	.195	---	4.28	---	4.28	**6.46**	3.62	**5.47**
2" x 4" - 10', 24" oc	Inst	LF	2C	.235	5.64	6.34	.69	12.67	**16.80**	10.58	**14.10**
2" x 4" - 12', 24" oc	Demo	LF	LB	.198	---	4.35	---	4.35	**6.56**	3.69	**5.57**
2" x 4" - 12', 24" oc	Inst	LF	2C	.239	6.35	6.44	.70	13.49	**17.80**	11.25	**14.90**
2" x 6" - 8', 12" oc	Demo	LF	LB	.286	---	6.28	---	6.28	**9.48**	5.31	**8.02**
2" x 6" - 8', 12" oc	Inst	LF	2C	.340	12.00	9.17	.99	22.16	**28.70**	18.50	**24.10**
2" x 6" - 10', 12" oc	Demo	LF	LB	.296	---	6.50	---	6.50	**9.81**	5.58	**8.42**
2" x 6" - 10', 12" oc	Inst	LF	2C	.356	14.30	9.60	1.04	24.94	**32.00**	20.72	**26.70**
2" x 6" - 12', 12" oc	Demo	LF	LB	.314	---	6.89	---	6.89	**10.40**	5.86	**8.85**
2" x 6" - 12', 12" oc	Inst	LF	2C	.372	16.60	10.00	1.09	27.69	**35.40**	23.16	**29.70**
2" x 6" - 8', 16" oc	Demo	LF	LB	.254	---	5.58	---	5.58	**8.42**	4.74	**7.16**
2" x 6" - 8', 16" oc	Inst	LF	2C	.308	9.66	8.30	.90	18.86	**24.60**	15.74	**20.60**
2" x 6" - 10', 16" oc	Demo	LF	LB	.267	---	5.86	---	5.86	**8.85**	4.94	**7.46**
2" x 6" - 10', 16" oc	Inst	LF	2C	.320	11.40	8.63	.93	20.96	**27.10**	17.42	**22.60**
2" x 6" - 12', 16" oc	Demo	LF	LB	.271	---	5.95	---	5.95	**8.98**	5.09	**7.69**
2" x 6" - 12', 16" oc	Inst	LF	2C	.327	13.10	8.82	.95	22.87	**29.40**	19.04	**24.50**
2" x 6" - 8', 24" oc	Demo	LF	LB	.216	---	4.74	---	4.74	**7.16**	4.04	**6.10**
2" x 6" - 8', 24" oc	Inst	LF	2C	.262	7.29	7.06	.76	15.11	**19.90**	12.62	**16.60**
2" x 6" - 10', 24" oc	Demo	LF	LB	.222	---	4.87	---	4.87	**7.36**	4.13	**6.23**
2" x 6" - 10', 24" oc	Inst	LF	2C	.267	8.50	7.20	.78	16.48	**21.50**	13.69	**17.90**
2" x 6" - 12', 24" oc	Demo	LF	LB	.225	---	4.94	---	4.94	**7.46**	4.17	**6.30**
2" x 6" - 12', 24" oc	Inst	LF	2C	.267	9.66	7.20	.78	17.64	**22.80**	14.76	**19.20**
2" x 8" - 8', 12" oc	Demo	LF	LB	.320	---	7.02	---	7.02	**10.60**	5.95	**8.98**
2" x 8" - 8', 12" oc	Inst	LF	2C	.381	17.10	10.30	1.11	28.51	**36.40**	23.77	**30.50**
2" x 8" - 10', 12" oc	Demo	LF	LB	.333	---	7.31	---	7.31	**11.00**	6.28	**9.48**
2" x 8" - 10', 12" oc	Inst	LF	2C	.400	20.30	10.80	1.17	32.27	**40.90**	26.86	**34.10**
2" x 8" - 12', 12" oc	Demo	LF	LB	.356	---	7.81	---	7.81	**11.80**	6.63	**10.00**
2" x 8" - 12', 12" oc	Inst	LF	2C	.421	23.70	11.40	1.23	36.33	**45.80**	30.04	**38.00**

Description	Oper	Unit	Crew Size	Man-Hours Per Unit	Avg Mat'l Unit Cost	Avg Labor Unit Cost	Avg Equip Unit Cost	Avg Total Unit Cost	Avg Price Incl O&P	Large Volume Avg Total Unit Cost	Large Volume Avg Price Incl O&P
2" x 8" - 8', 16" oc	Demo	LF	LB	.286	---	6.28	---	6.28	**9.48**	5.31	**8.02**
2" x 8" - 8', 16" oc	Inst	LF	2C	.340	13.70	9.17	.99	23.86	**30.70**	20.00	**25.70**
2" x 8" - 10', 16" oc	Demo	LF	LB	.296	---	6.50	---	6.50	**9.81**	5.58	**8.42**
2" x 8" - 10', 16" oc	Inst	LF	2C	.356	16.20	9.60	1.04	26.84	**34.30**	22.32	**28.60**
2" x 8" - 12', 16" oc	Demo	LF	LB	.308	---	6.76	---	6.76	**10.20**	5.75	**8.68**
2" x 8" - 12', 16" oc	Inst	LF	2C	.364	18.80	9.81	1.06	29.67	**37.60**	24.60	**31.20**
2" x 8" - 8', 24" oc	Demo	LF	LB	.242	---	5.31	---	5.31	**8.02**	4.50	**6.79**
2" x 8" - 8', 24" oc	Inst	LF	2C	.291	10.40	7.85	.85	19.10	**24.80**	15.91	**20.70**
2" x 8" - 10', 24" oc	Demo	LF	LB	.246	---	5.40	---	5.40	**8.15**	4.63	**6.99**
2" x 8" - 10', 24" oc	Inst	LF	2C	.296	12.10	7.98	.86	20.94	**26.90**	17.39	**22.40**
2" x 8" - 12', 24" oc	Demo	LF	LB	.254	---	5.58	---	5.58	**8.42**	4.74	**7.16**
2" x 8" - 12', 24" oc	Inst	LF	2C	.302	13.70	8.14	.88	22.72	**29.00**	19.01	**24.30**
Sheathing, walls											
Boards, 1" x 8"											
Horizontal	Demo	SF	LB	.025	---	.55	---	.55	**.83**	.46	**.70**
Horizontal	Inst	SF	2C	.027	1.00	.73	.08	1.81	**2.34**	1.50	**1.94**
Diagonal	Demo	SF	LB	.028	---	.61	---	.61	**.93**	.53	**.80**
Diagonal	Inst	SF	2C	.030	1.09	.81	.09	1.99	**2.57**	1.66	**2.16**
Plywood											
3/8"	Demo	SF	LB	.016	---	.35	---	.35	**.53**	.31	**.46**
3/8"	Inst	SF	2C	.018	.55	.49	.05	1.09	**1.42**	.90	**1.18**
1/2"	Demo	SF	LB	.016	---	.35	---	.35	**.53**	.31	**.46**
1/2"	Inst	SF	2C	.018	.75	.49	.05	1.29	**1.65**	1.06	**1.37**
5/8"	Demo	SF	LB	.016	---	.35	---	.35	**.53**	.31	**.46**
5/8"	Inst	SF	2C	.018	.92	.49	.05	1.46	**1.85**	1.19	**1.52**
Gypsum board 1/2"	Demo	SF	LB	.016	---	.35	---	.35	**.53**	.31	**.46**
1/2"	Inst	SF	2C	.018	.01	.49	.05	.55	**.80**	.45	**.67**
Wood fiber board, 1/2"											
With vapor barrier	Demo	SF	LB	.016	---	.35	---	.35	**.53**	.31	**.46**
With vapor barrier	Inst	SF	2C	.018	.01	.49	.05	.55	**.80**	.45	**.67**
Without vapor barrier	Demo	SF	LB	.016	---	.35	---	.35	**.53**	.31	**.46**
Without vapor barrier	Inst	SF	2C	.018	.01	.49	.05	.55	**.80**	.45	**.67**
Asphalt impregnated	Demo	SF	LB	.016	---	.35	---	.35	**.53**	.31	**.46**
Asphalt impregnated	Inst	SF	2C	.018	.01	.49	.05	.55	**.80**	.45	**.67**
Sheathing, roof											
Boards, 1" x 8"											
Horizontal	Demo	SF	LB	.021	---	.46	---	.46	**.70**	.40	**.60**
Horizontal	Inst	SF	2C	.023	1.00	.62	.07	1.69	**2.16**	1.41	**1.81**
Diagonal	Demo	SF	LB	.024	---	.53	---	.53	**.80**	.46	**.70**
Diagonal	Inst	SF	2C	.026	1.09	.70	.08	1.87	**2.40**	1.54	**1.98**

			Costs Based On Small Volume							Large Volume	
Description	Oper	Unit	Crew Size	Man-Hours Per Unit	Avg Mat'l Unit Cost	Avg Labor Unit Cost	Avg Equip Unit Cost	Avg Total Unit Cost	Avg Price Incl O&P	Avg Total Unit Cost	Avg Price Incl O&P
Plywood											
1/2"	Demo	SF	LB	.015	---	.33	---	.33	**.50**	.29	**.43**
1/2"	Inst	SF	2C	.016	.72	.43	.05	1.20	**1.53**	.98	**1.25**
5/8"	Demo	SF	LB	.015	---	.33	---	.33	**.50**	.29	**.43**
5/8"	Inst	SF	2C	.016	.88	.43	.05	1.36	**1.72**	1.11	**1.40**
3/4"	Demo	SF	LB	.016	---	.35	---	.35	**.53**	.29	**.43**
3/4"	Inst	SF	2C	.017	1.01	.46	.05	1.52	**1.91**	1.27	**1.61**
Sheathing, subfloor											
Boards, 1" x 8"											
Horizontal	Demo	SF	LB	.023	---	.50	---	.50	**.76**	.42	**.63**
Horizontal	Inst	SF	2C	.025	1.02	.67	.07	1.76	**2.27**	1.46	**1.88**
Diagonal	Demo	SF	LB	.026	---	.57	---	.57	**.86**	.48	**.73**
Diagonal	Inst	SF	2C	.028	1.11	.75	.08	1.94	**2.50**	1.63	**2.10**
Plywood											
5/8"	Demo	SF	LB	.015	---	.33	---	.33	**.50**	.29	**.43**
5/8"	Inst	SF	2C	.017	.92	.46	.05	1.43	**1.80**	1.17	**1.48**
3/4"	Demo	SF	LB	.017	---	.37	---	.37	**.56**	.31	**.46**
3/4"	Inst	SF	2C	.018	1.06	.49	.05	1.60	**2.01**	1.31	**1.66**
1-1/8"	Demo	SF	LB	.022	---	.48	---	.48	**.73**	.42	**.63**
1-1/8"	Inst	SF	2C	.024	1.55	.65	.07	2.27	**2.84**	1.88	**2.35**
Particleboard											
5/8"	Demo	SF	LB	.015	---	.33	---	.33	**.50**	.29	**.43**
5/8"	Inst	SF	2C	.017	.44	.46	.05	.95	**1.25**	.78	**1.03**
3/4"	Demo	SF	LB	.017	---	.37	---	.37	**.56**	.31	**.46**
3/4"	Inst	SF	2C	.018	.51	.49	.05	1.05	**1.37**	.86	**1.14**
Underlayment											
Plywood, 3/8"	Demo	SF	LB	.015	---	.33	---	.33	**.50**	.29	**.43**
Plywood, 3/8"	Inst	SF	2C	.017	.50	.46	.05	1.01	**1.32**	.83	**1.09**
Hardboard, 0.215"	Demo	SF	LB	.015	---	.33	---	.33	**.50**	.29	**.43**
Hardboard, 0.215"	Inst	SF	2C	.017	.43	.46	.05	.94	**1.24**	.77	**1.02**
Trusses, shop fabricated, wood "W" type											
1/8 pitch											
3" rise in 12" run											
20' span	Demo	Ea	LJ	1.032	---	22.70	---	22.70	**34.20**	19.00	**28.70**
20' span	Inst	Ea	CX	1.10	54.40	29.70	1.61	85.71	**109.00**	71.27	**91.10**
22' span	Demo	Ea	LJ	1.032	---	22.70	---	22.70	**34.20**	19.00	**28.70**
22' span	Inst	Ea	CX	1.10	60.60	29.70	1.61	91.91	**116.00**	76.27	**96.90**
24' span	Demo	Ea	LJ	1.032	---	22.70	---	22.70	**34.20**	19.00	**28.70**
24' span	Inst	Ea	CX	1.10	63.40	29.70	1.61	94.71	**120.00**	78.67	**99.60**
26' span	Demo	Ea	LJ	1.032	---	22.70	---	22.70	**34.20**	19.00	**28.70**
26' span	Inst	Ea	CX	1.10	69.60	29.70	1.61	100.91	**127.00**	83.67	**105.00**
28' span	Demo	Ea	LJ	1.032	---	22.70	---	22.70	**34.20**	19.00	**28.70**
28' span	Inst	Ea	CX	1.10	72.50	29.70	1.61	103.81	**130.00**	86.07	**108.00**

			Costs Based On Small Volume						Large Volume	
Description	Oper	Unit	Crew Size	Man-Hours Per Unit	Avg Mat'l Unit Cost	Avg Labor Unit Cost	Avg Equip Unit Cost	Avg Total Unit Cost	Avg Price Incl O&P	Avg Total Unit Cost	Avg Price Incl O&P
30' span	Demo	Ea	LJ	1.07	---	23.50	---	23.50	**35.50**	20.10	**30.30**
30' span	Inst	Ea	CX	1.19	78.70	32.10	1.73	112.53	**141.00**	92.76	**116.00**
32' span	Demo	Ea	LJ	1.07	---	23.50	---	23.50	**35.50**	20.10	**30.30**
32' span	Inst	Ea	CX	1.19	80.70	32.10	1.73	114.53	**143.00**	94.36	**118.00**
34' span	Demo	Ea	LJ	1.07	---	23.50	---	23.50	**35.50**	20.10	**30.30**
34' span	Inst	Ea	CX	1.19	84.40	32.10	1.73	118.23	**148.00**	97.46	**122.00**
36' span	Demo	Ea	LJ	1.14	---	25.00	---	25.00	**37.80**	21.30	**32.20**
36' span	Inst	Ea	CX	1.23	90.60	33.20	1.79	125.59	**156.00**	103.31	**129.00**
5/24 pitch											
5" rise in 12" run											
20' span	Demo	Ea	LJ	1.032	---	22.70	---	22.70	**34.20**	19.00	**28.70**
20' span	Inst	Ea	CX	1.10	43.40	29.70	1.61	74.71	**96.60**	62.27	**80.70**
22' span	Demo	Ea	LJ	1.032	---	22.70	---	22.70	**34.20**	19.00	**28.70**
22' span	Inst	Ea	CX	1.10	48.70	29.70	1.61	80.01	**103.00**	66.57	**85.70**
24' span	Demo	Ea	LJ	1.032	---	22.70	---	22.70	**34.20**	19.00	**28.70**
24' span	Inst	Ea	CX	1.10	51.40	29.70	1.61	82.71	**106.00**	68.77	**88.30**
26' span	Demo	Ea	LJ	1.032	---	22.70	---	22.70	**34.20**	19.00	**28.70**
26' span	Inst	Ea	CX	1.10	57.50	29.70	1.61	88.81	**113.00**	73.77	**94.00**
28' span	Demo	Ea	LJ	1.032	---	22.70	---	22.70	**34.20**	19.00	**28.70**
28' span	Inst	Ea	CX	1.10	68.60	29.70	1.61	99.91	**126.00**	82.87	**104.00**
30' span	Demo	Ea	LJ	1.07	---	23.50	---	23.50	**35.50**	20.10	**30.30**
30' span	Inst	Ea	CX	1.19	73.90	32.10	1.73	107.73	**135.00**	88.86	**112.00**
32' span	Demo	Ea	LJ	1.07	---	23.50	---	23.50	**35.50**	20.10	**30.30**
32' span	Inst	Ea	CX	1.19	75.80	32.10	1.73	109.63	**138.00**	90.46	**114.00**
34' span	Demo	Ea	LJ	1.07	---	23.50	---	23.50	**35.50**	20.10	**30.30**
34' span	Inst	Ea	CX	1.19	83.60	32.10	1.73	117.43	**147.00**	96.76	**121.00**
36' span	Demo	Ea	LJ	1.14	---	25.00	---	25.00	**37.80**	21.30	**32.20**
36' span	Inst	Ea	CX	1.23	86.30	33.20	1.79	121.29	**151.00**	99.81	**125.00**
1/4 pitch											
6" rise in 12" run											
20' span	Demo	Ea	LJ	1.032	---	22.70	---	22.70	**34.20**	19.00	**28.70**
20' span	Inst	Ea	CX	1.10	44.20	29.70	1.61	75.51	**97.60**	62.97	**81.50**
22' span	Demo	Ea	LJ	1.032	---	22.70	---	22.70	**34.20**	19.00	**28.70**
22' span	Inst	Ea	CX	1.10	49.50	29.70	1.61	80.81	**104.00**	67.27	**86.50**
24' span	Demo	Ea	LJ	1.032	---	22.70	---	22.70	**34.20**	19.00	**28.70**
24' span	Inst	Ea	CX	1.10	55.60	29.70	1.61	86.91	**111.00**	72.17	**92.20**
26' span	Demo	Ea	LJ	1.032	---	22.70	---	22.70	**34.20**	19.00	**28.70**
26' span	Inst	Ea	CX	1.10	57.50	29.70	1.61	88.81	**113.00**	73.77	**94.00**
28' span	Demo	Ea	LJ	1.032	---	22.70	---	22.70	**34.20**	19.00	**28.70**
28' span	Inst	Ea	CX	1.10	69.50	29.70	1.61	100.81	**127.00**	83.57	**105.00**

			Costs Based On Small Volume							Large Volume	
Description	Oper	Unit	Crew Size	Man-Hours Per Unit	Avg Mat'l Unit Cost	Avg Labor Unit Cost	Avg Equip Unit Cost	Avg Total Unit Cost	Avg Price Incl O&P	Avg Total Unit Cost	Avg Price Incl O&P
30' span	Demo	Ea	LJ	1.07	---	23.50	---	23.50	**35.50**	20.10	**30.30**
30' span	Inst	Ea	CX	1.19	72.20	32.10	1.73	106.03	**134.00**	87.46	**110.00**
32' span	Demo	Ea	LJ	1.07	---	23.50	---	23.50	**35.50**	20.10	**30.30**
32' span	Inst	Ea	CX	1.19	78.30	32.10	1.73	112.13	**141.00**	92.46	**116.00**
34' span	Demo	Ea	LJ	1.07	---	23.50	---	23.50	**35.50**	20.10	**30.30**
34' span	Inst	Ea	CX	1.19	84.40	32.10	1.73	118.23	**148.00**	97.46	**122.00**
36' span	Demo	Ea	LJ	1.14	---	25.00	---	25.00	**37.80**	21.30	**32.20**
36' span	Inst	Ea	CX	1.23	88.80	33.20	1.79	123.79	**154.00**	101.91	**127.00**
Garage doors											
Aluminum frame with plastic skin bonded to polystyrene foam core											
Jamb type with hardware and deluxe lock											
8' x 7', single	Inst	Ea	2C	5.71	469.00	154.00	---	623.00	**773.00**	540.00	**661.00**
8' x 8', single	Inst	Ea	2C	5.71	610.00	154.00	---	764.00	**936.00**	670.00	**810.00**
9' x 7', single	Inst	Ea	2C	5.71	512.00	154.00	---	666.00	**822.00**	579.00	**706.00**
9' x 8', single	Inst	Ea	2C	5.71	658.00	154.00	---	812.00	**990.00**	714.00	**860.00**
16' x 7', double	Inst	Ea	2C	7.62	884.00	205.00	---	1089.00	**1330.00**	958.00	**1150.00**
16' x 8', double	Inst	Ea	2C	7.62	1140.00	205.00	---	1345.00	**1630.00**	1194.00	**1430.00**
Track type with hardware and deluxe lock											
8' x 7', single	Inst	Ea	2C	5.71	554.00	154.00	---	708.00	**871.00**	618.00	**751.00**
9' x 7', single	Inst	Ea	2C	5.71	608.00	154.00	---	762.00	**934.00**	668.00	**808.00**
16' x 7', double	Inst	Ea	2C	7.62	1020.00	205.00	---	1225.00	**1480.00**	1080.00	**1300.00**
Sectional type with hardware and key lock											
8' x 7', single	Inst	Ea	2C	5.71	577.00	154.00	---	731.00	**898.00**	640.00	**775.00**
9' x 7', single	Inst	Ea	2C	5.71	613.00	154.00	---	767.00	**939.00**	673.00	**813.00**
16' x 7', double	Inst	Ea	2C	7.62	1050.00	205.00	---	1255.00	**1520.00**	1109.00	**1330.00**
Fiberglass											
Jamb type with hardware and deluxe lock											
8' x 7', single	Inst	Ea	2C	5.71	413.00	154.00	---	567.00	**709.00**	489.00	**602.00**
8' x 8', single	Inst	Ea	2C	5.71	581.00	154.00	---	735.00	**902.00**	643.00	**779.00**
9' x 7', single	Inst	Ea	2C	5.71	495.00	154.00	---	649.00	**804.00**	564.00	**688.00**
9' x 8', single	Inst	Ea	2C	5.71	645.00	154.00	---	799.00	**975.00**	701.00	**846.00**
16' x 7', double	Inst	Ea	2C	7.62	902.00	205.00	---	1107.00	**1350.00**	975.00	**1170.00**
16' x 8', double	Inst	Ea	2C	7.62	1200.00	205.00	---	1405.00	**1690.00**	1254.00	**1490.00**
Track type with hardware and deluxe lock											
8' x 7', single	Inst	Ea	2C	5.71	554.00	154.00	---	708.00	**871.00**	618.00	**751.00**
9' x 7', single	Inst	Ea	2C	5.71	608.00	154.00	---	762.00	**934.00**	668.00	**808.00**
16' x 7', double	Inst	Ea	2C	7.62	977.00	205.00	---	1182.00	**1440.00**	1044.00	**1250.00**
Sectional type with hardware and key lock											
8' x 7', single	Inst	Ea	2C	5.71	492.00	154.00	---	646.00	**800.00**	561.00	**685.00**
9' x 7', single	Inst	Ea	2C	5.71	549.00	154.00	---	703.00	**866.00**	614.00	**746.00**
16' x 7', double	Inst	Ea	2C	7.62	871.00	205.00	---	1076.00	**1310.00**	946.00	**1140.00**
Steel											
Jamb type with hardware and deluxe lock											
8' x 7', single	Inst	Ea	2C	5.71	359.00	154.00	---	513.00	**647.00**	439.00	**544.00**
8' x 8', single	Inst	Ea	2C	5.71	479.00	154.00	---	633.00	**785.00**	549.00	**671.00**
9' x 7', single	Inst	Ea	2C	5.71	394.00	154.00	---	548.00	**687.00**	470.00	**581.00**
9' x 8', single	Inst	Ea	2C	5.71	540.00	154.00	---	694.00	**854.00**	605.00	**735.00**
16' x 7', double	Inst	Ea	2C	7.62	674.00	205.00	---	879.00	**1090.00**	765.00	**932.00**
16' x 8', double	Inst	Ea	2C	7.62	935.00	205.00	---	1140.00	**1390.00**	1005.00	**1210.00**

Description	Oper	Unit	Crew Size	Man-Hours Per Unit	Avg Mat'l Unit Cost	Avg Labor Unit Cost	Avg Equip Unit Cost	Avg Total Unit Cost	Avg Price Incl O&P	Avg Total Unit Cost	Avg Price Incl O&P
			Costs Based On Small Volume							Large Volume	
Track type with hardware and deluxe lock											
8' x 7', single	Inst	Ea	2C	5.71	410.00	154.00	---	564.00	**706.00**	486.00	**598.00**
9' x 7', single	Inst	Ea	2C	5.71	456.00	154.00	---	610.00	**758.00**	528.00	**647.00**
16' x 7', double	Inst	Ea	2C	7.62	758.00	205.00	---	963.00	**1180.00**	842.00	**1020.00**
Sectional type with hardware and key lock											
8' x 7', single	Inst	Ea	2C	5.71	441.00	154.00	---	595.00	**741.00**	514.00	**631.00**
9' x 7', single	Inst	Ea	2C	5.71	479.00	154.00	---	633.00	**785.00**	549.00	**671.00**
16' x 7', double	Inst	Ea	2C	7.62	830.00	205.00	---	1035.00	**1270.00**	908.00	**1100.00**
Wood											
Jamb type with hardware and deluxe lock											
8' x 7', single	Inst	Ea	2C	5.71	499.00	154.00	---	653.00	**807.00**	567.00	**692.00**
8' x 8', single	Inst	Ea	2C	5.71	517.00	154.00	---	671.00	**828.00**	584.00	**711.00**
9' x 7', single	Inst	Ea	2C	5.71	543.00	154.00	---	697.00	**858.00**	608.00	**739.00**
9' x 8', single	Inst	Ea	2C	5.71	672.00	154.00	---	826.00	**1010.00**	727.00	**876.00**
16' x 7', double	Inst	Ea	2C	7.62	913.00	205.00	---	1118.00	**1360.00**	985.00	**1190.00**
16' x 8', double	Inst	Ea	2C	7.62	1180.00	205.00	---	1385.00	**1670.00**	1234.00	**1470.00**
Track type with hardware and deluxe lock											
8' x 7', single	Inst	Ea	2C	5.71	567.00	154.00	---	721.00	**887.00**	630.00	**765.00**
9' x 7', single	Inst	Ea	2C	5.71	615.00	154.00	---	769.00	**941.00**	674.00	**815.00**
16' x 7', double	Inst	Ea	2C	7.62	1020.00	205.00	---	1225.00	**1480.00**	1082.00	**1300.00**
Sectional type with hardware and key lock											
8' x 7', single	Inst	Ea	2C	5.71	556.00	154.00	---	710.00	**873.00**	620.00	**753.00**
9' x 7', single	Inst	Ea	2C	5.71	610.00	154.00	---	764.00	**936.00**	670.00	**810.00**
16' x 7', double	Inst	Ea	2C	7.62	917.00	205.00	---	1122.00	**1370.00**	988.00	**1190.00**

Garage door operators

Radio controlled for single or double doors. Labor includes wiring, connection and installation.

Description	Oper	Unit	Crew Size	Man-Hours Per Unit	Avg Mat'l Unit Cost	Avg Labor Unit Cost	Avg Equip Unit Cost	Avg Total Unit Cost	Avg Price Incl O&P	Avg Total Unit Cost	Avg Price Incl O&P
Chain drive, 1/4 hp, with receiver and one transmitter	Inst	LS	ED	7.11	189.00	198.00	---	387.00	**511.00**	317.00	**413.00**
Screw-worm drive, 1/3 hp, with receiver and one transmitter	Inst	LS	ED	7.11	259.00	198.00	---	457.00	**592.00**	379.00	**485.00**
Deluxe models, 1/2 hp, with receiver, transmitter, and time delay light											
Chain drive, not for vault-type garages	Inst	LS	ED	7.11	378.00	198.00	---	576.00	**728.00**	485.00	**606.00**
Screw drive with threaded worm screw	Inst	LS	ED	7.11	408.00	198.00	---	606.00	**762.00**	511.00	**637.00**
Additional transmitters, ADD	Inst	Ea	EA	---	44.60	---	---	44.60	**51.20**	39.60	**45.50**
Exterior key switch	Inst	Ea	EA	.889	13.50	25.60	---	39.10	**53.50**	31.20	**42.30**
To remove and replace unit and receiver	Inst	LS	2C	3.56	---	96.00	---	96.00	**146.00**	72.00	**109.00**

Garbage disposers

In-Sink-Erator Products, with wall switch. Labor includes rough-in.

See also Trash compactors, page 254

Description	Oper	Unit	Crew Size	Man-Hours Per Unit	Avg Mat'l Unit Cost	Avg Labor Unit Cost	Avg Equip Unit Cost	Avg Total Unit Cost	Avg Price Incl O&P	Avg Total Unit Cost	Avg Price Incl O&P
Model "Badger 1," 1/3 HP, continuous feed, 1 year parts protection	Inst	Ea	SC	9.09	92.30	270.00	---	362.30	**508.00**	292.00	**409.00**
Model "Badger V," 1/2 HP, continuous feed, 1 year parts protection	Inst	Ea	SC	9.09	105.00	270.00	---	375.00	**522.00**	302.40	**421.00**

Description	Oper	Unit	Crew Size	Man-Hours Per Unit	Avg Mat'l Unit Cost	Avg Labor Unit Cost	Avg Equip Unit Cost	Avg Total Unit Cost	Avg Price Incl O&P	Large Volume Avg Total Unit Cost	Large Volume Avg Price Incl O&P
Model 333, 1/2 HP, continuous feed, 4 year parts protection											
	Inst	Ea	SC	9.09	160.00	270.00	---	430.00	**586.00**	348.00	**473.00**
With stainless steel construction	Inst	Ea	SC	9.09	214.00	270.00	---	484.00	**647.00**	392.00	**523.00**
Model 77, 3/4 HP, automatic reversing feed, 5 year parts protection, stainless steel construction											
	Inst	Ea	SC	9.09	265.00	270.00	---	535.00	**707.00**	434.00	**572.00**
Classic, 1 HP, continuous feed, auto reversing, 7 year parts protection, stainless steel construction											
	Inst	Ea	SC	9.09	308.00	270.00	---	578.00	**756.00**	470.00	**613.00**
Model 17, 3/4 HP, batch feed, auto reversing, 5 year parts protection, stainless steel construction											
	Inst	Ea	SC	9.09	348.00	270.00	---	618.00	**801.00**	502.00	**650.00**
Classic/LC, 1 HP, batch feed, auto reversing, 7 year parts protection, stainless steel construction											
	Inst	Ea	SC	9.09	439.00	270.00	---	709.00	**906.00**	578.00	**737.00**
Adjustments											
To only remove and reset garbage disposer	Reset	Ea	SA	1.25	---	38.10	---	38.10	**56.70**	30.50	**45.40**
Parts and accessories											
Stainless steel stopper	Inst	Ea	---	---	9.04	---	---	9.04	**9.04**	7.45	**7.45**
Dishwasher connector kit	Inst	Ea	---	---	8.09	---	---	8.09	**8.09**	6.66	**6.66**
Splash guard	Inst	Ea	---	---	4.52	---	---	4.52	**4.52**	3.72	**3.72**
Flexible tail pipe	Inst	Ea	---	---	10.40	---	---	10.40	**10.40**	8.53	**8.53**
Power cord accessory kit	Inst	Ea	---	---	8.81	---	---	8.81	**8.81**	7.25	**7.25**
Service wrench	Inst	Ea	---	---	3.93	---	---	3.93	**3.93**	3.23	**3.23**
Plastic stopper	Inst	Ea	---	---	3.81	---	---	3.81	**3.81**	3.14	**3.14**
Deluxe mounting gasket	Inst	Ea	---	---	6.19	---	---	6.19	**6.19**	5.10	**5.10**
Economy mounting kit	Inst	Ea	---	---	5.47	---	---	5.47	**5.47**	4.51	**4.51**

Girders. See Framing, page 106

Glass and glazing

Description	Oper	Unit	Crew Size	Man-Hours Per Unit	Avg Mat'l Unit Cost	Avg Labor Unit Cost	Avg Equip Unit Cost	Avg Total Unit Cost	Avg Price Incl O&P	Large Volume Avg Total Unit Cost	Large Volume Avg Price Incl O&P
3/16" T float with putty in wood sash											
8" x 12"	Demo	SF	GA	.533	---	13.40	---	13.40	**20.10**	8.04	**12.10**
8" x 12"	Inst	SF	GA	.444	6.13	11.20	---	17.33	**23.50**	12.07	**16.00**
12" x 16"	Demo	SF	GA	.296	---	7.43	---	7.43	**11.20**	4.47	**6.71**
12" x 16"	Inst	SF	GA	.242	5.03	6.08	---	11.11	**14.70**	8.04	**10.30**
14" x 20"	Demo	SF	GA	.242	---	6.08	---	6.08	**9.12**	3.64	**5.46**
14" x 20"	Inst	SF	GA	.190	4.58	4.77	---	9.35	**12.20**	6.87	**8.71**
16" x 24"	Demo	SF	GA	.190	---	4.77	---	4.77	**7.16**	2.86	**4.29**
16" x 24"	Inst	SF	GA	.148	4.28	3.72	---	8.00	**10.30**	5.97	**7.47**
24" x 26"	Demo	SF	GA	.140	---	3.52	---	3.52	**5.27**	2.11	**3.16**
24" x 26"	Inst	SF	GA	.111	3.84	2.79	---	6.63	**8.41**	5.04	**6.22**
36" x 24"	Demo	SF	GA	.107	---	2.69	---	2.69	**4.03**	1.61	**2.41**
36" x 24"	Inst	SF	GA	.086	3.66	2.16	---	5.82	**7.27**	4.51	**5.48**

Description	Oper	Unit	Crew Size	Man-Hours Per Unit	Avg Mat'l Unit Cost	Avg Labor Unit Cost	Avg Equip Unit Cost	Avg Total Unit Cost	Avg Price Incl O&P	Large Volume Avg Total Unit Cost	Large Volume Avg Price Incl O&P
			Costs Based On Small Volume							Large Volume	
1/8" T float with putty in steel sash											
12" x 16"	Demo	SF	GA	.296	---	7.43	---	7.43	**11.20**	4.47	**6.71**
12" x 16"	Inst	SF	GA	.242	7.07	6.08	---	13.15	**16.90**	9.83	**12.30**
16" x 20"	Demo	SF	GA	.205	---	5.15	---	5.15	**7.72**	3.09	**4.63**
16" x 20"	Inst	SF	GA	.167	6.11	4.19	---	10.30	**13.00**	7.85	**9.64**
16" x 24"	Demo	SF	GA	.190	---	4.77	---	4.77	**7.16**	2.86	**4.29**
16" x 24"	Inst	SF	GA	.148	5.88	3.72	---	9.60	**12.00**	7.37	**9.01**
24" x 24"	Demo	SF	GA	.140	---	3.52	---	3.52	**5.27**	2.11	**3.16**
24" x 24"	Inst	SF	GA	.111	5.18	2.79	---	7.97	**9.88**	6.22	**7.52**
28" x 32"	Demo	SF	GA	.103	---	2.59	---	2.59	**3.88**	1.56	**2.34**
28" x 32"	Inst	SF	GA	.083	4.81	2.08	---	6.89	**8.42**	5.47	**6.51**
36" x 36"	Demo	SF	GA	.083	---	2.08	---	2.08	**3.13**	1.26	**1.88**
36" x 36"	Inst	SF	GA	.065	4.48	1.63	---	6.11	**7.38**	4.90	**5.78**
36" x 48"	Demo	SF	GA	.070	---	1.76	---	1.76	**2.64**	1.05	**1.58**
36" x 48"	Inst	SF	GA	.053	4.26	1.33	---	5.59	**6.68**	4.53	**5.31**
1/4" T float											
With putty and points in wood sash											
72" x 48"	Demo	SF	GA	.072	---	1.81	---	1.81	**2.71**	1.08	**1.62**
72" x 48"	Inst	SF	GB	.087	4.23	2.18	---	6.41	**7.93**	5.01	**6.03**
With aluminum channel and rigid neoprene rubber in aluminum sash											
48" x 96"	Demo	SF	GA	.076	---	1.91	---	1.91	**2.86**	1.16	**1.73**
48" x 96"	Inst	SF	GB	.076	3.76	1.91	---	5.67	**7.00**	4.45	**5.35**
96" x 96"	Demo	SF	GA	.074	---	1.86	---	1.86	**2.79**	1.10	**1.66**
96" x 96"	Inst	SF	GC	.062	3.72	1.56	---	5.28	**6.43**	4.18	**4.97**
1" T insulating glass (2 pieces 1/4" T float with 1/2" air space) with putty and points in wood sash											
To 6.0 SF	Demo	SF	GA	.267	---	6.70	---	6.70	**10.10**	4.02	**6.03**
To 6.0 SF	Inst	SF	GA	.222	10.20	5.57	---	15.77	**19.60**	12.26	**14.80**
6.1 SF to 12.0 SF	Demo	SF	GA	.121	---	3.04	---	3.04	**4.56**	1.83	**2.75**
6.1 SF to 12.0 SF	Inst	SF	GA	.099	9.19	2.49	---	11.68	**13.80**	9.52	**11.10**
12.1 SF to 18.0 SF	Demo	SF	GA	.089	---	2.23	---	2.23	**3.35**	1.33	**2.00**
12.1 SF to 18.0 SF	Inst	SF	GA	.074	8.77	1.86	---	10.63	**12.40**	8.77	**10.10**
18.1 SF to 24.0 SF	Demo	SF	GA	.092	---	2.31	---	2.31	**3.47**	1.38	**2.07**
18.1 SF to 24.0 SF	Inst	SF	GB	.107	8.48	2.69	---	11.17	**13.40**	9.03	**10.60**
Aluminum sliding door glass with aluminum channel and rigid neoprene rubber											
3/16" T tempered glass											
34" W x 76" H	Demo	SF	GA	.068	---	1.71	---	1.71	**2.56**	1.03	**1.54**
34" W x 76" H	Inst	SF	GA	.054	3.10	1.36	---	4.46	**5.44**	3.54	**4.22**
46" W x 76" H	Demo	SF	GA	.054	---	1.36	---	1.36	**2.03**	.83	**1.24**
46" W x 76" H	Inst	SF	GB	.081	3.07	2.03	---	5.10	**6.43**	3.89	**4.76**
5/8" T insulating glass (2 pieces 3/16" T tempered with 1/4" T air space)											
34" W x 76" H	Demo	SF	GA	.078	---	1.96	---	1.96	**2.94**	1.18	**1.77**
34" W x 76" H	Inst	SF	GA	.058	6.21	1.46	---	7.67	**9.02**	6.31	**7.29**
46" W x 76" H	Demo	SF	GA	.062	---	1.56	---	1.56	**2.34**	.93	**1.39**
46" W x 76" H	Inst	SF	GB	.086	6.12	2.16	---	8.28	**9.97**	6.67	**7.85**

Grading. See Concrete, page 60

Glu-lam products

Beams

3-1/8" thick, SF pricing based on 16' oc

Description	Oper	Unit	Crew Size	Man-Hours Per Unit	Avg Mat'l Unit Cost	Avg Labor Unit Cost	Avg Equip Unit Cost	Avg Total Unit Cost	Avg Price Incl O&P	Avg Total Unit Cost	Avg Price Incl O&P
				Costs Based On Small Volume						Large Volume	
9" deep, 20' long	Demo	LF	LK	.205	---	5.15	.64	5.79	**8.42**	3.76	**5.47**
9" deep, 20' long	Inst	LF	CY	.239	10.10	6.07	.64	16.81	**21.50**	12.33	**15.60**
9" deep, 20' long	Demo	BF	LK	.068	---	1.71	.21	1.92	**2.79**	1.25	**1.81**
9" deep, 20' long	Inst	BF	CY	.080	3.35	2.03	.21	5.59	**7.15**	4.10	**5.18**
9" deep, 20' long	Demo	SF	LK	.011	---	.28	.04	.32	**.46**	.20	**.29**
9" deep, 20' long	Inst	SF	CY	.013	.55	.33	.04	.92	**1.17**	.68	**.86**
10-1/2" deep, 20' long	Demo	LF	LK	.205	---	5.15	.64	5.79	**8.42**	3.76	**5.47**
10-1/2" deep, 20' long	Inst	LF	CY	.239	11.70	6.07	.64	18.41	**23.30**	13.59	**17.00**
10-1/2" deep, 20' long	Demo	BF	LK	.059	---	1.48	.18	1.66	**2.42**	1.08	**1.56**
10-1/2" deep, 20' long	Inst	BF	CY	.068	3.34	1.73	.18	5.25	**6.65**	3.87	**4.84**
10-1/2" deep, 20' long	Demo	SF	LK	.011	---	.28	.04	.32	**.46**	.20	**.29**
10-1/2" deep, 20' long	Inst	SF	CY	.013	.64	.33	.04	1.01	**1.28**	.75	**.94**
12" deep, 20' long	Demo	LF	LK	.205	---	5.15	.64	5.79	**8.42**	3.76	**5.47**
12" deep, 20' long	Inst	LF	CY	.239	13.00	6.07	.64	19.71	**24.80**	14.58	**18.20**
12" deep, 20' long	Demo	BF	LK	.051	---	1.28	.16	1.44	**2.10**	.93	**1.35**
12" deep, 20' long	Inst	BF	CY	.060	3.24	1.52	.16	4.92	**6.20**	3.64	**4.54**
12" deep, 20' long	Demo	SF	LK	.011	---	.28	.04	.32	**.46**	.20	**.29**
12" deep, 20' long	Inst	SF	CY	.013	.71	.33	.04	1.08	**1.36**	.81	**1.01**
13-1/2" deep, 20' long	Demo	LF	LK	.205	---	5.15	.64	5.79	**8.42**	3.76	**5.47**
13-1/2" deep, 20' long	Inst	LF	CY	.239	14.50	6.07	.64	21.21	**26.50**	15.78	**19.50**
13-1/2" deep, 20' long	Demo	BF	LK	.046	---	1.16	.14	1.30	**1.89**	.84	**1.23**
13-1/2" deep, 20' long	Inst	BF	CY	.053	3.21	1.35	.14	4.70	**5.88**	3.51	**4.35**
13-1/2" deep, 20' long	Demo	SF	LK	.011	---	.28	.04	.32	**.46**	.20	**.29**
13-1/2" deep, 20' long	Inst	SF	CY	.013	.80	.33	.04	1.17	**1.46**	.88	**1.09**
15" deep, 20' long	Demo	LF	LK	.205	---	5.15	.64	5.79	**8.42**	3.76	**5.47**
15" deep, 20' long	Inst	LF	CY	.239	15.90	6.07	.64	22.61	**28.20**	16.88	**20.90**
15" deep, 20' long	Demo	BF	LK	.041	---	1.03	.13	1.16	**1.69**	.76	**1.10**
15" deep, 20' long	Inst	BF	CY	.048	3.19	1.22	.13	4.54	**5.65**	3.38	**4.16**
15" deep, 20' long	Demo	SF	LK	.011	---	.28	.04	.32	**.46**	.20	**.29**
15" deep, 20' long	Inst	SF	CY	.013	.86	.33	.04	1.23	**1.53**	.93	**1.15**
16-1/2" deep, 20' long	Demo	LF	LK	.205	---	5.15	.64	5.79	**8.42**	3.76	**5.47**
16-1/2" deep, 20' long	Inst	LF	CY	.239	17.30	6.07	.64	24.01	**29.80**	18.08	**22.10**
16-1/2" deep, 20' long	Demo	BF	LK	.037	---	.93	.12	1.05	**1.52**	.68	**.99**
16-1/2" deep, 20' long	Inst	BF	CY	.044	3.15	1.12	.12	4.39	**5.44**	3.27	**4.01**
16-1/2" deep, 20' long	Demo	SF	LK	.011	---	.28	.04	.32	**.46**	.20	**.29**
16-1/2" deep, 20' long	Inst	SF	CY	.013	.94	.33	.04	1.31	**1.62**	.99	**1.22**
18" deep, 20' long	Demo	LF	LK	.205	---	5.15	.64	5.79	**8.42**	3.76	**5.47**
18" deep, 20' long	Inst	LF	CY	.239	18.80	6.07	.64	25.51	**31.50**	19.18	**23.50**
18" deep, 20' long	Demo	BF	LK	.034	---	.85	.11	.96	**1.40**	.62	**.91**
18" deep, 20' long	Inst	BF	CY	.040	3.14	1.02	.11	4.27	**5.27**	3.20	**3.91**
18" deep, 20' long	Demo	SF	LK	.011	---	.28	.04	.32	**.46**	.20	**.29**
18" deep, 20' long	Inst	SF	CY	.013	1.03	.33	.04	1.40	**1.73**	1.06	**1.30**

Description	Oper	Unit	Crew Size	Man-Hours Per Unit	Avg Mat'l Unit Cost	Avg Labor Unit Cost	Avg Equip Unit Cost	Avg Total Unit Cost	Avg Price Incl O&P	Avg Total Unit Cost	Avg Price Incl O&P
				Costs Based On Small Volume						Large Volume	
19-1/2" deep, 30' long	Demo	LF	LK	.205	---	5.15	.64	5.79	**8.42**	3.76	**5.47**
19-1/2" deep, 30' long	Inst	LF	CY	.239	19.40	6.07	.64	26.11	**32.10**	19.68	**24.00**
19-1/2" deep, 30' long	Demo	BF	LK	.032	---	.80	.10	.90	**1.31**	.59	**.86**
19-1/2" deep, 30' long	Inst	BF	CY	.037	2.97	.94	.10	4.01	**4.94**	3.01	**3.68**
19-1/2" deep, 30' long	Demo	SF	LK	.011	---	.28	.04	.32	**.46**	.20	**.29**
19-1/2" deep, 30' long	Inst	SF	CY	.013	1.07	.33	.04	1.44	**1.77**	1.09	**1.33**
21" deep, 30' long	Demo	LF	LK	.205	---	5.15	.64	5.79	**8.42**	3.76	**5.47**
21" deep, 30' long	Inst	LF	CY	.239	20.80	6.07	.64	27.51	**33.80**	20.78	**25.30**
21" deep, 30' long	Demo	BF	LK	.029	---	.73	.09	.82	**1.19**	.54	**.78**
21" deep, 30' long	Inst	BF	CY	.034	2.97	.86	.09	3.92	**4.82**	2.96	**3.60**
21" deep, 30' long	Demo	SF	LK	.011	---	.28	.04	.32	**.46**	.20	**.29**
21" deep, 30' long	Inst	SF	CY	.013	1.13	.33	.04	1.50	**1.84**	1.14	**1.39**
22-1/2" deep, 30' long	Demo	LF	LK	.205	---	5.15	.64	5.79	**8.42**	3.76	**5.47**
22-1/2" deep, 30' long	Inst	LF	CY	.239	22.20	6.07	.64	28.91	**35.40**	21.88	**26.50**
22-1/2" deep, 30' long	Demo	BF	LK	.027	---	.68	.09	.77	**1.11**	.51	**.74**
22-1/2" deep, 30' long	Inst	BF	CY	.032	2.96	.81	.09	3.86	**4.73**	2.92	**3.55**
22-1/2" deep, 30' long	Demo	SF	LK	.011	---	.28	.04	.32	**.46**	.20	**.29**
22-1/2" deep, 30' long	Inst	SF	CY	.013	1.21	.33	.04	1.58	**1.93**	1.20	**1.46**
24" deep, 30' long	Demo	LF	LK	.205	---	5.15	.64	5.79	**8.42**	3.76	**5.47**
24" deep, 30' long	Inst	LF	CY	.239	23.60	6.07	.64	30.31	**37.00**	22.98	**27.80**
24" deep, 30' long	Demo	BF	LK	.026	---	.65	.08	.73	**1.07**	.48	**.70**
24" deep, 30' long	Inst	BF	CY	.030	2.95	.76	.08	3.79	**4.63**	2.85	**3.45**
24" deep, 30' long	Demo	SF	LK	.011	---	.28	.04	.32	**.46**	.20	**.29**
24" deep, 30' long	Inst	SF	CY	.013	1.28	.33	.04	1.65	**2.01**	1.26	**1.53**
25-1/2" deep, 30' long	Demo	LF	LK	.205	---	5.15	.64	5.79	**8.42**	3.76	**5.47**
25-1/2" deep, 30' long	Inst	LF	CY	.239	25.10	6.07	.64	31.81	**38.70**	24.18	**29.20**
25-1/2" deep, 30' long	Demo	BF	LK	.024	---	.60	.08	.68	**.99**	.45	**.66**
25-1/2" deep, 30' long	Inst	BF	CY	.028	2.95	.71	.08	3.74	**4.55**	2.83	**3.41**
25-1/2" deep, 30' long	Demo	SF	LK	.011	---	.28	.04	.32	**.46**	.20	**.29**
25-1/2" deep, 30' long	Inst	SF	CY	.013	1.37	.33	.04	1.74	**2.12**	1.33	**1.61**
27" deep, 30' long	Demo	LF	LK	.205	---	5.15	.64	5.79	**8.42**	3.76	**5.47**
27" deep, 30' long	Inst	LF	CY	.239	26.50	6.07	.64	33.21	**40.30**	25.28	**30.40**
27" deep, 30' long	Demo	BF	LK	.023	---	.58	.07	.65	**.94**	.43	**.62**
27" deep, 30' long	Inst	BF	CY	.027	2.95	.69	.07	3.71	**4.50**	2.80	**3.37**
27" deep, 30' long	Demo	SF	LK	.011	---	.28	.04	.32	**.46**	.20	**.29**
27" deep, 30' long	Inst	SF	CY	.013	1.45	.33	.04	1.82	**2.21**	1.39	**1.68**
5-1/8" thick, SF pricing based on 16' oc											
12" deep, 20' long	Demo	LF	LK	.205	---	5.15	.64	5.79	**8.42**	3.76	**5.47**
12" deep, 20' long	Inst	LF	CY	.239	15.90	6.07	.64	22.61	**28.10**	16.88	**20.80**
12" deep, 20' long	Demo	BF	LK	.034	---	.85	.11	.96	**1.40**	.62	**.91**
12" deep, 20' long	Inst	BF	CY	.040	2.65	1.02	.11	3.78	**4.70**	2.82	**3.48**
12" deep, 20' long	Demo	SF	LK	.011	---	.28	.04	.32	**.46**	.20	**.29**
12" deep, 20' long	Inst	SF	CY	.013	.86	.33	.04	1.23	**1.53**	.93	**1.15**
13-1/2" deep, 20' long	Demo	LF	LK	.205	---	5.15	.64	5.79	**8.42**	3.76	**5.47**
13-1/2" deep, 20' long	Inst	LF	CY	.239	17.60	6.07	.64	24.31	**30.20**	18.28	**22.40**
13-1/2" deep, 20' long	Demo	BF	LK	.030	---	.75	.09	.84	**1.23**	.56	**.82**
13-1/2" deep, 20' long	Inst	BF	CY	.035	2.60	.89	.09	3.58	**4.43**	2.69	**3.31**
13-1/2" deep, 20' long	Demo	SF	LK	.011	---	.28	.04	.32	**.46**	.20	**.29**
13-1/2" deep, 20' long	Inst	SF	CY	.013	.95	.33	.04	1.32	**1.63**	1.00	**1.23**

			Costs Based On Small Volume						Large Volume	
Description	Oper	Unit	Crew Size	Man-Hours Per Unit	Avg Mat'l Unit Cost	Avg Labor Unit Cost	Avg Equip Unit Cost	Avg Total Unit Cost	Avg Price Incl O&P	Avg Total Unit Cost	Avg Price Incl O&P
15" deep, 20' long	Demo	LF	LK	.205	---	5.15	.64	5.79	**8.42**	3.76	**5.47**
15" deep, 20' long	Inst	LF	CY	.239	19.40	6.07	.64	26.11	**32.20**	19.68	**24.00**
15" deep, 20' long	Demo	BF	LK	.027	---	.68	.09	.77	**1.11**	.51	**.74**
15" deep, 20' long	Inst	BF	CY	.032	2.59	.81	.09	3.49	**4.30**	2.63	**3.22**
15" deep, 20' long	Demo	SF	LK	.011	---	.28	.04	.32	**.46**	.20	**.29**
15" deep, 20' long	Inst	SF	CY	.013	1.07	.33	.04	1.44	**1.77**	1.09	**1.33**
16-1/2" deep, 20' long	Demo	LF	LK	.205	---	5.15	.64	5.79	**8.42**	3.76	**5.47**
16-1/2" deep, 20' long	Inst	LF	CY	.239	21.20	6.07	.64	27.91	**34.20**	21.08	**25.60**
16-1/2" deep, 20' long	Demo	BF	LK	.025	---	.63	.08	.71	**1.03**	.45	**.66**
16-1/2" deep, 20' long	Inst	BF	CY	.029	2.57	.74	.08	3.39	**4.16**	2.55	**3.11**
16-1/2" deep, 20' long	Demo	SF	LK	.011	---	.28	.04	.32	**.46**	.20	**.29**
16-1/2" deep, 20' long	Inst	SF	CY	.013	1.14	.33	.04	1.51	**1.85**	1.15	**1.40**
18" deep, 20' long	Demo	LF	LK	.205	---	5.15	.64	5.79	**8.42**	3.76	**5.47**
18" deep, 20' long	Inst	LF	CY	.239	22.90	6.07	.64	29.61	**36.20**	22.48	**27.20**
18" deep, 20' long	Demo	BF	LK	.023	---	.58	.07	.65	**.94**	.43	**.62**
18" deep, 20' long	Inst	BF	CY	.027	2.55	.69	.07	3.31	**4.04**	2.49	**3.02**
18" deep, 20' long	Demo	SF	LK	.011	---	.28	.04	.32	**.46**	.20	**.29**
18" deep, 20' long	Inst	SF	CY	.013	1.26	.33	.04	1.63	**1.99**	1.24	**1.51**
19-1/2" deep, 20' long	Demo	LF	LK	.205	---	5.15	.64	5.79	**8.42**	3.76	**5.47**
19-1/2" deep, 20' long	Inst	LF	CY	.239	24.60	6.07	.64	31.31	**38.20**	23.78	**28.70**
19-1/2" deep, 20' long	Demo	BF	LK	.021	---	.53	.07	.60	**.87**	.39	**.57**
19-1/2" deep, 20' long	Inst	BF	CY	.025	2.51	.64	.07	3.22	**3.92**	2.43	**2.93**
19-1/2" deep, 20' long	Demo	SF	LK	.011	---	.28	.04	.32	**.46**	.20	**.29**
19-1/2" deep, 20' long	Inst	SF	CY	.013	1.35	.33	.04	1.72	**2.09**	1.31	**1.59**
21" deep, 30' long	Demo	LF	LK	.205	---	5.15	.64	5.79	**8.42**	3.76	**5.47**
21" deep, 30' long	Inst	LF	CY	.239	25.50	6.07	.64	32.21	**39.20**	24.48	**29.50**
21" deep, 30' long	Demo	BF	LK	.020	---	.50	.06	.56	**.82**	.37	**.53**
21" deep, 30' long	Inst	BF	CY	.023	2.43	.58	.06	3.07	**3.74**	2.33	**2.82**
21" deep, 30' long	Demo	SF	LK	.011	---	.28	.04	.32	**.46**	.20	**.29**
21" deep, 30' long	Inst	SF	CY	.013	1.40	.33	.04	1.77	**2.15**	1.35	**1.63**
22-1/2" deep, 30' long	Demo	LF	LK	.205	---	5.15	.64	5.79	**8.42**	3.76	**5.47**
22-1/2" deep, 30' long	Inst	LF	CY	.239	27.20	6.07	.64	33.91	**41.20**	25.78	**31.10**
22-1/2" deep, 30' long	Demo	BF	LK	.018	---	.45	.06	.51	**.74**	.34	**.50**
22-1/2" deep, 30' long	Inst	BF	CY	.021	2.41	.53	.06	3.00	**3.64**	2.30	**2.77**
22-1/2" deep, 30' long	Demo	SF	LK	.011	---	.28	.04	.32	**.46**	.20	**.29**
22-1/2" deep, 30' long	Inst	SF	CY	.013	1.49	.33	.04	1.86	**2.26**	1.42	**1.71**
24" deep, 30' long	Demo	LF	LK	.205	---	5.15	.64	5.79	**8.42**	3.76	**5.47**
24" deep, 30' long	Inst	LF	CY	.239	28.90	6.07	.64	35.61	**43.10**	27.18	**32.60**
24" deep, 30' long	Demo	BF	LK	.017	---	.43	.05	.48	**.70**	.31	**.45**
24" deep, 30' long	Inst	BF	CY	.020	2.41	.51	.05	2.97	**3.59**	2.26	**2.72**
24" deep, 30' long	Demo	SF	LK	.011	---	.28	.04	.32	**.46**	.20	**.29**
24" deep, 30' long	Inst	SF	CY	.013	1.57	.33	.04	1.94	**2.35**	1.49	**1.79**
25-1/2" deep, 30' long	Demo	LF	LK	.205	---	5.15	.64	5.79	**8.42**	3.76	**5.47**
25-1/2" deep, 30' long	Inst	LF	CY	.239	30.70	6.07	.64	37.41	**45.10**	28.48	**34.20**
25-1/2" deep, 30' long	Demo	BF	LK	.016	---	.40	.05	.45	**.66**	.28	**.41**
25-1/2" deep, 30' long	Inst	BF	CY	.019	2.40	.48	.05	2.93	**3.54**	2.22	**2.67**
25-1/2" deep, 30' long	Demo	SF	LK	.011	---	.28	.04	.32	**.46**	.20	**.29**
25-1/2" deep, 30' long	Inst	SF	CY	.013	1.68	.33	.04	2.05	**2.47**	1.57	**1.89**

			Costs Based On Small Volume							Large Volume	
Description	Oper	Unit	Crew Size	Man-Hours Per Unit	Avg Mat'l Unit Cost	Avg Labor Unit Cost	Avg Equip Unit Cost	Avg Total Unit Cost	Avg Price Incl O&P	Avg Total Unit Cost	Avg Price Incl O&P
27" deep, 40' long	Demo	LF	LK	.205	---	5.15	.64	5.79	**8.42**	3.76	**5.47**
27" deep, 40' long	Inst	LF	CY	.239	31.80	6.07	.64	38.51	**46.40**	29.38	**35.20**
27" deep, 40' long	Demo	BF	LK	.015	---	.38	.05	.43	**.62**	.28	**.41**
27" deep, 40' long	Inst	BF	CY	.018	2.36	.46	.05	2.87	**3.46**	2.19	**2.63**
27" deep, 40' long	Demo	SF	LK	.011	---	.28	.04	.32	**.46**	.20	**.29**
27" deep, 40' long	Inst	SF	CY	.013	1.74	.33	.04	2.11	**2.54**	1.62	**1.94**
28-1/2" deep, 40' long	Demo	LF	LK	.205	---	5.15	.64	5.79	**8.42**	3.76	**5.47**
28-1/2" deep, 40' long	Inst	LF	CY	.239	33.40	6.07	.64	40.11	**48.30**	30.68	**36.70**
28-1/2" deep, 40' long	Demo	BF	LK	.014	---	.35	.04	.39	**.57**	.26	**.37**
28-1/2" deep, 40' long	Inst	BF	CY	.017	2.34	.43	.04	2.81	**3.39**	2.15	**2.57**
28-1/2" deep, 40' long	Demo	SF	LK	.011	---	.28	.04	.32	**.46**	.20	**.29**
28-1/2" deep, 40' long	Inst	SF	CY	.013	1.83	.33	.04	2.20	**2.65**	1.69	**2.02**
30" deep, 40' long	Demo	LF	LK	.205	---	5.15	.64	5.79	**8.42**	3.76	**5.47**
30" deep, 40' long	Inst	LF	CY	.239	35.20	6.07	.64	41.91	**50.30**	32.08	**38.30**
30" deep, 40' long	Demo	BF	LK	.014	---	.35	.04	.39	**.57**	.26	**.37**
30" deep, 40' long	Inst	BF	CY	.016	2.34	.41	.04	2.79	**3.35**	2.12	**2.53**
30" deep, 40' long	Demo	SF	LK	.011	---	.28	.04	.32	**.46**	.20	**.29**
30" deep, 40' long	Inst	SF	CY	.013	1.92	.33	.04	2.29	**2.75**	1.76	**2.10**
31-1/2" deep, 40' long	Demo	LF	LK	.205	---	5.15	.64	5.79	**8.42**	3.76	**5.47**
31-1/2" deep, 40' long	Inst	LF	CY	.239	36.80	6.07	.64	43.51	**52.20**	33.38	**39.80**
31-1/2" deep, 40' long	Demo	BF	LK	.013	---	.33	.04	.37	**.53**	.23	**.33**
31-1/2" deep, 40' long	Inst	BF	CY	.015	2.34	.38	.04	2.76	**3.31**	2.12	**2.53**
31-1/2" deep, 40' long	Demo	SF	LK	.011	---	.28	.04	.32	**.46**	.20	**.29**
31-1/2" deep, 40' long	Inst	SF	CY	.013	2.02	.33	.04	2.39	**2.86**	1.84	**2.20**
33" deep, 40' long	Demo	LF	LK	.205	---	5.15	.64	5.79	**8.42**	3.76	**5.47**
33" deep, 40' long	Inst	LF	CY	.239	38.50	6.07	.64	45.21	**54.20**	34.68	**41.30**
33" deep, 40' long	Demo	BF	LK	.012	---	.30	.04	.34	**.50**	.23	**.33**
33" deep, 40' long	Inst	BF	CY	.015	2.32	.38	.04	2.74	**3.29**	2.09	**2.48**
33" deep, 40' long	Demo	SF	LK	.011	---	.28	.04	.32	**.46**	.20	**.29**
33" deep, 40' long	Inst	SF	CY	.013	2.11	.33	.04	2.48	**2.97**	1.91	**2.28**
34-1/2" deep, 40' long	Demo	LF	LK	.205	---	5.15	.64	5.79	**8.42**	3.76	**5.47**
34-1/2" deep, 40' long	Inst	LF	CY	.239	40.20	6.07	.64	46.91	**56.10**	36.08	**42.90**
34-1/2" deep, 40' long	Demo	BF	LK	.012	---	.30	.04	.34	**.50**	.22	**.32**
34-1/2" deep, 40' long	Inst	BF	CY	.014	2.32	.36	.04	2.72	**3.25**	2.08	**2.47**
34-1/2" deep, 40' long	Demo	SF	LK	.011	---	.28	.04	.32	**.46**	.20	**.29**
34-1/2" deep, 40' long	Inst	SF	CY	.013	2.20	.33	.04	2.57	**3.07**	1.98	**2.36**
36" deep, 50' long	Demo	LF	LK	.205	---	5.15	.64	5.79	**8.42**	3.76	**5.47**
36" deep, 50' long	Inst	LF	CY	.239	41.40	6.07	.64	48.11	**57.50**	36.98	**44.00**
36" deep, 50' long	Demo	BF	LK	.011	---	.28	.04	.32	**.46**	.20	**.29**
36" deep, 50' long	Inst	BF	CY	.013	2.30	.33	.04	2.67	**3.19**	2.06	**2.45**
36" deep, 50' long	Demo	SF	LK	.011	---	.28	.04	.32	**.46**	.20	**.29**
36" deep, 50' long	Inst	SF	CY	.013	2.27	.33	.04	2.64	**3.15**	2.04	**2.43**

Description	Oper	Unit	Crew Size	Man-Hours Per Unit	Avg Mat'l Unit Cost	Avg Labor Unit Cost	Avg Equip Unit Cost	Avg Total Unit Cost	Avg Price Incl O&P	Large Volume Avg Total Unit Cost	Large Volume Avg Price Incl O&P
				Costs Based On Small Volume						Large Volume	
37-1/2" deep, 50' long	Demo	LF	LK	.205	---	5.15	.64	5.79	**8.42**	3.76	**5.47**
37-1/2" deep, 50' long	Inst	LF	CY	.239	43.20	6.07	.64	49.91	**59.50**	38.38	**45.50**
37-1/2" deep, 50' long	Demo	BF	LK	.011	---	.28	.03	.31	**.45**	.20	**.29**
37-1/2" deep, 50' long	Inst	BF	CY	.013	2.30	.33	.03	2.66	**3.18**	2.03	**2.41**
37-1/2" deep, 50' long	Demo	SF	LK	.011	---	.28	.04	.32	**.46**	.20	**.29**
37-1/2" deep, 50' long	Inst	SF	CY	.013	2.36	.33	.04	2.73	**3.26**	2.11	**2.51**
39" deep, 50' long	Demo	LF	LK	.205	---	5.15	.64	5.79	**8.42**	3.76	**5.47**
39" deep, 50' long	Inst	LF	CY	.239	44.90	6.07	.64	51.61	**61.50**	39.78	**47.10**
39" deep, 50' long	Demo	BF	LK	.011	---	.28	.03	.31	**.45**	.20	**.29**
39" deep, 50' long	Inst	BF	CY	.012	2.30	.30	.03	2.63	**3.14**	2.03	**2.41**
39" deep, 50' long	Demo	SF	LK	.011	---	.28	.04	.32	**.46**	.20	**.29**
39" deep, 50' long	Inst	SF	CY	.013	2.45	.33	.04	2.82	**3.36**	2.18	**2.59**
6-3/4" thick, SF pricing based on 16' oc											
30" deep, 30' long	Demo	LF	LK	.205	---	5.15	.64	5.79	**8.42**	3.76	**5.47**
30" deep, 30' long	Inst	LF	CY	.239	42.20	6.07	.64	48.91	**58.40**	37.68	**44.70**
30" deep, 30' long	Demo	BF	LK	.010	---	.25	.03	.28	**.41**	.20	**.29**
30" deep, 30' long	Inst	BF	CY	.012	2.11	.30	.03	2.44	**2.92**	1.88	**2.24**
30" deep, 30' long	Demo	SF	LK	.011	---	.28	.04	.32	**.46**	.20	**.29**
30" deep, 30' long	Inst	SF	CY	.013	2.31	.33	.04	2.68	**3.20**	2.07	**2.46**
31-1/2" deep, 30' long	Demo	LF	LK	.205	---	5.15	.64	5.79	**8.42**	3.76	**5.47**
31-1/2" deep, 30' long	Inst	LF	CY	.239	44.30	6.07	.64	51.01	**60.80**	39.28	**46.50**
31-1/2" deep, 30' long	Demo	BF	LK	.010	---	.25	.03	.28	**.41**	.17	**.25**
31-1/2" deep, 30' long	Inst	BF	CY	.011	2.11	.28	.03	2.42	**2.88**	1.86	**2.20**
31-1/2" deep, 30' long	Demo	SF	LK	.011	---	.28	.04	.32	**.46**	.20	**.29**
31-1/2" deep, 30' long	Inst	SF	CY	.013	2.41	.33	.04	2.78	**3.31**	2.15	**2.55**
33" deep, 30' long	Demo	LF	LK	.205	---	5.15	.64	5.79	**8.42**	3.76	**5.47**
33" deep, 30' long	Inst	LF	CY	.239	46.30	6.07	.64	53.01	**63.10**	40.88	**48.40**
33" deep, 30' long	Demo	BF	LK	.009	---	.23	.03	.26	**.37**	.17	**.25**
33" deep, 30' long	Inst	BF	CY	.011	2.11	.28	.03	2.42	**2.88**	1.86	**2.20**
33" deep, 30' long	Demo	SF	LK	.011	---	.28	.04	.32	**.46**	.20	**.29**
33" deep, 30' long	Inst	SF	CY	.013	2.54	.33	.04	2.91	**3.46**	2.25	**2.67**
34-1/2" deep, 40' long	Demo	LF	LK	.205	---	5.15	.64	5.79	**8.42**	3.76	**5.47**
34-1/2" deep, 40' long	Inst	LF	CY	.239	47.50	6.07	.64	54.21	**64.50**	41.78	**49.50**
34-1/2" deep, 40' long	Demo	BF	LK	.009	---	.23	.03	.26	**.37**	.17	**.25**
34-1/2" deep, 40' long	Inst	BF	CY	.010	2.06	.25	.03	2.34	**2.79**	1.82	**2.15**
34-1/2" deep, 40' long	Demo	SF	LK	.011	---	.28	.04	.32	**.46**	.20	**.29**
34-1/2" deep, 40' long	Inst	SF	CY	.013	2.59	.33	.04	2.96	**3.52**	2.29	**2.71**
36" deep, 40' long	Demo	LF	LK	.205	---	5.15	.64	5.79	**8.42**	3.76	**5.47**
36" deep, 40' long	Inst	LF	CY	.239	49.50	6.07	.64	56.21	**66.80**	43.38	**51.30**
36" deep, 40' long	Demo	BF	LK	.009	---	.23	.03	.26	**.37**	.17	**.25**
36" deep, 40' long	Inst	BF	CY	.010	2.06	.25	.03	2.34	**2.79**	1.79	**2.11**
36" deep, 40' long	Demo	SF	LK	.011	---	.28	.04	.32	**.46**	.20	**.29**
36" deep, 40' long	Inst	SF	CY	.013	2.71	.33	.04	3.08	**3.66**	2.38	**2.82**
37-1/2" deep, 40' long	Demo	LF	LK	.205	---	5.15	.64	5.79	**8.42**	3.76	**5.47**
37-1/2" deep, 40' long	Inst	LF	CY	.239	51.50	6.07	.64	58.21	**69.10**	44.98	**53.10**
37-1/2" deep, 40' long	Demo	BF	LK	.008	---	.20	.03	.23	**.33**	.15	**.21**
37-1/2" deep, 40' long	Inst	BF	CY	.010	2.06	.25	.03	2.34	**2.79**	1.79	**2.11**
37-1/2" deep, 40' long	Demo	SF	LK	.011	---	.28	.04	.32	**.46**	.20	**.29**
37-1/2" deep, 40' long	Inst	SF	CY	.013	2.82	.33	.04	3.19	**3.78**	2.47	**2.92**

Description	Oper	Unit	Crew Size	Man-Hours Per Unit	Avg Mat'l Unit Cost	Avg Labor Unit Cost	Avg Equip Unit Cost	Avg Total Unit Cost	Avg Price Incl O&P	Avg Total Unit Cost	Avg Price Incl O&P
				Costs Based On Small Volume						Large Volume	
39" deep, 50' long	Demo	LF	LK	.205	---	5.15	.64	5.79	**8.42**	3.76	**5.47**
39" deep, 50' long	Inst	LF	CY	.239	53.00	6.07	.64	59.71	**70.80**	46.18	**54.50**
39" deep, 50' long	Demo	BF	LK	.008	---	.20	.02	.22	**.32**	.15	**.21**
39" deep, 50' long	Inst	BF	CY	.009	2.03	.23	.02	2.28	**2.70**	1.77	**2.09**
39" deep, 50' long	Demo	SF	LK	.011	---	.28	.04	.32	**.46**	.20	**.29**
39" deep, 50' long	Inst	SF	CY	.013	2.91	.33	.04	3.28	**3.89**	2.54	**3.00**
40-1/2" deep, 50' long	Demo	LF	LK	.205	---	5.15	.64	5.79	**8.42**	3.76	**5.47**
40-1/2" deep, 50' long	Inst	LF	CY	.239	55.00	6.07	.64	61.71	**73.10**	47.68	**56.30**
40-1/2" deep, 50' long	Demo	BF	LK	.008	---	.20	.02	.22	**.32**	.15	**.21**
40-1/2" deep, 50' long	Inst	BF	CY	.009	2.03	.23	.02	2.28	**2.70**	1.77	**2.09**
40-1/2" deep, 50' long	Demo	SF	LK	.011	---	.28	.04	.32	**.46**	.20	**.29**
40-1/2" deep, 50' long	Inst	SF	CY	.013	3.01	.33	.04	3.38	**4.00**	2.62	**3.09**
42" deep, 50' long	Demo	LF	LK	.205	---	5.15	.64	5.79	**8.42**	3.76	**5.47**
42" deep, 50' long	Inst	LF	CY	.239	57.00	6.07	.64	63.71	**75.40**	49.28	**58.10**
42" deep, 50' long	Demo	BF	LK	.007	---	.18	.02	.20	**.29**	.14	**.20**
42" deep, 50' long	Inst	BF	CY	.009	2.03	.23	.02	2.28	**2.70**	1.76	**2.08**
42" deep, 50' long	Demo	SF	LK	.011	---	.28	.04	.32	**.46**	.20	**.29**
42" deep, 50' long	Inst	SF	CY	.013	3.11	.33	.04	3.48	**4.12**	2.70	**3.18**
43-1/2" deep, 50' long	Demo	LF	LK	.205	---	5.15	.64	5.79	**8.42**	3.76	**5.47**
43-1/2" deep, 50' long	Inst	LF	CY	.239	59.00	6.07	.64	65.71	**77.70**	50.78	**59.80**
43-1/2" deep, 50' long	Demo	BF	LK	.007	---	.18	.02	.20	**.29**	.14	**.20**
43-1/2" deep, 50' long	Inst	BF	CY	.008	2.03	.20	.02	2.25	**2.66**	1.74	**2.04**
43-1/2" deep, 50' long	Demo	SF	LK	.011	---	.28	.04	.32	**.46**	.20	**.29**
43-1/2" deep, 50' long	Inst	SF	CY	.013	3.23	.33	.04	3.60	**4.26**	2.79	**3.29**
45" deep, 50' long	Demo	LF	LK	.205	---	5.15	.64	5.79	**8.42**	3.76	**5.47**
45" deep, 50' long	Inst	LF	CY	.239	61.00	6.07	.64	67.71	**80.00**	52.38	**61.60**
45" deep, 50' long	Demo	BF	LK	.007	---	.18	.02	.20	**.29**	.11	**.16**
45" deep, 50' long	Inst	BF	CY	.008	2.03	.20	.02	2.25	**2.66**	1.74	**2.04**
45" deep, 50' long	Demo	SF	LK	.011	---	.28	.04	.32	**.46**	.20	**.29**
45" deep, 50' long	Inst	SF	CY	.013	3.33	.33	.04	3.70	**4.37**	2.87	**3.38**
8-3/4" thick, SF pricing based on 16' oc											
36" deep, 30' long	Demo	LF	LK	.205	---	5.15	.64	5.79	**8.42**	3.76	**5.47**
36" deep, 30' long	Inst	LF	CY	.239	57.80	6.07	.64	64.51	**76.30**	49.88	**58.80**
36" deep, 30' long	Demo	BF	LK	.007	---	.18	.02	.20	**.29**	.11	**.16**
36" deep, 30' long	Inst	BF	CY	.008	1.93	.20	.02	2.15	**2.55**	1.66	**1.95**
36" deep, 30' long	Demo	SF	LK	.011	---	.28	.04	.32	**.46**	.20	**.29**
36" deep, 30' long	Inst	SF	CY	.013	3.15	.33	.04	3.52	**4.16**	2.73	**3.22**
37-1/2" deep, 30' long	Demo	LF	LK	.205	---	5.15	.64	5.79	**8.42**	3.76	**5.47**
37-1/2" deep, 30' long	Inst	LF	CY	.239	60.00	6.07	.64	66.71	**78.90**	51.68	**60.80**
37-1/2" deep, 30' long	Demo	BF	LK	.007	---	.18	.02	.20	**.29**	.11	**.16**
37-1/2" deep, 30' long	Inst	BF	CY	.008	1.92	.20	.02	2.14	**2.54**	1.65	**1.94**
37-1/2" deep, 30' long	Demo	SF	LK	.011	---	.28	.04	.32	**.46**	.20	**.29**
37-1/2" deep, 30' long	Inst	SF	CY	.013	3.28	.33	.04	3.65	**4.31**	2.83	**3.33**
39" deep, 30' long	Demo	LF	LK	.205	---	5.15	.64	5.79	**8.42**	3.76	**5.47**
39" deep, 30' long	Inst	LF	CY	.239	62.50	6.07	.64	69.21	**81.70**	53.58	**63.00**
39" deep, 30' long	Demo	BF	LK	.006	---	.15	.02	.17	**.25**	.11	**.16**
39" deep, 30' long	Inst	BF	CY	.007	1.92	.18	.02	2.12	**2.50**	1.65	**1.94**
39" deep, 30' long	Demo	SF	LK	.011	---	.28	.04	.32	**.46**	.20	**.29**
39" deep, 30' long	Inst	SF	CY	.013	3.40	.33	.04	3.77	**4.45**	2.93	**3.45**

Description	Oper	Unit	Crew Size	Man-Hours Per Unit	Avg Mat'l Unit Cost	Avg Labor Unit Cost	Avg Equip Unit Cost	Avg Total Unit Cost	Avg Price Incl O&P	Large Volume Avg Total Unit Cost	Large Volume Avg Price Incl O&P
40-1/2" deep, 30' long	Demo	LF	LK	.205	---	5.15	.64	5.79	**8.42**	3.76	**5.47**
40-1/2" deep, 30' long	Inst	LF	CY	.239	64.80	6.07	.64	71.51	**84.40**	55.38	**65.10**
40-1/2" deep, 30' long	Demo	BF	LK	.006	---	.15	.02	.17	**.25**	.11	**.16**
40-1/2" deep, 30' long	Inst	BF	CY	.007	1.92	.18	.02	2.12	**2.50**	1.65	**1.94**
40-1/2" deep, 30' long	Demo	SF	LK	.011	---	.28	.04	.32	**.46**	.20	**.29**
40-1/2" deep, 30' long	Inst	SF	CY	.013	3.53	.33	.04	3.90	**4.60**	3.03	**3.56**
42" deep, 30' long	Demo	LF	LK	.205	---	5.15	.64	5.79	**8.42**	3.76	**5.47**
42" deep, 30' long	Inst	LF	CY	.239	67.00	6.07	.64	73.71	**87.00**	57.18	**67.20**
42" deep, 30' long	Demo	BF	LK	.006	---	.15	.02	.17	**.25**	.11	**.16**
42" deep, 30' long	Inst	BF	CY	.007	1.92	.18	.02	2.12	**2.50**	1.62	**1.90**
42" deep, 30' long	Demo	SF	LK	.011	---	.28	.04	.32	**.46**	.20	**.29**
42" deep, 30' long	Inst	SF	CY	.013	3.66	.33	.04	4.03	**4.75**	3.13	**3.68**
43-1/2" deep, 40' long	Demo	LF	LK	.205	---	5.15	.64	5.79	**8.42**	3.76	**5.47**
43-1/2" deep, 40' long	Inst	LF	CY	.239	68.60	6.07	.64	75.31	**88.80**	58.38	**68.60**
43-1/2" deep, 40' long	Demo	BF	LK	.006	---	.15	.02	.17	**.25**	.11	**.16**
43-1/2" deep, 40' long	Inst	BF	CY	.007	1.88	.18	.02	2.08	**2.45**	1.59	**1.87**
43-1/2" deep, 40' long	Demo	SF	LK	.011	---	.28	.04	.32	**.46**	.20	**.29**
43-1/2" deep, 40' long	Inst	SF	CY	.013	3.75	.33	.04	4.12	**4.85**	3.20	**3.76**
45" deep, 40' long	Demo	LF	LK	.205	---	5.15	.64	5.79	**8.42**	3.76	**5.47**
45" deep, 40' long	Inst	LF	CY	.239	70.90	6.07	.64	77.61	**91.40**	60.18	**70.70**
45" deep, 40' long	Demo	BF	LK	.005	---	.13	.02	.15	**.21**	.11	**.16**
45" deep, 40' long	Inst	BF	CY	.006	1.88	.15	.02	2.05	**2.41**	1.59	**1.87**
45" deep, 40' long	Demo	SF	LK	.011	---	.28	.04	.32	**.46**	.20	**.29**
45" deep, 40' long	Inst	SF	CY	.013	3.87	.33	.04	4.24	**4.99**	3.30	**3.87**
46-1/2" deep, 40' long	Demo	LF	LK	.205	---	5.15	.64	5.79	**8.42**	3.76	**5.47**
46-1/2" deep, 40' long	Inst	LF	CY	.239	73.20	6.07	.64	79.91	**94.10**	62.08	**72.70**
46-1/2" deep, 40' long	Demo	BF	LK	.005	---	.13	.02	.15	**.21**	.09	**.12**
46-1/2" deep, 40' long	Inst	BF	CY	.006	1.88	.15	.02	2.05	**2.41**	1.59	**1.87**
46-1/2" deep, 40' long	Demo	SF	LK	.011	---	.28	.04	.32	**.46**	.20	**.29**
46-1/2" deep, 40' long	Inst	SF	CY	.013	3.99	.33	.04	4.36	**5.13**	3.39	**3.98**
48" deep, 50' long	Demo	LF	LK	.205	---	5.15	.64	5.79	**8.42**	3.76	**5.47**
48" deep, 50' long	Inst	LF	CY	.239	75.00	6.07	.64	81.71	**96.10**	63.38	**74.30**
48" deep, 50' long	Demo	BF	LK	.005	---	.13	.02	.15	**.21**	.09	**.12**
48" deep, 50' long	Inst	BF	CY	.006	1.87	.15	.02	2.04	**2.40**	1.58	**1.85**
48" deep, 50' long	Demo	SF	LK	.011	---	.28	.04	.32	**.46**	.20	**.29**
48" deep, 50' long	Inst	SF	CY	.013	4.09	.33	.04	4.46	**5.25**	3.47	**4.07**
49-1/2" deep, 50' long	Demo	LF	LK	.205	---	5.15	.64	5.79	**8.42**	3.76	**5.47**
49-1/2" deep, 50' long	Inst	LF	CY	.239	77.30	6.07	.64	84.01	**98.70**	65.18	**76.40**
49-1/2" deep, 50' long	Demo	BF	LK	.005	---	.13	.02	.15	**.21**	.09	**.12**
49-1/2" deep, 50' long	Inst	BF	CY	.006	1.87	.15	.02	2.04	**2.40**	1.58	**1.85**
49-1/2" deep, 50' long	Demo	SF	LK	.011	---	.28	.04	.32	**.46**	.20	**.29**
49-1/2" deep, 50' long	Inst	SF	CY	.013	4.22	.33	.04	4.59	**5.39**	3.57	**4.19**
51" deep, 50' long	Demo	LF	LK	.205	---	5.15	.64	5.79	**8.42**	3.76	**5.47**
51" deep, 50' long	Inst	LF	CY	.239	79.60	6.07	.64	86.31	**101.00**	67.08	**78.50**
51" deep, 50' long	Demo	BF	LK	.005	---	.13	.02	.15	**.21**	.09	**.12**
51" deep, 50' long	Inst	BF	CY	.006	1.87	.15	.02	2.04	**2.40**	1.58	**1.85**
51" deep, 50' long	Demo	SF	LK	.011	---	.28	.04	.32	**.46**	.20	**.29**
51" deep, 50' long	Inst	SF	CY	.013	4.34	.33	.04	4.71	**5.53**	3.67	**4.30**

Description	Oper	Unit	Crew Size	Man-Hours Per Unit	Avg Mat'l Unit Cost	Avg Labor Unit Cost	Avg Equip Unit Cost	Avg Total Unit Cost	Avg Price Incl O&P	Large Volume Avg Total Unit Cost	Large Volume Avg Price Incl O&P
				Costs Based On Small Volume						Large Volume	
52-1/2" deep, 50' long	Demo	LF	LK	.205	---	5.15	.64	5.79	**8.42**	3.76	**5.47**
52-1/2" deep, 50' long	Inst	LF	CY	.239	81.90	6.07	.64	88.61	**104.00**	68.88	**80.60**
52-1/2" deep, 50' long	Demo	BF	LK	.005	---	.13	.01	.14	**.20**	.09	**.12**
52-1/2" deep, 50' long	Inst	BF	CY	.005	1.87	.13	.01	2.01	**2.35**	1.58	**1.85**
52-1/2" deep, 50' long	Demo	SF	LK	.011	---	.28	.04	.32	**.46**	.20	**.29**
52-1/2" deep, 50' long	Inst	SF	CY	.013	4.46	.33	.04	4.83	**5.67**	3.76	**4.40**
54" deep, 50' long	Demo	LF	LK	.205	---	5.15	.64	5.79	**8.42**	3.76	**5.47**
54" deep, 50' long	Inst	LF	CY	.239	84.20	6.07	.64	90.91	**107.00**	70.68	**82.60**
54" deep, 50' long	Demo	BF	LK	.005	---	.13	.01	.14	**.20**	.09	**.12**
54" deep, 50' long	Inst	BF	CY	.005	1.87	.13	.01	2.01	**2.35**	1.56	**1.82**
54" deep, 50' long	Demo	SF	LK	.011	---	.28	.04	.32	**.46**	.20	**.29**
54" deep, 50' long	Inst	SF	CY	.013	4.58	.33	.04	4.95	**5.81**	3.86	**4.52**
55-1/2" deep, 50' long	Demo	LF	LK	.205	---	5.15	.64	5.79	**8.42**	3.76	**5.47**
55-1/2" deep, 50' long	Inst	LF	CY	.239	86.50	6.07	.64	93.21	**109.00**	72.48	**84.70**
55-1/2" deep, 50' long	Demo	BF	LK	.004	---	.10	.01	.11	**.16**	.09	**.12**
55-1/2" deep, 50' long	Inst	BF	CY	.005	1.87	.13	.01	2.01	**2.35**	1.56	**1.82**
55-1/2" deep, 50' long	Demo	SF	LK	.011	---	.28	.04	.32	**.46**	.20	**.29**
55-1/2" deep, 50' long	Inst	SF	CY	.013	4.71	.33	.04	5.08	**5.96**	3.96	**4.63**
57" deep, 50' long	Demo	LF	LK	.205	---	5.15	.64	5.79	**8.42**	3.76	**5.47**
57" deep, 50' long	Inst	LF	CY	.239	88.80	6.07	.64	95.51	**112.00**	74.28	**86.80**
57" deep, 50' long	Demo	BF	LK	.004	---	.10	.01	.11	**.16**	.09	**.12**
57" deep, 50' long	Inst	BF	CY	.005	1.87	.13	.01	2.01	**2.35**	1.56	**1.82**
57" deep, 50' long	Demo	SF	LK	.011	---	.28	.04	.32	**.46**	.20	**.29**
57" deep, 50' long	Inst	SF	CY	.013	4.85	.33	.04	5.22	**6.12**	4.07	**4.76**
10-3/4" thick, SF pricing based on 16' oc											
42" deep, 50' long	Demo	LF	LK	.205	---	5.15	.64	5.79	**8.42**	3.76	**5.47**
42" deep, 50' long	Inst	LF	CY	.239	74.80	6.07	.64	81.51	**95.90**	63.28	**74.20**
42" deep, 50' long	Demo	BF	LK	.005	---	.13	.02	.15	**.21**	.09	**.12**
42" deep, 50' long	Inst	BF	CY	.006	1.78	.15	.02	1.95	**2.30**	1.51	**1.77**
42" deep, 50' long	Demo	SF	LK	.011	---	.28	.04	.32	**.46**	.20	**.29**
42" deep, 50' long	Inst	SF	CY	.013	4.08	.33	.04	4.45	**5.23**	3.46	**4.06**
43-1/2" deep, 50' long	Demo	LF	LK	.205	---	5.15	.64	5.79	**8.42**	3.76	**5.47**
43-1/2" deep, 50' long	Inst	LF	CY	.239	77.40	6.07	.64	84.11	**98.90**	65.38	**76.60**
43-1/2" deep, 50' long	Demo	BF	LK	.005	---	.13	.01	.14	**.20**	.09	**.12**
43-1/2" deep, 50' long	Inst	BF	CY	.006	1.78	.15	.01	1.94	**2.29**	1.51	**1.77**
43-1/2" deep, 50' long	Demo	SF	LK	.011	---	.28	.04	.32	**.46**	.20	**.29**
43-1/2" deep, 50' long	Inst	SF	CY	.013	4.23	.33	.04	4.60	**5.41**	3.58	**4.20**
45" deep, 50' long	Demo	LF	LK	.205	---	5.15	.64	5.79	**8.42**	3.76	**5.47**
45" deep, 50' long	Inst	LF	CY	.239	80.10	6.07	.64	86.81	**102.00**	67.38	**78.90**
45" deep, 50' long	Demo	BF	LK	.005	---	.13	.01	.14	**.20**	.09	**.12**
45" deep, 50' long	Inst	BF	CY	.005	1.78	.13	.01	1.92	**2.25**	1.49	**1.74**
45" deep, 50' long	Demo	SF	LK	.011	---	.28	.04	.32	**.46**	.20	**.29**
45" deep, 50' long	Inst	SF	CY	.013	4.38	.33	.04	4.75	**5.58**	3.70	**4.33**
46-1/2" deep, 50' long	Demo	LF	LK	.205	---	5.15	.64	5.79	**8.42**	3.76	**5.47**
46-1/2" deep, 50' long	Inst	LF	CY	.239	82.70	6.07	.64	89.41	**105.00**	69.48	**81.30**
46-1/2" deep, 50' long	Demo	BF	LK	.004	---	.10	.01	.11	**.16**	.09	**.12**
46-1/2" deep, 50' long	Inst	BF	CY	.005	1.78	.13	.01	1.92	**2.25**	1.49	**1.74**
46-1/2" deep, 50' long	Demo	SF	LK	.011	---	.28	.04	.32	**.46**	.20	**.29**
46-1/2" deep, 50' long	Inst	SF	CY	.013	4.51	.33	.04	4.88	**5.73**	3.80	**4.45**

Description	Oper	Unit	Crew Size	Man-Hours Per Unit	Avg Mat'l Unit Cost	Avg Labor Unit Cost	Avg Equip Unit Cost	Avg Total Unit Cost	Avg Price Incl O&P	Large Volume Avg Total Unit Cost	Large Volume Avg Price Incl O&P
48" deep, 50' long	Demo	LF	LK	.205	---	5.15	.64	5.79	**8.42**	3.76	**5.47**
48" deep, 50' long	Inst	LF	CY	.239	85.30	6.07	.64	92.01	**108.00**	71.48	**83.70**
48" deep, 50' long	Demo	BF	LK	.004	---	.10	.01	.11	**.16**	.09	**.12**
48" deep, 50' long	Inst	BF	CY	.005	1.78	.13	.01	1.92	**2.25**	1.49	**1.74**
48" deep, 50' long	Demo	SF	LK	.011	---	.28	.04	.32	**.46**	.20	**.29**
48" deep, 50' long	Inst	SF	CY	.013	4.66	.33	.04	5.03	**5.90**	3.92	**4.59**
49-1/2" deep, 50' long	Demo	LF	LK	.205	---	5.15	.64	5.79	**8.42**	3.76	**5.47**
49-1/2" deep, 50' long	Inst	LF	CY	.239	87.90	6.07	.64	94.61	**111.00**	73.58	**86.00**
49-1/2" deep, 50' long	Demo	BF	LK	.004	---	.10	.01	.11	**.16**	.09	**.12**
49-1/2" deep, 50' long	Inst	BF	CY	.005	1.77	.13	.01	1.91	**2.24**	1.48	**1.72**
49-1/2" deep, 50' long	Demo	SF	LK	.011	---	.28	.04	.32	**.46**	.20	**.29**
49-1/2" deep, 50' long	Inst	SF	CY	.013	4.80	.33	.04	5.17	**6.06**	4.03	**4.71**
51" deep, 50' long	Demo	LF	LK	.205	---	5.15	.64	5.79	**8.42**	3.76	**5.47**
51" deep, 50' long	Inst	LF	CY	.239	90.40	6.07	.64	97.11	**114.00**	75.58	**88.30**
51" deep, 50' long	Demo	BF	LK	.004	---	.10	.01	.11	**.16**	.09	**.12**
51" deep, 50' long	Inst	BF	CY	.005	1.77	.13	.01	1.91	**2.24**	1.48	**1.72**
51" deep, 50' long	Demo	SF	LK	.011	---	.28	.04	.32	**.46**	.20	**.29**
51" deep, 50' long	Inst	SF	CY	.013	4.94	.33	.04	5.31	**6.22**	4.14	**4.84**
52-1/2" deep, 60' long	Demo	LF	LK	.205	---	5.15	.64	5.79	**8.42**	3.76	**5.47**
52-1/2" deep, 60' long	Inst	LF	CY	.239	92.90	6.07	.64	99.61	**117.00**	77.48	**90.60**
52-1/2" deep, 60' long	Demo	BF	LK	.004	---	.10	.01	.11	**.16**	.09	**.12**
52-1/2" deep, 60' long	Inst	BF	CY	.005	1.77	.13	.01	1.91	**2.24**	1.48	**1.72**
52-1/2" deep, 60' long	Demo	SF	LK	.011	---	.28	.04	.32	**.46**	.20	**.29**
52-1/2" deep, 60' long	Inst	SF	CY	.013	5.07	.33	.04	5.44	**6.37**	4.24	**4.96**
54" deep, 60' long	Demo	LF	LK	.205	---	5.15	.64	5.79	**8.42**	3.76	**5.47**
54" deep, 60' long	Inst	LF	CY	.239	95.50	6.07	.64	102.21	**120.00**	79.58	**92.90**
54" deep, 60' long	Demo	BF	LK	.004	---	.10	.01	.11	**.16**	.06	**.09**
54" deep, 60' long	Inst	BF	CY	.004	1.77	.10	.01	1.88	**2.20**	1.48	**1.72**
54" deep, 60' long	Demo	SF	LK	.011	---	.28	.04	.32	**.46**	.20	**.29**
54" deep, 60' long	Inst	SF	CY	.013	5.22	.33	.04	5.59	**6.54**	4.36	**5.09**
55-1/2" deep, 60' long	Demo	LF	LK	.205	---	5.15	.64	5.79	**8.42**	3.76	**5.47**
55-1/2" deep, 60' long	Inst	LF	CY	.239	98.10	6.07	.64	104.81	**123.00**	81.58	**95.30**
55-1/2" deep, 60' long	Demo	BF	LK	.004	---	.10	.01	.11	**.16**	.06	**.09**
55-1/2" deep, 60' long	Inst	BF	CY	.004	1.77	.10	.01	1.88	**2.20**	1.48	**1.72**
55-1/2" deep, 60' long	Demo	SF	LK	.011	---	.28	.04	.32	**.46**	.20	**.29**
55-1/2" deep, 60' long	Inst	SF	CY	.013	5.35	.33	.04	5.72	**6.69**	4.46	**5.21**
57" deep, 60' long	Demo	LF	LK	.205	---	5.15	.64	5.79	**8.42**	3.76	**5.47**
57" deep, 60' long	Inst	LF	CY	.239	101.00	6.07	.64	107.71	**126.00**	83.68	**97.60**
57" deep, 60' long	Demo	BF	LK	.004	---	.10	.01	.11	**.16**	.06	**.09**
57" deep, 60' long	Inst	BF	CY	.004	1.77	.10	.01	1.88	**2.20**	1.48	**1.72**
57" deep, 60' long	Demo	SF	LK	.011	---	.28	.04	.32	**.46**	.20	**.29**
57" deep, 60' long	Inst	SF	CY	.013	5.50	.33	.04	5.87	**6.87**	4.58	**5.35**

Description	Oper	Unit	Crew Size	Man-Hours Per Unit	Avg Mat'l Unit Cost	Avg Labor Unit Cost	Avg Equip Unit Cost	Avg Total Unit Cost	Avg Price Incl O&P	Avg Total Unit Cost	Avg Price Incl O&P
				Costs Based On Small Volume						Large Volume	
58-1/2" deep, 60' long	Demo	LF	LK	.205	---	5.15	.64	5.79	**8.42**	3.76	**5.47**
58-1/2" deep, 60' long	Inst	LF	CY	.239	103.00	6.07	.64	109.71	**129.00**	85.68	**100.00**
58-1/2" deep, 60' long	Demo	BF	LK	.004	---	.10	.01	.11	**.16**	.06	**.09**
58-1/2" deep, 60' long	Inst	BF	CY	.004	1.77	.10	.01	1.88	**2.20**	1.48	**1.72**
58-1/2" deep, 60' long	Demo	SF	LK	.011	---	.28	.04	.32	**.46**	.20	**.29**
58-1/2" deep, 60' long	Inst	SF	CY	.013	5.64	.33	.04	6.01	**7.03**	4.69	**5.47**
60" deep, 60' long	Demo	LF	LK	.205	---	5.15	.64	5.79	**8.42**	3.76	**5.47**
60" deep, 60' long	Inst	LF	CY	.239	106.00	6.07	.64	112.71	**132.00**	87.78	**102.00**
60" deep, 60' long	Demo	BF	LK	.003	---	.08	.01	.09	**.12**	.06	**.09**
60" deep, 60' long	Inst	BF	CY	.004	1.77	.10	.01	1.88	**2.20**	1.48	**1.72**
60" deep, 60' long	Demo	SF	LK	.011	---	.28	.04	.32	**.46**	.20	**.29**
60" deep, 60' long	Inst	SF	CY	.013	5.78	.33	.04	6.15	**7.19**	4.80	**5.60**
Purlins											
16' long, STR #1, SF pricing based on 8' oc											
2" x 8"	Demo	LF	LL	.046	---	1.12	.14	1.26	**1.83**	.82	**1.19**
2" x 8"	Inst	LF	CZ	.103	2.82	2.59	.21	5.62	**7.40**	4.05	**5.26**
2" x 8"	Demo	BF	LL	.034	---	.83	.11	.94	**1.36**	.60	**.88**
2" x 8"	Inst	BF	CZ	.077	2.12	1.94	.16	4.22	**5.55**	3.03	**3.94**
2" x 8"	Demo	SF	LL	.005	---	.12	.02	.14	**.20**	.08	**.12**
2" x 8"	Inst	SF	CZ	.011	.30	.28	.02	.60	**.79**	.44	**.56**
2" x 10"	Demo	LF	LL	.046	---	1.12	.14	1.26	**1.83**	.82	**1.19**
2" x 10"	Inst	LF	CZ	.103	3.33	2.59	.21	6.13	**7.98**	4.45	**5.72**
2" x 10"	Demo	BF	LL	.027	---	.66	.09	.75	**1.08**	.50	**.72**
2" x 10"	Inst	BF	CZ	.061	1.40	1.54	.13	3.07	**4.08**	2.19	**2.88**
2" x 10"	Demo	SF	LL	.005	---	.12	.02	.14	**.20**	.08	**.12**
2" x 10"	Inst	SF	CZ	.011	.37	.28	.02	.67	**.87**	.49	**.62**
2" x 12"	Demo	LF	LL	.046	---	1.12	.14	1.26	**1.83**	.82	**1.19**
2" x 12"	Inst	LF	CZ	.103	3.72	2.59	.21	6.52	**8.43**	4.76	**6.07**
2" x 12"	Demo	BF	LL	.023	---	.56	.07	.63	**.91**	.41	**.60**
2" x 12"	Inst	BF	CZ	.051	1.21	1.28	.11	2.60	**3.45**	1.85	**2.43**
2" x 12"	Demo	SF	LL	.005	---	.12	.02	.14	**.20**	.08	**.12**
2" x 12"	Inst	SF	CZ	.011	.41	.28	.02	.71	**.91**	.52	**.66**
3" x 8"	Demo	LF	LL	.046	---	1.12	.14	1.26	**1.83**	.82	**1.19**
3" x 8"	Inst	LF	CZ	.103	3.51	2.59	.21	6.31	**8.19**	4.59	**5.88**
3" x 8"	Demo	BF	LL	.023	---	.56	.07	.63	**.91**	.41	**.60**
3" x 8"	Inst	BF	CZ	.051	1.09	1.28	.11	2.48	**3.32**	1.76	**2.32**
3" x 8"	Demo	SF	LL	.005	---	.12	.02	.14	**.20**	.08	**.12**
3" x 8"	Inst	SF	CZ	.011	.38	.28	.02	.68	**.88**	.50	**.63**
3" x 10"	Demo	LF	LL	.046	---	1.12	.14	1.26	**1.83**	.82	**1.19**
3" x 10"	Inst	LF	CZ	.103	4.04	2.59	.21	6.84	**8.80**	5.01	**6.36**
3" x 10"	Demo	BF	LL	.018	---	.44	.06	.50	**.72**	.33	**.48**
3" x 10"	Inst	BF	CZ	.041	.95	1.03	.09	2.07	**2.75**	1.49	**1.96**
3" x 10"	Demo	SF	LL	.005	---	.12	.02	.14	**.20**	.08	**.12**
3" x 10"	Inst	SF	CZ	.011	.44	.28	.02	.74	**.95**	.55	**.69**
3" x 12"	Demo	LF	LL	.046	---	1.12	.14	1.26	**1.83**	.82	**1.19**
3" x 12"	Inst	LF	CZ	.103	4.62	2.59	.21	7.42	**9.47**	5.47	**6.89**
3" x 12"	Demo	BF	LL	.015	---	.36	.05	.41	**.60**	.27	**.40**
3" x 12"	Inst	BF	CZ	.034	.89	.86	.07	1.82	**2.40**	1.30	**1.70**

			Costs Based On Small Volume							Large Volume	
Description	Oper	Unit	Crew Size	Man-Hours Per Unit	Avg Mat'l Unit Cost	Avg Labor Unit Cost	Avg Equip Unit Cost	Avg Total Unit Cost	Avg Price Incl O&P	Avg Total Unit Cost	Avg Price Incl O&P
3" x 12"	Demo	SF	LL	.005	---	.12	.02	.14	**.20**	.08	**.12**
3" x 12"	Inst	SF	CZ	.011	.50	.28	.02	.80	**1.02**	.59	**.74**
3" x 14"	Demo	LF	LL	.046	---	1.12	.14	1.26	**1.83**	.82	**1.19**
3" x 14"	Inst	LF	CZ	.103	5.51	2.59	.21	8.31	**10.50**	6.17	**7.70**
3" x 14"	Demo	BF	LL	.013	---	.32	.04	.36	**.52**	.22	**.32**
3" x 14"	Inst	BF	CZ	.029	1.59	.73	.06	2.38	**3.00**	1.77	**2.21**
3" x 14"	Demo	SF	LL	.005	---	.12	.02	.14	**.20**	.08	**.12**
3" x 14"	Inst	SF	CZ	.011	.61	.28	.02	.91	**1.14**	.68	**.84**
Sub-purlins											
8' long, STR #1, SF pricing based on 2' oc											
2" x 4"	Demo	LF	LL	.020	---	.49	.06	.55	**.79**	.36	**.52**
2" x 4"	Inst	LF	CZ	.046	.39	1.16	.10	1.65	**2.31**	1.13	**1.57**
2" x 4"	Demo	BF	LL	.030	---	.73	.09	.82	**1.19**	.55	**.79**
2" x 4"	Inst	BF	CZ	.068	.58	1.71	.14	2.43	**3.41**	1.66	**2.30**
2" x 4"	Demo	SF	LL	.009	---	.22	.03	.25	**.36**	.17	**.24**
2" x 4"	Inst	SF	CZ	.021	.18	.53	.04	.75	**1.05**	.50	**.69**
2" x 6"	Demo	LF	LL	.020	---	.49	.06	.55	**.79**	.36	**.52**
2" x 6"	Inst	LF	CZ	.046	.58	1.16	.10	1.84	**2.53**	1.28	**1.74**
2" x 6"	Demo	BF	LL	.020	---	.49	.06	.55	**.79**	.36	**.52**
2" x 6"	Inst	BF	CZ	.046	.58	1.16	.10	1.84	**2.53**	1.28	**1.74**
2" x 6"	Demo	SF	LL	.009	---	.22	.03	.25	**.36**	.17	**.24**
2" x 6"	Inst	SF	CZ	.021	.27	.53	.04	.84	**1.15**	.57	**.77**
3" x 6"	Demo	LF	LL	.020	---	.49	.06	.55	**.79**	.36	**.52**
3" x 6"	Inst	LF	CZ	.046	1.00	1.16	.10	2.26	**3.01**	1.61	**2.12**
3" x 6"	Demo	BF	LL	.014	---	.34	.04	.38	**.55**	.25	**.36**
3" x 6"	Inst	BF	CZ	.031	.66	.78	.06	1.50	**2.01**	1.06	**1.40**
3" x 6"	Demo	SF	LL	.009	---	.22	.03	.25	**.36**	.17	**.24**
3" x 6"	Inst	SF	CZ	.021	.44	.53	.04	1.01	**1.35**	.71	**.93**
3" x 8"	Demo	LF	LL	.020	---	.49	.06	.55	**.79**	.36	**.52**
3" x 8"	Inst	LF	CZ	.046	1.32	1.16	.10	2.58	**3.38**	1.86	**2.40**
3" x 8"	Demo	BF	LL	.010	---	.24	.03	.27	**.40**	.19	**.28**
3" x 8"	Inst	BF	CZ	.023	.66	.58	.05	1.29	**1.69**	.93	**1.20**
3" x 8"	Demo	SF	LL	.009	---	.22	.03	.25	**.36**	.17	**.24**
3" x 8"	Inst	SF	CZ	.021	.57	.53	.04	1.14	**1.50**	.81	**1.05**
2" x 8"	Demo	LF	LL	.020	---	.49	.06	.55	**.79**	.36	**.52**
2" x 8"	Inst	LF	CZ	.046	.89	1.16	.10	2.15	**2.88**	1.52	**2.01**
2" x 8"	Demo	BF	LL	.015	---	.36	.05	.41	**.60**	.27	**.40**
2" x 8"	Inst	BF	CZ	.034	.66	.86	.07	1.59	**2.13**	1.12	**1.49**
2" x 8"	Demo	SF	LL	.009	---	.22	.03	.25	**.36**	.17	**.24**
2" x 8"	Inst	SF	CZ	.021	.38	.53	.04	.95	**1.28**	.66	**.87**
3" x 4"	Demo	LF	LL	.020	---	.49	.06	.55	**.79**	.36	**.52**
3" x 4"	Inst	LF	CZ	.046	.66	1.16	.10	1.92	**2.62**	1.34	**1.81**
3" x 4"	Demo	BF	LL	.020	---	.49	.06	.55	**.79**	.36	**.52**
3" x 4"	Inst	BF	CZ	.046	.66	1.16	.10	1.92	**2.62**	1.34	**1.81**
3" x 4"	Demo	SF	LL	.009	---	.22	.03	.25	**.36**	.17	**.24**
3" x 4"	Inst	SF	CZ	.021	.29	.53	.04	.86	**1.18**	.59	**.79**

Description	Oper	Unit	Crew Size	Man-Hours Per Unit	Avg Mat'l Unit Cost	Avg Labor Unit Cost	Avg Equip Unit Cost	Avg Total Unit Cost	Avg Price Incl O&P	Large Volume Avg Total Unit Cost	Large Volume Avg Price Incl O&P
Ledgers											
Bolts 2' oc											
3" x 8"	Demo	LF	LC	.057	---	1.35	.36	1.71	**2.39**	1.10	**1.55**
3" x 8"	Inst	LF	CS	.085	2.58	2.15	.53	5.26	**6.76**	3.80	**4.84**
3" x 8"	Demo	BF	LC	.029	---	.68	.18	.86	**1.21**	.57	**.80**
3" x 8"	Inst	BF	CS	.043	1.30	1.09	.27	2.66	**3.42**	1.90	**2.42**
3" x 8"	Demo	SF	LC	.003	---	.07	.02	.09	**.13**	.06	**.08**
3" x 8"	Inst	SF	CS	.004	.13	.10	.03	.26	**.33**	.20	**.25**
3" x 10"	Demo	LF	LC	.057	---	1.35	.36	1.71	**2.39**	1.10	**1.55**
3" x 10"	Inst	LF	CS	.085	2.95	2.15	.53	5.63	**7.19**	4.09	**5.17**
3" x 10"	Demo	BF	LC	.023	---	.54	.14	.68	**.96**	.44	**.63**
3" x 10"	Inst	BF	CS	.034	1.18	.86	.21	2.25	**2.87**	1.63	**2.06**
3" x 10"	Demo	SF	LC	.003	---	.07	.02	.09	**.13**	.06	**.08**
3" x 10"	Inst	SF	CS	.004	.15	.10	.03	.28	**.36**	.22	**.27**
3" x 12"	Demo	LF	LC	.057	---	1.35	.36	1.71	**2.39**	1.10	**1.55**
3" x 12"	Inst	LF	CS	.085	3.31	2.15	.53	5.99	**7.60**	4.38	**5.50**
3" x 12"	Demo	BF	LC	.019	---	.45	.12	.57	**.80**	.36	**.51**
3" x 12"	Inst	BF	CS	.029	1.10	.73	.18	2.01	**2.56**	1.47	**1.85**
3" x 12"	Demo	SF	LC	.003	---	.07	.02	.09	**.13**	.06	**.08**
3" x 12"	Inst	SF	CS	.004	.17	.10	.03	.30	**.38**	.23	**.28**
4" x 8"	Demo	LF	LC	.057	---	1.35	.36	1.71	**2.39**	1.10	**1.55**
4" x 8"	Inst	LF	CS	.085	3.28	2.15	.53	5.96	**7.57**	4.35	**5.47**
4" x 8"	Demo	BF	LC	.021	---	.50	.13	.63	**.88**	.42	**.59**
4" x 8"	Inst	BF	CS	.032	1.22	.81	.20	2.23	**2.83**	1.62	**2.04**
4" x 8"	Demo	SF	LC	.003	---	.07	.02	.09	**.13**	.06	**.08**
4" x 8"	Inst	SF	CS	.004	.17	.10	.03	.30	**.38**	.23	**.28**
4" x 10"	Demo	LF	LC	.057	---	1.35	.36	1.71	**2.39**	1.10	**1.55**
4" x 10"	Inst	LF	CS	.085	3.80	2.15	.53	6.48	**8.17**	4.76	**5.94**
4" x 10"	Demo	BF	LC	.017	---	.40	.11	.51	**.72**	.33	**.46**
4" x 10"	Inst	BF	CS	.026	1.14	.66	.16	1.96	**2.47**	1.43	**1.79**
4" x 10"	Demo	SF	LC	.003	---	.07	.02	.09	**.13**	.06	**.08**
4" x 10"	Inst	SF	CS	.004	.19	.10	.03	.32	**.40**	.25	**.31**
4" x 12"	Demo	LF	LC	.057	---	1.35	.36	1.71	**2.39**	1.10	**1.55**
4" x 12"	Inst	LF	CS	.085	4.34	2.15	.53	7.02	**8.79**	5.19	**6.44**
4" x 12"	Demo	BF	LC	.014	---	.33	.09	.42	**.59**	.27	**.38**
4" x 12"	Inst	BF	CS	.021	1.09	.53	.13	1.75	**2.19**	1.30	**1.62**
4" x 12"	Demo	SF	LC	.003	---	.07	.02	.09	**.13**	.06	**.08**
4" x 12"	Inst	SF	CS	.004	.20	.10	.03	.33	**.41**	.26	**.32**

Description	Oper	Unit	Crew Size	Man-Hours Per Unit	Avg Mat'l Unit Cost	Avg Labor Unit Cost	Avg Equip Unit Cost	Avg Total Unit Cost	Avg Price Incl O&P	Avg Total Unit Cost	Avg Price Incl O&P
			Costs Based On Small Volume							Large Volume	

Gutters and downspouts

Aluminum, baked on painted finish (white or brown)

Description	Oper	Unit	Crew Size	Man-Hours Per Unit	Avg Mat'l Unit Cost	Avg Labor Unit Cost	Avg Equip Unit Cost	Avg Total Unit Cost	Avg Price Incl O&P	Avg Total Unit Cost (LV)	Avg Price Incl O&P (LV)
Gutter, 5" box type											
Heavyweight gauge	Inst	LF	UB	.114	1.38	3.39	---	4.77	**6.68**	3.48	**4.78**
Standard weight gauge	Inst	LF	UB	.114	1.10	3.39	---	4.49	**6.35**	3.23	**4.49**
End caps	Inst	Ea	UB	---	.79	---	---	.79	**.79**	.74	**.74**
Drop outlet for downspouts	Inst	Ea	UB	---	4.52	---	---	4.52	**4.52**	4.21	**4.21**
Inside/outside corner	Inst	Ea	UB	---	4.84	---	---	4.84	**4.84**	4.51	**4.51**
Joint connector (4 each package)	Inst	Pkg	UB	---	6.14	---	---	6.14	**6.14**	5.72	**5.72**
Strap hanger (10 each package)	Inst	Pkg	UB	---	8.89	---	---	8.89	**8.89**	8.29	**8.29**
Fascia bracket (4 each package)	Inst	Pkg	UB	---	8.89	---	---	8.89	**8.89**	8.29	**8.29**
Spike and ferrule (10 each package)	Inst	Pkg	UB	---	7.27	---	---	7.27	**7.27**	6.78	**6.78**
Downspout, corrugated square											
Standard weight gauge	Inst	LF	UB	.098	1.13	2.92	---	4.05	**5.67**	2.96	**4.08**
Regular/side elbow	Inst	Ea	UB	---	1.93	---	---	1.93	**1.93**	1.80	**1.80**
Downspout holder (4 each package)	Inst	Pkg	UB	---	5.65	---	---	5.65	**5.65**	5.27	**5.27**

Steel, natural finish

Description	Oper	Unit	Crew Size	Man-Hours Per Unit	Avg Mat'l Unit Cost	Avg Labor Unit Cost	Avg Equip Unit Cost	Avg Total Unit Cost	Avg Price Incl O&P	Avg Total Unit Cost (LV)	Avg Price Incl O&P (LV)
Gutter, 4" box type											
Heavyweight gauge	Inst	LF	UB	.124	.89	3.69	---	4.58	**6.56**	3.03	**4.26**
End caps	Inst	Ea	UB	---	.79	---	---	.79	**.79**	.74	**.74**
Drop outlet for downspouts	Inst	Ea	UB	---	2.90	---	---	2.90	**2.90**	2.70	**2.70**
Inside/outside corner	Inst	Ea	UB	---	4.36	---	---	4.36	**4.36**	4.06	**4.06**
Joint connector (4 each package)	Inst	Pkg	UB	---	4.52	---	---	4.52	**4.52**	4.21	**4.21**
Strap hanger (10 each package)	Inst	Pkg	UB	---	6.14	---	---	6.14	**6.14**	5.72	**5.72**
Fascia bracket (4 each package)	Inst	Pkg	UB	---	4.84	---	---	4.84	**4.84**	4.51	**4.51**
Spike and ferrule (10 each package)	Inst	Pkg	UB	---	5.33	---	---	5.33	**5.33**	4.97	**4.97**
Downspout, 30 gauge galvanized steel											
Standard weight gauge	Inst	LF	UB	.124	1.22	3.69	---	4.91	**6.94**	3.33	**4.60**
Regular/side elbow	Inst	Ea	UB	---	1.93	---	---	1.93	**1.93**	1.80	**1.80**
Downspout holder (4 each package)	Inst	Pkg	UB	---	5.17	---	---	5.17	**5.17**	4.82	**4.82**

Vinyl, extruded 5" PVC

Description	Oper	Unit	Crew Size	Man-Hours Per Unit	Avg Mat'l Unit Cost	Avg Labor Unit Cost	Avg Equip Unit Cost	Avg Total Unit Cost	Avg Price Incl O&P	Avg Total Unit Cost (LV)	Avg Price Incl O&P (LV)
White											
Gutter	Inst	LF	UB	.124	1.05	3.69	---	4.74	**6.74**	3.18	**4.43**
End caps	Inst	Ea	UB	---	1.93	---	---	1.93	**1.93**	1.80	**1.80**
Drop outlet for downspouts	Inst	Ea	UB	---	8.08	---	---	8.08	**8.08**	7.53	**7.53**
Inside/outside corner	Inst	Ea	UB	---	8.08	---	---	8.08	**8.08**	7.53	**7.53**

Description	Oper	Unit	Crew Size	Man-Hours Per Unit	Avg Mat'l Unit Cost	Avg Labor Unit Cost	Avg Equip Unit Cost	Avg Total Unit Cost	Avg Price Incl O&P	Large Volume Avg Total Unit Cost	Large Volume Avg Price Incl O&P
			Costs Based On Small Volume							Large Volume	
Joint connector	Inst	Ea	UB	---	2.41	---	---	2.41	**2.41**	2.25	**2.25**
Fascia bracket (4 each package)	Inst	Pkg	UB	---	10.50	---	---	10.50	**10.50**	9.80	**9.80**
Downspout	Inst	LF	UB	.107	1.54	3.18	---	4.72	**6.55**	3.33	**4.50**
Downspout driplet	Inst	Ea	UB	---	10.50	---	---	10.50	**10.50**	9.80	**9.80**
Downspout joiner	Inst	Ea	UB	---	3.22	---	---	3.22	**3.22**	3.00	**3.00**
Downspout holder (2 each package)	Inst	Pkg	UB	---	10.50	---	---	10.50	**10.50**	9.80	**9.80**
Regular elbow	Inst	Ea	UB	---	3.55	---	---	3.55	**3.55**	3.31	**3.31**
Well cap	Inst	Ea	UB	---	4.52	---	---	4.52	**4.52**	4.21	**4.21**
Well outlet	Inst	Ea	UB	---	6.14	---	---	6.14	**6.14**	5.72	**5.72**
Expansion joint connector	Inst	Ea	UB	---	9.70	---	---	9.70	**9.70**	9.04	**9.04**
Rafter adapter (4 each package)	Inst	Pkg	UB	---	6.46	---	---	6.46	**6.46**	6.02	**6.02**
Brown											
Gutter	Inst	LF	UB	.124	1.30	3.69	---	4.99	**7.03**	3.41	**4.69**
End caps	Inst	Ea	UB	---	2.41	---	---	2.41	**2.41**	2.25	**2.25**
Drop outlet for downspouts	Inst	Ea	UB	---	9.70	---	---	9.70	**9.70**	9.04	**9.04**
Inside/outside corner	Inst	Ea	UB	---	10.50	---	---	10.50	**10.50**	9.80	**9.80**
Joint connector	Inst	Ea	UB	---	3.22	---	---	3.22	**3.22**	3.00	**3.00**
Fascia bracket (4 each package)	Inst	Pkg	UB	---	12.10	---	---	12.10	**12.10**	11.30	**11.30**
Downspout	Inst	LF	UB	.107	1.70	3.18	---	4.88	**6.73**	3.49	**4.69**
Downspout driplet	Inst	Ea	UB	---	12.10	---	---	12.10	**12.10**	11.30	**11.30**
Downspout joiner	Inst	Ea	UB	---	3.87	---	---	3.87	**3.87**	3.61	**3.61**
Downspout holder (2 each package)	Inst	Pkg	UB	---	12.90	---	---	12.90	**12.90**	12.10	**12.10**
Regular elbow	Inst	Ea	UB	---	4.52	---	---	4.52	**4.52**	4.21	**4.21**
Well cap	Inst	Ea	UB	---	5.33	---	---	5.33	**5.33**	4.97	**4.97**
Well outlet	Inst	Ea	UB	---	7.76	---	---	7.76	**7.76**	7.23	**7.23**
Expansion joint connector	Inst	Ea	UB	---	12.10	---	---	12.10	**12.10**	11.30	**11.30**
Rafter adapter (4 each package)	Inst	Pkg	UB	---	6.46	---	---	6.46	**6.46**	6.02	**6.02**

Hardboard. See Paneling, page 193

Hardwood flooring

1. Strip flooring is nailed into place over wood subflooring or over wood sleeper strips. Using 3¼" W strips leaves 25% cutting and fitting waste; 2¼" W strips leave 33% waste. Nails and the respective cutting and fitting waste have been included in the unit costs.

2. Block flooring is laid in mastic applied to felt-covered wood subfloor. Mastic, 5% block waste and felt are included in material unit costs for block or parquet flooring.

Two types of wood block flooring

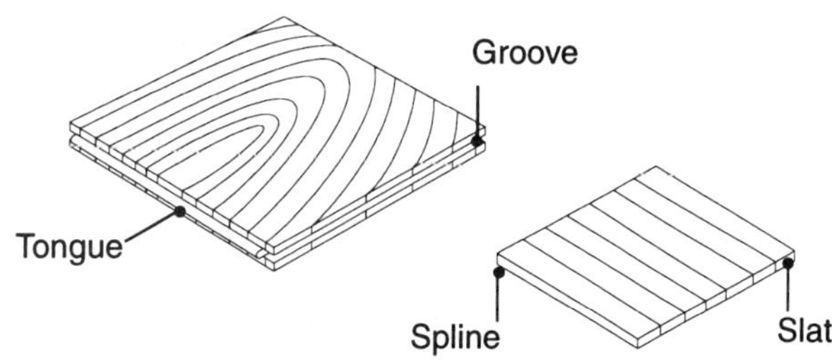

Strip flooring:
A Side and end matched
B Side matched
C Square edged

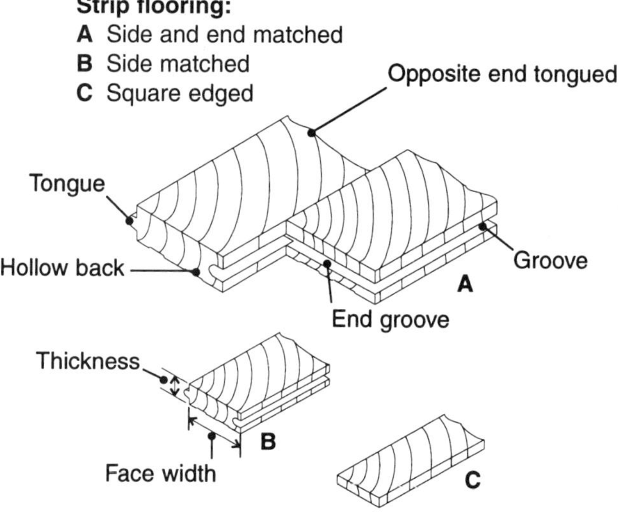

Installation of first strip of flooring

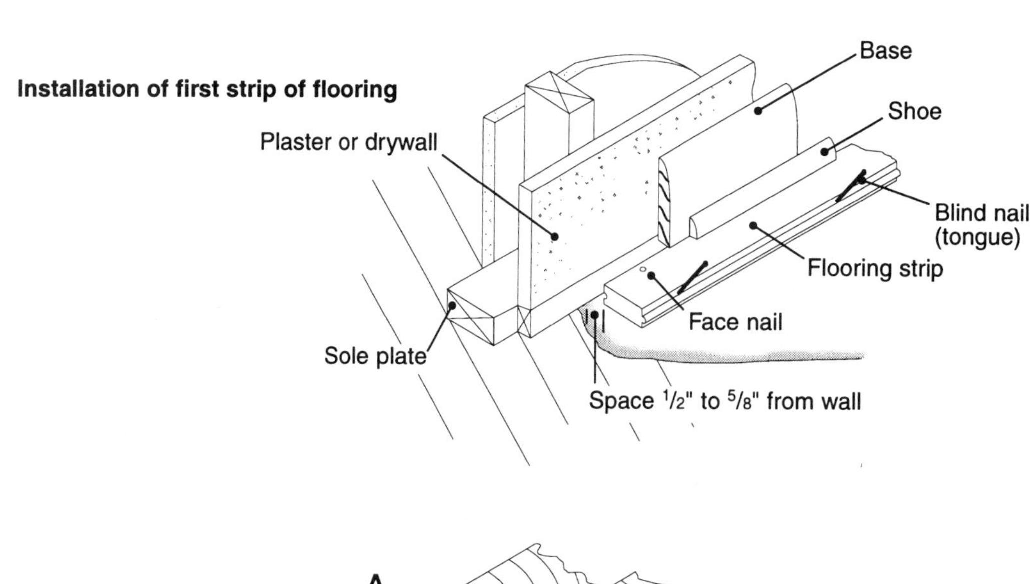

Nailing of flooring:
A Angle of nailing
B Setting the nail without damage to the flooring

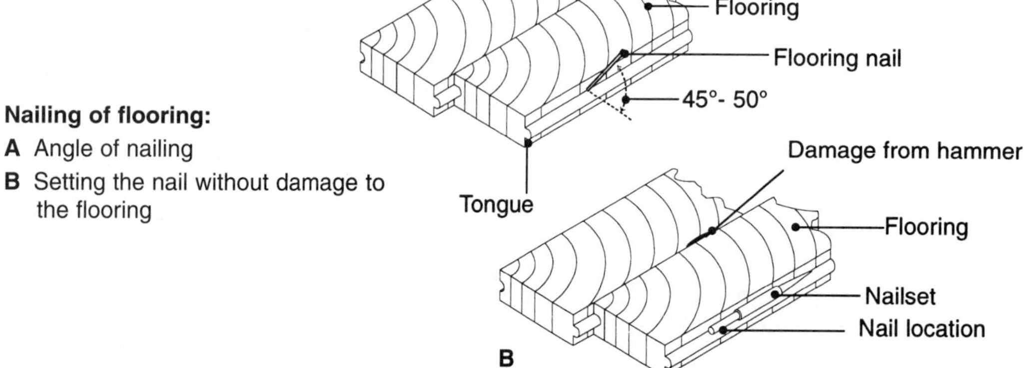

Description	Oper	Unit	Crew Size	Man-Hours Per Unit	Avg Mat'l Unit Cost	Avg Labor Unit Cost	Avg Equip Unit Cost	Avg Total Unit Cost	Avg Price Incl O&P	Large Volume Avg Total Unit Cost	Large Volume Avg Price Incl O&P
Hardwood flooring											
Includes waste and nails											
Strip; installed over wood subfloor											
Prefinished oak, prime											
25/32" x 3-1/4"											
Lay floor	Inst	SF	FD	.075	10.10	1.99	---	12.09	**14.60**	10.21	**12.20**
Wax, polish, machine buff	Inst	SF	FC	.016	.04	.43	.04	.51	**.73**	.31	**.44**
25/32" x 2-1/4"											
Lay floor	Inst	SF	FD	.100	9.11	2.66	---	11.77	**14.40**	9.70	**11.70**
Wax, polish, machine buff	Inst	SF	FC	.016	.04	.43	.04	.51	**.73**	.31	**.44**
Unfinished											
25/32" x 3-1/4", lay floor only											
Fir											
Vertical grain	Inst	SF	FD	.063	3.13	1.68	---	4.81	**6.09**	3.80	**4.71**
Flat grain	Inst	SF	FD	.063	3.56	1.68	---	5.24	**6.59**	4.17	**5.14**
Yellow pine	Inst	SF	FD	.063	3.05	1.68	---	4.73	**6.00**	3.72	**4.62**
25/32" x 2-1/4", lay floor only											
Maple	Inst	SF	FD	.094	4.58	2.50	---	7.08	**8.99**	5.55	**6.89**
Oak	Inst	SF	FD	.094	5.11	2.50	---	7.61	**9.60**	6.03	**7.44**
Yellow pine	Inst	SF	FD	.078	3.10	2.07	---	5.17	**6.65**	4.01	**5.04**
Related materials and operations											
Machine sand, fill and finish											
New floors		SF	FC	.031	.12	.84	.09	1.05	**1.48**	.67	**.93**
Damaged floors		SF	FC	.047	.16	1.28	.13	1.57	**2.22**	.98	**1.37**
Wax, polish and machine buff		SF	FC	.016	.04	.43	.04	.51	**.73**	.31	**.44**
Block											
Laid in mastic over wood subfloor covered with felt											
Oak, 5/16" x 12" x 12"											
Lay floor only											
Prefinished	Inst	SF	FD	.058	11.20	1.54	---	12.74	**15.20**	10.91	**12.90**
Unfinished	Inst	SF	FD	.055	6.00	1.46	---	7.46	**9.08**	6.22	**7.45**
Teak, 5/16" x 12" x 12"											
Lay floor only											
Prefinished	Inst	SF	FD	.058	13.50	1.54	---	15.04	**17.80**	12.93	**15.10**
Unfinished	Inst	SF	FD	.055	8.23	1.46	---	9.69	**11.60**	8.19	**9.71**
Oak, 13/16" x 12" x 12"											
Lay floor only											
Prefinished	Inst	SF	FD	.058	20.20	1.54	---	21.74	**25.50**	18.83	**22.00**
Unfinished	Inst	SF	FD	.060	14.90	1.60	---	16.50	**19.60**	14.26	**16.70**
Machine sand, fill and finish											
New floors	Inst	SF	FC	.031	.12	.84	.09	1.05	**1.48**	.67	**.93**
Damaged floors	Inst	SF	FC	.047	.16	1.28	.13	1.57	**2.22**	.98	**1.37**
Wax, polish and machine buff	Inst	SF	FC	.016	.04	.43	.04	.51	**.73**	.31	**.44**

			Costs Based On Small Volume							Large Volume	
Description	Oper	Unit	Crew Size	Man-Hours Per Unit	Avg Mat'l Unit Cost	Avg Labor Unit Cost	Avg Equip Unit Cost	Avg Total Unit Cost	Avg Price Incl O&P	Avg Total Unit Cost	Avg Price Incl O&P
Parquetry, 5/16" x 9" x 9"											
Lay floor only											
Oak	Inst	SF	FD	.060	6.00	1.60	---	7.60	**9.28**	6.30	**7.57**
Walnut	Inst	SF	FD	.060	15.50	1.60	---	17.10	**20.10**	14.66	**17.20**
Teak	Inst	SF	FD	.060	8.23	1.60	---	9.83	**11.80**	8.27	**9.83**
Machine sand, fill and finish											
New floors	Inst	SF	FC	.031	.12	.84	.09	1.05	**1.48**	.67	**.93**
Damaged floors	Inst	SF	FC	.047	.16	1.28	.13	1.57	**2.22**	.98	**1.37**
Wax, polish and machine buff	Inst	SF	FC	.016	.04	.43	.04	.51	**.73**	.31	**.44**
Acrylic wood parquet blocks											
5/16" x 12" x 12" set in epoxy	Inst	SF	FD	.060	7.28	1.60	---	8.88	**10.80**	7.43	**8.87**
Wax, polish and machine buff	Inst	SF	FC	.016	.04	.43	.04	.51	**.73**	.31	**.44**

Heating

Boilers

Electric fired heaters, includes standard controls and trim, ASME

Description	Oper	Unit	Crew Size	Man-Hours Per Unit	Avg Mat'l Unit Cost	Avg Labor Unit Cost	Avg Equip Unit Cost	Avg Total Unit Cost	Avg Price Incl O&P	Avg Total Unit Cost	Avg Price Incl O&P
Hot water											
12 KW, 40 MBHP	Inst	Ea	SD	26.7	3960.00	723.00	---	4683.00	**5630.00**	4032.00	**4820.00**
24 KW, 82 MBHP	Inst	Ea	SD	27.9	4220.00	756.00	---	4976.00	**5980.00**	4296.00	**5130.00**
45 KW, 154 MBHP	Inst	Ea	SD	28.6	4640.00	774.00	---	5414.00	**6490.00**	4680.00	**5570.00**
60 KW, 205 MBHP	Inst	Ea	SD	28.9	5500.00	783.00	---	6283.00	**7500.00**	5450.00	**6460.00**
240 KW, 820 MBHP	Inst	Ea	SD	32.0	11900	867.00	---	12767	**15000**	11150	**13000**
480 KW, 1,635 MBHP	Inst	Ea	SD	33.8	20500	915.00	---	21415	**24900**	18685	**21800**
1,200 KW, 4,095 MBHP	Inst	Ea	SD	35.3	37300	956.00	---	38256	**44300**	33623	**38900**
2,400 KW, 8,190 MBHP	Inst	Ea	SD	37.5	63600	1020.00	---	64620	**74700**	56864	**65700**
Steam											
6 KW, 20 MBHP	Inst	Ea	SD	28.9	9400.00	783.00	---	10183	**12000**	8890.00	**10400**
60 KW, 205 MBHP	Inst	Ea	SD	35.3	11400	956.00	---	12356	**14500**	10823	**12600**
240 KW, 815 MBHP	Inst	Ea	SD	80.0	18000	2170.00	---	20170	**23900**	17520	**20700**
600 KW, 2,047 MBHP	Inst	Ea	SE	213	27700	5950.00	---	33650	**40700**	28970	**34800**
Minimum Job Charge	Inst	Job	SD	17.1	---	463.00	---	463.00	**690.00**	325.00	**484.00**

Gas fired, natural or propane, heaters; includes standard controls; MBHP gross output

Cast iron with insulated jacket

Description	Oper	Unit	Crew Size	Man-Hours Per Unit	Avg Mat'l Unit Cost	Avg Labor Unit Cost	Avg Equip Unit Cost	Avg Total Unit Cost	Avg Price Incl O&P	Avg Total Unit Cost	Avg Price Incl O&P
Hot water											
80 MBHP	Inst	Ea	SF	22.9	1250.00	633.00	---	1883.00	**2380.00**	1572.00	**1970.00**
100 MBHP	Inst	Ea	SF	23.3	1430.00	644.00	---	2074.00	**2600.00**	1743.00	**2170.00**
122 MBHP	Inst	Ea	SF	23.8	1580.00	657.00	---	2237.00	**2790.00**	1882.00	**2330.00**
163 MBHP	Inst	Ea	SF	24.2	1930.00	668.00	---	2598.00	**3220.00**	2213.00	**2710.00**
440 MBHP	Inst	Ea	SF	24.5	3780.00	677.00	---	4457.00	**5350.00**	3841.00	**4600.00**
2,000 MBHP	Inst	Ea	SF	32.0	12300	884.00	---	13184	**15400**	11463	**13400**
6,970 MBHP	Inst	Ea	SF	80.0	83300	2210.00	---	85510	**99100**	75160	**87000**
Steam											
80 MBHP	Inst	Ea	SF	22.9	1250.00	633.00	---	1883.00	**2380.00**	1572.00	**1970.00**
163 MBHP	Inst	Ea	SF	24.2	1930.00	668.00	---	2598.00	**3220.00**	2213.00	**2710.00**
440 MBHP	Inst	Ea	SF	24.5	3780.00	677.00	---	4457.00	**5350.00**	3841.00	**4600.00**
3,570 MBHP	Inst	Ea	SF	35.3	19300	975.00	---	20275	**23600**	17737	**20700**
6,970 MBHP	Inst	Ea	SF	80.0	62500	2210.00	---	64710	**75100**	56760	**65900**

Description	Oper	Unit	Crew Size	Man-Hours Per Unit	Avg Mat'l Unit Cost	Avg Labor Unit Cost	Avg Equip Unit Cost	Avg Total Unit Cost	Avg Price Incl O&P	Large Volume Avg Total Unit Cost	Large Volume Avg Price Incl O&P
Steel with insulated jacket; includes burner and one zone valve											
Hot water											
50 MBHP	Inst	Ea	SF	14.5	1870.00	400.00	---	2270.00	**2750.00**	1951.00	**2350.00**
70 MBHP	Inst	Ea	SF	14.5	2080.00	400.00	---	2480.00	**2990.00**	2141.00	**2560.00**
90 MBHP	Inst	Ea	SF	15.2	2110.00	420.00	---	2530.00	**3050.00**	2175.00	**2610.00**
105 MBHP	Inst	Ea	SF	16.0	2380.00	442.00	---	2822.00	**3400.00**	2431.00	**2910.00**
130 MBHP	Inst	Ea	SF	16.8	2710.00	464.00	---	3174.00	**3800.00**	2738.00	**3270.00**
150 MBHP	Inst	Ea	SF	17.8	3150.00	492.00	---	3642.00	**4360.00**	3147.00	**3750.00**
185 MBHP	Inst	Ea	SF	19.4	3810.00	536.00	---	4346.00	**5180.00**	3760.00	**4460.00**
235 MBHP	Inst	Ea	SF	21.2	4820.00	586.00	---	5406.00	**6420.00**	4692.00	**5550.00**
290 MBHP	Inst	Ea	SF	22.9	5440.00	633.00	---	6073.00	**7200.00**	5272.00	**6230.00**
480 MBHP	Inst	Ea	SF	35.3	7850.00	975.00	---	8825.00	**10500**	7667.00	**9070.00**
640 MBHP	Inst	Ea	SF	45.3	9340.00	1250.00	---	10590	**12600**	9187.00	**10900**
800 MBHP	Inst	Ea	SF	54.5	10900	1510.00	---	12410	**14800**	10750	**12800**
960 MBHP	Inst	Ea	SF	58.5	13600	1620.00	---	15220	**18000**	13200	**15600**
Minimum Job Charge	Inst	Job	SF	17.1	---	472.00	---	472.00	**704.00**	331.00	**494.00**
Oil fired heaters; includes standard controls; flame retention burner; MBHP gross output											
Cast iron with insulated jacket											
Hot water											
110 MBHP	Inst	Ea	SF	32.0	1400.00	884.00	---	2284.00	**2930.00**	1893.00	**2410.00**
200 MBHP	Inst	Ea	SF	45.3	1930.00	1250.00	---	3180.00	**4090.00**	2657.00	**3370.00**
1,080 MBHP	Inst	Ea	SF	104	9880.00	2870.00	---	12750	**15600**	10930	**13300**
1,320 MBHP	Inst	Ea	SF	126	8950.00	3480.00	---	12430	**15500**	10550	**13000**
2,100 MBHP	Inst	Ea	SF	160	11900	4420.00	---	16320	**20200**	13810	**17000**
4,360 MBHP	Inst	Ea	SF	218	21300	6020.00	---	27320	**33500**	23220	**28200**
6,970 MBHP	Inst	Ea	SF	400	61900	11000	---	72900	**87600**	62890	**75100**
Steam											
110 MBHP	Inst	Ea	SF	32.0	1400.00	884.00	---	2284.00	**2930.00**	1893.00	**2410.00**
205 MBHP	Inst	Ea	SF	45.3	1930.00	1250.00	---	3180.00	**4090.00**	2657.00	**3370.00**
1,085 MBHP	Inst	Ea	SF	104.3	9880.00	2880.00	---	12760	**15700**	10930	**13300**
1,360 MBHP	Inst	Ea	SF	126	8950.00	3480.00	---	12430	**15500**	10550	**13000**
2,175 MBHP	Inst	Ea	SF	160	11900	4420.00	---	16320	**20200**	13810	**17000**
4,360 MBHP	Inst	Ea	SF	218	21300	6020.00	---	27320	**33500**	23220	**28200**
6,970 MBHP	Inst	Ea	SF	400	61900	11000	---	72900	**87600**	62890	**75100**
Steel insulated jacket burner											
Hot water											
105 MBHP	Inst	Ea	SF	17.8	2470.00	492.00	---	2962.00	**3570.00**	2547.00	**3050.00**
120 MBHP	Inst	Ea	SF	20.0	2500.00	552.00	---	3052.00	**3700.00**	2624.00	**3150.00**
140 MBHP	Inst	Ea	SF	22.9	2620.00	633.00	---	3253.00	**3950.00**	2782.00	**3360.00**
170 MBHP	Inst	Ea	SF	26.7	3120.00	737.00	---	3857.00	**4690.00**	3312.00	**3990.00**
225 MBHP	Inst	Ea	SF	35.3	3720.00	975.00	---	4695.00	**5730.00**	4017.00	**4870.00**
315 MBHP	Inst	Ea	SF	40.0	5470.00	1100.00	---	6570.00	**7940.00**	5659.00	**6790.00**
420 MBHP	Inst	Ea	SF	45.3	6160.00	1250.00	---	7410.00	**8950.00**	6377.00	**7660.00**
Minimum Job Charge	Inst	Job	SF	17.1	---	472.00	---	472.00	**704.00**	331.00	**494.00**

Description	Oper	Unit	Crew Size	Man-Hours Per Unit	Avg Mat'l Unit Cost	Avg Labor Unit Cost	Avg Equip Unit Cost	Avg Total Unit Cost	Avg Price Incl O&P	Large Volume Avg Total Unit Cost	Large Volume Avg Price Incl O&P
Boiler accessories											
Burners											
Conversion, gas fired, LP or natural											
Residential, gun type, atmospheric input											
72 to 200 MBHP	Inst	Ea	SB	10.7	779.00	280.00	---	1059.00	**1310.00**	898.00	**1100.00**
120 to 360 MBHP	Inst	Ea	SB	12.2	863.00	320.00	---	1183.00	**1470.00**	1001.00	**1230.00**
280 to 800 MBHP	Inst	Ea	SB	14.2	1670.00	372.00	---	2042.00	**2470.00**	1750.00	**2110.00**
Flame retention, oil fired assembly											
2.0 to 5.0 GPH	Inst	Ea	SB	12.2	595.00	320.00	---	915.00	**1160.00**	765.00	**961.00**
Forced warm air systems											
Duct furnaces											
Furnace includes burner, controls, stainless steel heat exchanger.											
Gas fired with an electric ignition											
Outdoor installation, includes vent cap											
225 MBHP output	Inst	Ea	SB	10.7	3330.00	280.00	---	3610.00	**4250.00**	3150.00	**3690.00**
375 MBHP output	Inst	Ea	SB	14.2	4820.00	372.00	---	5192.00	**6100.00**	4530.00	**5310.00**
450 MBHP output	Inst	Ea	SB	16.3	5060.00	427.00	---	5487.00	**6450.00**	4782.00	**5610.00**
Furnaces, hot air heating with blowers and standard controls											
Gas or oil lines and couplings not included (see next page), flue piping not included (see page 254)											
Electric fired, UL listed, heat staging, 240 volt run and connection											
30 MBHP	Inst	Ea	UE	7.41	482.00	196.00	---	678.00	**848.00**	572.00	**710.00**
75 MBHP	Inst	Ea	UE	7.60	613.00	201.00	---	814.00	**1010.00**	692.00	**848.00**
85 MBHP	Inst	Ea	UE	8.33	684.00	220.00	---	904.00	**1120.00**	769.00	**942.00**
90 MBHP	Inst	Ea	UE	8.89	857.00	235.00	---	1092.00	**1340.00**	932.00	**1130.00**
Minimum Job Charge	Inst	Job	UD	11.4	---	295.00	---	295.00	**442.00**	207.00	**310.00**
Gas fired, AGA certified, direct drive models											
40 MBHP	Inst	Ea	UD	5.93	518.00	153.00	---	671.00	**825.00**	572.00	**697.00**
65 MBHP	Inst	Ea	UD	6.08	690.00	157.00	---	847.00	**1030.00**	727.00	**878.00**
80 MBHP	Inst	Ea	UD	6.27	732.00	162.00	---	894.00	**1080.00**	768.00	**925.00**
85 MBHP	Inst	Ea	UD	6.67	756.00	172.00	---	928.00	**1130.00**	796.00	**961.00**
105 MBHP	Inst	Ea	UD	7.11	779.00	184.00	---	963.00	**1170.00**	826.00	**998.00**
125 MBHP	Inst	Ea	UD	7.62	898.00	197.00	---	1095.00	**1330.00**	941.00	**1130.00**
160 MBHP	Inst	Ea	UD	8.21	922.00	212.00	---	1134.00	**1380.00**	973.00	**1170.00**
200 MBHP	Inst	Ea	UD	8.89	2200.00	230.00	---	2430.00	**2880.00**	2112.00	**2490.00**
Minimum Job Charge	Inst	Job	UD	11.4	---	295.00	---	295.00	**442.00**	207.00	**310.00**
Gas line with couplings	Inst	LF	SB	.474	---	12.40	---	12.40	**18.50**	9.33	**13.90**
Oil fired, UL listed, gun type burner											
55 MBHP	Inst	Ea	UD	6.08	887.00	157.00	---	1044.00	**1260.00**	900.00	**1080.00**
100 MBHP	Inst	Ea	UD	7.11	940.00	184.00	---	1124.00	**1360.00**	968.00	**1160.00**
125 MBHP	Inst	Ea	UD	7.62	1310.00	197.00	---	1507.00	**1800.00**	1308.00	**1550.00**
150 MBHP	Inst	Ea	UD	8.21	1460.00	212.00	---	1672.00	**1990.00**	1449.00	**1720.00**
200 MBHP	Inst	Ea	UD	8.89	2440.00	230.00	---	2670.00	**3150.00**	2322.00	**2730.00**
Minimum Job Charge	Inst	Job	UD	11.4	---	295.00	---	295.00	**442.00**	207.00	**310.00**
Oil line with couplings	Inst	LF	SB	.267	---	7.00	---	7.00	**10.40**	5.24	**7.81**

Description	Oper	Unit	Crew Size	Man-Hours Per Unit	Avg Mat'l Unit Cost	Avg Labor Unit Cost	Avg Equip Unit Cost	Avg Total Unit Cost	Avg Price Incl O&P	Large Volume Avg Total Unit Cost	Large Volume Avg Price Incl O&P
Combo fired (wood, coal, oil combination) complete with burner											
115 MBHP (based on oil)	Inst	Ea	UD	7.34	2650.00	190.00	---	2840.00	**3330.00**	2483.00	**2900.00**
140 MBHP (based on oil)	Inst	Ea	UD	7.88	4250.00	204.00	---	4454.00	**5200.00**	3903.00	**4550.00**
150 MBHP (based on oil)	Inst	Ea	UD	8.21	4250.00	212.00	---	4462.00	**5210.00**	3909.00	**4560.00**
170 MBHP (based on oil)	Inst	Ea	UD	8.51	4490.00	220.00	---	4710.00	**5500.00**	4126.00	**4810.00**
Minimum Job Charge	Inst	Job	UD	11.4	---	295.00	---	295.00	**442.00**	207.00	**310.00**
Oil line with couplings	Inst	LF	SB	.267	---	7.00	---	7.00	**10.40**	5.24	**7.81**
Space heaters, gas fired											
Unit includes cabinet, grilles, fan, controls, burner and thermostat; no flue piping included (see page 254)											
Floor mounted											
60 MBHP	Inst	Ea	SB	2.67	637.00	70.00	---	707.00	**836.00**	614.40	**724.00**
180 MBHP	Inst	Ea	SB	5.33	946.00	140.00	---	1086.00	**1300.00**	940.00	**1120.00**
Suspension mounted, propeller fan											
20 MBHP	Inst	Ea	SB	3.56	476.00	93.30	---	569.30	**686.00**	490.00	**587.00**
60 MBHP	Inst	Ea	SB	4.27	589.00	112.00	---	701.00	**844.00**	603.90	**723.00**
130 MBHP	Inst	Ea	SB	5.33	869.00	140.00	---	1009.00	**1210.00**	872.00	**1040.00**
320 MBHP	Inst	Ea	SB	10.7	1790.00	280.00	---	2070.00	**2470.00**	1790.00	**2120.00**
Powered venter, adapter	ADD	Ea	SB	2.67	302.00	70.00	---	372.00	**452.00**	319.40	**385.00**
Wall furnace, self-contained thermostat											
Single capacity, recessed or surface mounted, 1-speed fan											
15 MBHP	Inst	Ea	SB	4.27	524.00	112.00	---	636.00	**769.00**	545.90	**656.00**
25 MBHP	Inst	Ea	SB	5.33	541.00	140.00	---	681.00	**831.00**	583.00	**706.00**
35 MBHP	Inst	Ea	SB	7.11	726.00	186.00	---	912.00	**1110.00**	781.00	**945.00**
Dual capacity, recessed or surface mounted, 2 speed blowers											
50 MBHP (direct vent)	Inst	Ea	SB	10.7	726.00	280.00	---	1006.00	**1250.00**	851.00	**1050.00**
60 MBHP (up vent)	Inst	Ea	SB	10.7	643.00	280.00	---	923.00	**1160.00**	777.00	**964.00**
Register kit for circulating heat to second room	Inst	Ea	SB	2.67	65.50	70.00	---	135.50	**180.00**	110.20	**145.00**
Minimum Job Charge	Inst	Job	SB	11.4	---	299.00	---	299.00	**445.00**	210.00	**312.00**
Bathroom heaters, electric fired											
Ceiling Heat-A-Ventlite, includes grille, blower, 4" round duct 13" x 7" dia., 3-way switch											
1,500 watt	Inst	Ea	EA	5.33	282.00	154.00	---	436.00	**552.00**	364.00	**457.00**
1,800 watt	Inst	Ea	EA	5.33	313.00	154.00	---	467.00	**587.00**	391.00	**488.00**
Ceiling Heat-A-Lite, includes grille, airotor wheel, 13" x 7" dia., 2-way switch, 1500 watt	Inst	Ea	EA	5.33	208.00	154.00	---	362.00	**467.00**	299.00	**382.00**
Ceiling radiant heating using infrared lamps											
Recessed, Heat-A-Lamp											
One bulb, 250 watt lamp	Inst	Ea	EA	4.44	53.60	128.00	---	181.60	**251.00**	143.30	**196.00**
Two bulb, 500 watt lamp	Inst	Ea	EA	4.44	91.60	128.00	---	219.60	**295.00**	176.90	**235.00**
Three bulb, 750 watt lamp	Inst	Ea	EA	4.44	159.00	128.00	---	287.00	**373.00**	237.00	**304.00**
Recessed, Heat-A-Vent											
One bulb, 250 watt lamp	Inst	Ea	EA	4.44	111.00	128.00	---	239.00	**317.00**	193.70	**254.00**
Two bulb, 500 watt lamp	Inst	Ea	EA	4.44	129.00	128.00	---	257.00	**337.00**	209.00	**273.00**

Description	Oper	Unit	Crew Size	Man-Hours Per Unit	Avg Mat'l Unit Cost	Avg Labor Unit Cost	Avg Equip Unit Cost	Avg Total Unit Cost	Avg Price Incl O&P	Large Volume Avg Total Unit Cost	Large Volume Avg Price Incl O&P
Wall heaters, recessed											
Fan forced											
1250 watt heating element	Inst	Ea	EA	3.33	148.00	96.00	---	244.00	**312.00**	202.10	**256.00**
Radiant heating											
1200 watt heating element	Inst	Ea	EA	3.33	117.00	96.00	---	213.00	**276.00**	175.10	**225.00**
1500 watt heating element	Inst	Ea	EA	3.33	123.00	96.00	---	219.00	**283.00**	180.10	**231.00**
Wiring, connection, and installation in closed wall or ceiling structure											
ADD	Inst	Ea	EA	3.46	---	99.80	---	99.80	**148.00**	69.80	**103.00**

Insulation

Batt or roll

Description	Oper	Unit	Crew Size	Man-Hours Per Unit	Avg Mat'l Unit Cost	Avg Labor Unit Cost	Avg Equip Unit Cost	Avg Total Unit Cost	Avg Price Incl O&P	Large Volume Avg Total Unit Cost	Large Volume Avg Price Incl O&P
With wall or ceiling finish already removed											
Joists, 16" or 24" oc	Demo	SF	LB	.007	---	.15	---	.15	**.23**	.11	**.17**
Rafters, 16" or 24" oc	Demo	SF	LB	.008	---	.18	---	.18	**.27**	.13	**.20**
Studs, 16" or 24" oc	Demo	SF	LB	.006	---	.13	---	.13	**.20**	.11	**.17**
Place and/or staple, Johns-Manville fiberglass; allowance made for joists, rafters, studs											
Joists											
Unfaced											
3-1/2" T (R-13)											
16" oc	Inst	SF	CA	.011	.36	.30	---	.66	**.86**	.53	**.68**
6-1/2" T (R-19)											
16" oc	Inst	SF	CA	.011	.43	.30	---	.73	**.95**	.59	**.75**
24" oc	Inst	SF	CA	.007	.43	.19	---	.62	**.78**	.50	**.63**
7" T (R-22)											
16" oc	Inst	SF	CA	.011	.57	.30	---	.87	**1.11**	.71	**.89**
24" oc	Inst	SF	CA	.007	.57	.19	---	.76	**.94**	.62	**.77**
9-1/4" T (R-30)											
16" oc	Inst	SF	CA	.011	.64	.30	---	.94	**1.19**	.78	**.97**
24" oc	Inst	SF	CA	.007	.64	.19	---	.83	**1.02**	.69	**.85**
Kraft-faced											
3-1/2" T (R-11)											
16" oc	Inst	SF	CA	.011	.29	.30	---	.59	**.78**	.47	**.62**
24" oc	Inst	SF	CA	.007	.32	.19	---	.51	**.65**	.40	**.52**
3-1/2" T (R-13)											
16" oc	Inst	SF	CA	.011	.38	.30	---	.68	**.89**	.55	**.71**
6-1/2" T (R-19)											
16" oc	Inst	SF	CA	.011	.46	.30	---	.76	**.98**	.62	**.79**
24" oc	Inst	SF	CA	.007	.46	.19	---	.65	**.82**	.53	**.66**
7" T (R-22)											
16" oc	Inst	SF	CA	.011	.62	.30	---	.92	**1.16**	.76	**.95**
24" oc	Inst	SF	CA	.007	.62	.19	---	.81	**1.00**	.67	**.83**
9-1/4" T (R-30)											
16" oc	Inst	SF	CA	.011	.71	.30	---	1.01	**1.27**	.84	**1.04**
24" oc	Inst	SF	CA	.007	.71	.19	---	.90	**1.10**	.75	**.92**
Foil-faced											
4" T (R-11)											
16" oc	Inst	SF	CA	.011	.32	.30	---	.62	**.82**	.49	**.64**
24" oc	Inst	SF	CA	.007	.32	.19	---	.51	**.65**	.40	**.52**

Description	Oper	Unit	Crew Size	Man-Hours Per Unit	Avg Mat'l Unit Cost	Avg Labor Unit Cost	Avg Equip Unit Cost	Avg Total Unit Cost	Avg Price Incl O&P	Large Volume Avg Total Unit Cost	Large Volume Avg Price Incl O&P
			Costs Based On Small Volume							Large Volume	
6-1/2" T (R-19)											
16" oc	Inst	SF	CA	.011	.46	.30	---	.76	**.98**	.62	**.79**
24" oc	Inst	SF	CA	.007	.46	.19	---	.65	**.82**	.53	**.66**
Rafters											
Unfaced											
3-1/2" T (R-13)											
16" oc	Inst	SF	CA	.016	.36	.43	---	.79	**1.07**	.63	**.85**
6-1/2" T (R-19)											
16" oc	Inst	SF	CA	.016	.43	.43	---	.86	**1.15**	.69	**.92**
24" oc	Inst	SF	CA	.011	.43	.30	---	.73	**.95**	.59	**.75**
7" T (R-22)											
16" oc	Inst	SF	CA	.016	.55	.43	---	.98	**1.29**	.80	**1.04**
24" oc	Inst	SF	CA	.011	.55	.30	---	.85	**1.08**	.70	**.88**
9-1/4" T (R-30)											
16" oc	Inst	SF	CA	.016	.64	.43	---	1.07	**1.39**	.88	**1.14**
24" oc	Inst	SF	CA	.011	.64	.30	---	.94	**1.19**	.78	**.97**
Kraft-faced											
3-1/2" T (R-11)											
16" oc	Inst	SF	CA	.016	.29	.43	---	.72	**.99**	.57	**.78**
24" oc	Inst	SF	CA	.011	.29	.30	---	.59	**.78**	.47	**.62**
3-1/2" T (R-13)											
16" oc	Inst	SF	CA	.016	.38	.43	---	.81	**1.09**	.65	**.87**
6-1/2" T (R-19)											
16" oc	Inst	SF	CA	.016	.46	.43	---	.89	**1.18**	.72	**.95**
24" oc	Inst	SF	CA	.011	.46	.30	---	.76	**.98**	.62	**.79**
7" T (R-22)											
16" oc	Inst	SF	CA	.016	.62	.43	---	1.05	**1.37**	.86	**1.11**
24" oc	Inst	SF	CA	.011	.62	.30	---	.92	**1.16**	.76	**.95**
9-1/4" T (R-30)											
16" oc	Inst	SF	CA	.016	.71	.43	---	1.14	**1.47**	.94	**1.20**
24" oc	Inst	SF	CA	.011	.71	.30	---	1.01	**1.27**	.84	**1.04**
Foil-faced											
4" T (R-11)											
16" oc	Inst	SF	CA	.016	.32	.43	---	.75	**1.02**	.59	**.80**
24" oc	Inst	SF	CA	.011	.32	.30	---	.62	**.82**	.49	**.64**
6-1/2" T (R-19)											
16" oc	Inst	SF	CA	.016	.46	.43	---	.89	**1.18**	.72	**.95**
24" oc	Inst	SF	CA	.011	.46	.30	---	.76	**.98**	.62	**.79**
Studs											
Unfaced											
3-1/2" T (R-13)											
16" oc	Inst	SF	CA	.013	.36	.35	---	.71	**.95**	.58	**.77**
6-1/2" T (R-19)											
16" oc	Inst	SF	CA	.013	.43	.35	---	.78	**1.03**	.64	**.84**
24" oc	Inst	SF	CA	.009	.43	.24	---	.67	**.86**	.56	**.71**
7" T (R-22)											
16" oc	Inst	SF	CA	.013	.57	.35	---	.92	**1.19**	.76	**.97**
24" oc	Inst	SF	CA	.009	.57	.24	---	.81	**1.02**	.68	**.85**

Description	Oper	Unit	Crew Size	Man-Hours Per Unit	Avg Mat'l Unit Cost	Avg Labor Unit Cost	Avg Equip Unit Cost	Avg Total Unit Cost	Avg Price Incl O&P	Large Volume Avg Total Unit Cost	Large Volume Avg Price Incl O&P
9-1/4" T (R-30)											
16" oc	Inst	SF	CA	.013	.64	.35	---	.99	**1.27**	.83	**1.05**
24" oc	Inst	SF	CA	.009	.64	.24	---	.88	**1.10**	.75	**.93**
Kraft-faced											
3-1/2" T (R-11)											
16" oc	Inst	SF	CA	.013	.29	.35	---	.64	**.87**	.52	**.70**
24" oc	Inst	SF	CA	.009	.32	.24	---	.56	**.74**	.46	**.60**
3-1/2" T (R-13)											
16" oc	Inst	SF	CA	.013	.38	.35	---	.73	**.97**	.60	**.79**
6-1/2" T (R-19)											
16" oc	Inst	SF	CA	.013	.46	.35	---	.81	**1.06**	.67	**.87**
24" oc	Inst	SF	CA	.009	.46	.24	---	.70	**.90**	.59	**.75**
7" T (R-22)											
16" oc	Inst	SF	CA	.013	.62	.35	---	.97	**1.25**	.81	**1.03**
24" oc	Inst	SF	CA	.009	.62	.24	---	.86	**1.08**	.73	**.91**
9-1/4" T (R-30)											
16" oc	Inst	SF	CA	.013	.71	.35	---	1.06	**1.35**	.89	**1.12**
24" oc	Inst	SF	CA	.009	.71	.24	---	.95	**1.19**	.81	**1.00**
Foil-faced											
4" T (R-11)											
16" oc	Inst	SF	CA	.013	.32	.35	---	.67	**.90**	.54	**.72**
24" oc	Inst	SF	CA	.009	.32	.24	---	.56	**.74**	.46	**.60**
6-1/2" T (R-19)											
16" oc	Inst	SF	CA	.013	.46	.35	---	.81	**1.06**	.67	**.87**
24" oc	Inst	SF	CA	.009	.46	.24	---	.70	**.90**	.59	**.75**

Loose fill

With ceiling finish already removed

Description	Oper	Unit	Crew Size	Man-Hours Per Unit	Avg Mat'l Unit Cost	Avg Labor Unit Cost	Avg Equip Unit Cost	Avg Total Unit Cost	Avg Price Incl O&P	Large Volume Avg Total Unit Cost	Large Volume Avg Price Incl O&P
Joists, 16" or 24" oc											
4" T	Demo	SF	LB	.005	---	.11	---	.11	**.17**	.09	**.13**
6" T	Demo	SF	LB	.009	---	.20	---	.20	**.30**	.15	**.23**

Allowance made for joists, cavities, and cores

Insulating wood, granule or pellet (40 lbs/bag, 4 CF/bag)

Description	Oper	Unit	Crew Size	Man-Hours Per Unit	Avg Mat'l Unit Cost	Avg Labor Unit Cost	Avg Equip Unit Cost	Avg Total Unit Cost	Avg Price Incl O&P	Large Volume Avg Total Unit Cost	Large Volume Avg Price Incl O&P
Joists, @ 7 lbs/CF density											
16" oc											
4" T	Inst	SF	CH	.017	.69	.43	---	1.12	**1.45**	.93	**1.19**
6" T	Inst	SF	CH	.022	1.04	.56	---	1.60	**2.04**	1.33	**1.69**
24" oc											
4" T	Inst	SF	CH	.017	.71	.43	---	1.14	**1.47**	.95	**1.21**
6" T	Inst	SF	CH	.023	1.07	.58	---	1.65	**2.11**	1.36	**1.72**

Vermiculite/Perlite (4 CF/bag, approximately 10 lbs/bag)

Description	Oper	Unit	Crew Size	Man-Hours Per Unit	Avg Mat'l Unit Cost	Avg Labor Unit Cost	Avg Equip Unit Cost	Avg Total Unit Cost	Avg Price Incl O&P	Large Volume Avg Total Unit Cost	Large Volume Avg Price Incl O&P
Joists											
16" oc											
4" T	Inst	SF	CH	.012	.51	.30	---	.81	**1.05**	.67	**.85**
6" T	Inst	SF	CH	.016	.75	.40	---	1.15	**1.48**	.95	**1.21**
24" oc											
4" T	Inst	SF	CH	.012	.52	.30	---	.82	**1.06**	.68	**.86**
6" T	Inst	SF	CH	.017	.77	.43	---	1.20	**1.54**	1.00	**1.27**
Cavity walls											
1" T	Inst	SF	CH	.005	.14	.13	---	.27	**.35**	.22	**.29**
2" T	Inst	SF	CH	.009	.25	.23	---	.48	**.63**	.40	**.52**

Description	Oper	Unit	Crew Size	Man-Hours Per Unit	Avg Mat'l Unit Cost	Avg Labor Unit Cost	Avg Equip Unit Cost	Avg Total Unit Cost	Avg Price Incl O&P	Large Volume Avg Total Unit Cost	Large Volume Avg Price Incl O&P
Block walls (2 cores/block)											
8" T block	Inst	SF	CH	.020	.42	.51	---	.93	**1.25**	.74	**.99**
12" T block	Inst	SF	CH	.026	.78	.66	---	1.44	**1.90**	1.19	**1.55**
Rigid											
Roofs											
1/2" T	Demo	SQ	LB	1.23	---	27.00	---	27.00	**40.80**	20.70	**31.20**
1" T	Demo	SQ	LB	1.45	---	31.80	---	31.80	**48.10**	23.50	**35.50**
Walls, 1/2" T	Demo	SF	LB	.010	---	.22	---	.22	**.33**	.15	**.23**
Rigid insulating board											
Roofs, over wood decks, 5% waste included											
Normal (dry) moisture conditions within building											
Nail one ply 15 lb felt, set and nail:											
2' x 8' x 1/2" T&G asphalt sheathing	Inst	SF	CN	.026	.31	.67	---	.98	**1.38**	.78	**1.09**
4' x 8' x 1/2" asphalt sheathing	Inst	SF	2C	.019	.31	.51	---	.82	**1.14**	.64	**.87**
4' x 8' x 1/2" building block	Inst	SF	2C	.019	.36	.51	---	.87	**1.19**	.69	**.93**
Excessive (humid) moisture conditions within building											
Nail and overlap three plies 15 lb felt, mop laps and surface one coat and embed:											
2' x 8' x 1/2" T&G asphalt sheathing	Inst	SF	RT	.041	.54	1.16	---	1.70	**2.43**	1.34	**1.91**
4' x 8' x 1/2" asphalt sheathing	Inst	SF	RT	.039	.54	1.10	---	1.64	**2.35**	1.32	**1.87**
4' x 8' x 1/2" building block	Inst	SF	RT	.039	.57	1.10	---	1.67	**2.38**	1.34	**1.89**
Over noncombustible decks, 5% waste included											
Normal (dry) moisture conditions within building											
Mop one coat and embed											
2' x 8' x 1/2" T&G asphalt sheathing	Inst	SF	RT	.029	.40	.82	---	1.22	**1.74**	.96	**1.36**
4' x 8' x 1/2" asphalt sheathing	Inst	SF	RT	.028	.40	.79	---	1.19	**1.70**	.93	**1.32**
4' x 8' x 1/2" building block	Inst	SF	RT	.028	.42	.79	---	1.21	**1.72**	.95	**1.34**
Excessive (humid) moisture conditions within building											
Mop one coat, embedded two plies 15 lb. felt, mop and embed:											
2' x 8' x 1/2" T&G asphalt sheathing	Inst	SF	RT	.051	.54	1.44	---	1.98	**2.88**	1.54	**2.22**
4' x 8' x 1/2" asphalt sheathing	Inst	SF	RT	.049	.54	1.38	---	1.92	**2.79**	1.51	**2.18**
4' x 8' x 1/2" building block	Inst	SF	RT	.049	.57	1.38	---	1.95	**2.82**	1.53	**2.20**
Walls, nailed, 5% waste included											
4' x 8' x 1/2" asphalt sheathing											
Straight wall	Inst	SF	2C	.015	.28	.40	---	.68	**.94**	.55	**.74**
Cut-up wall	Inst	SF	2C	.018	.28	.49	---	.77	**1.06**	.60	**.82**
4' x 8' x 1/2" building board											
Straight wall	Inst	SF	2C	.015	.32	.40	---	.72	**.98**	.57	**.76**
Cut-up wall	Inst	SF	2C	.018	.32	.49	---	.81	**1.11**	.62	**.84**

Intercom systems. See Electrical, page 95

Jacuzzi whirlpools. See Spas, page 245

Lath & plaster. See Plaster, page 196

			Costs Based On Small Volume							Large Volume	
Description	Oper	Unit	Crew Size	Man-Hours Per Unit	Avg Mat'l Unit Cost	Avg Labor Unit Cost	Avg Equip Unit Cost	Avg Total Unit Cost	Avg Price Incl O&P	Avg Total Unit Cost	Avg Price Incl O&P

Lighting fixtures

Labor includes hanging and connecting fixtures

Indoor lighting

Fluorescent

Description	Oper	Unit	Crew Size	Man-Hours Per Unit	Avg Mat'l Unit Cost	Avg Labor Unit Cost	Avg Equip Unit Cost	Avg Total Unit Cost	Avg Price Incl O&P	Avg Total Unit Cost (Large Volume)	Avg Price Incl O&P (Large Volume)
Pendant mounted worklights											
4' L, two 40 watt RS	Inst	Ea	EA	1.23	50.30	35.50	---	85.80	**110.00**	66.50	**84.10**
4' L, two 60 watt RS	Inst	Ea	EA	1.23	81.90	35.50	---	117.40	**147.00**	93.80	**115.00**
8' L, two 75 watt RS	Inst	Ea	EA	1.54	94.80	44.40	---	139.20	**175.00**	110.60	**137.00**
Recessed mounted light fixture with acrylic diffuser											
1' W x 4' L, two 40 watt RS	Inst	Ea	EA	1.23	53.80	35.50	---	89.30	**114.00**	69.60	**87.60**
2' W x 2' L, two U 40 watt RS	Inst	Ea	EA	1.23	58.50	35.50	---	94.00	**120.00**	73.60	**92.20**
2' W x 4' L, four 40 watt RS	Inst	Ea	EA	1.54	65.50	44.40	---	109.90	**141.00**	85.40	**108.00**
Strip lighting fixtures											
4' L, one 40 watt RS	Inst	Ea	EA	1.23	30.40	35.50	---	65.90	**87.50**	49.40	**64.30**
4' L, two 40 watt RS	Inst	Ea	EA	1.23	32.80	35.50	---	68.30	**90.20**	51.40	**66.70**
8' L, one 75 watt SL	Inst	Ea	EA	1.54	45.60	44.40	---	90.00	**118.00**	68.20	**88.00**
8' L, two 75 watt SL	Inst	Ea	EA	1.54	55.00	44.40	---	99.40	**129.00**	76.30	**97.30**
Surface mounted, acrylic diffuser											
1' W x 4' L, two 40 watt RS	Inst	Ea	EA	1.23	79.60	35.50	---	115.10	**144.00**	91.80	**113.00**
2' W x 2' L, two U 40 watt RS	Inst	Ea	EA	1.23	99.50	35.50	---	135.00	**167.00**	109.00	**133.00**
2' W x 4' L, four 40 watt RS	Inst	Ea	EA	1.54	102.00	44.40	---	146.40	**183.00**	116.70	**144.00**
White enameled circline steel ceiling fixtures											
8" W, 22 watt RS	Inst	Ea	EA	1.03	28.50	29.70	---	58.20	**76.70**	43.80	**56.80**
12" W, 22 to 32 watt RS	Inst	Ea	EA	1.03	42.80	29.70	---	72.50	**93.10**	56.10	**70.90**
16" W, 22 to 40 watt RS	Inst	Ea	EA	1.03	57.00	29.70	---	86.70	**110.00**	68.40	**85.10**
Decorative circline fixtures											
12" W, 22 to 32 watt RS	Inst	Ea	EA	1.03	111.00	29.70	---	140.70	**172.00**	115.20	**139.00**
16" W, 22 to 40 watt RS	Inst	Ea	EA	1.03	192.00	29.70	---	221.70	**265.00**	185.20	**219.00**

Incandescent

Description	Oper	Unit	Crew Size	Man-Hours Per Unit	Avg Mat'l Unit Cost	Avg Labor Unit Cost	Avg Equip Unit Cost	Avg Total Unit Cost	Avg Price Incl O&P	Avg Total Unit Cost (Large Volume)	Avg Price Incl O&P (Large Volume)
Ceiling fixture surface mounted											
15" x 5" white bent glass	Inst	Ea	EA	1.03	32.80	29.70	---	62.50	**81.70**	47.50	**61.00**
10" x 7" two light circular	Inst	Ea	EA	1.03	41.30	29.70	---	71.00	**91.50**	54.90	**69.50**
8" x 8" one light screw-in	Inst	Ea	EA	1.03	27.10	29.70	---	56.80	**75.10**	42.60	**55.30**
Ceiling fixture recessed											
Square fixture drop or flat lens											
8" frame	Inst	Ea	EA	1.54	41.30	44.40	---	85.70	**113.00**	64.50	**83.70**
10" frame	Inst	Ea	EA	1.54	47.00	44.40	---	91.40	**120.00**	69.40	**89.40**
12" frame	Inst	Ea	EA	1.54	49.90	44.40	---	94.30	**123.00**	71.90	**92.20**
Round fixture for concentrated light over small areas											
7" shower fixture frame	Inst	Ea	EA	1.54	41.30	44.40	---	85.70	**113.00**	64.50	**83.70**
8" spotlight fixture	Inst	Ea	EA	1.54	39.90	44.40	---	84.30	**112.00**	63.20	**82.30**
8" flat lens or stepped baffle frame	Inst	Ea	EA	1.54	41.30	44.40	---	85.70	**113.00**	64.50	**83.70**

Description	Oper	Unit	Crew Size	Man-Hours Per Unit	Avg Mat'l Unit Cost	Avg Labor Unit Cost	Avg Equip Unit Cost	Avg Total Unit Cost	Avg Price Incl O&P	Large Volume Avg Total Unit Cost	Large Volume Avg Price Incl O&P
Track lighting for highlighting effects from a ceiling or wall											
Swivel track heads with the ability to slide head along track to a new position											
Track heads											
Large cylinder	Inst	Ea	EA	---	48.50	---	---	48.50	**48.50**	41.80	**41.80**
Small cylinder	Inst	Ea	EA	---	39.90	---	---	39.90	**39.90**	34.40	**34.40**
Sphere cylinder	Inst	Ea	EA	---	48.50	---	---	48.50	**48.50**	41.80	**41.80**
Track; 1-7/16" W x 3/4" D											
2' track	Inst	Ea	EA	.513	20.00	14.80	---	34.80	**44.80**	26.80	**34.00**
4' track	Inst	Ea	EA	.769	49.90	22.20	---	72.10	**90.20**	57.50	**70.90**
8' track	Inst	Ea	EA	1.03	74.10	29.70	---	103.80	**129.00**	83.20	**102.00**
Straight connector; joins two track sections end to end; 7-5/8" L	Inst	Ea	EA	---	14.30	---	---	14.30	**14.30**	12.30	**12.30**
L - connector, joins two track sections for 90 degree angle turns; 7-5/8" L	Inst	Ea	EA	---	14.30	---	---	14.30	**14.30**	12.30	**12.30**
Feed in unit, attaches to ceiling or wall outlet box, supplies electrical current to all heads on track(s)	Inst	Ea	EA	1.03	15.70	29.70	---	45.40	**62.00**	32.70	**44.00**
Wall fixtures											
White glass with on/off switch											
1 light, 5" W x 5" H	Inst	Ea	EA	1.03	15.70	29.70	---	45.40	**62.00**	32.70	**44.00**
2 light 14" W x 5" H	Inst	Ea	EA	1.03	24.20	29.70	---	53.90	**71.80**	40.10	**52.50**
4 light, 24" W x 4" H	Inst	Ea	EA	1.03	35.60	29.70	---	65.30	**84.90**	50.00	**63.80**
Swivel wall fixture											
1 light, 4" W x 8" H	Inst	Ea	EA	1.03	24.60	29.70	---	54.30	**72.20**	40.40	**52.90**
2 light, 9" W x 9" H	Inst	Ea	EA	1.03	42.80	29.70	---	72.50	**93.10**	56.10	**70.90**
Outdoor lighting											
Ceiling fixture for porch											
7" W x 13" L, one 75 watt RS	Inst	Ea	EA	1.03	51.30	29.70	---	81.00	**103.00**	63.50	**79.40**
11" W x 4" L, two 60 watt RS	Inst	Ea	EA	1.03	62.70	29.70	---	92.40	**116.00**	73.30	**90.70**
15" W x 4" L, three 60 watt RS	Inst	Ea	EA	1.03	64.10	29.70	---	93.80	**118.00**	74.60	**92.10**
Wall fixture for porch											
6" W x 10" L, one 75 watt RS	Inst	Ea	EA	1.03	55.60	29.70	---	85.30	**108.00**	67.20	**83.60**
6" W x 16" L, one 75 watt RS	Inst	Ea	EA	1.03	69.80	29.70	---	99.50	**124.00**	79.50	**97.80**
Post lantern fixture											
Aluminum cast posts											
84" H post with lantern, set in concrete, includes 50' of conduit, circuit, and cement	Inst	Ea	EA	11.4	328.00	329.00	---	657.00	**863.00**	514.00	**667.00**
Urethane (in form of simulated redwood) over steel post											
Post with matching lantern, set in concrete, includes 50' of conduit, circuit, and cement	Inst	Ea	EA	11.4	392.00	329.00	---	721.00	**937.00**	569.00	**730.00**

Linoleum. See Resilient flooring, page 200

Lumber. See Framing, page 106

Description	Oper	Unit	Crew Size	Man-Hours Per Unit	Avg Mat'l Unit Cost	Avg Labor Unit Cost	Avg Equip Unit Cost	Avg Total Unit Cost	Avg Price Incl O&P	Large Volume Avg Total Unit Cost	Large Volume Avg Price Incl O&P
Mantels, fireplace											
Ponderosa pine, kiln-dried, unfinished, assembled											
Versailles, ornate, French design 59" W x 46" H	Inst	Ea	CJ	5.71	1170.00	140.00	---	1310.00	**1560.00**	1147.80	**1350.00**
Victorian, ornate, English design 63" W x 52" H	Inst	Ea	CJ	5.71	910.00	140.00	---	1050.00	**1260.00**	909.80	**1080.00**
Chelsea, plain, English design 70" W x 52" H	Inst	Ea	CJ	5.71	823.00	140.00	---	963.00	**1160.00**	831.80	**993.00**
Jamestown, plain, Early American 68" W x 53" H	Inst	Ea	CJ	5.71	417.00	140.00	---	557.00	**692.00**	469.80	**576.00**
Marlite paneling											
Panels, 4' x 8', adhesive set	Demo	SF	LB	.012	---	.26	---	.26	**.40**	.20	**.30**
Plastic coated masonite panels; 4' x 8', 5' x 5'; screw applied; channel molding around perimeter; 1/8" T											
Solid colors	Inst	SF	CJ	.067	1.69	1.64	---	3.33	**4.43**	2.65	**3.47**
Patterned panels	Inst	SF	CJ	.067	2.03	1.64	---	3.67	**4.82**	2.96	**3.83**
Molding, 1/8" panels; corners, divisions, or edging; nailed to framing or sheathing											
Bright anodized	Inst	LF	CJ	.090	.19	2.20	---	2.39	**3.56**	1.71	**2.54**
Gold anodized	Inst	LF	CJ	.090	.22	2.20	---	2.42	**3.60**	1.73	**2.56**
Colors	Inst	LF	CJ	.090	.27	2.20	---	2.47	**3.66**	1.78	**2.62**

Masonry

Brick. All material costs include mortar and waste, 5% on brick and 30% on mortar.

1. **Dimensions**
   a. Standard or regular: 8" L x $2^1/_4$" H x $3^3/_4$" W.
   b. Modular: $7^5/_8$" L x $2^1/_4$" H x $3^5/_8$" W.
   c. Norman: $11^5/_8$" L x $2^2/_3$" H x $3^5/_8$" W.
   d. Roman: $11^5/_8$" L x $1^5/_8$" H x $3^5/_8$" W.
2. **Installation**
   a. Mortar joints are $^3/_8$" thick both horizontally and vertically.
   b. Mortar mix is 1:3, 1 part masonry cement and 3 parts sand.
   c. Galvanized, corrugated wall ties are used on veneers at 1 tie per SF wall area. The tie is $^7/_8$" x 7" x 16 ga.
   d. Running bond used on veneers and walls.
3. **Notes on Labor.** Output is based on a crew composed of bricklayers and bricktenders at a 1:1 ratio.
4. **Estimating Technique.** Chimneys and columns figured per vertical linear foot with allowances already made for brick waste and mortar waste. Veneers and walls are computed per square foot of wall area.

Concrete (Masonry) Block. All material costs include 3% block waste and 30% mortar waste.

1. **Dimension.** All blocks are two core.
   a. Heavyweight: 8" T blocks weigh approximately 46 lbs/block; 12" T blocks weigh approximately 65 lbs/block.
   b. Lightweight: Also known as haydite blocks. 8" T blocks weigh approximately 30 lbs/block; 12" T blocks weigh approximately 41 lbs/block.
2. **Installation**
   a. Mortar joints are $^3/_8$" T both horizontally and vertically.
   b. Mortar mix is 1:3, 1 part masonry cement and 3 parts sand.
   c. Reinforcing: Lateral metal is regular truss with 9 gauge sides and ties. Vertical steel is #4 ($^1/_2$" dia.) rods, at 0.668 lbs/LF.
3. **Notes on Labor.** Output is based on a crew composed of bricklayers and bricktenders.
4. **Estimating Technique.** Figure chimneys and columns per vertical linear foot, with allowances already made for block waste and mortar waste. Veneers and walls are computed per square foot of wall area.

Quarry Tile (on Floor). Includes 5% tile waste.

1. **Dimensions:** 6" square tile is $^1/_2$" T and 9" square tile is $^3/_4$" T.
2. **Installation**
   a. Conventional mortar set utilizes portland cement, mortar mix, sand, and water. The mortar dry-cures and bonds to tile.
   b. Dry-set mortar utilizes dry-set portland cement, mortar mix, sand, and water. The mortar dry-cures and bonds to tile.
3. **Notes on Labor.** Output is based on a crew composed of bricklayers and bricktenders.
4. **Estimating Technique.** Compute square feet of floor area.

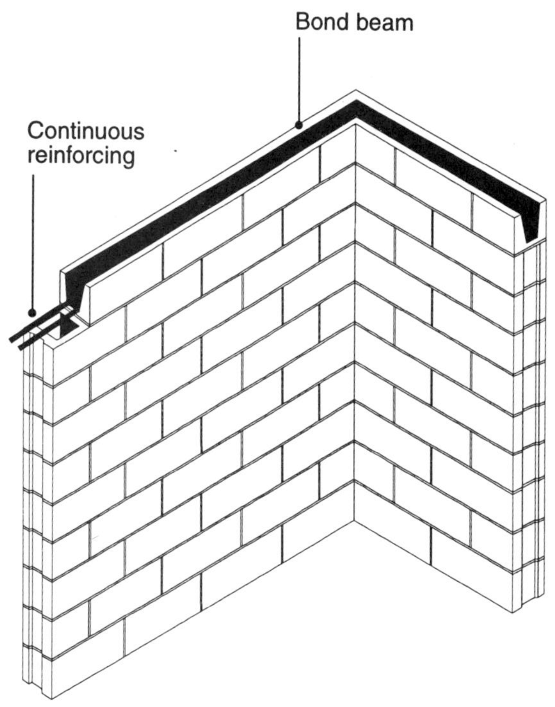

Typical concrete block wall

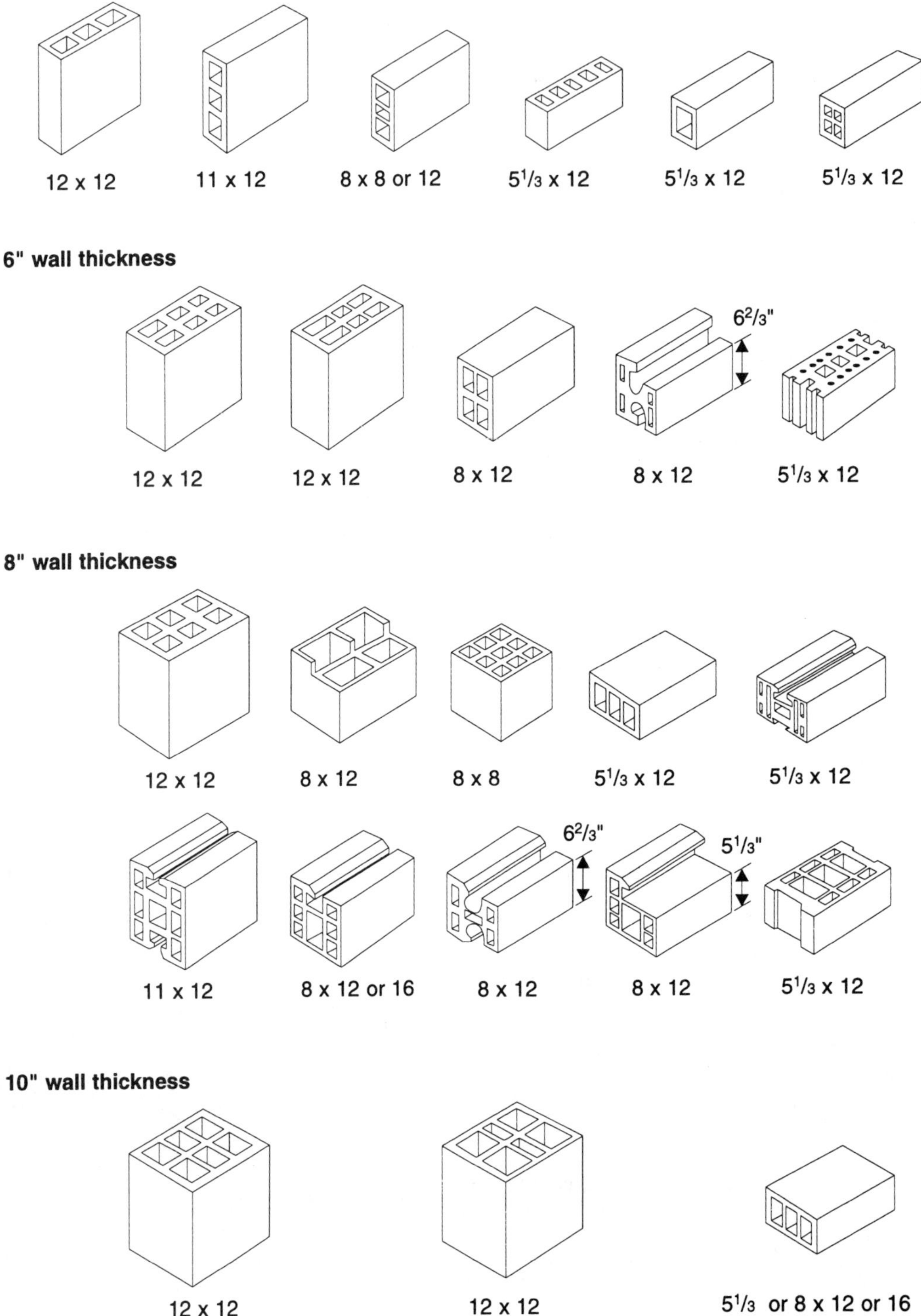

Common sizes and shapes for clay tile

Description	Oper	Unit	Crew Size	Man-Hours Per Unit	Avg Mat'l Unit Cost	Avg Labor Unit Cost	Avg Equip Unit Cost	Avg Total Unit Cost	Avg Price Incl O&P	Large Volume Avg Total Unit Cost	Large Volume Avg Price Incl O&P
Masonry											
Brick, standard											
Running bond, 3/8" mortar joints											
Chimneys											
Flue lining included; no scaffolding included											
4" T wall with standard brick											
16" x 16" with one 8" x 8" flue	Demo	VLF	LB	1.48	---	32.50	---	32.50	**49.10**	27.77	**40.10**
20" x 16" with one 12" x 8" flue	Demo	VLF	LB	1.72	---	37.80	---	37.80	**57.00**	32.08	**46.30**
20" x 20" with one 12" x 12" flue	Demo	VLF	LB	1.87	---	41.10	---	41.10	**62.00**	35.03	**50.50**
28" x 16" with two 8" x 8" flues	Demo	VLF	LB	2.29	---	50.30	---	50.30	**75.90**	42.83	**61.70**
32" x 16" with one 8" x 8" and one 12" x 8" flue	Demo	VLF	LB	2.43	---	53.30	---	53.30	**80.50**	45.46	**65.50**
36" x 16" with two 12" x 8" flues	Demo	VLF	LB	2.60	---	57.10	---	57.10	**86.20**	48.72	**70.20**
36" x 20" with two 12" x 12" flues	Demo	VLF	LB	2.89	---	63.40	---	63.40	**95.80**	54.27	**78.20**
16" x 16" with one 8" x 8" flue	Inst	VLF	BK	1.49	15.90	36.90	---	52.80	**74.00**	41.60	**57.80**
20" x 16" with one 12" x 8" flue	Inst	VLF	BK	1.69	19.30	41.80	---	61.10	**85.30**	48.20	**66.70**
20" x 20" with one 12" x 12" flue	Inst	VLF	BK	1.89	22.50	46.70	---	69.20	**96.40**	54.60	**75.40**
28" x 16" with two 8" x 8" flues	Inst	VLF	BK	2.27	28.00	56.10	---	84.10	**117.00**	66.40	**91.50**
32" x 16" with one 8" x 8" and one 12" x 8" flue	Inst	VLF	BK	2.42	31.00	59.90	---	90.90	**126.00**	71.90	**98.90**
36" x 16" with two 12" x 8" flues	Inst	VLF	BK	2.60	34.30	64.30	---	98.60	**137.00**	78.00	**107.00**
36" x 20" with two 12" x 12" flues	Inst	VLF	BK	2.88	41.10	71.20	---	112.30	**155.00**	89.00	**122.00**
8" T wall with standard brick											
24" x 24" with one 8" x 8" flue	Demo	VLF	LB	3.17	---	69.60	---	69.60	**105.00**	59.39	**85.60**
28" x 24" with one 12" x 8" flue	Demo	VLF	LB	3.41	---	74.90	---	74.90	**113.00**	63.91	**92.10**
28" x 28" with one 12" x 12" flue	Demo	VLF	LB	3.75	---	82.30	---	82.30	**124.00**	70.05	**101.00**
36" x 24" with two 8" x 8" flues	Demo	VLF	LB	4.16	---	91.30	---	91.30	**138.00**	77.85	**112.00**
40" x 24" with one 8" x 8" and one 12" x 8" flue	Demo	VLF	LB	4.57	---	100.00	---	100.00	**151.00**	85.60	**123.00**
44" x 24" with two 12" x 8" flues	Demo	VLF	LB	4.76	---	104.00	---	104.00	**158.00**	89.10	**128.00**
44" x 28" with two 12" x 12" flues	Demo	VLF	LB	5.08	---	112.00	---	112.00	**168.00**	95.20	**137.00**
24" x 24" with one 8" x 8" flue	Inst	VLF	BK	3.33	35.60	82.40	---	118.00	**165.00**	92.80	**129.00**
28" x 24" with one 12" x 8" flue	Inst	VLF	BK	3.68	40.60	91.00	---	131.60	**184.00**	103.70	**144.00**
28" x 28" with one 12" x 12" flue	Inst	VLF	BK	3.95	45.90	97.70	---	143.60	**200.00**	113.10	**156.00**
36" x 24" with two 8" x 8" flue	Inst	VLF	BK	4.44	53.10	110.00	---	163.10	**227.00**	128.60	**177.00**
40" x 24" with one 8" x 8" and one 12" x 8" flue	Inst	VLF	BK	4.95	60.30	122.00	---	182.30	**254.00**	144.50	**199.00**
44" x 24" with two 12" x 8" flues	Inst	VLF	BK	5.08	63.70	126.00	---	189.70	**263.00**	149.60	**206.00**
44" x 28" with two 12" x 12" flues	Inst	VLF	BK	5.46	72.50	135.00	---	207.50	**287.00**	164.10	**226.00**
Columns											
Outside dimension; no shoring; solid centers; no scaffolding included											
8" x 8"	Demo	VLF	LB	.473	---	10.40	---	10.40	**15.70**	8.85	**12.80**
12" x 8"	Demo	VLF	LB	.780	---	17.10	---	17.10	**25.90**	14.62	**21.10**
16" x 8"	Demo	VLF	LB	1.02	---	22.40	---	22.40	**33.80**	19.04	**27.40**
20" x 8"	Demo	VLF	LB	1.24	---	27.20	---	27.20	**41.10**	23.24	**33.50**
24" x 8"	Demo	VLF	LB	1.46	---	32.10	---	32.10	**48.40**	27.31	**39.30**

			Costs Based On Small Volume						Large Volume	
Description	Oper	Unit	Crew Size	Man-Hours Per Unit	Avg Mat'l Unit Cost	Avg Labor Unit Cost	Avg Equip Unit Cost	Avg Total Unit Cost	**Avg Price Incl O&P**	Avg Total Unit Cost	**Avg Price Incl O&P**
12" x 12"	Demo	VLF	LB	1.13	---	24.80	---	24.80	**37.50**	21.19	**30.50**
16" x 12"	Demo	VLF	LB	1.46	---	32.10	---	32.10	**48.40**	27.31	**39.30**
20" x 12"	Demo	VLF	LB	1.76	---	38.60	---	38.60	**58.30**	32.87	**47.40**
24" x 12"	Demo	VLF	LB	2.04	---	44.80	---	44.80	**67.60**	38.29	**55.10**
28" x 12"	Demo	VLF	LB	2.31	---	50.70	---	50.70	**76.60**	43.28	**62.40**
32" x 12"	Demo	VLF	LB	2.57	---	56.40	---	56.40	**85.20**	48.16	**69.40**
16" x 16"	Demo	VLF	AB	1.86	---	43.90	5.85	49.75	**71.70**	34.79	**50.10**
20" x 16"	Demo	VLF	AB	2.22	---	52.40	6.98	59.38	**85.60**	41.49	**59.80**
24" x 16"	Demo	VLF	AB	2.57	---	60.70	8.08	68.78	**99.10**	48.16	**69.40**
28" x 16"	Demo	VLF	AB	2.86	---	67.50	8.99	76.49	**110.00**	53.49	**77.10**
32" x 16"	Demo	VLF	AB	3.17	---	74.80	9.99	84.79	**122.00**	59.39	**85.60**
36" x 16"	Demo	VLF	AB	3.41	---	80.50	10.70	91.20	**131.00**	63.91	**92.10**
20" x 20"	Demo	VLF	AB	2.63	---	62.10	8.27	70.37	**101.00**	49.19	**70.90**
24" x 20"	Demo	VLF	AB	3.09	---	72.90	9.72	82.62	**119.00**	57.80	**83.30**
28" x 20"	Demo	VLF	AB	3.41	---	80.50	10.70	91.20	**131.00**	63.91	**92.10**
32" x 20"	Demo	VLF	AB	3.75	---	88.50	11.80	100.30	**145.00**	70.05	**101.00**
36" x 20"	Demo	VLF	AB	4.01	---	94.60	12.60	107.20	**155.00**	75.13	**108.00**
24" x 24"	Demo	VLF	AB	3.41	---	80.50	10.70	91.20	**131.00**	63.91	**92.10**
28" x 24"	Demo	VLF	AB	3.81	---	89.90	12.00	101.90	**147.00**	71.39	**103.00**
32" x 24"	Demo	VLF	AB	4.16	---	98.20	13.10	111.30	**160.00**	77.85	**112.00**
36" x 24"	Demo	VLF	AB	4.48	---	106.00	14.10	120.10	**173.00**	83.97	**121.00**
28" x 28"	Demo	VLF	AB	4.23	---	99.80	13.30	113.10	**163.00**	79.22	**114.00**
32" x 28"	Demo	VLF	AB	4.66	---	110.00	14.70	124.70	**180.00**	87.50	**126.00**
36" x 28"	Demo	VLF	AB	5.08	---	120.00	16.00	136.00	**196.00**	95.20	**137.00**
32" x 32"	Demo	VLF	AB	5.08	---	120.00	16.00	136.00	**196.00**	95.20	**137.00**
36" x 32"	Demo	VLF	AB	5.57	---	131.00	17.50	148.50	**215.00**	104.30	**150.00**
36" x 36"	Demo	VLF	AB	6.02	---	142.00	18.90	160.90	**232.00**	112.70	**162.00**
8" x 8" (9.33 Brick/VLF)	Inst	VLF	BK	.553	3.87	13.70	---	17.57	**25.10**	13.67	**19.40**
12" x 8" (14.00 Brick/VLF)	Inst	VLF	BK	.808	5.90	20.00	---	25.90	**37.00**	20.14	**28.50**
16" x 8" (18.67 Brick/VLF)	Inst	VLF	BK	1.05	7.84	26.00	---	33.84	**48.20**	26.23	**37.10**
20" x 8" (23.33 Brick/VLF)	Inst	VLF	BK	1.28	9.77	31.70	---	41.47	**59.00**	32.21	**45.60**
24" x 8" (28.00 Brick/VLF)	Inst	VLF	BK	1.49	11.70	36.90	---	48.60	**69.10**	37.90	**53.60**
12" x 12" (21.00 Brick/VLF)	Inst	VLF	BK	1.17	8.75	28.90	---	37.65	**53.80**	29.23	**41.40**
16" x 12" (28.00 Brick/VLF)	Inst	VLF	BK	1.49	11.70	36.90	---	48.60	**69.10**	37.90	**53.60**
20" x 12" (35.00 Brick/VLF)	Inst	VLF	BK	1.79	14.70	44.30	---	59.00	**83.70**	45.90	**64.70**
24" x 12" (42.00 Brick/VLF)	Inst	VLF	BK	2.09	17.50	51.70	---	69.20	**98.20**	54.10	**76.20**
28" x 12" (49.00 Brick/VLF)	Inst	VLF	BK	2.34	20.50	57.90	---	78.40	**111.00**	61.30	**86.20**
32" x 12" (56.00 Brick/VLF)	Inst	VLF	BK	2.60	23.40	64.30	---	87.70	**124.00**	68.60	**96.30**
16" x 16" (37.33 Brick/VLF)	Inst	VLF	BK	1.89	15.60	46.70	---	62.30	**88.50**	48.70	**68.60**
20" x 16" (46.67 Brick/VLF)	Inst	VLF	BK	2.27	19.50	56.10	---	75.60	**107.00**	59.00	**83.10**
24" x 16" (56.00 Brick/VLF)	Inst	VLF	BK	2.60	23.40	64.30	---	87.70	**124.00**	68.60	**96.30**
28" x 16" (65.33 Brick/VLF)	Inst	VLF	BK	2.88	27.30	71.20	---	98.50	**139.00**	77.20	**108.00**
32" x 16" (74.67 Brick/VLF)	Inst	VLF	BK	3.18	31.20	78.60	---	109.80	**155.00**	86.30	**121.00**
36" x 16" (84.00 Brick/VLF)	Inst	VLF	BK	3.44	35.10	85.10	---	120.20	**169.00**	94.40	**132.00**
20" x 20" (58.33 Brick/VLF)	Inst	VLF	BK	2.67	24.40	66.00	---	90.40	**128.00**	70.80	**99.20**
24" x 20" (70.00 Brick/VLF)	Inst	VLF	BK	3.09	29.20	76.40	---	105.60	**149.00**	82.80	**116.00**
28" x 20" (81.67 Brick/VLF)	Inst	VLF	BK	3.44	34.10	85.10	---	119.20	**168.00**	93.50	**130.00**
32" x 20" (93.33 Brick/VLF)	Inst	VLF	BK	3.74	39.00	92.50	---	131.50	**184.00**	103.50	**144.00**
36" x 20" (105.00 Brick/VLF)	Inst	VLF	BK	4.02	43.90	99.40	---	143.30	**201.00**	112.90	**157.00**

Description	Oper	Unit	Crew Size	Man-Hours Per Unit	Avg Mat'l Unit Cost	Avg Labor Unit Cost	Avg Equip Unit Cost	Avg Total Unit Cost	Avg Price Incl O&P	Large Volume Avg Total Unit Cost	Large Volume Avg Price Incl O&P
			Costs Based On Small Volume							Large Volume	
24" x 24" (84.00 Brick/VLF)	Inst	VLF	BK	3.44	35.10	85.10	---	120.20	**169.00**	94.40	**132.00**
28" x 24" (98.00 Brick/VLF)	Inst	VLF	BK	3.81	40.90	94.20	---	135.10	**189.00**	106.30	**148.00**
32" x 24" (112.00 Brick/VLF)	Inst	VLF	BK	4.10	46.80	101.00	---	147.80	**207.00**	117.00	**162.00**
36" x 24" (126.00 Brick/VLF)	Inst	VLF	BK	4.44	52.60	110.00	---	162.60	**226.00**	128.20	**177.00**
28" x 28" (114.33 Brick/VLF)	Inst	VLF	BK	4.27	47.80	106.00	---	153.80	**214.00**	120.80	**167.00**
32" x 28" (130.67 Brick/VLF)	Inst	VLF	BK	4.64	54.70	115.00	---	169.70	**236.00**	133.70	**185.00**
36" x 28" (147.00 Brick/VLF)	Inst	VLF	BK	5.19	61.50	128.00	---	189.50	**264.00**	150.10	**207.00**
32" x 32" (149.33 Brick/VLF)	Inst	VLF	BK	5.33	62.40	132.00	---	194.40	**271.00**	153.30	**212.00**
36" x 32" (168.00 Brick/VLF)	Inst	VLF	BK	5.93	70.20	147.00	---	217.20	**302.00**	171.20	**236.00**
36" x 36" (189.00 Brick/VLF)	Inst	VLF	BK	6.67	79.00	165.00	---	244.00	**340.00**	192.80	**266.00**
Veneers											
4" T, with air tools	Demo	SF	AB	.070	---	1.65	.22	1.87	**2.70**	1.31	**1.88**
4" T, with wall ties; no scaffolding included											
Common, 8" x 2-2/3" x 4"	Inst	SF	BO	.182	2.65	4.50	.20	7.35	**10.00**	6.74	**9.29**
Standard face, 8" x 2-2/3" x 4"	Inst	SF	BO	.194	2.85	4.80	.22	7.87	**10.70**	7.26	**9.99**
Glazed, 8" x 2-2/3" x 4"	Inst	SF	BO	.203	8.18	5.02	.23	13.43	**17.20**	12.06	**15.60**
Other brick types/sizes											
8" x 4" x 4"	Inst	SF	BO	.135	2.77	3.34	.15	6.26	**8.38**	5.72	**7.73**
8" x 3-1/5" x 4"	Inst	SF	BO	.161	4.11	3.98	.18	8.27	**10.90**	7.50	**9.99**
12" x 4" x 6"	Inst	SF	BO	.097	4.40	2.40	.11	6.91	**8.79**	6.17	**7.91**
12" x 2-2/3" x 4"	Inst	SF	BO	.131	3.04	3.24	.15	6.43	**8.54**	5.85	**7.84**
12" x 3-1/5" x 4"	Inst	SF	BO	.114	2.77	2.82	.13	5.72	**7.57**	5.20	**6.94**
12" x 2" x 4"	Inst	SF	BO	.171	4.90	4.23	.19	9.32	**12.20**	8.41	**11.10**
12" x 2-2/3" x 6"	Inst	SF	BO	.135	4.16	3.34	.15	7.65	**9.97**	6.94	**9.13**
12" x 4" x 4"	Inst	SF	BO	.095	3.29	2.35	.11	5.75	**7.44**	5.16	**6.73**
Walls											
With air tools											
8" T	Demo	SF	AB	.115	---	2.71	.36	3.07	**4.43**	2.30	**3.31**
12" T	Demo	SF	AB	.164	---	3.87	.52	4.39	**6.33**	3.29	**4.74**
16" T	Demo	SF	AB	.213	---	5.03	.67	5.70	**8.21**	4.28	**6.16**
24" T	Demo	SF	AB	.251	---	5.92	.79	6.71	**9.68**	5.03	**7.25**
With common brick; no scaffolding included											
8" T (13.50 Brick/SF)	Inst	SF	ML	.337	5.20	8.70	.50	14.40	**19.60**	13.19	**18.10**
12" T (20.25 Brick/SF)	Inst	SF	ML	.457	7.85	11.80	.68	20.33	**27.50**	18.57	**25.40**
16" T (27.00 Brick/SF)	Inst	SF	ML	.582	10.50	15.00	.86	26.36	**35.60**	24.10	**32.80**
24" T (40.50 Brick/SF)	Inst	SF	ML	.800	15.70	20.70	1.19	37.59	**50.40**	34.19	**46.30**
Brick, adobe											
Running bond, 3/8" mortar joints											
Walls, with air tools											
4" T	Demo	SF	AB	.059	---	1.39	.18	1.57	**2.27**	1.10	**1.58**
6" T	Demo	SF	AB	.060	---	1.42	.19	1.61	**2.31**	1.12	**1.62**
8" T	Demo	SF	AB	.062	---	1.46	.19	1.65	**2.38**	1.15	**1.66**
12" T	Demo	SF	AB	.065	---	1.53	.21	1.74	**2.51**	1.23	**1.77**
Walls; no scaffolding included											
4" x 4" x 16"	Inst	SF	ML	.139	5.16	3.59	.21	8.96	**11.60**	8.06	**10.50**
6" x 4" x 16"	Inst	SF	ML	.145	5.16	3.74	.22	9.12	**11.80**	9.48	**12.20**
8" x 4" x 16"	Inst	SF	ML	.152	7.97	3.92	.23	12.12	**15.30**	10.80	**13.80**
12" x 4" x 16"	Inst	SF	ML	.168	10.30	4.34	.25	14.89	**18.60**	13.22	**16.70**

Description	Oper	Unit	Crew Size	Man-Hours Per Unit	Avg Mat'l Unit Cost	Avg Labor Unit Cost	Avg Equip Unit Cost	Avg Total Unit Cost	Avg Price Incl O&P	Large Volume Avg Total Unit Cost	Large Volume Avg Price Incl O&P
Concrete block											
Lightweight (haydite) or heavyweight blocks; 2 cores/block, solid face; includes allowances for lintels, bond beams											
Foundations and retaining walls											
No excavation included											
Without reinforcing or with lateral reinforcing only											
With air tools											
8" W x 8" H x 16" L	Demo	SF	AB	.068	---	1.60	.21	1.81	**2.62**	1.28	**1.85**
10" W x 8" H x 16" L	Demo	SF	AB	.074	---	1.75	.23	1.98	**2.85**	1.39	**2.00**
12" W x 8" H x 16" L	Demo	SF	AB	.079	---	1.86	.25	2.11	**3.05**	1.47	**2.12**
Without air tools											
8" W x 8" H x 16" L	Demo	SF	LB	.085	---	1.87	---	1.87	**2.82**	1.30	**1.96**
10" W x 8" H x 16" L	Demo	SF	LB	.091	---	2.00	---	2.00	**3.02**	1.40	**2.12**
12" W x 8" H x 16" L	Demo	SF	LB	.099	---	2.17	---	2.17	**3.28**	1.54	**2.32**
With vertical reinforcing in every other core (2 core blocks) with core concrete filled											
With air tools											
8" W x 8" H x 16" L	Demo	SF	AB	.111	---	2.62	.35	2.97	**4.28**	2.09	**3.01**
10" W x 8" H x 16" L	Demo	SF	AB	.120	---	2.83	.38	3.21	**4.63**	2.24	**3.23**
12" W x 8" H x 16" L	Demo	SF	AB	.131	---	3.09	.41	3.50	**5.05**	2.44	**3.51**
Foundations and retaining walls											
No reinforcing											
Heavyweight blocks											
8" x 8" x 16"	Inst	SF	BO	.135	2.25	3.34	.15	5.74	**7.78**	5.28	**7.22**
10" x 8" x 16"	Inst	SF	BO	.147	2.92	3.64	.16	6.72	**9.01**	6.15	**8.32**
12" x 8" x 16"	Inst	SF	BO	.161	3.40	3.98	.18	7.56	**10.10**	6.89	**9.29**
Lightweight blocks											
8" x 8" x 16"	Inst	SF	BO	.122	2.62	3.02	.14	5.78	**7.71**	5.25	**7.07**
10" x 8" x 16"	Inst	SF	BO	.133	3.30	3.29	.15	6.74	**8.91**	6.12	**8.17**
12" x 8" x 16"	Inst	SF	BO	.145	3.77	3.59	.16	7.52	**9.91**	6.82	**9.07**
Lateral reinforcing every second course											
Heavyweight blocks											
8" x 8" x 16"	Inst	SF	BO	.142	2.25	3.51	.16	5.92	**8.05**	5.45	**7.46**
10" x 8" x 16"	Inst	SF	BO	.155	2.92	3.83	.17	6.92	**9.32**	6.34	**8.60**
12" x 8" x 16"	Inst	SF	BO	.171	3.40	4.23	.19	7.82	**10.50**	7.13	**9.64**
Lightweight blocks											
8" x 8" x 16"	Inst	SF	BO	.127	2.62	3.14	.14	5.90	**7.90**	5.39	**7.27**
10" x 8" x 16"	Inst	SF	BO	.140	3.30	3.46	.16	6.92	**9.18**	6.28	**8.41**
12" x 8" x 16"	Inst	SF	BO	.152	3.77	3.76	.17	7.70	**10.20**	7.01	**9.35**
Vertical reinforcing (No. 4 rod) every second core with core concrete filled											
Heavyweight blocks											
8" x 8" x 16"	Inst	SF	BD	.242	3.00	6.14	.22	9.36	**12.90**	7.37	**10.10**
10" x 8" x 16"	Inst	SF	BD	.267	3.77	6.78	.24	10.79	**14.80**	8.51	**11.60**
12" x 8" x 16"	Inst	SF	BD	.288	4.37	7.31	.26	11.94	**16.30**	9.44	**12.80**
Lightweight blocks											
8" x 8" x 16"	Inst	SF	BD	.222	3.37	5.63	.20	9.20	**12.60**	7.29	**9.88**
10" x 8" x 16"	Inst	SF	BD	.242	4.14	6.14	.22	10.50	**14.30**	8.35	**11.20**
12" x 8" x 16"	Inst	SF	BD	.267	4.74	6.78	.24	11.76	**15.90**	9.35	**12.60**

Description	Oper	Unit	Crew Size	Man-Hours Per Unit	Avg Mat'l Unit Cost	Avg Labor Unit Cost	Avg Equip Unit Cost	Avg Total Unit Cost	Avg Price Incl O&P	Large Volume Avg Total Unit Cost	Large Volume Avg Price Incl O&P
			Costs Based On Small Volume							Large Volume	
Exterior walls (above grade)											
No bracing or shoring included											
Without reinforcing or with lateral reinforcing only											
With air tools											
8" W x 8" H x 16" L	Demo	SF	AB	.068	---	1.60	.21	1.81	**2.62**	1.28	**1.85**
10" W x 8" H x 16" L	Demo	SF	AB	.075	---	1.77	.24	2.01	**2.90**	1.40	**2.01**
12" W x 8" H x 16" L	Demo	SF	AB	.083	---	1.96	.26	2.22	**3.20**	1.55	**2.23**
Without air tools											
8" W x 8" H x 16" L	Demo	SF	LB	.086	---	1.89	---	1.89	**2.85**	1.32	**1.99**
10" W x 8" H x 16" L	Demo	SF	LB	.095	---	2.09	---	2.09	**3.15**	1.47	**2.22**
12" W x 8" H x 16" L	Demo	SF	LB	.104	---	2.28	---	2.28	**3.45**	1.60	**2.42**
Exterior walls (above grade)											
No reinforcing											
Heavyweight blocks											
8" x 8" x 16"	Inst	SF	BO	.145	2.25	3.59	.16	6.00	**8.16**	5.52	**7.57**
10" x 8" x 16"	Inst	SF	BO	.158	2.92	3.91	.18	7.01	**9.44**	6.42	**8.71**
12" x 8" x 16"	Inst	SF	BO	.171	3.40	4.23	.19	7.82	**10.50**	7.13	**9.64**
Lightweight blocks											
8" x 8" x 16"	Inst	SF	BO	.129	2.62	3.19	.14	5.95	**7.97**	5.44	**7.35**
10" x 8" x 16"	Inst	SF	BO	.140	2.85	3.46	.16	6.47	**8.67**	6.28	**8.41**
12" x 8" x 16"	Inst	SF	BO	.152	3.77	3.76	.17	7.70	**10.20**	7.01	**9.35**
Lateral reinforcing every second course											
Heavyweight blocks											
8" x 8" x 16"	Inst	SF	BO	.152	2.25	3.76	.17	6.18	**8.43**	5.71	**7.85**
10" x 8" x 16"	Inst	SF	BO	.164	2.92	4.06	.18	7.16	**9.66**	6.58	**8.95**
12" x 8" x 16"	Inst	SF	BO	.178	3.40	4.40	.20	8.00	**10.80**	7.32	**9.92**
Lightweight blocks											
8" x 8" x 16"	Inst	SF	BO	.135	2.62	3.34	.15	6.11	**8.20**	5.59	**7.58**
10" x 8" x 16"	Inst	SF	BO	.147	2.85	3.64	.16	6.65	**8.93**	6.78	**9.04**
12" x 8" x 16"	Inst	SF	BO	.161	3.77	3.98	.18	7.93	**10.50**	7.21	**9.65**
Partitions (above grade)											
No bracing or shoring included											
Without reinforcing or with lateral reinforcing only											
With air tools											
4" W x 8" H x 16" L	Demo	SF	AB	.069	---	1.63	.22	1.85	**2.66**	1.28	**1.85**
6" W x 8" H x 16" L	Demo	SF	AB	.074	---	1.75	.23	1.98	**2.85**	1.39	**2.00**
8" W x 8" H x 16" L	Demo	SF	AB	.077	---	1.82	.24	2.06	**2.97**	1.44	**2.08**
10" W x 8" H x 16" L	Demo	SF	AB	.086	---	2.03	.27	2.30	**3.31**	1.61	**2.31**
12" W x 8" H x 16" L	Demo	SF	AB	.097	---	2.29	.31	2.60	**3.74**	1.81	**2.62**
Partitions (above grade)											
No reinforcing											
Heavyweight blocks											
4" W x 8" H x 16" L	Inst	SF	BO	.138	1.43	3.41	.15	4.99	**6.95**	4.62	**6.48**
6" W x 8" H x 16" L	Inst	SF	BO	.147	1.76	3.64	.16	5.56	**7.67**	5.14	**7.16**
8" W x 8" H x 16" L	Inst	SF	BO	.161	2.25	3.98	.18	6.41	**8.78**	5.91	**8.16**
10" W x 8" H x 16" L	Inst	SF	BO	.174	2.92	4.30	.19	7.41	**10.10**	6.82	**9.31**
12" W x 8" H x 16" L	Inst	SF	BO	.190	3.40	4.70	.21	8.31	**11.20**	7.61	**10.40**

Description	Oper	Unit	Crew Size	Man-Hours Per Unit	Avg Mat'l Unit Cost	Avg Labor Unit Cost	Avg Equip Unit Cost	Avg Total Unit Cost	Avg Price Incl O&P	Large Volume Avg Total Unit Cost	Large Volume Avg Price Incl O&P
Lightweight blocks											
4" W x 8" H x 16" L	Inst	SF	BO	.127	1.79	3.14	.14	5.07	**6.94**	4.68	**6.45**
6" W x 8" H x 16" L	Inst	SF	BO	.135	2.14	3.34	.15	5.63	**7.65**	5.17	**7.09**
8" W x 8" H x 16" L	Inst	SF	BO	.145	2.62	3.59	.16	6.37	**8.59**	5.83	**7.93**
10" W x 8" H x 16" L	Inst	SF	BO	.158	2.85	3.91	.18	6.94	**9.36**	6.73	**9.07**
12" W x 8" H x 16" L	Inst	SF	BO	.171	3.77	4.23	.19	8.19	**10.90**	7.45	**10.00**
Lateral reinforcing every second course											
Heavyweight blocks											
4" W x 8" H x 16" L	Inst	SF	BO	.145	1.43	3.59	.16	5.18	**7.22**	4.80	**6.75**
6" W x 8" H x 16" L	Inst	SF	BO	.155	1.76	3.83	.17	5.76	**7.98**	5.33	**7.43**
8" W x 8" H x 16" L	Inst	SF	BO	.171	2.25	4.23	.19	6.67	**9.16**	6.15	**8.51**
10" W x 8" H x 16" L	Inst	SF	BO	.186	2.92	4.60	.21	7.73	**10.50**	7.11	**9.74**
12" W x 8" H x 16" L	Inst	SF	BO	.203	3.40	5.02	.23	8.65	**11.70**	7.92	**10.80**
Lightweight blocks											
4" W x 8" H x 16" L	Inst	SF	BO	.133	1.79	3.29	.15	5.23	**7.17**	4.83	**6.68**
6" W x 8" H x 16" L	Inst	SF	BO	.142	2.14	3.51	.16	5.81	**7.92**	5.34	**7.33**
8" W x 8" H x 16" L	Inst	SF	BO	.152	2.62	3.76	.17	6.55	**8.86**	6.02	**8.21**
10" W x 8" H x 16" L	Inst	SF	BO	.164	2.85	4.06	.18	7.09	**9.58**	6.89	**9.31**
12" W x 8" H x 16" L	Inst	SF	BO	.178	3.77	4.40	.20	8.37	**11.20**	7.64	**10.30**
Fences											
Without reinforcing or with lateral reinforcing only											
With air tools											
6" W x 4" H x 16" L	Demo	SF	AB	.059	---	1.39	.18	1.57	**2.27**	1.10	**1.58**
6" W x 6" H x 16" L	Demo	SF	AB	.062	---	1.46	.19	1.65	**2.38**	1.15	**1.66**
8" W x 8" H x 16" L	Demo	SF	AB	.065	---	1.53	.21	1.74	**2.51**	1.23	**1.77**
10" W x 8" H x 16" L	Demo	SF	AB	.070	---	1.65	.22	1.87	**2.70**	1.31	**1.88**
12" W x 8" H x 16" L	Demo	SF	AB	.076	---	1.79	.24	2.03	**2.93**	1.42	**2.05**
Without air tools											
6" W x 4" H x 16" L	Demo	SF	LB	.074	---	1.62	---	1.62	**2.45**	1.14	**1.72**
6" W x 6" H x 16" L	Demo	SF	LB	.077	---	1.69	---	1.69	**2.55**	1.19	**1.79**
8" W x 8" H x 16" L	Demo	SF	LB	.082	---	1.80	---	1.80	**2.72**	1.25	**1.89**
10" W x 8" H x 16" L	Demo	SF	LB	.104	---	2.28	---	2.28	**3.45**	1.60	**2.42**
12" W x 8" H x 16" L	Demo	SF	LB	.095	---	2.09	---	2.09	**3.15**	1.47	**2.22**
Fences, lightweight blocks											
No reinforcing											
4" W x 8" H x 16" L	Inst	SF	BO	.115	1.79	2.84	.13	4.76	**6.48**	4.39	**6.02**
6" W x 4" H x 16" L	Inst	SF	BO	.237	3.91	5.86	.26	10.03	**13.60**	9.19	**12.60**
6" W x 6" H x 16" L	Inst	SF	BO	.164	2.75	4.06	.18	6.99	**9.47**	6.42	**8.77**
6" W x 8" H x 16" L	Inst	SF	BO	.122	2.14	3.02	.14	5.30	**7.16**	4.83	**6.58**
8" W x 8" H x 16" L	Inst	SF	BO	.129	2.62	3.19	.14	5.95	**7.97**	5.44	**7.35**
10" W x 8" H x 16" L	Inst	SF	BO	.140	2.85	3.46	.16	6.47	**8.67**	6.28	**8.41**
12" W x 8" H x 16" L	Inst	SF	BO	.152	3.77	3.76	.17	7.70	**10.20**	7.01	**9.35**
Lateral reinforcing every third course											
4" W x 8" H x 16" L	Inst	SF	BO	.120	1.79	2.97	.13	4.89	**6.67**	4.52	**6.21**
6" W x 4" H x 16" L	Inst	SF	BO	.244	3.91	6.03	.27	10.21	**13.90**	9.37	**12.80**
6" W x 6" H x 16" L	Inst	SF	BO	.174	2.75	4.30	.19	7.24	**9.85**	6.66	**9.12**
6" W x 8" H x 16" L	Inst	SF	BO	.127	2.14	3.14	.14	5.42	**7.34**	4.97	**6.79**
8" W x 8" H x 16" L	Inst	SF	BO	.135	2.62	3.34	.15	6.11	**8.20**	5.59	**7.58**
10" W x 8" H x 16" L	Inst	SF	BO	.147	2.85	3.64	.16	6.65	**8.93**	6.46	**8.67**
12" W x 8" H x 16" L	Inst	SF	BO	.161	3.77	3.98	.18	7.93	**10.50**	7.21	**9.65**

Description	Oper	Unit	Crew Size	Man-Hours Per Unit	Avg Mat'l Unit Cost	Avg Labor Unit Cost	Avg Equip Unit Cost	Avg Total Unit Cost	Avg Price Incl O&P	Avg Total Unit Cost	Avg Price Incl O&P
			Costs Based On Small Volume							Large Volume	
Concrete slump block											
Running bond, 3/8" mortar joints											
Walls, with air tools											
4" T	Demo	SF	AB	.062	---	1.46	.19	1.65	**2.38**	1.15	**1.66**
6" T	Demo	SF	AB	.065	---	1.53	.21	1.74	**2.51**	1.23	**1.77**
8" T	Demo	SF	AB	.068	---	1.60	.21	1.81	**2.62**	1.28	**1.85**
12" T	Demo	SF	AB	.083	---	1.96	.26	2.22	**3.20**	1.55	**2.23**
Walls; no scaffolding included											
4" x 4" x 16"	Inst	SF	ML	.128	4.54	3.30	.19	8.03	**10.40**	7.22	**9.42**
6" x 4" x 16"	Inst	SF	ML	.133	5.60	3.43	.20	9.23	**11.80**	8.27	**10.70**
6" x 6" x 16"	Inst	SF	ML	.119	2.97	3.07	.18	6.22	**8.23**	5.64	**7.53**
8" x 4" x 16"	Inst	SF	ML	.139	6.97	3.59	.21	10.77	**13.60**	9.60	**12.30**
8" x 6" x 16"	Inst	SF	ML	.123	3.72	3.17	.18	7.07	**9.25**	6.39	**8.42**
12" x 4" x 16"	Inst	SF	ML	.152	11.70	3.92	.23	15.85	**19.60**	14.03	**17.40**
12" x 6" x 16"	Inst	SF	ML	.133	5.96	3.43	.20	9.59	**12.20**	8.57	**11.00**
Concrete slump brick											
Running bond, 3/8" mortar joints											
Walls, with air tools											
4" T	Demo	SF	AB	.062	---	1.46	.19	1.65	**2.38**	1.15	**1.66**
Walls; no scaffolding included											
4" x 4" x 8"	Inst	SF	ML	.145	13.40	3.74	.22	17.36	**21.30**	15.25	**18.90**
4" x 4" x 12"	Inst	SF	ML	.133	6.90	3.43	.20	10.53	**13.30**	9.40	**12.00**
Concrete textured screen block											
Running bond, 3/8" mortar joints											
Walls, with air tools											
4" T	Demo	SF	AB	.048	---	1.13	.15	1.28	**1.85**	.88	**1.27**
Walls; no scaffolding included											
4" x 6" x 6"	Inst	SF	ML	.200	11.90	5.16	.30	17.36	**21.80**	15.47	**19.50**
4" x 8" x 8"	Inst	SF	ML	.133	9.02	3.43	.20	12.65	**15.80**	11.22	**14.10**
4" x 12" x 12"	Inst	SF	ML	.089	2.63	2.30	.13	5.06	**6.62**	4.57	**6.03**
Glass block											
Plain, 4" thick											
6" x 6"	Demo	SF	LB	.091	---	2.00	---	2.00	**3.02**	1.40	**2.12**
8" x 8"	Demo	SF	LB	.082	---	1.80	---	1.80	**2.72**	1.25	**1.89**
12" x 12"	Demo	SF	LB	.074	---	1.62	---	1.62	**2.45**	1.14	**1.72**
6" x 6"	Inst	SF	BK	.284	21.40	7.02	---	28.42	**35.20**	26.66	**33.70**
8" x 8"	Inst	SF	BK	.194	15.40	4.80	---	20.20	**25.00**	18.93	**23.80**
12" x 12"	Inst	SF	BK	.129	19.10	3.19	---	22.29	**26.80**	20.24	**24.60**

Description	Oper	Unit	Crew Size	Man-Hours Per Unit	Avg Mat'l Unit Cost	Avg Labor Unit Cost	Avg Equip Unit Cost	Avg Total Unit Cost	Avg Price Incl O&P	Large Volume Avg Total Unit Cost	Large Volume Avg Price Incl O&P
Glazed tile											
6-T Series. Includes normal allowance for special shapes											
All sizes	Demo	SF	LB	.074	---	1.62	---	1.62	**2.45**	1.14	**1.72**
Glazed 1 side											
2" x 5-1/3" x 12"	Inst	SF	BK	.379	3.60	9.37	---	12.97	**18.30**	10.47	**14.70**
4" x 5-1/3" x 12"	Inst	SF	BK	.398	5.42	9.84	---	15.26	**21.10**	12.42	**17.10**
6" x 5-1/3" x 12"	Inst	SF	BK	.419	7.08	10.40	---	17.48	**23.80**	14.27	**19.30**
8" x 5-1/3" x 12"	Inst	SF	BK	.468	8.43	11.60	---	20.03	**27.20**	16.37	**22.10**
Glazed 2 sides											
4" x 5-1/3" x 12"	Inst	SF	BK	.498	8.18	12.30	---	20.48	**28.00**	16.74	**22.70**
6" x 5-1/3" x 12"	Inst	SF	BK	.531	10.00	13.10	---	23.10	**31.30**	18.93	**25.50**
8" x 5-1/3" x 12"	Inst	SF	BK	.612	12.10	15.10	---	27.20	**36.70**	22.30	**30.00**
Quarry tile											
Floors											
Conventional mortar set	Demo	SF	LB	.043	---	.94	---	.94	**1.43**	.66	**.99**
Dry-set mortar	Demo	SF	LB	.037	---	.81	---	.81	**1.23**	.57	**.86**
Conventional mortar set with unmounted tile											
6" x 6" x 1/2" T	Inst	SF	ML	.164	1.31	4.23	.24	5.78	**8.14**	4.31	**6.03**
9" x 9" x 3/4" T	Inst	SF	ML	.138	1.84	3.56	.21	5.61	**7.70**	4.26	**5.79**
Dry-set mortar with unmounted tile											
6" x 6" x 1/2" T	Inst	SF	ML	.135	4.47	3.48	.20	8.15	**10.60**	6.49	**8.36**
9" x 9" x 3/4" T	Inst	SF	ML	.109	5.01	2.81	.16	7.98	**10.20**	6.44	**8.12**

Molding

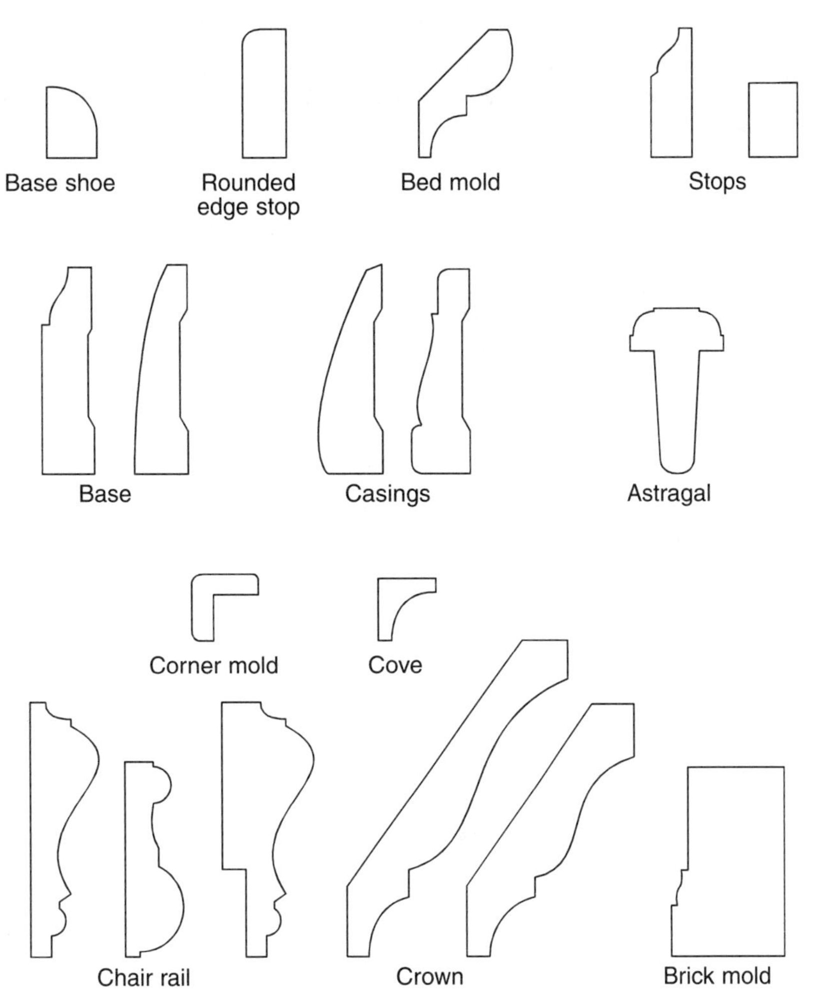

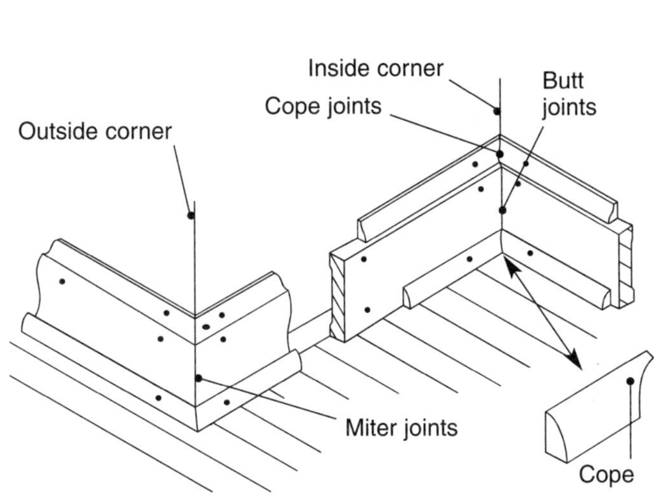

Base molding installation

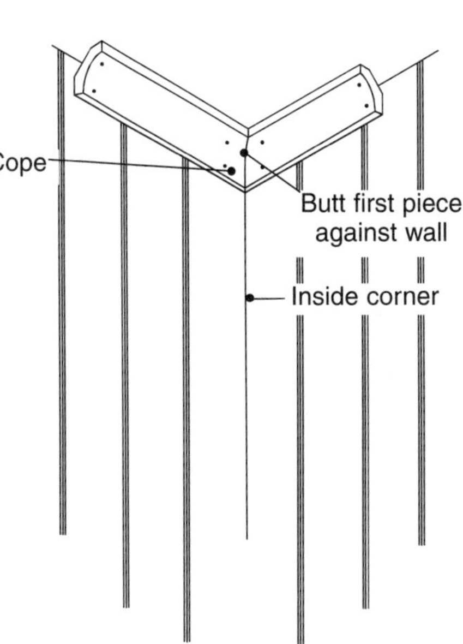

Ceiling molding installation

Description	Oper	Unit	Crew Size	Man-Hours Per Unit	Avg Mat'l Unit Cost	Avg Labor Unit Cost	Avg Equip Unit Cost	Avg Total Unit Cost	Avg Price Incl O&P	Large Volume Avg Total Unit Cost	Large Volume Avg Price Incl O&P
Molding and trim											
Removal											
At base (floor)	Demo	LF	LB	.023	---	.50	---	.50	**.76**	.33	**.50**
At ceiling	Demo	LF	LB	.021	---	.46	---	.46	**.70**	.29	**.43**
On wall or cabinets	Demo	LF	LB	.015	---	.33	---	.33	**.50**	.22	**.33**
Solid pine											
Unfinished											
Casing											
Flat streamline											
1/2" x 1-5/8"	Inst	LF	CA	.046	.86	1.24	---	2.10	**2.87**	1.56	**2.09**
5/8" x 1-5/8"	Inst	LF	CA	.046	.98	1.24	---	2.22	**3.01**	1.66	**2.21**
5/8" x 2-1/2"	Inst	LF	CA	.046	1.33	1.24	---	2.57	**3.41**	1.97	**2.56**
Oval streamline											
1/2" x 1-5/8"	Inst	LF	CA	.046	.92	1.24	---	2.16	**2.94**	1.61	**2.15**
5/8" x 1-5/8"	Inst	LF	CA	.046	.98	1.24	---	2.22	**3.01**	1.66	**2.21**
5/8" x 2-1/2"	Inst	LF	CA	.046	1.44	1.24	---	2.68	**3.54**	2.06	**2.67**
711											
5/8" x 1-5/8"	Inst	LF	CA	.046	.98	1.24	---	2.22	**3.01**	1.66	**2.21**
5/8" x 2-1/2"	Inst	LF	CA	.046	1.44	1.24	---	2.68	**3.54**	2.06	**2.67**
3 step											
5/8" x 1-5/8"	Inst	LF	CA	.046	.87	1.24	---	2.11	**2.89**	1.57	**2.10**
5/8" x 2-1/2"	Inst	LF	CA	.046	1.29	1.24	---	2.53	**3.37**	1.93	**2.52**
2 round edge											
5/8" x 2-1/2"	Inst	LF	CA	.046	.95	1.24	---	2.19	**2.98**	1.64	**2.18**
B & B											
1/2" x 1-5/8"	Inst	LF	CA	.046	.87	1.24	---	2.11	**2.89**	1.57	**2.10**
4 bead											
1/2" x 1-5/8"	Inst	LF	CA	.046	.83	1.24	---	2.07	**2.84**	1.53	**2.06**
1550											
1/2" x 1-5/8"	Inst	LF	CA	.046	.72	1.24	---	1.96	**2.71**	1.44	**1.95**
Mullion											
1/4" x 3-1/2"	Inst	LF	CA	.046	.92	1.24	---	2.16	**2.94**	1.61	**2.15**
Universal											
5/8" x 2-1/2"	Inst	LF	CA	.046	1.02	1.24	---	2.26	**3.06**	1.70	**2.25**
Colonial											
11/16" x 2-1/2"	Inst	LF	CA	.046	1.17	1.24	---	2.41	**3.23**	1.83	**2.40**
11/16" x 3-1/2"	Inst	LF	CA	.046	1.47	1.24	---	2.71	**3.58**	2.09	**2.70**
11/16" x 4-1/2"	Inst	LF	CA	.046	2.70	1.24	---	3.94	**4.99**	3.16	**3.93**
11/16" x 2-7/8"	Inst	LF	CA	.046	1.45	1.24	---	2.69	**3.55**	2.07	**2.68**
356											
11/16" x 2-1/4"	Inst	LF	CA	.046	1.18	1.24	---	2.42	**3.24**	1.84	**2.41**
Back band											
11/16" x 1-1/8"	Inst	LF	CA	.046	.67	1.24	---	1.91	**2.66**	1.39	**1.90**
Cape Cod											
5/8" x 2-1/2"	Inst	LF	CA	.046	1.29	1.24	---	2.53	**3.37**	1.93	**2.52**
11/16" x 3-1/2"	Inst	LF	CA	.046	2.00	1.24	---	3.24	**4.19**	2.55	**3.23**

Description	Oper	Unit	Crew Size	Man-Hours Per Unit	Avg Mat'l Unit Cost	Avg Labor Unit Cost	Avg Equip Unit Cost	Avg Total Unit Cost	Avg Price Incl O&P	Avg Total Unit Cost	Avg Price Incl O&P
			Costs Based On Small Volume							Large Volume	
Rosette plinth block w/ edge detail											
3/4" x 2-3/4"	Inst	Ea	CA	.381	3.51	10.30	---	13.81	**19.70**	9.79	**13.80**
3/4" x 3-1/2"	Inst	Ea	CA	.381	3.88	10.30	---	14.18	**20.10**	10.11	**14.10**
3/4" x 3-3/4"	Inst	Ea	CA	.381	4.49	10.30	---	14.79	**20.80**	10.64	**14.70**
Base & base shoe											
Shoe											
3/8" x 3/4"	Inst	LF	CA	.046	.28	1.24	---	1.52	**2.21**	1.05	**1.51**
Flat streamline											
7/16" x 1-1/2"	Inst	LF	CA	.046	.52	1.24	---	1.76	**2.48**	1.26	**1.75**
Streamline											
7/16" x 1-5/8"	Inst	LF	CA	.046	.81	1.24	---	2.05	**2.82**	1.51	**2.03**
1/2" x 1-5/8"	Inst	LF	CA	.046	.84	1.24	---	2.08	**2.85**	1.54	**2.07**
3/8" x 2-1/4"	Inst	LF	CA	.046	.78	1.24	---	2.02	**2.78**	1.49	**2.01**
7/16" x 2-1/2"	Inst	LF	CA	.046	.83	1.24	---	2.07	**2.84**	1.53	**2.06**
7/16" x 3-1/2"	Inst	LF	CA	.046	1.32	1.24	---	2.56	**3.40**	1.96	**2.55**
1/2" x 2-1/2"	Inst	LF	CA	.046	1.21	1.24	---	2.45	**3.28**	1.86	**2.44**
Combination											
1/2" x 1-5/8"	Inst	LF	CA	.046	.84	1.24	---	2.08	**2.85**	1.54	**2.07**
1/2" x 2-1/2"	Inst	LF	CA	.046	1.24	1.24	---	2.48	**3.31**	1.89	**2.47**
711											
1/2" x 2-1/2"	Inst	LF	CA	.046	1.21	1.24	---	2.45	**3.28**	1.86	**2.44**
7/16" x 3-1/2"	Inst	LF	CA	.046	1.32	1.24	---	2.56	**3.40**	1.96	**2.55**
3 step											
1/2" x 2-1/2"	Inst	LF	CA	.046	.98	1.24	---	2.22	**3.01**	1.66	**2.21**
1/2" x 3-1/2"	Inst	LF	CA	.046	1.41	1.24	---	2.65	**3.51**	2.04	**2.64**
1550											
7/16" x 2-1/2"	Inst	LF	CA	.046	.97	1.24	---	2.21	**3.00**	1.65	**2.20**
Cape Cod											
9/16" x 4"	Inst	LF	CA	.046	1.61	1.24	---	2.85	**3.74**	2.21	**2.84**
Colonial											
11/16" x 3-1/2"	Inst	LF	CA	.046	1.98	1.24	---	3.22	**4.16**	2.53	**3.21**
11/16" x 4-1/2"	Inst	LF	CA	.046	2.24	1.24	---	3.48	**4.46**	2.76	**3.47**
Base cap											
11/16" x 1-3/8"	Inst	LF	CA	.046	.93	1.24	---	2.17	**2.95**	1.62	**2.16**
Rosette plinth block w/ edge detail											
3/4" x 2-3/4" x 6"	Inst	Ea	CA	.381	3.51	10.30	---	13.81	**19.70**	9.79	**13.80**
3/4" x 3-1/2" x 6"	Inst	Ea	CA	.381	3.88	10.30	---	14.18	**20.10**	10.11	**14.10**
3/4" x 3-3/4" x 6"	Inst	Ea	CA	.381	4.49	10.30	---	14.79	**20.80**	10.64	**14.70**
Colonial plinth block w/ edge detail											
1" x 2-3/4" x 6"	Inst	Ea	CA	.381	3.51	10.30	---	13.81	**19.70**	9.79	**13.80**
1" x 3-3/4" x 6"	Inst	Ea	CA	.381	4.49	10.30	---	14.79	**20.80**	10.64	**14.70**
Stops											
Round edge											
3/8" x 1/2"	Inst	LF	CA	.046	.29	1.24	---	1.53	**2.22**	1.06	**1.52**
3/8" x 3/4"	Inst	LF	CA	.046	.35	1.24	---	1.59	**2.29**	1.11	**1.57**
3/8" x 1"	Inst	LF	CA	.046	.46	1.24	---	1.70	**2.41**	1.21	**1.69**
3/8" x 1-1/4"	Inst	LF	CA	.046	.52	1.24	---	1.76	**2.48**	1.26	**1.75**
3/8" x 1-5/8"	Inst	LF	CA	.046	.63	1.24	---	1.87	**2.61**	1.36	**1.86**
1/2" x 3/4"	Inst	LF	CA	.046	.40	1.24	---	1.64	**2.35**	1.16	**1.63**
1/2" x 1"	Inst	LF	CA	.046	.37	1.24	---	1.61	**2.31**	1.13	**1.60**

Description	Oper	Unit	Crew Size	Man-Hours Per Unit	Avg Mat'l Unit Cost	Avg Labor Unit Cost	Avg Equip Unit Cost	Avg Total Unit Cost	Avg Price Incl O&P	Large Volume Avg Total Unit Cost	Large Volume Avg Price Incl O&P
1/2" x 1-1/4"	Inst	LF	CA	.046	.81	1.24	---	2.05	**2.82**	1.51	**2.03**
1/2" x 1-5/8"	Inst	LF	CA	.046	.98	1.24	---	2.22	**3.01**	1.66	**2.21**
Ogee stop											
7/16" x 1-5/8"	Inst	LF	CA	.046	.51	1.24	---	1.75	**2.47**	1.25	**1.74**
Crown & cornice											
Crown											
9/16" x 1-5/8"	Inst	LF	CA	.062	.98	1.67	---	2.65	**3.67**	1.93	**2.62**
9/16" x 2-1/4"	Inst	LF	CA	.062	1.09	1.67	---	2.76	**3.79**	2.03	**2.73**
9/16" x 3-1/2"	Inst	LF	CA	.062	1.73	1.67	---	3.40	**4.53**	2.58	**3.36**
9/16" x 4-1/4"	Inst	LF	CA	.062	2.15	1.67	---	3.82	**5.01**	2.95	**3.79**
9/16" x 5-1/2"	Inst	LF	CA	.062	3.45	1.67	---	5.12	**6.51**	4.08	**5.09**
Colonial crown											
9/16" x 3-3/8"	Inst	LF	CA	.062	1.36	1.67	---	3.03	**4.10**	2.26	**3.00**
9/16" x 4-1/2"	Inst	LF	CA	.062	1.37	1.67	---	3.04	**4.12**	2.27	**3.01**
Cornice											
1-1/8" x 4-1/2"	Inst	LF	CA	.062	5.69	1.67	---	7.36	**9.08**	6.03	**7.33**
1-1/4" x 6"	Inst	LF	CA	.062	6.21	1.67	---	7.88	**9.68**	6.48	**7.85**
Chair rail											
Chair rail											
1/2" x 1-1/2"	Inst	LF	CA	.046	.82	1.24	---	2.06	**2.83**	1.52	**2.05**
9/16" x 2-1/2"	Inst	LF	CA	.046	1.79	1.24	---	3.03	**3.94**	2.37	**3.02**
Colonial chair rail											
11/16" x 2-1/2"	Inst	LF	CA	.046	1.38	1.24	---	2.62	**3.47**	2.01	**2.61**
Outside corner guard											
3/4" x 3/4"	Inst	LF	CA	.046	.60	1.24	---	1.84	**2.58**	1.33	**1.83**
1" x 1"	Inst	LF	CA	.046	.99	1.24	---	2.23	**3.02**	1.67	**2.22**
1-3/8" x 1-3/8"	Inst	LF	CA	.046	1.22	1.24	---	2.46	**3.29**	1.87	**2.45**
1-1/8" x 1-1/8"	Inst	LF	CA	.046	1.48	1.24	---	2.72	**3.59**	2.10	**2.71**
Astragal											
Flat astragal											
3/4" x 1-5/8"	Inst	LF	CA	.046	1.31	1.24	---	2.55	**3.39**	1.95	**2.54**
T astragal											
1-1/4" x 2-1/4"	Inst	LF	CA	.046	2.13	1.24	---	3.37	**4.33**	2.66	**3.36**
Cap											
Panel cap											
1/2" x 1-1/2"	Inst	LF	CA	.046	.81	1.24	---	2.05	**2.82**	1.51	**2.03**
Wainscot cap											
5/8" x 1-1/4"	Inst	LF	CA	.046	.81	1.24	---	2.05	**2.82**	1.51	**2.03**
Glass bead & screen											
Insert											
3/16" x 1/2"	Inst	LF	CA	.046	.29	1.24	---	1.53	**2.22**	1.06	**1.52**
Flat											
1/4" x 3/4"	Inst	LF	CA	.046	.29	1.24	---	1.53	**2.22**	1.06	**1.52**
Beaded											
1/4" x 3/4"	Inst	LF	CA	.046	.29	1.24	---	1.53	**2.22**	1.06	**1.52**
Scribe											
3/16" x 3/4"	Inst	LF	CA	.046	.17	1.24	---	1.41	**2.08**	.96	**1.40**

Description	Oper	Unit	Crew Size	Man-Hours Per Unit	Avg Mat'l Unit Cost	Avg Labor Unit Cost	Avg Equip Unit Cost	Avg Total Unit Cost	Avg Price Incl O&P	Large Volume Avg Total Unit Cost	Large Volume Avg Price Incl O&P
			Costs Based On Small Volume							Large Volume	
Cloverleaf											
3/8" x 3/4"	Inst	LF	CA	.046	.35	1.24	---	1.59	**2.29**	1.11	**1.57**
Glass bead											
1/4" x 7/16"	Inst	LF	CA	.046	.16	1.24	---	1.40	**2.07**	.95	**1.39**
3/8" x 3/8"	Inst	LF	CA	.046	.16	1.24	---	1.40	**2.07**	.95	**1.39**
Quarter round											
1/4" x 1/4"	Inst	LF	CA	.046	.23	1.24	---	1.47	**2.15**	1.01	**1.46**
3/8" x 3/8"	Inst	LF	CA	.046	.29	1.24	---	1.53	**2.22**	1.06	**1.52**
1/2" x 1/2"	Inst	LF	CA	.046	.35	1.24	---	1.59	**2.29**	1.11	**1.57**
5/8" x 5/8"	Inst	LF	CA	.046	.40	1.24	---	1.64	**2.35**	1.16	**1.63**
3/4" x 3/4"	Inst	LF	CA	.046	.52	1.24	---	1.76	**2.48**	1.26	**1.75**
1" x 1"	Inst	LF	CA	.046	1.04	1.24	---	2.28	**3.08**	1.71	**2.26**
Cove											
Cove											
3/8" x 3/8"	Inst	LF	CA	.062	.29	1.67	---	1.96	**2.87**	1.33	**1.93**
1/2" x 1/2"	Inst	LF	CA	.062	.35	1.67	---	2.02	**2.94**	1.38	**1.98**
5/8" x 5/8"	Inst	LF	CA	.062	.52	1.67	---	2.19	**3.14**	1.53	**2.16**
3/4" x 3/4"	Inst	LF	CA	.062	.52	1.67	---	2.19	**3.14**	1.53	**2.16**
1" x 1"	Inst	LF	CA	.062	1.04	1.67	---	2.71	**3.74**	1.98	**2.67**
Sprung cove											
5/8" x 1-5/8"	Inst	LF	CA	.062	1.00	1.67	---	2.67	**3.69**	1.95	**2.64**
5/8" x 2-1/4"	Inst	LF	CA	.062	1.22	1.67	---	2.89	**3.94**	2.14	**2.86**
Linoleum cove											
7/16" x 1-1/4"	Inst	LF	CA	.062	.59	1.67	---	2.26	**3.22**	1.59	**2.23**
Half round											
1/4" x 1/2"	Inst	LF	CA	.046	.29	1.24	---	1.53	**2.22**	1.06	**1.52**
5/16" x 5/8"	Inst	LF	CA	.046	.29	1.24	---	1.53	**2.22**	1.06	**1.52**
3/8" x 3/4"	Inst	LF	CA	.046	.35	1.24	---	1.59	**2.29**	1.11	**1.57**
1/2" x 1"	Inst	LF	CA	.046	.63	1.24	---	1.87	**2.61**	1.36	**1.86**
Lattice - batts - S4S											
Lattice											
5/16" x 1-1/8"	Inst	LF	CA	.046	.41	1.24	---	1.65	**2.36**	1.17	**1.64**
5/16" x 1-1/4"	Inst	LF	CA	.046	.46	1.24	---	1.70	**2.41**	1.21	**1.69**
5/16" x 1-5/8"	Inst	LF	CA	.046	.52	1.24	---	1.76	**2.48**	1.26	**1.75**
Batts											
5/16" x 2-1/2"	Inst	LF	CA	.046	.75	1.24	---	1.99	**2.75**	1.46	**1.98**
5/16" x 3-1/2"	Inst	LF	CA	.046	1.09	1.24	---	2.33	**3.14**	1.76	**2.32**
S4S											
1/2" x 1/2"	Inst	LF	CA	.046	.35	1.24	---	1.59	**2.29**	1.11	**1.57**
1/2" x 3/4"	Inst	LF	CA	.046	.40	1.24	---	1.64	**2.35**	1.16	**1.63**
1/2" x 1-1/2"	Inst	LF	CA	.046	.83	1.24	---	2.07	**2.84**	1.53	**2.06**
1/2" x 2-1/2"	Inst	LF	CA	.046	1.17	1.24	---	2.41	**3.23**	1.83	**2.40**
1/2" x 3-1/2"	Inst	LF	CA	.046	1.62	1.24	---	2.86	**3.75**	2.22	**2.85**
1/2" x 5-1/2"	Inst	LF	CA	.046	2.84	1.24	---	4.08	**5.15**	3.28	**4.07**
3/4" x 3/4"	Inst	LF	CA	.046	.58	1.24	---	1.82	**2.55**	1.31	**1.80**

Description	Oper	Unit	Crew Size	Man-Hours Per Unit	Avg Mat'l Unit Cost	Avg Labor Unit Cost	Avg Equip Unit Cost	Avg Total Unit Cost	Avg Price Incl O&P	Large Volume Avg Total Unit Cost	Large Volume Avg Price Incl O&P
11/16" x 1-1/4"	Inst	LF	CA	.046	.92	1.24	---	2.16	**2.94**	1.61	**2.15**
11/16" x 1-5/8"	Inst	LF	CA	.046	1.14	1.24	---	2.38	**3.20**	1.80	**2.37**
11/16" x 2-1/2"	Inst	LF	CA	.046	1.46	1.24	---	2.70	**3.56**	2.08	**2.69**
11/16" x 3-1/2"	Inst	LF	CA	.046	2.00	1.24	---	3.24	**4.19**	2.55	**3.23**
1" x 1"	Inst	LF	CA	.046	1.04	1.24	---	2.28	**3.08**	1.71	**2.26**
1-9/16" x 1-9/16"	Inst	LF	CA	.046	1.97	1.24	---	3.21	**4.15**	2.52	**3.20**
S4S 4EE											
1-1/2" x 1-1/2"	Inst	LF	CA	.046	1.91	1.24	---	3.15	**4.08**	2.47	**3.14**
Window stool											
Flat stool											
11/16" x 5-1/2"	Inst	LF	CA	.046	3.57	1.24	---	4.81	**5.99**	3.91	**4.79**
11/16" x 4-3/4"	Inst	LF	CA	.046	3.28	1.24	---	4.52	**5.66**	3.66	**4.51**
11/16" x 7-1/4"	Inst	LF	CA	.046	5.31	1.24	---	6.55	**7.99**	5.43	**6.54**
Rabbeted stool											
7/8" x 2-1/2"	Inst	LF	CA	.046	2.55	1.24	---	3.79	**4.82**	3.03	**3.78**
7/8" x 3-1/2"	Inst	LF	CA	.046	3.57	1.24	---	4.81	**5.99**	3.91	**4.79**
7/8" x 5-1/2"	Inst	LF	CA	.046	4.31	1.24	---	5.55	**6.84**	4.56	**5.54**
Panel & decorative molds											
Panel											
3/8" x 5/8"	Inst	LF	CA	.046	.52	1.24	---	1.76	**2.48**	1.26	**1.75**
5/8" x 3/4"	Inst	LF	CA	.046	.77	1.24	---	2.01	**2.77**	1.48	**2.00**
3/4" x 1"	Inst	LF	CA	.046	1.12	1.24	---	2.36	**3.17**	1.78	**2.34**
Panel or brick											
1-1/16" x 1-1/2"	Inst	LF	CA	.046	1.45	1.24	---	2.69	**3.55**	2.07	**2.68**
Panel											
3/8" x 1-1/8"	Inst	LF	CA	.046	.77	1.24	---	2.01	**2.77**	1.48	**2.00**
3/4" x 1-5/8"	Inst	LF	CA	.046	1.46	1.24	---	2.70	**3.56**	2.08	**2.69**
9/16" x 1-3/8"	Inst	LF	CA	.046	.74	1.24	---	1.98	**2.74**	1.45	**1.97**
Beauty M-340											
3/4" x 1-5/8"	Inst	LF	CA	.046	1.10	1.24	---	2.34	**3.15**	1.77	**2.33**
Beauty M-400											
3/4" x 1-7/8"	Inst	LF	CA	.046	1.74	1.24	---	2.98	**3.89**	2.32	**2.97**
Decorative											
1-3/8" x 3-1/4"	Inst	LF	CA	.046	4.36	1.24	---	5.60	**6.90**	4.60	**5.59**
Base cap											
11/16" x 1-3/8"	Inst	LF	CA	.046	1.27	1.24	---	2.51	**3.35**	1.91	**2.49**
M340 corner arcs											
3/4" x 1-5/8"	Inst	LF	CA	.381	2.76	10.30	---	13.06	**18.80**	9.14	**13.00**
Miscellaneous											
Band mold											
3/4" x 1-5/8"	Inst	LF	CA	.046	.98	1.24	---	2.22	**3.01**	1.66	**2.21**
Parting bead											
3/8" x 3/4"	Inst	LF	CA	.046	1.35	1.24	---	2.59	**3.44**	1.98	**2.57**
Apron											
3/8" x 1-1/2"	Inst	LF	CA	.046	.69	1.24	---	1.93	**2.68**	1.41	**1.92**
Flat Hoffco											
3/8" x 1"	Inst	LF	CA	.046	.48	1.24	---	1.72	**2.44**	1.23	**1.71**
Drip											
3/4" x 1-1/4"	Inst	LF	CA	.046	1.00	1.24	---	2.24	**3.04**	1.68	**2.23**

Description	Oper	Unit	Crew Size	Man-Hours Per Unit	Avg Mat'l Unit Cost	Avg Labor Unit Cost	Avg Equip Unit Cost	Avg Total Unit Cost	Avg Price Incl O&P	Large Volume Avg Total Unit Cost	Large Volume Avg Price Incl O&P
1139 picture											
3/4" x 1-5/8"	Inst	LF	CA	.046	1.07	1.24	---	2.31	**3.12**	1.74	**2.30**
Chamfer											
1/2" x 1/2"	Inst	LF	CA	.046	.26	1.24	---	1.50	**2.18**	1.04	**1.49**
3/4" x 3/4"	Inst	LF	CA	.046	.35	1.24	---	1.59	**2.29**	1.11	**1.57**
1" x 1"	Inst	LF	CA	.046	.76	1.24	---	2.00	**2.76**	1.47	**1.99**
Nose & cove											
1" x 1-5/8"	Inst	LF	CA	.046	1.79	1.24	---	3.03	**3.94**	2.37	**3.02**
Cabinet crown											
5/16" x 1-1/4"	Inst	LF	CA	.046	.38	1.24	---	1.62	**2.32**	1.14	**1.61**

Finger joint pine

Paint grade

Casing

Description	Oper	Unit	Crew Size	Man-Hours Per Unit	Avg Mat'l Unit Cost	Avg Labor Unit Cost	Avg Equip Unit Cost	Avg Total Unit Cost	Avg Price Incl O&P	Large Volume Avg Total Unit Cost	Large Volume Avg Price Incl O&P
Prefit											
9/16" x 1-1/2"	Inst	LF	CA	.046	.38	1.24	---	1.62	**2.32**	1.14	**1.61**
Flat streamline											
1/2" x 1-5/8"	Inst	LF	CA	.046	.51	1.24	---	1.75	**2.47**	1.25	**1.74**
5/8" x 1-5/8"	Inst	LF	CA	.046	.52	1.24	---	1.76	**2.48**	1.26	**1.75**
11/16" x 2-1/4"	Inst	LF	CA	.046	.74	1.24	---	1.98	**2.74**	1.45	**1.97**
711											
5/8" x 1-5/8"	Inst	LF	CA	.046	.53	1.24	---	1.77	**2.49**	1.27	**1.76**
5/8" x 2-1/2"	Inst	LF	CA	.046	.79	1.24	---	2.03	**2.79**	1.50	**2.02**
Monterey											
5/8" x 2-1/2"	Inst	LF	CA	.046	.85	1.24	---	2.09	**2.86**	1.55	**2.08**
3 step											
5/8" x 2-1/2"	Inst	LF	CA	.046	.79	1.24	---	2.03	**2.79**	1.50	**2.02**
Universal											
5/8" x 2-1/2"	Inst	LF	CA	.046	.81	1.24	---	2.05	**2.82**	1.51	**2.03**
Colonial											
9/16" x 4-1/4"	Inst	LF	CA	.046	1.33	1.24	---	2.57	**3.41**	1.97	**2.56**
11/16" x 2-7/8"	Inst	LF	CA	.046	.95	1.24	---	2.19	**2.98**	1.64	**2.18**
Clam shell											
11/16" x 2"	Inst	LF	CA	.046	.64	1.24	---	1.88	**2.62**	1.37	**1.87**
356											
11/16" x 2-1/4"	Inst	LF	CA	.046	.70	1.24	---	1.94	**2.69**	1.42	**1.93**
9/16" x 2"	Inst	LF	CA	.046	.64	1.24	---	1.88	**2.62**	1.37	**1.87**
WM 366											
11/16" x 2-1/4"	Inst	LF	CA	.046	.70	1.24	---	1.94	**2.69**	1.42	**1.93**
3 fluted											
1/2" x 3-1/4"	Inst	LF	CA	.046	1.01	1.24	---	2.25	**3.05**	1.69	**2.24**
5 fluted											
1/2" x 3-1/2"	Inst	LF	CA	.046	1.08	1.24	---	2.32	**3.13**	1.75	**2.31**
Cape Cod											
5/8" x 2-1/2"	Inst	LF	CA	.046	.84	1.24	---	2.08	**2.85**	1.54	**2.07**
11/16" x 3-1/2"	Inst	LF	CA	.046	1.40	1.24	---	2.64	**3.50**	2.03	**2.63**
Chesapeake											
3/4" x 3-1/2"	Inst	LF	CA	.046	1.41	1.24	---	2.65	**3.51**	2.04	**2.64**
Coronado											
1-1/4" x 2-1/2"	Inst	LF	CA	.046	1.66	1.24	---	2.90	**3.79**	2.25	**2.89**

			Costs Based On Small Volume							Large Volume	
Description	Oper	Unit	Crew Size	Man-Hours Per Unit	Avg Mat'l Unit Cost	Avg Labor Unit Cost	Avg Equip Unit Cost	Avg Total Unit Cost	Avg Price Incl O&P	Avg Total Unit Cost	Avg Price Incl O&P
Bell court											
1-1/4" x 2-1/2"	Inst	LF	CA	.046	1.66	1.24	---	2.90	**3.79**	2.25	**2.89**
Hermosa casing											
1-1/4" x 3-1/4"	Inst	LF	CA	.046	2.24	1.24	---	3.48	**4.46**	2.76	**3.47**
Cambridge											
11/16" x 3-1/4"	Inst	LF	CA	.046	1.38	1.24	---	2.62	**3.47**	2.01	**2.61**
275											
5/8" x 3"	Inst	LF	CA	.046	1.04	1.24	---	2.28	**3.08**	1.71	**2.26**
Stops											
2 round edge											
3/8" x 1-1/4"	Inst	LF	CA	.046	.30	1.24	---	1.54	**2.23**	1.07	**1.53**
Round edge											
1/2" x 1-1/4"	Inst	LF	CA	.046	.43	1.24	---	1.67	**2.38**	1.18	**1.65**
1/2" x 1-1/2"	Inst	LF	CA	.046	.55	1.24	---	1.79	**2.52**	1.29	**1.78**
Ogee											
1/2" x 1-3/4"	Inst	LF	CA	.046	.59	1.24	---	1.83	**2.56**	1.32	**1.82**
Base											
Base shoe											
3/8" x 3/4"	Inst	LF	CA	.046	.28	1.24	---	1.52	**2.21**	1.05	**1.51**
Flat											
7/16" x 1-1/2"	Inst	LF	CA	.046	.35	1.24	---	1.59	**2.29**	1.11	**1.57**
Streamline or reversible											
3/8" x 2-1/4"	Inst	LF	CA	.046	.46	1.24	---	1.70	**2.41**	1.21	**1.69**
Streamline											
7/16" x 2-1/2"	Inst	LF	CA	.046	.56	1.24	---	1.80	**2.53**	1.30	**1.79**
7/16" x 3-1/2"	Inst	LF	CA	.046	.86	1.24	---	2.10	**2.87**	1.56	**2.09**
711											
7/16" x 2-1/2"	Inst	LF	CA	.062	.60	1.67	---	2.27	**3.23**	1.60	**2.24**
7/16" x 3-1/2"	Inst	LF	CA	.062	.81	1.67	---	2.48	**3.47**	1.78	**2.44**
3 step											
7/16" x 2-1/2"	Inst	LF	CA	.062	.66	1.67	---	2.33	**3.30**	1.65	**2.29**
3 step base											
1/2" x 4-1/2"	Inst	LF	CA	.062	1.48	1.67	---	3.15	**4.24**	2.37	**3.12**
WM 623											
9/16" x 3-1/4"	Inst	LF	CA	.062	1.05	1.67	---	2.72	**3.75**	1.99	**2.69**
WM 618											
9/16" x 5-1/4"	Inst	LF	CA	.062	1.62	1.67	---	3.29	**4.40**	2.49	**3.26**
Monterey											
7/16" x 4-1/4"	Inst	LF	CA	.062	1.07	1.67	---	2.74	**3.77**	2.01	**2.71**
Cape Cod											
9/16" x 4"	Inst	LF	CA	.062	1.35	1.67	---	3.02	**4.09**	2.25	**2.98**
9/16" x 5-1/4"	Inst	LF	CA	.062	1.90	1.67	---	3.57	**4.73**	2.73	**3.54**
"B" Cape Cod base											
9/16" x 5-1/4"	Inst	LF	CA	.062	1.90	1.67	---	3.57	**4.73**	2.73	**3.54**

Description	Oper	Unit	Crew Size	Man-Hours Per Unit	Avg Mat'l Unit Cost	Avg Labor Unit Cost	Avg Equip Unit Cost	Avg Total Unit Cost	Avg Price Incl O&P	Large Volume Avg Total Unit Cost	Large Volume Avg Price Incl O&P
356 base											
1/2" x 2-1/2"	Inst	LF	CA	.046	.76	1.24	---	2.00	**2.76**	1.47	**1.99**
1/2" x 3-1/4"	Inst	LF	CA	.046	.99	1.24	---	2.23	**3.02**	1.67	**2.22**
175 base											
9/16" x 4-1/4"	Inst	LF	CA	.062	1.39	1.67	---	3.06	**4.14**	2.29	**3.03**
Colonial											
9/16" x 4-1/4"	Inst	LF	CA	.062	2.04	1.67	---	3.71	**4.89**	2.85	**3.67**
Crown & cornice											
Colonial crown											
9/16" x 2-1/4"	Inst	LF	CA	.062	.74	1.67	---	2.41	**3.39**	1.72	**2.38**
9/16" x 3-3/8"	Inst	LF	CA	.062	1.18	1.67	---	2.85	**3.90**	2.11	**2.82**
9/16" x 4-1/2"	Inst	LF	CA	.062	1.62	1.67	---	3.29	**4.40**	2.49	**3.26**
Cornice											
1-1/4" x 5-7/8"	Inst	LF	CA	.062	4.57	1.67	---	6.24	**7.80**	5.05	**6.20**
13/16" x 4-5/8"	Inst	LF	CA	.062	2.97	1.67	---	4.64	**5.96**	3.66	**4.61**
Crown											
3/4" x 6"	Inst	LF	CA	.062	3.11	1.67	---	4.78	**6.12**	3.78	**4.74**
Cornice											
1-1/8" x 3-1/2"	Inst	LF	CA	.062	2.24	1.67	---	3.91	**5.12**	3.03	**3.88**
Georgian crown											
11/16" x 4-1/4"	Inst	LF	CA	.062	1.81	1.67	---	3.48	**4.62**	2.65	**3.44**
Cove											
11/16" x 11/16"	Inst	LF	CA	.046	.36	1.24	---	1.60	**2.30**	1.12	**1.59**
Chair rail											
Chair rail											
9/16" x 2-1/2"	Inst	LF	CA	.046	.83	1.24	---	2.07	**2.84**	1.53	**2.06**
Colonial chair rail											
3/4" x 3"	Inst	LF	CA	.046	1.25	1.24	---	2.49	**3.32**	1.90	**2.48**
Exterior mold											
Stucco mold											
7/8" x 1-1/4"	Inst	LF	CA	.046	.66	1.24	---	1.90	**2.64**	1.38	**1.88**
WM 180 eastern brick											
1-1/4" x 2"	Inst	LF	CA	.046	1.41	1.24	---	2.65	**3.51**	2.04	**2.64**
Stool & apron											
Flat stool											
11/16" x 5-1/2"	Inst	LF	CA	.046	2.76	1.24	---	4.00	**5.06**	3.21	**3.99**
11/16" x 4-3/4"	Inst	LF	CA	.046	2.07	1.24	---	3.31	**4.27**	2.61	**3.30**
11/16" x 7-1/4"	Inst	LF	CA	.046	3.45	1.24	---	4.69	**5.85**	3.81	**4.68**
Apron											
3/8" x 1-1/2"	Inst	LF	CA	.046	.43	1.24	---	1.67	**2.38**	1.18	**1.65**

Description	Oper	Unit	Crew Size	Man-Hours Per Unit	Avg Mat'l Unit Cost	Avg Labor Unit Cost	Avg Equip Unit Cost	Avg Total Unit Cost	Avg Price Incl O&P	Large Volume Avg Total Unit Cost	Large Volume Avg Price Incl O&P
Oak											
Hardwood											
Casing											
Streamline casing											
5/8" x 1-5/8"	Inst	LF	CA	.046	1.52	1.24	---	2.76	**3.63**	2.13	**2.75**
711 casing											
1/2" x 1-5/8"	Inst	LF	CA	.046	1.44	1.24	---	2.68	**3.54**	2.06	**2.67**
5/8" x 2-3/8"	Inst	LF	CA	.046	1.79	1.24	---	3.03	**3.94**	2.37	**3.02**
Universal											
5/8" x 2-1/2"	Inst	LF	CA	.046	2.16	1.24	---	3.40	**4.37**	2.69	**3.39**
Colonial											
11/16" x 3-1/2"	Inst	LF	CA	.046	2.33	1.24	---	3.57	**4.56**	2.84	**3.56**
11/16" x 2-7/8"	Inst	LF	CA	.046	1.67	1.24	---	2.91	**3.81**	2.26	**2.90**
11/16" x 4-1/4"	Inst	LF	CA	.046	2.58	1.24	---	3.82	**4.85**	3.05	**3.81**
Mull casing											
5/16" x 3-1/4"	Inst	LF	CA	.046	1.98	1.24	---	3.22	**4.16**	2.53	**3.21**
356 casing											
9/16" x 2-1/4"	Inst	LF	CA	.046	1.21	1.24	---	2.45	**3.28**	1.86	**2.44**
Cape Cod											
5/8" x 2-1/2"	Inst	LF	CA	.046	1.28	1.24	---	2.52	**3.36**	1.92	**2.51**
Fluted											
1/2" x 3-1/2"	Inst	LF	CA	.046	1.82	1.24	---	3.06	**3.98**	2.39	**3.05**
Victorian											
11/16" x 4-1/4"	Inst	LF	CA	.046	5.00	1.24	---	6.24	**7.64**	5.16	**6.23**
Cambridge											
11/16" x 3-1/4"	Inst	LF	CA	.046	2.93	1.24	---	4.17	**5.25**	3.36	**4.16**
Rosette plinth block w/ edge detail											
3/4" x 2-3/4"	Inst	Ea	CA	.381	4.31	10.30	---	14.61	**20.60**	10.49	**14.60**
3/4" x 3-3/4"	Inst	Ea	CA	.381	5.18	10.30	---	15.48	**21.60**	11.24	**15.40**
Base											
Base shoe											
1/2" x 3/4"	Inst	LF	CA	.046	.46	1.24	---	1.70	**2.41**	1.21	**1.69**
Colonial											
11/16" x 4-1/4"	Inst	LF	CA	.046	3.05	1.24	---	4.29	**5.39**	3.46	**4.28**
Streamline base											
1/2" x 2-1/2"	Inst	LF	CA	.046	1.16	1.24	---	2.40	**3.22**	1.82	**2.39**
1/2" x 3-1/2"	Inst	LF	CA	.046	2.31	1.24	---	3.55	**4.54**	2.82	**3.54**
711											
1/2" x 2-1/2"	Inst	LF	CA	.046	1.21	1.24	---	2.45	**3.28**	1.86	**2.44**
1/2" x 3-1/2"	Inst	LF	CA	.046	1.61	1.24	---	2.85	**3.74**	2.21	**4.06**
Cape Cod base											
7/16" x 4-1/4"	Inst	LF	CA	.046	2.83	1.24	---	4.07	**5.14**	3.27	**4.06**
Monterey base											
7/16" x 4-1/4"	Inst	LF	CA	.046	2.83	1.24	---	4.07	**5.14**	3.27	**4.06**
Rosette plinth block w/ edge detail											
3/4" x 2-3/4"	Inst	Ea	CA	.381	6.64	10.30	---	16.94	**23.30**	12.51	**16.90**
3/4" x 3-1/2"	Inst	Ea	CA	.381	7.94	10.30	---	18.24	**24.70**	13.64	**18.20**

Description	Oper	Unit	Crew Size	Man-Hours Per Unit	Avg Mat'l Unit Cost	Avg Labor Unit Cost	Avg Equip Unit Cost	Avg Total Unit Cost	Avg Price Incl O&P	Large Volume Avg Total Unit Cost	Large Volume Avg Price Incl O&P
Colonial plinth block w/ edge detail											
1" x 2-3/4" x 6"	Inst	Ea	CA	.381	5.00	10.30	---	15.30	**21.40**	11.09	**15.30**
1" x 3-3/4" x 6"	Inst	Ea	CA	.381	6.12	10.30	---	16.42	**22.70**	12.06	**16.40**
Stops											
Round edge											
3/8" x 1-1/4"	Inst	LF	CA	.046	.90	1.24	---	2.14	**2.92**	1.59	**2.13**
Ogee											
1/2" x 1-3/4"	Inst	LF	CA	.046	1.67	1.24	---	2.91	**3.81**	2.26	**2.90**
Crown & cornice											
Crown											
1/2" x 1-5/8"	Inst	LF	CA	.062	1.41	1.67	---	3.08	**4.16**	2.31	**3.05**
1/2" x 2-1/4"	Inst	LF	CA	.062	1.98	1.67	---	3.65	**4.82**	2.80	**3.62**
1/2" x 3-1/2"	Inst	LF	CA	.062	2.84	1.67	---	4.51	**5.81**	3.55	**4.48**
Colonial crown											
5/8" x 3-1/2"	Inst	LF	CA	.062	2.93	1.67	---	4.60	**5.91**	3.63	**4.57**
5/8" x 4-1/2"	Inst	LF	CA	.062	4.49	1.67	---	6.16	**7.70**	4.98	**6.12**
Cornice											
1-1/4" x 6"	Inst	LF	CA	.062	9.14	1.67	---	10.81	**13.10**	9.03	**10.80**
13/16" x 4-5/8"	Inst	LF	CA	.062	4.57	1.67	---	6.24	**7.80**	5.05	**6.20**
Quarter round											
1/2" x 1/2"	Inst	LF	CA	.046	.72	1.24	---	1.96	**2.71**	1.44	**1.95**
3/4 x 3/4"	Inst	LF	CA	.046	1.28	1.24	---	2.52	**3.36**	1.92	**2.51**
Cove											
1/2" x 1/2"	Inst	LF	CA	.062	.72	1.67	---	2.39	**3.37**	1.71	**2.36**
3/4 x 3/4"	Inst	LF	CA	.062	1.28	1.67	---	2.95	**4.01**	2.19	**2.92**
Half round											
3/8" x 3/4"	Inst	LF	CA	.046	.77	1.24	---	2.01	**2.77**	1.48	**2.00**
Screen mold											
Flat screen											
3/8" x 3/4"	Inst	LF	CA	.046	.81	1.24	---	2.05	**2.82**	1.51	**2.03**
Scribe											
3/16" x 3/4"	Inst	LF	CA	.046	.70	1.24	---	1.94	**2.69**	1.42	**1.93**
Cloverleaf											
3/8" x 3/4"	Inst	LF	CA	.046	.97	1.24	---	2.21	**3.00**	1.65	**2.20**
Outside corner guard											
3/4" x 3/4"	Inst	LF	CA	.046	1.29	1.24	---	2.53	**3.37**	1.93	**2.52**
1" x 1"	Inst	LF	CA	.046	1.48	1.24	---	2.72	**3.59**	2.10	**2.71**
1-1/8" x 1-1/8"	Inst	LF	CA	.046	2.21	1.24	---	3.45	**4.43**	2.73	**3.44**
Chair rail											
Chair rail											
5/8" x 2-1/2"	Inst	LF	CA	.046	2.32	1.24	---	3.56	**4.55**	2.83	**3.55**
Colonial chair rail											
5/8" x 2-1/2"	Inst	LF	CA	.046	2.76	1.24	---	4.00	**5.06**	3.21	**3.99**
Victorian											
11/16" x 4-1/4"	Inst	LF	CA	.046	6.18	1.24	---	7.42	**8.99**	6.18	**7.40**

Description	Oper	Unit	Crew Size	Man-Hours Per Unit	Avg Mat'l Unit Cost	Avg Labor Unit Cost	Avg Equip Unit Cost	Avg Total Unit Cost	Avg Price Incl O&P	Large Volume Avg Total Unit Cost	Large Volume Avg Price Incl O&P
Panel cap											
1/2" x 1-1/2"	Inst	LF	CA	.046	1.59	1.24	---	2.83	**3.71**	2.19	**2.82**
Astragal											
Flat astragal											
3/4" x 1-5/8"	Inst	LF	CA	.046	2.31	1.24	---	3.55	**4.54**	2.82	**3.54**
1-3/4 T astragal											
1-1/4" x 2-1/4"	Inst	LF	CA	.046	8.45	1.24	---	9.69	**11.60**	8.16	**9.68**
T astragal mahogany											
1-1/4" x 2-1/4" (1-3/8" door)	Inst	LF	CA	.046	4.30	1.24	---	5.54	**6.83**	4.55	**5.53**
Nose & cove											
1" x 1-5/8"	Inst	LF	CA	.062	2.84	1.67	---	4.51	**5.81**	3.55	**4.48**
Hand rail											
Hand rail											
1-1/2" x 2-1/2"	Inst	LF	CA	.046	5.00	1.24	---	6.24	**7.64**	5.16	**6.23**
Plowed 1/4" x 1-1/4"											
1-1/2" x 2-1/2"	Inst	LF	CA	.046	5.52	1.24	---	6.76	**8.23**	5.61	**6.75**
Panel mold											
Panel mold											
3/8" x 5/8"	Inst	LF	CA	.046	.83	1.24	---	2.07	**2.84**	1.53	**2.06**
5/8" x 3/4"	Inst	LF	CA	.046	1.18	1.24	---	2.42	**3.24**	1.84	**2.41**
3/4" x 1"	Inst	LF	CA	.046	1.63	1.24	---	2.87	**3.76**	2.23	**2.86**
3/4" x 1-5/8"	Inst	LF	CA	.046	2.38	1.24	---	3.62	**4.62**	2.88	**3.61**
Beauty mold											
9/16" x 1-3/8"	Inst	LF	CA	.046	1.98	1.24	---	3.22	**4.16**	2.53	**3.21**
3/4" x 1-7/8"	Inst	LF	CA	.046	3.04	1.24	---	4.28	**5.38**	3.45	**4.27**
Raised panel											
1-1/8" x 1-1/4"	Inst	LF	CA	.046	2.38	1.24	---	3.62	**4.62**	2.88	**3.61**
Full round											
1-3/8"	Inst	LF	CA	.046	3.36	1.24	---	4.60	**5.75**	3.73	**4.59**
S4S											
1 x 4	Inst	LF	CA	.062	4.57	1.67	---	6.24	**7.80**	5.05	**6.20**
1 x 6	Inst	LF	CA	.062	6.90	1.67	---	8.57	**10.50**	7.08	**8.54**
1 x 8	Inst	LF	CA	.062	9.14	1.67	---	10.81	**13.10**	9.03	**10.80**
1 x 10	Inst	LF	CA	.062	11.80	1.67	---	13.47	**16.10**	11.38	**13.50**
1 x 12	Inst	LF	CA	.062	14.70	1.67	---	16.37	**19.40**	13.88	**16.30**
Oak bar nosing											
1-1/4" x 3-1/2"	Inst	LF	CA	.046	8.28	1.24	---	9.52	**11.40**	8.01	**9.51**
1-5/8" x 5"	Inst	LF	CA	.046	11.80	1.24	---	13.04	**15.50**	11.11	**13.00**
Redwood											
Lattice											
Lattice											
5/16" x 1-1/4"	Inst	LF	CA	.046	.38	1.24	---	1.62	**2.32**	1.14	**1.61**
5/16" x 1-5/8"	Inst	LF	CA	.046	.45	1.24	---	1.69	**2.40**	1.20	**1.68**
Batts											
5/16" x 2-1/2"	Inst	LF	CA	.046	.70	1.24	---	1.94	**2.69**	1.42	**1.93**

Description	Oper	Unit	Crew Size	Man-Hours Per Unit	Avg Mat'l Unit Cost	Avg Labor Unit Cost	Avg Equip Unit Cost	Avg Total Unit Cost	Avg Price Incl O&P	Large Volume Avg Total Unit Cost	Large Volume Avg Price Incl O&P
5/16" x 3-1/2"	Inst	LF	CA	.046	.98	1.24	---	2.22	**3.01**	1.66	**2.21**
5/16" x 5-1/2"	Inst	LF	CA	.046	1.67	1.24	---	2.91	**3.81**	2.26	**2.90**
Miscellaneous exterior molds											
Bricks 1-1/2" x 1-1/2"	Inst	LF	CA	.062	1.70	1.67	---	3.37	**4.50**	2.56	**3.34**
Siding 7/8" x 1-5/8"	Inst	LF	CA	.062	1.25	1.67	---	2.92	**3.98**	2.17	**2.89**
Stucco 7/8" x 1-1/2"	Inst	LF	CA	.062	.74	1.67	---	2.41	**3.39**	1.72	**2.38**
Watertable without lip 1-1/2" x 2-3/8"	Inst	LF	CA	.062	2.73	1.67	---	4.40	**5.68**	3.45	**4.36**
Watertable with lip 1-1/2" x 2-3/8"	Inst	LF	CA	.062	2.73	1.67	---	4.40	**5.68**	3.45	**4.36**
Corrugated 3/4" x 1-5/8"	Inst	LF	CA	.062	.52	1.67	---	2.19	**3.14**	1.53	**2.16**
Crest 3/4" x 1-5/8"	Inst	LF	CA	.062	.52	1.67	---	2.19	**3.14**	1.53	**2.16**
Resin flexible molding											
Primed paint grade											
Diameter casing											
711 casing 5/8" x 2-1/2"	Inst	LF	CA	.046	12.20	1.24	---	13.44	**16.00**	11.41	**13.50**
Universal casing 5/8" x 2-1/2"	Inst	LF	CA	.046	12.20	1.24	---	13.44	**16.00**	11.41	**13.50**
356 casing 5/8" x 2-1/2"	Inst	LF	CA	.046	12.20	1.24	---	13.44	**16.00**	11.41	**13.50**
Cape Cod casing 5/8" x 2-1/2"	Inst	LF	CA	.046	12.20	1.24	---	13.44	**16.00**	11.41	**13.50**
Bases & round corner blocks											
Base shoe 3/8" x 3/4"	Inst	LF	CA	.046	2.31	1.24	---	3.55	**4.54**	2.82	**3.54**
Streamline base 3/8" x 2-1/4"	Inst	LF	CA	.046	3.77	1.24	---	5.01	**6.22**	4.09	**5.00**
7/16" x 3-1/2"	Inst	LF	CA	.046	6.08	1.24	---	7.32	**8.88**	6.10	**7.31**
711 base 7/16" x 2-1/2"	Inst	LF	CA	.046	3.63	1.24	---	4.87	**6.06**	3.97	**4.86**
7/16" x 3-1/2"	Inst	LF	CA	.046	5.49	1.24	---	6.73	**8.20**	5.58	**6.71**
WM 623 base 9/16" x 3-1/4"	Inst	LF	CA	.046	5.03	1.24	---	6.27	**7.67**	5.18	**6.25**
WM 618 9/16" x 5-1/4"	Inst	LF	CA	.046	11.60	1.24	---	12.84	**15.30**	10.91	**12.90**
Cape Cod base 9/16" x 5-1/4"	Inst	LF	CA	.046	9.26	1.24	---	10.50	**12.50**	8.86	**10.50**
9/16" x 4"	Inst	LF	CA	.046	6.61	1.24	---	7.85	**9.49**	6.56	**7.84**
"B" Cape Cod base 9/16" x 5-1/4"	Inst	LF	CA	.046	10.60	1.24	---	11.84	**14.10**	10.01	**11.80**
Colonial base 11/16" x 4-1/4"	Inst	LF	CA	.046	7.94	1.24	---	9.18	**11.00**	7.71	**9.16**

Description	Oper	Unit	Crew Size	Man-Hours Per Unit	Avg Mat'l Unit Cost	Avg Labor Unit Cost	Avg Equip Unit Cost	Avg Total Unit Cost	Avg Price Incl O&P	Large Volume Avg Total Unit Cost	Large Volume Avg Price Incl O&P
Monterey base											
7/16" x 4-1/4"	Inst	LF	CA	.046	7.94	1.24	---	9.18	**11.00**	7.71	**9.16**
356 base											
1-1/2" x 2-1/2"	Inst	LF	CA	.046	3.97	1.24	---	5.21	**6.45**	4.26	**5.20**
1-1/2" x 3-1/4"	Inst	LF	CA	.046	6.08	1.24	---	7.32	**8.88**	6.10	**7.31**
Step base											
7/16" x 2-1/2"	Inst	LF	CA	.046	4.23	1.24	---	5.47	**6.75**	4.49	**5.46**
1/2" x 4-1/2"	Inst	LF	CA	.046	8.34	1.24	---	9.58	**11.50**	8.06	**9.57**
Oak grain											
Diameter casing											
711 casing											
5/8" x 2-1/2"	Inst	LF	CA	.046	13.20	1.24	---	14.44	**17.10**	12.31	**14.50**
356 casing											
11/16" x 2-1/2"	Inst	LF	CA	.046	13.20	1.24	---	14.44	**17.10**	12.31	**14.50**
Bases & base shoe											
711 base											
3/8" x 2-1/2"	Inst	LF	CA	.046	5.29	1.24	---	6.53	**7.97**	5.41	**6.52**
3/8" x 3-1/2"	Inst	LF	CA	.046	7.94	1.24	---	9.18	**11.00**	7.71	**9.16**
Base shoe											
3/8" x 3/4"	Inst	LF	CA	.046	2.65	1.24	---	3.89	**4.93**	3.11	**3.87**

Spindles

Western hemlock, clear, kiln dried, turned for decorative applications

Description	Oper	Unit	Crew Size	Man-Hours Per Unit	Avg Mat'l Unit Cost	Avg Labor Unit Cost	Avg Equip Unit Cost	Avg Total Unit Cost	Avg Price Incl O&P	Large Volume Avg Total Unit Cost	Large Volume Avg Price Incl O&P
Planter design											
1-11/16" x 1-11/16"											
3'-0" H	Inst	Ea	CA	.500	4.90	13.50	---	18.40	**26.10**	13.24	**18.60**
4'-0" H	Inst	Ea	CA	.500	6.28	13.50	---	19.78	**27.70**	14.44	**19.90**
2-3/8" x 2-3/8"											
3'-0" H	Inst	Ea	CA	.500	10.20	13.50	---	23.70	**32.20**	17.84	**23.80**
4'-0" H	Inst	Ea	CA	.500	14.00	13.50	---	27.50	**36.50**	21.08	**27.60**
5'-0" H	Inst	Ea	CA	.500	16.90	13.50	---	30.40	**40.00**	23.68	**30.60**
6'-0" H	Inst	Ea	CA	.615	21.60	16.60	---	38.20	**50.00**	29.60	**38.00**
8'-0" H	Inst	Ea	CA	.615	41.90	16.60	---	58.50	**73.40**	47.30	**58.30**
3-1/4" x 3-1/4"											
3'-0" H	Inst	Ea	CA	.500	19.30	13.50	---	32.80	**42.70**	25.78	**33.00**
4'-0" H	Inst	Ea	CA	.500	25.80	13.50	---	39.30	**50.20**	31.38	**39.40**
5'-0" H	Inst	Ea	CA	.500	32.70	13.50	---	46.20	**58.10**	37.38	**46.30**
6'-0" H	Inst	Ea	CA	.615	39.10	16.60	---	55.70	**70.10**	44.80	**55.50**
8'-0" H	Inst	Ea	CA	.615	56.40	16.60	---	73.00	**90.10**	59.90	**72.80**
Colonial design											
1-11/16" x 1-11/16"											
1'-0" H	Inst	Ea	CA	.444	1.59	12.00	---	13.59	**20.00**	9.09	**13.30**
1'-6" H	Inst	Ea	CA	.444	2.84	12.00	---	14.84	**21.50**	10.18	**14.60**
2'-0" H	Inst	Ea	CA	.444	3.63	12.00	---	15.63	**22.40**	10.87	**15.40**
2'-4" H	Inst	Ea	CA	.500	3.77	13.50	---	17.27	**24.80**	12.26	**17.40**
2'-8" H	Inst	Ea	CA	.500	4.23	13.50	---	17.73	**25.40**	12.66	**17.90**
3'-0" H	Inst	Ea	CA	.500	5.03	13.50	---	18.53	**26.30**	13.35	**18.70**

Description	Oper	Unit	Crew Size	Man-Hours Per Unit	Avg Mat'l Unit Cost	Avg Labor Unit Cost	Avg Equip Unit Cost	Avg Total Unit Cost	Avg Price Incl O&P	Avg Total Unit Cost	Avg Price Incl O&P
			Costs Based On Small Volume							Large Volume	
2-3/8" x 2-3/8"											
1'-0" H	Inst	Ea	CA	.444	3.24	12.00	---	15.24	**21.90**	10.53	**15.00**
1'-6" H	Inst	Ea	CA	.444	4.50	12.00	---	16.50	**23.40**	11.62	**16.20**
2'-0" H	Inst	Ea	CA	.444	6.81	12.00	---	18.81	**26.00**	13.63	**18.50**
2'-4" H	Inst	Ea	CA	.500	7.28	13.50	---	20.78	**28.90**	15.31	**20.90**
2'-8" H	Inst	Ea	CA	.500	9.26	13.50	---	22.76	**31.10**	17.03	**22.90**
3'-0" H	Inst	Ea	CA	.500	10.20	13.50	---	23.70	**32.20**	17.84	**23.80**
3-1/4" x 3-1/4"											
1'-6" H	Inst	Ea	CA	.444	7.60	12.00	---	19.60	**26.90**	14.32	**19.30**
2'-0" H	Inst	Ea	CA	.444	10.30	12.00	---	22.30	**30.10**	16.68	**22.00**
2'-4" H	Inst	Ea	CA	.500	12.10	13.50	---	25.60	**34.40**	19.48	**25.70**
2'-8" H	Inst	Ea	CA	.500	13.70	13.50	---	27.20	**36.20**	20.88	**27.30**
3'-0" H	Inst	Ea	CA	.500	19.30	13.50	---	32.80	**42.70**	25.78	**33.00**
8'-0" H	Inst	Ea	CA	.615	56.40	16.60	---	73.00	**90.10**	59.90	**72.80**
Mediterranean design											
1-11/16" x 1-11/16"											
1'-0" H	Inst	Ea	CA	.444	1.66	12.00	---	13.66	**20.10**	9.15	**13.40**
1'-6" H	Inst	Ea	CA	.444	2.71	12.00	---	14.71	**21.30**	10.07	**14.40**
2'-0" H	Inst	Ea	CA	.444	2.98	12.00	---	14.98	**21.60**	10.30	**14.70**
2'-4" H	Inst	Ea	CA	.500	3.51	13.50	---	17.01	**24.50**	12.03	**17.20**
2'-8" H	Inst	Ea	CA	.500	3.90	13.50	---	17.40	**25.00**	12.37	**17.50**
3'-0" H	Inst	Ea	CA	.500	4.90	13.50	---	18.40	**26.10**	13.24	**18.60**
4'-0" H	Inst	Ea	CA	.500	6.28	13.50	---	19.78	**27.70**	14.44	**19.90**
5'-0" H	Inst	Ea	CA	.500	11.30	13.50	---	24.80	**33.50**	18.81	**25.00**
2-3/8" x 2-3/8"											
1'-0" H	Inst	Ea	CA	.444	2.98	12.00	---	14.98	**21.60**	10.30	**14.70**
1'-6" H	Inst	Ea	CA	.444	4.30	12.00	---	16.30	**23.10**	11.45	**16.00**
2'-0" H	Inst	Ea	CA	.444	5.43	12.00	---	17.43	**24.40**	12.43	**17.20**
2'-4" H	Inst	Ea	CA	.500	6.95	13.50	---	20.45	**28.50**	15.02	**20.60**
2'-8" H	Inst	Ea	CA	.500	8.66	13.50	---	22.16	**30.50**	16.51	**22.30**
3'-0" H	Inst	Ea	CA	.500	9.92	13.50	---	23.42	**31.90**	17.61	**23.60**
4'-0" H	Inst	Ea	CA	.500	11.80	13.50	---	25.30	**34.00**	19.18	**25.40**
5'-0" H	Inst	Ea	CA	.500	16.20	13.50	---	29.70	**39.10**	23.08	**29.90**
6'-0" H	Inst	Ea	CA	.615	22.80	16.60	---	39.40	**51.40**	30.60	**39.10**
8'-0" H	Inst	Ea	CA	.615	41.90	16.60	---	58.50	**73.40**	47.30	**58.30**
3-1/4" x 3-1/4"											
3'-0" H	Inst	Ea	CA	.500	18.70	13.50	---	32.20	**42.00**	25.28	**32.40**
4'-0" H	Inst	Ea	CA	.500	22.00	13.50	---	35.50	**45.80**	28.18	**35.70**
5'-0" H	Inst	Ea	CA	.500	29.60	13.50	---	43.10	**54.60**	34.78	**43.30**
6'-0" H	Inst	Ea	CA	.615	41.70	16.60	---	58.30	**73.10**	47.00	**58.10**
8'-0" H	Inst	Ea	CA	.615	65.70	16.60	---	82.30	**101.00**	68.00	**82.10**
Spindle rails, 8'-0" H pieces											
For 1-11/16" spindles	Inst	Ea	CA	---	16.70	---	---	16.70	**19.20**	14.50	**16.70**
For 2-3/8" spindles	Inst	Ea	CA	---	24.10	---	---	24.10	**27.70**	20.90	**24.10**
For 3-1/4" spindles	Inst	Ea	CA	---	28.00	---	---	28.00	**32.20**	24.30	**28.00**
Wood bullnose round corners											
Pine base R/C											
Colonial base											
11/16" x 4-1/4"	Inst	LF	CA	.046	1.35	1.24	---	2.59	**3.44**	1.98	**2.57**
Streamline base											
3/8" x 2-1/4"	Inst	LF	CA	.046	.70	1.24	---	1.94	**2.69**	1.42	**1.93**
7/16" x 3-1/2"	Inst	LF	CA	.046	1.00	1.24	---	2.24	**3.04**	1.68	**2.23**

Description	Oper	Unit	Crew Size	Man-Hours Per Unit	Avg Mat'l Unit Cost	Avg Labor Unit Cost	Avg Equip Unit Cost	Avg Total Unit Cost	Avg Price Incl O&P	Large Volume Avg Total Unit Cost	Large Volume Avg Price Incl O&P
			Costs Based On Small Volume							**Large Volume**	
711 base											
7/16" x 2-1/2"	Inst	LF	CA	.046	.83	1.24	---	2.07	**2.84**	1.53	**2.06**
7/16" x 3-1/2"	Inst	LF	CA	.046	1.17	1.24	---	2.41	**3.23**	1.83	**2.40**
3 step base											
7/16" x 2-1/2"	Inst	LF	CA	.046	.74	1.24	---	1.98	**2.74**	1.45	**1.97**
1/2" x 4-1/2"	Inst	LF	CA	.046	2.17	1.24	---	3.41	**4.38**	2.70	**3.40**
WM 623 base											
9/16" x 3-1/4"	Inst	LF	CA	.046	1.02	1.24	---	2.26	**3.06**	1.70	**2.25**
WM 618 base											
9/16" x 5-1/4"	Inst	LF	CA	.046	1.58	1.24	---	2.82	**3.70**	2.18	**2.80**
Monterey base											
7/16" x 4-1/4"	Inst	LF	CA	.046	1.09	1.24	---	2.33	**3.14**	1.76	**2.32**
Cape Cod base											
9/16" x 5-1/4"	Inst	LF	CA	.046	1.59	1.24	---	2.83	**3.71**	2.19	**2.82**
9/16" x 4"	Inst	LF	CA	.046	1.32	1.24	---	2.56	**3.40**	1.96	**2.55**
"B" Cape Cod base											
9/16" x 5-1/4"	Inst	LF	CA	.046	1.59	1.24	---	2.83	**3.71**	2.19	**2.82**
356 base											
1/2" x 2-1/4"	Inst	LF	CA	.046	.97	1.24	---	2.21	**3.00**	1.65	**2.20**
1/2" x 3-1/4"	Inst	LF	CA	.046	1.22	1.24	---	2.46	**3.29**	1.87	**2.45**
Base shoe											
3/8" x 3/4"	Inst	LF	CA	.046	.25	1.24	---	1.49	**2.17**	1.03	**1.48**
Oak base R/C											
Base shoe											
3/8" x 3/4"	Inst	LF	CA	.046	.41	1.24	---	1.65	**2.36**	1.17	**1.64**
711 base											
3/8" x 2-1/2"	Inst	LF	CA	.046	1.21	1.24	---	2.45	**3.28**	1.86	**2.44**
3/8" x 3-1/2"	Inst	LF	CA	.046	1.52	1.24	---	2.76	**3.63**	2.13	**2.75**
Pine chair rail R/C											
Chair rail											
9/16" x 2-1/2"	Inst	LF	CA	.046	1.07	1.24	---	2.31	**3.12**	1.74	**2.30**
Colonial chair rail											
3/4" x 3"	Inst	LF	CA	.046	1.24	1.24	---	2.48	**3.31**	1.89	**2.47**
11/16" x 2-1/2"	Inst	LF	CA	.046	1.08	1.24	---	2.32	**3.13**	1.75	**2.31**
Pine crown & cornice R/C											
Colonial crown											
9/16" x 2-1/4"	Inst	LF	CA	.046	.97	1.24	---	2.21	**3.00**	1.65	**2.20**
11/16" x 3-3/8"	Inst	LF	CA	.046	1.30	1.24	---	2.54	**3.38**	1.94	**2.53**
11/16" x 4-1/2"	Inst	LF	CA	.046	2.27	1.24	---	3.51	**4.50**	2.78	**3.49**
Cornice											
13/16" x 4-5/8"	Inst	LF	CA	.046	2.56	1.24	---	3.80	**4.83**	3.04	**3.79**
Georgian crown											
11/16" x 4-1/4"	Inst	LF	CA	.046	2.81	1.24	---	4.05	**5.12**	3.25	**4.04**

Painting

Interior and Exterior. There is a paint for almost every type of surface and surface condition. The large variety makes it impractical to consider each paint individually. For this reason, average output and average material cost/unit are based on the paints and prices listed below.

1. **Installation.** Paint can be applied by brush, roller or spray gun. Only application by brush or roller is considered in this section.
2. **Notes on Labor.** Average Manhours per Unit, for both roller and brush, is based on what one painter can do in one day. The output for cleaning is also based on what one painter can do in one day.
3. **Estimating Technique.** Use these techniques to determine quantities for interior and exterior painting before you apply unit costs.

Interior

a. Floors, walls and ceilings. Figure actual area. No deductions for openings.

b. Doors and windows. **Only openings to be painted:** Figure 36 SF or 4 SY for each side of each door and 27 SF for each side of each window. Based on doors 3'-0" x 7'-0" or smaller and windows 3'-0" x 4'-0" or smaller. **Openings to be painted with walls:** Figure wall area plus 27 SF or 3 SY for each side of each door and 18 SF or 2 SY for each side of each window. Based on doors 3'-0" x 7'-0" or smaller and windows 3'-0" x 4'-0" or smaller.

For larger doors and windows, add 1'-0" to height and width and figure area.

c. Base or picture moldings and chair rails. Less than 1'-0" wide, figure one SF/LF. On 1'-0" or larger, figure actual area.

d. Stairs (including treads, risers, cove and stringers). Add 2'-0" width (for treads, risers, etc.) times length plus 2'-0".

e. Balustrades. Add 1'-0" to height, figure two times area to paint two sides.

Exterior

a. Walls. (No deductions for openings.)

Siding. Figure actual area plus 10%.

Shingles. Figure actual area plus 40%.

Characteristics - Interior			Characteristics - Exterior		
Type	**Coverage SF/Gal.**	**Surface**	**Type**	**Coverage SF/Gal.**	**Surface**
Latex, flat	450	Plaster/drywall	Oil base*	300	Plain siding & stucco
Latex, enamel	450	Doors, windows, trim	Oil base*	450	Door, windows, trim
Shellac	500	Woodwork	Oil base*	300	Shingle siding
Varnish	500	Woodwork	Stain	200	Shingle siding
Stain	500	Woodwork	Latex, masonry	400	Stucco & masonry

*Certain latex paints may also be used on exterior work.

Brick, stucco, concrete and smooth wood surfaces. Figure actual area.

b. Doors and windows. See Interior, doors and windows.

c. Eaves (including soffit or exposed rafter ends and fascia). **Enclosed.** If sidewalls are to be painted the same color, figure 1.5 times actual area. If sidewalls are to be painted a different color, figure 2.0 times actual area. If sidewalls are not to be painted, figure 2.5 times actual area. **Rafter ends exposed:** If sidewalls are to be painted same color, figure 2.5 times actual area. If sidewalls are to be painted a different color, figure 3.0 times actual area. If sidewalls are not to be painted, figure 4.0 times actual area.

d. Porch rails. See Interior, balustrades (previous page).

e. Gutters and downspouts. Figure 2.0 SF/LF or $^2/_9$ SY/LF.

f. Latticework. Figure 2.0 times actual area for each side.

g. Fences. **Solid fence:** Figure actual area of each side to be painted. **Basketweave:** Figure 1.5 times actual area for each side to be painted.

Calculating Square Foot Coverage

Triangle

To find the number of square feet in any shape triangle or 3 sided surface, multiply the height by the width and divide the total by 2.

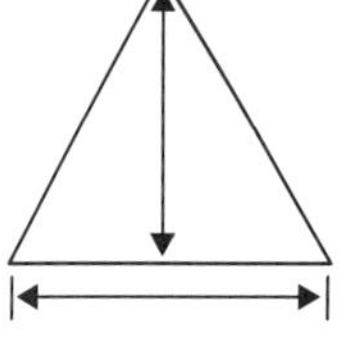

Square

Multiply the base measurement in feet times the height in feet.

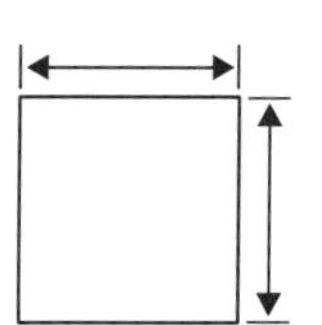

Rectangle

Multiply the base measurement in feet times the height in feet.

Arch Roof

Multiply length (B) by width (A) and add one-half the total.

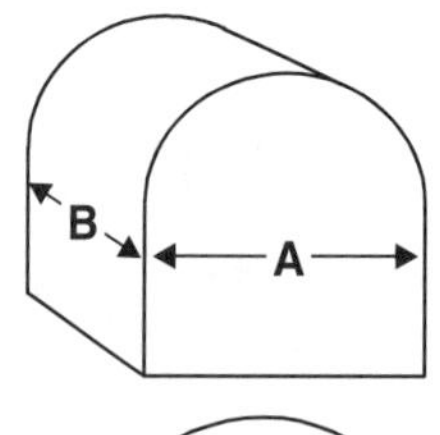

Circle

To find the number of square feet in a circle multiply the diameter (distance across) by itself and them multiply this total by .7854.

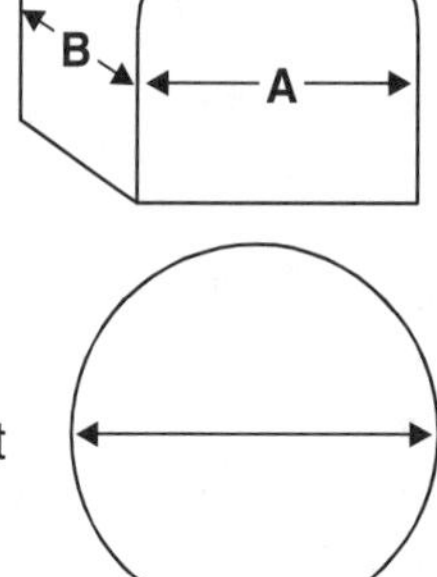

Cylinder

When the circumference (distance around the cylinder) is known, multiply height by circumference. When the diameter (distance across) is known, multiply diameter by 3.1416. This gives circumference. Then multiply by height.

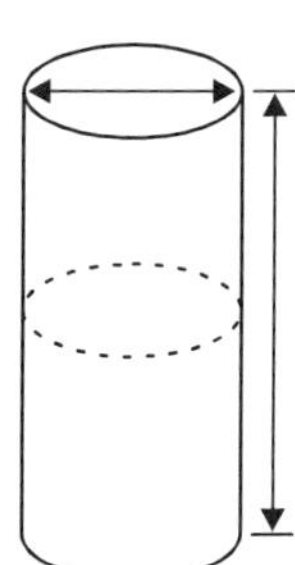

Gambrel Roof

Multiply length (B) by width (A) and add one-third of the total.

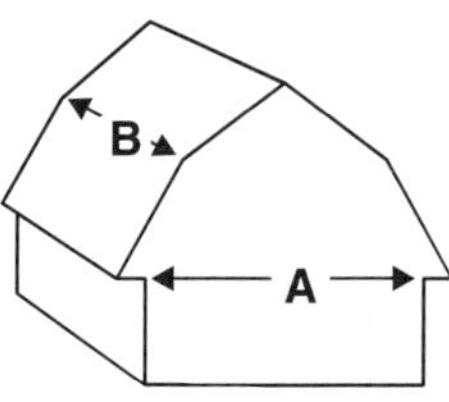

Cone

Determine area of base by multiplying 3.1416 times radius (A) in feet.

Determine the surface area of a cone by multiplying circumference of base (in feet) times one-half of the slant height (B) in feet.

Add the square foot area of the base to the square foot area of the cone for total square foot area.

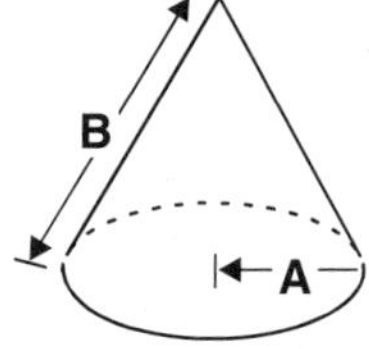

Description	Oper	Unit	Crew Size	Man-Hours Per Unit	Avg Mat'l Unit Cost	Avg Labor Unit Cost	Avg Equip Unit Cost	Avg Total Unit Cost	Avg Price Incl O&P	Avg Total Unit Cost	Avg Price Incl O&P
			Costs Based On Small Volume							Large Volume	
Painting and cleaning											
Interior											
Time calculations include normal materials handling and protection of furniture and other property not to be painted.											
Preparation											
Excluding openings unless otherwise indicated											
Cleaning, wet											
Smooth finishes											
Plaster and drywall	Inst	SF	NA	.007	.01	.20	---	.21	**.31**	.15	**.22**
Paneling	Inst	SF	NA	.007	.01	.20	---	.21	**.31**	.12	**.18**
Millwork and trim	Inst	SF	NA	.008	.01	.22	---	.23	**.35**	.15	**.22**
Floors	Inst	SF	NA	.005	.01	.14	---	.15	**.22**	.12	**.18**
Sand finishes	Inst	SF	NA	.011	.01	.31	---	.32	**.48**	.21	**.31**
Sheetrock or drywall; tape, fill, and finish only; tape kit includes 25 lb. bag taping compound and 250 ft. roll perforated tape; 735 SF per kit	Inst	SF	NA	.014	.02	.39	---	.41	**.61**	.27	**.40**
Sheetrock or drywall; thin coat plaster in lieu of taping (125 SF per 50 lb. bag)	Inst	SF	NA	.018	.11	.51	---	.62	**.89**	.45	**.63**
Light sanding											
Before first coat	Inst	SF	NA	.008	.01	.22	---	.23	**.35**	.15	**.22**
Before second coat	Inst	SF	NA	.007	.01	.20	---	.21	**.31**	.15	**.22**
Before third coat	Inst	SF	NA	.006	.01	.17	---	.18	**.26**	.12	**.18**
Liquid removal of paint or varnish											
Paneling (170 SF/gal)	Inst	SF	NA	.038	.13	1.07	---	1.20	**1.75**	.81	**1.18**
Millwork & trim (170 SF/gal)	Inst	SF	NA	.044	.13	1.24	---	1.37	**2.00**	.92	**1.35**
Floors (170 SF/gal)	Inst	SF	NA	.026	.13	.73	---	.86	**1.25**	.59	**.84**
Burning off paint	Inst	SF	NA	.062	.01	1.74	---	1.75	**2.62**	1.13	**1.70**
One coat application											
Excluding openings unless otherwise indicated											
Sizing, on sheetrock or plaster											
Smooth finish											
Brush (650 SF/gal)	Inst	SF	NA	.005	.02	.14	---	.16	**.23**	.10	**.15**
Roller (625 SF/gal)	Inst	SF	NA	.004	.02	.11	---	.13	**.19**	.10	**.15**
Sand finish											
Brush (550 SF/gal)	Inst	SF	NA	.007	.03	.20	---	.23	**.33**	.16	**.23**
Roller (525 SF/gal)	Inst	SF	NA	.005	.03	.14	---	.17	**.25**	.10	**.15**
Sealer											
Sheetrock or plaster											
Smooth finish											
Brush (300 SF/gal)	Inst	SF	NA	.009	.06	.25	---	.31	**.45**	.22	**.31**
Roller (285 SF/gal)	Inst	SF	NA	.007	.06	.20	---	.26	**.36**	.16	**.23**
Spray (250 SF/gal)	Inst	SF	NC	.005	.07	.14	.01	.22	**.31**	.16	**.21**
Sand finish											
Brush (250 SF/gal)	Inst	SF	NA	.012	.07	.34	---	.41	**.59**	.28	**.41**
Roller (235 SF/gal)	Inst	SF	NA	.009	.08	.25	---	.33	**.47**	.24	**.33**
Spray (210 SF/gal)	Inst	SF	NC	.005	.09	.14	.01	.24	**.33**	.17	**.22**

Description	Oper	Unit	Crew Size	Man-Hours Per Unit	Avg Mat'l Unit Cost	Avg Labor Unit Cost	Avg Equip Unit Cost	Avg Total Unit Cost	Avg Price Incl O&P	Large Volume Avg Total Unit Cost	Large Volume Avg Price Incl O&P
Acoustical tile or panels											
Brush (225 SF/gal)	Inst	SF	NA	.010	.08	.28	---	.36	**.51**	.24	**.33**
Roller (200 SF/gal)	Inst	SF	NA	.008	.09	.22	---	.31	**.44**	.22	**.30**
Spray (160 SF/gal)	Inst	SF	NC	.006	.11	.17	.01	.29	**.40**	.23	**.30**
Latex											
Drywall or plaster, latex flat											
Smooth finish											
Brush (300 SF/gal)	Inst	SF	NA	.009	.07	.25	---	.32	**.46**	.23	**.32**
Roller (285 SF/gal)	Inst	SF	NA	.007	.08	.20	---	.28	**.39**	.21	**.29**
Spray (260 SF/gal)	Inst	SF	NC	.005	.08	.14	.01	.23	**.32**	.17	**.22**
Sand finish											
Brush (250 SF/gal)	Inst	SF	NA	.013	.09	.37	---	.46	**.65**	.29	**.42**
Roller (235 SF/gal)	Inst	SF	NA	.010	.09	.28	---	.37	**.53**	.28	**.39**
Spray (210 SF/gal)	Inst	SF	NC	.005	.10	.14	.01	.25	**.34**	.19	**.24**
Texture or stipple applied to drywall, one coat											
Brush (125 SF/gal)	Inst	SF	NA	.010	.18	.28	---	.46	**.63**	.35	**.47**
Roller (120 SF/gal)	Inst	SF	NA	.008	.18	.22	---	.40	**.54**	.30	**.39**
Paneling, latex enamel											
Brush (300 SF/gal)	Inst	SF	NA	.009	.07	.25	---	.32	**.46**	.23	**.32**
Roller (285 SF/gal)	Inst	SF	NA	.007	.08	.20	---	.28	**.39**	.21	**.29**
Spray (260 SF/gal)	Inst	SF	NC	.005	.08	.14	.01	.23	**.32**	.17	**.22**
Acoustical tile or panels, latex flat											
Brush (225 SF/gal)	Inst	SF	NA	.011	.10	.31	---	.41	**.58**	.28	**.39**
Roller (210 SF/gal)	Inst	SF	NA	.009	.10	.25	---	.35	**.49**	.26	**.36**
Spray (185 SF/gal)	Inst	SF	NC	.006	.12	.17	.01	.30	**.41**	.23	**.30**
Millwork and trim, latex enamel											
Doors and windows											
Roller and/or brush (360 SF/gal)	Inst	SF	NA	.019	.06	.53	---	.59	**.87**	.42	**.61**
Spray, doors only (325 SF / gal)	Inst	SF	NC	.009	.07	.26	.02	.35	**.49**	.25	**.35**
Cabinets											
Roller and/or brush (360 SF/gal)	Inst	SF	NA	.021	.06	.59	---	.65	**.95**	.42	**.61**
Spray, doors only (325 SF / gal)	Inst	SF	NC	.009	.07	.26	.02	.35	**.49**	.25	**.35**
Louvers, spray (300 SF/gal)	Inst	SF	NC	.028	.07	.81	.05	.93	**1.35**	.63	**.90**
Picture molding, chair rail, base, ceiling mold etc., less than 6" high											
Note: SF equals LF on trim less than 6" high											
Brush (900 SF/gal)	Inst	SF	NA	.013	.02	.37	---	.39	**.57**	.24	**.36**
Floors, wood											
Brush (405 SF/gal)	Inst	SF	NA	.006	.05	.17	---	.22	**.31**	.16	**.23**
Roller (385 SF/gal)	Inst	SF	NA	.005	.06	.14	---	.20	**.28**	.13	**.18**
For custom colors, ADD	Inst	%	---	---	10.0	---	---	---	**---**	---	**---**
Floor seal											
Brush (450 SF/gal)	Inst	SF	NA	.004	.06	.11	---	.17	**.24**	.13	**.18**
Roller (430 SF/gal)	Inst	SF	NA	.004	.06	.11	---	.17	**.24**	.13	**.18**
Penetrating stainwax (hardwood floors)											
Brush (450 SF/gal)	Inst	SF	NA	.006	.06	.17	---	.23	**.32**	.16	**.23**
Roller (425 SF/gal)	Inst	SF	NA	.005	.06	.14	---	.20	**.28**	.13	**.18**

Description	Oper	Unit	Crew Size	Man-Hours Per Unit	Avg Mat'l Unit Cost	Avg Labor Unit Cost	Avg Equip Unit Cost	Avg Total Unit Cost	Avg Price Incl O&P	Large Volume Avg Total Unit Cost	Large Volume Avg Price Incl O&P
Natural finishes											
Paneling, brush work unless otherwise indicated											
Stain, brush on - wipe off (360 SF/gal)	Inst	SF	NA	.022	.07	.62	---	.69	**1.01**	.45	**.66**
Varnish (380 SF/gal)	Inst	SF	NA	.009	.07	.25	---	.32	**.46**	.23	**.32**
Shellac (630 SF/gal)	Inst	SF	NA	.008	.05	.22	---	.27	**.39**	.18	**.26**
Lacquer											
Brush (450 SF/gal)	Inst	SF	NA	.006	.06	.17	---	.23	**.32**	.16	**.23**
Spray (300 SF/gal)	Inst	SF	NC	.005	.09	.14	.01	.24	**.33**	.17	**.22**
Doors and windows, brush work unless otherwise indicated											
Stain, brush on - wipe off (450 SF/gal)	Inst	SF	NA	.049	.06	1.38	---	1.44	**2.13**	.95	**1.41**
Varnish (550 SF/gal)	Inst	SF	NA	.036	.05	1.01	---	1.06	**1.57**	.71	**1.06**
Shellac (550 SF/gal)	Inst	SF	NA	.034	.05	.96	---	1.01	**1.49**	.67	**.98**
Lacquer											
Brush (550 SF/gal)	Inst	SF	NA	.030	.05	.84	---	.89	**1.32**	.60	**.89**
Spray doors (300 SF/gal)	Inst	SF	NC	.015	.09	.43	.03	.55	**.78**	.39	**.54**
Cabinets, brush work unless otherwise indicated											
Stain, brush on - wipe off (450 SF/gal)	Inst	SF	NA	.052	.06	1.46	---	1.52	**2.26**	1.01	**1.49**
Varnish (550 SF/gal)	Inst	SF	NA	.038	.05	1.07	---	1.12	**1.66**	.74	**1.10**
Shellac (550 SF/gal)	Inst	SF	NA	.036	.05	1.01	---	1.06	**1.57**	.70	**1.03**
Lacquer											
Brush (550 SF/gal)	Inst	SF	NA	.032	.05	.90	---	.95	**1.41**	.63	**.93**
Spray (300 SF/gal)	Inst	SF	NC	.016	.09	.46	.03	.58	**.83**	.42	**.59**
Louvers, lacquer, spray (300 SF/gal)	Inst	SF	NC	.026	.09	.75	.05	.89	**1.28**	.61	**.87**
Picture molding, chair rail, base, ceiling mold etc., less than 6" high											
Note: SF equals LF on trim less than 6" high											
Varnish, brush (900 SF/gal)	Inst	SF	NA	.012	.03	.34	---	.37	**.54**	.25	**.37**
Shellac, brush (900 SF/gal)	Inst	SF	NA	.012	.03	.34	---	.37	**.54**	.25	**.37**
Lacquer, spray (700 SF/gal)	Inst	SF	NA	.010	.04	.28	---	.32	**.47**	.20	**.29**
Floors, wood, brush work unless otherwise indicated											
Shellac (450 SF/gal)	Inst	SF	NA	.006	.07	.17	---	.24	**.33**	.17	**.24**
Varnish (500 SF/gal)	Inst	SF	NA	.006	.06	.17	---	.23	**.32**	.16	**.23**
Buffing, by machine	Inst	SF	NA	.004	.01	.11	---	.12	**.18**	.09	**.14**
Waxing and polishing, by hand (1,000 SF/gal)	Inst	SF	NA	.008	.01	.22	---	.23	**.35**	.15	**.22**

Two coat application

Excluding openings unless otherwise indicated

For sizing or sealer, see One coat application, page 187

Latex

No spray work included, see One coat application, page 188

Description	Oper	Unit	Crew Size	Man-Hours Per Unit	Avg Mat'l Unit Cost	Avg Labor Unit Cost	Avg Equip Unit Cost	Avg Total Unit Cost	Avg Price Incl O&P	Large Volume Avg Total Unit Cost	Large Volume Avg Price Incl O&P
Drywall or plaster, latex flat											
Smooth finish											
Brush (170 SF/gal)	Inst	SF	NA	.014	.13	.39	---	.52	**.74**	.36	**.51**
Roller (160 SF/gal)	Inst	SF	NA	.012	.14	.34	---	.48	**.67**	.34	**.48**

Description	Oper	Unit	Crew Size	Man-Hours Per Unit	Avg Mat'l Unit Cost	Avg Labor Unit Cost	Avg Equip Unit Cost	Avg Total Unit Cost	Avg Price Incl O&P	Large Volume Avg Total Unit Cost	Large Volume Avg Price Incl O&P
Sand finish											
Brush (170 SF/gal)	Inst	SF	NA	.021	.13	.59	---	.72	**1.03**	.48	**.67**
Roller (160 SF/gal)	Inst	SF	NA	.016	.14	.45	---	.59	**.84**	.43	**.60**
Paneling, latex enamel											
Brush (170 SF/gal)	Inst	SF	NA	.014	.13	.39	---	.52	**.74**	.36	**.51**
Roller (160 SF/gal)	Inst	SF	NA	.012	.14	.34	---	.48	**.67**	.34	**.48**
Acoustical tile or panels, latex flat											
Brush (130 SF/gal)	Inst	SF	NA	.018	.17	.51	---	.68	**.95**	.48	**.67**
Roller (120 SF/gal)	Inst	SF	NA	.014	.18	.39	---	.57	**.80**	.41	**.56**
Millwork and trim, enamel											
Doors and windows											
Roller and/or brush (200 SF/gal)	Inst	SF	NA	.037	.11	1.04	---	1.15	**1.69**	.76	**1.12**
Cabinets											
Roller and/or brush (200 SF/gal)	Inst	SF	NA	.039	.11	1.10	---	1.21	**1.77**	.79	**1.16**
Louvers, see One coat application, page 188											
Picture molding, chair rail, base, ceiling mold etc., less than 6" high											
Note: SF equals LF on trim less than 6" high											
Brush (510 SF/gal)	Inst	SF	NA	.025	.04	.70	---	.74	**1.10**	.49	**.72**
Floors, wood											
Brush (230 SF/gal)	Inst	SF	NA	.011	.10	.31	---	.41	**.58**	.28	**.39**
Roller (220 SF/gal)	Inst	SF	NA	.010	.10	.28	---	.38	**.54**	.26	**.36**
For custom colors, ADD	Inst	%	---	---	10.0	---	---	---	**---**	---	**---**

For wood floor seal, penetrating stainwax, or natural finish, see One coat application, page 188

Exterior

Time calculations include normal materials handling and protection of property not to be painted

Preparation

Excluding openings unless otherwise indicated

Description	Oper	Unit	Crew Size	Man-Hours Per Unit	Avg Mat'l Unit Cost	Avg Labor Unit Cost	Avg Equip Unit Cost	Avg Total Unit Cost	Avg Price Incl O&P	Large Volume Avg Total Unit Cost	Large Volume Avg Price Incl O&P
Cleaning, wet											
Plain siding	Inst	SF	NA	.008	.02	.22	---	.24	**.36**	.16	**.23**
Exterior doors and trim											
Note: SF equals LF on trim less than 6" high	Inst	SF	NA	.009	.02	.25	---	.27	**.40**	.19	**.28**
Windows, wash and clean glass	Inst	SF	NA	.010	.02	.28	---	.30	**.44**	.19	**.28**
Porch floors and steps	Inst	SF	NA	.005	.02	.14	---	.16	**.23**	.13	**.19**
Acid wash											
Gutters and downspouts	Inst	SF	NA	.012	.02	.34	---	.36	**.53**	.24	**.36**
Sanding, light											
Porch floors and steps	Inst	SF	NA	.007	.02	.20	---	.22	**.32**	.16	**.23**
Sanding and puttying											
Plain siding	Inst	SF	NA	.009	.02	.25	---	.27	**.40**	.19	**.28**
Exterior doors and trim											
Note: SF equals LF on trim less than 6" high	Inst	SF	NA	.017	.02	.48	---	.50	**.74**	.33	**.49**

Description	Oper	Unit	Crew Size	Man-Hours Per Unit	Avg Mat'l Unit Cost	Avg Labor Unit Cost	Avg Equip Unit Cost	Avg Total Unit Cost	Avg Price Incl O&P	Large Volume Avg Total Unit Cost	Large Volume Avg Price Incl O&P
Puttying sash or reglazing											
Windows (30 SF glass/lb. glazing compound)	Inst	SF	NA	.051	.73	1.43	---	2.16	**2.99**	1.55	**2.10**
One coat application											
Excluding openings unless otherwise indicated											
Latex, flat (unless otherwise indicated)											
Plain siding											
Brush (300 SF/gal)	Inst	SF	NA	.015	.07	.42	---	.49	**.71**	.34	**.49**
Roller (275 SF/gal)	Inst	SF	NA	.012	.08	.34	---	.42	**.60**	.29	**.42**
Spray (325 SF/gal)	Inst	SF	NC	.007	.07	.20	.01	.28	**.39**	.21	**.30**
Shingle siding											
Brush (270 SF/gal)	Inst	SF	NA	.014	.08	.39	---	.47	**.68**	.32	**.46**
Roller (260 SF/gal)	Inst	SF	NA	.010	.08	.28	---	.36	**.51**	.27	**.38**
Spray (300 SF/gal)	Inst	SF	NC	.008	.07	.23	.01	.31	**.44**	.21	**.30**
Stucco											
Brush (135 SF/gal)	Inst	SF	NA	.015	.16	.42	---	.58	**.82**	.42	**.58**
Roller (125 SF/gal)	Inst	SF	NA	.010	.18	.28	---	.46	**.63**	.35	**.47**
Spray (150 SF/gal)	Inst	SF	NC	.008	.15	.23	.01	.39	**.53**	.27	**.37**
Cement walls, see Cement base paint, page 192											
Masonry block, brick, tile; masonry latex											
Brush (180 SF/gal)	Inst	SF	NA	.014	.12	.39	---	.51	**.73**	.35	**.49**
Roller (125 SF/gal)	Inst	SF	NA	.009	.18	.25	---	.43	**.59**	.32	**.43**
Spray (160 SF/gal)	Inst	SF	NC	.007	.14	.20	.01	.35	**.47**	.25	**.32**
Doors, exterior side only											
Brush (375 SF/gal)	Inst	SF	NA	.019	.06	.53	---	.59	**.87**	.42	**.61**
Roller (375 SF/gal)	Inst	SF	NA	.014	.06	.39	---	.45	**.66**	.30	**.44**
Windows, exterior side only, brush work (1500 SF/gal)	Inst	SF	NA	.019	.01	.53	---	.54	**.81**	.38	**.56**
Trim, less than 6" high, brush											
Note: SF equals LF on trim less than 6" high											
High gloss (300 SF/gal)	Inst	SF	NA	.014	.07	.39	---	.46	**.67**	.31	**.45**
Screens, full; high gloss											
Paint applied to wood only, brush work (700 SF/gal)	Inst	SF	NA	.023	.03	.65	---	.68	**1.00**	.45	**.67**
Paint applied to wood (brush) and wire (spray) (475 SF/gal)	Inst	SF	NC	.027	.05	.78	.05	.88	**1.28**	.56	**.83**
Storm windows and doors, 2 lites, brush work (340 SF/gal)	Inst	SF	NA	.036	.06	1.01	---	1.07	**1.59**	.73	**1.08**
Blinds or shutters											
Brush (120 SF/gal)	Inst	SF	NA	.094	.18	2.64	---	2.82	**4.17**	1.90	**2.80**
Spray (300 SF/gal)	Inst	SF	NC	.031	.07	.90	.06	1.03	**1.49**	.70	**1.00**
Gutters and downspouts, brush work (225 LF/gal), galvanized	Inst	SF	NA	.021	.10	.59	---	.69	**1.00**	.45	**.64**
Porch floors and steps, wood											
Brush (340 SF/gal)	Inst	SF	NA	.008	.06	.22	---	.28	**.41**	.20	**.28**
Roller (325 SF/gal)	Inst	SF	NA	.006	.07	.17	---	.24	**.33**	.17	**.24**
Shingle roofs											
Brush (135 SF/gal)	Inst	SF	NA	.012	.16	.34	---	.50	**.69**	.36	**.50**
Roller (125 SF/gal)	Inst	SF	NA	.009	.18	.25	---	.43	**.59**	.32	**.43**
Spray (150 SF/gal)	Inst	SF	NC	.006	.15	.17	.01	.33	**.44**	.25	**.32**
For custom colors, ADD	Inst	%		---	10.0	---	---	---	**---**	---	**---**

Description	Oper	Unit	Crew Size	Man-Hours Per Unit	Avg Mat'l Unit Cost	Avg Labor Unit Cost	Avg Equip Unit Cost	Avg Total Unit Cost	Avg Price Incl O&P	Large Volume Avg Total Unit Cost	Large Volume Avg Price Incl O&P
Cement base paint (epoxy concrete enamel)											
Cement walls, smooth finish											
Brush (120 SF/gal)	Inst	SF	NA	.009	.26	.25	---	.51	**.68**	.39	**.51**
Roller (110 SF/gal)	Inst	SF	NA	.006	.28	.17	---	.45	**.57**	.35	**.44**
Concrete porch floors and steps											
Brush (400 SF/gal)	Inst	SF	NA	.006	.08	.17	---	.25	**.34**	.18	**.25**
Roller (375 SF/gal)	Inst	SF	NA	.005	.08	.14	---	.22	**.30**	.15	**.21**
Stain											
Shingle siding											
Brush (180 SF/gal)	Inst	SF	NA	.015	.14	.42	---	.56	**.79**	.40	**.56**
Roller (170 SF/gal)	Inst	SF	NA	.011	.15	.31	---	.46	**.64**	.33	**.44**
Spray (200 SF/gal)	Inst	SF	NC	.008	.13	.23	.02	.38	**.52**	.27	**.36**
Shingle roofs											
Brush (180 SF/gal)	Inst	SF	NA	.013	.14	.37	---	.51	**.71**	.34	**.48**
Roller (170 SF/gal)	Inst	SF	NA	.010	.15	.28	---	.43	**.59**	.30	**.40**
Spray (200 SF/gal)	Inst	SF	NC	.006	.13	.17	.01	.31	**.42**	.24	**.31**
Two coat application											
Excluding openings unless otherwise indicated											
Latex, flat (unless otherwise indicated)											
Plain siding											
Brush (170 SF/gal)	Inst	SF	NA	.028	.13	.79	---	.92	**1.33**	.62	**.89**
Roller (155 SF/gal)	Inst	SF	NA	.022	.14	.62	---	.76	**1.09**	.51	**.73**
Spray (185 SF/gal)	Inst	SF	NC	.013	.12	.38	.02	.52	**.72**	.35	**.48**
Shingle siding											
Brush (150 SF/gal)	Inst	SF	NA	.026	.15	.73	---	.88	**1.27**	.60	**.85**
Roller (150 SF/gal)	Inst	SF	NA	.019	.15	.53	---	.68	**.97**	.49	**.69**
Spray (170 SF/gal)	Inst	SF	NC	.014	.13	.41	.03	.57	**.79**	.40	**.55**
Stucco											
Brush (90 SF/gal)	Inst	SF	NA	.026	.24	.73	---	.97	**1.37**	.69	**.96**
Roller (85 SF/gal)	Inst	SF	NA	.018	.26	.51	---	.77	**1.06**	.56	**.76**
Spray (100 SF/gal)	Inst	SF	NC	.014	.22	.41	.03	.66	**.89**	.48	**.64**
Cement wall, see Cement base paint, page 193											
Masonry block, brick, tile; masonry latex											
Brush (120 SF/gal)	Inst	SF	NA	.024	.18	.67	---	.85	**1.22**	.58	**.82**
Roller (85 SF/gal)	Inst	SF	NA	.016	.26	.45	---	.71	**.97**	.53	**.72**
Spray (105 SF/gal)	Inst	SF	NC	.012	.21	.35	.02	.58	**.78**	.43	**.57**
Doors, exterior side only											
Brush (215 SF/gal)	Inst	SF	NA	.037	.10	1.04	---	1.14	**1.67**	.76	**1.12**
Roller (215 SF/gal)	Inst	SF	NA	.027	.10	.76	---	.86	**1.25**	.60	**.86**
Windows, exterior side only, brush work (850 SF/gal)	Inst	SF	NA	.037	.03	1.04	---	1.07	**1.59**	.69	**1.03**
Trim, less than 6" high, brush											
Note: SF equals LF on trim less than 6" high											
High gloss (230 SF/gal)	Inst	SF	NA	.026	.10	.73	---	.83	**1.21**	.56	**.81**
Screens, full; high gloss											
Paint applied to wood only, brush work (400 SF/gal)	Inst	SF	NA	.043	.06	1.21	---	1.27	**1.88**	.84	**1.24**
Paint applied to wood (brush) and wire (spray) (270 SF/gal)	Inst	SF	NC	.051	.08	1.48	.10	1.66	**2.41**	1.13	**1.61**

Description	Oper	Unit	Crew Size	Man-Hours Per Unit	Avg Mat'l Unit Cost	Avg Labor Unit Cost	Avg Equip Unit Cost	Avg Total Unit Cost	Avg Price Incl O&P	Avg Total Unit Cost	Avg Price Incl O&P
			Costs Based On Small Volume							Large Volume	
Blinds or shutters											
Brush (65 SF/gal)	Inst	SF	NA	.174	.34	4.89	---	5.23	**7.73**	3.49	**5.14**
Spray (170 SF/gal)	Inst	SF	NC	.058	.13	1.68	.11	1.92	**2.78**	1.32	**1.89**
Gutters and downspouts, brush work (130 LF/gal), galvanized	Inst	SF	NA	.039	.17	1.10	---	1.27	**1.84**	.84	**1.21**
Porch floors and steps, wood											
Brush (195 SF/gal)	Inst	SF	NA	.015	.11	.42	---	.53	**.76**	.38	**.54**
Roller (185 SF/gal)	Inst	SF	NA	.012	.12	.34	---	.46	**.64**	.32	**.45**
Shingle roofs											
Brush (75 SF/gal)	Inst	SF	NA	.019	.29	.53	---	.82	**1.13**	.62	**.84**
Roller (70 SF/gal)	Inst	SF	NA	.015	.31	.42	---	.73	**.99**	.55	**.73**
Spray (85 SF/gal)	Inst	SF	NC	.012	.26	.35	.02	.63	**.84**	.47	**.62**
For custom colors, ADD	Inst	%	---	---	10.0	---	---	---	**---**	---	**---**
Cement base paint (epoxy concrete enamel)											
Cement walls, smooth finish											
Brush (80 SF/gal)	Inst	SF	NA	.017	.39	.48	---	.87	**1.17**	.64	**.84**
Roller (75 SF/gal)	Inst	SF	NA	.012	.41	.34	---	.75	**.98**	.57	**.74**
Concrete porch floors and steps											
Brush (225 SF/gal)	Inst	SF	NA	.012	.14	.34	---	.48	**.67**	.34	**.48**
Roller (210 SF/gal)	Inst	SF	NA	.012	.15	.34	---	.49	**.68**	.35	**.49**
Stain											
Shingle siding											
Brush (105 SF/gal)	Inst	SF	NA	.026	.25	.73	---	.98	**1.38**	.69	**.96**
Roller (100 SF/gal)	Inst	SF	NA	.021	.26	.59	---	.85	**1.18**	.59	**.80**
Spray (115 SF/gal)	Inst	SF	NC	.014	.23	.41	.03	.67	**.90**	.48	**.64**
Shingle roofs											
Brush (105 SF/gal)	Inst	SF	NA	.022	.25	.62	---	.87	**1.21**	.60	**.83**
Roller (100 SF/gal)	Inst	SF	NA	.016	.26	.45	---	.71	**.97**	.53	**.72**
Spray (115 SF/gal)	Inst	SF	NC	.011	.23	.32	.02	.57	**.76**	.41	**.54**

Paneling

Hardboard and plywood

Demolition

Description	Oper	Unit	Crew Size	Man-Hours Per Unit	Avg Mat'l Unit Cost	Avg Labor Unit Cost	Avg Equip Unit Cost	Avg Total Unit Cost	Avg Price Incl O&P	Avg Total Unit Cost	Avg Price Incl O&P
Sheets, plywood or hardboard	Demo	SF	LB	.013	---	.29	---	.29	**.43**	.20	**.30**
Boards, wood	Demo	SF	LB	.015	---	.33	---	.33	**.50**	.22	**.33**

Installation

Waste, nails, and adhesives not included

Economy hardboard

Description	Oper	Unit	Crew Size	Man-Hours Per Unit	Avg Mat'l Unit Cost	Avg Labor Unit Cost	Avg Equip Unit Cost	Avg Total Unit Cost	Avg Price Incl O&P	Avg Total Unit Cost	Avg Price Incl O&P
Presdwood, 4' x 8' sheets											
Standard											
1/8" T	Inst	SF	2C	.024	.22	.65	---	.87	**1.24**	.63	**.89**
1/4" T	Inst	SF	2C	.024	.35	.65	---	1.00	**1.39**	.72	**.99**
Tempered											
1/8" T	Inst	SF	2C	.024	.38	.65	---	1.03	**1.42**	.76	**1.04**
1/4" T	Inst	SF	2C	.024	.61	.65	---	1.26	**1.69**	.96	**1.27**

Description	Oper	Unit	Crew Size	Man-Hours Per Unit	Avg Mat'l Unit Cost	Avg Labor Unit Cost	Avg Equip Unit Cost	Avg Total Unit Cost	Avg Price Incl O&P	Large Volume Avg Total Unit Cost	Large Volume Avg Price Incl O&P
Duolux, 4' x 8' sheets											
Standard											
1/8" T	Inst	SF	2C	.024	.23	.65	---	.88	**1.25**	.64	**.90**
1/4" T	Inst	SF	2C	.024	.36	.65	---	1.01	**1.40**	.74	**1.01**
Tempered											
1/8" T	Inst	SF	2C	.024	.43	.65	---	1.08	**1.48**	.80	**1.08**
1/4" T	Inst	SF	2C	.024	.46	.65	---	1.11	**1.51**	.83	**1.12**
Particleboard, 40 lb. interior underlayment											
Nailed to floors											
3/8" T	Inst	SF	2C	.016	.43	.43	---	.86	**1.15**	.64	**.84**
1/2" T	Inst	SF	2C	.016	.49	.43	---	.92	**1.22**	.69	**.89**
5/8" T	Inst	SF	2C	.016	.56	.43	---	.99	**1.30**	.75	**.96**
3/4" T	Inst	SF	2C	.016	.63	.43	---	1.06	**1.38**	.82	**1.04**
Nailed to walls											
3/8" T	Inst	SF	2C	.017	.43	.46	---	.89	**1.19**	.67	**.88**
1/2" T	Inst	SF	2C	.017	.49	.46	---	.95	**1.26**	.72	**.93**
5/8" T	Inst	SF	2C	.017	.56	.46	---	1.02	**1.34**	.78	**1.00**
3/4" T	Inst	SF	2C	.017	.63	.46	---	1.09	**1.42**	.85	**1.08**
Masonite prefinished 4' x 8' panels											
1/4" T; oak and maple designs	Inst	SF	2C	.031	.57	.84	---	1.41	**1.93**	1.03	**1.38**
1/4" T; nutwood designs	Inst	SF	2C	.031	.65	.84	---	1.49	**2.02**	1.10	**1.46**
1/4" T; weathered white	Inst	SF	2C	.031	.65	.84	---	1.49	**2.02**	1.10	**1.46**
1/4" T; brick or stone designs	Inst	SF	2C	.031	.92	.84	---	1.76	**2.33**	1.35	**1.75**
Pegboard, 4' x 8' sheets											
Presdwood, tempered											
1/8" T	Inst	SF	2C	.024	.49	.65	---	1.14	**1.55**	.85	**1.14**
1/4" T	Inst	SF	2C	.024	.65	.65	---	1.30	**1.73**	1.00	**1.31**
Duolux, tempered											
1/8" T	Inst	SF	2C	.024	.72	.65	---	1.37	**1.81**	1.05	**1.37**
1/4" T	Inst	SF	2C	.024	1.03	.65	---	1.68	**2.17**	1.32	**1.68**
Unfinished hardwood plywood, applied with nails (nail heads filled)											
Ash-sen, flush face											
1/8" x 4' x 7', 8'	Inst	SF	CS	.041	1.09	1.04	---	2.13	**2.83**	1.63	**2.13**
3/16" x 4' x 8'	Inst	SF	CS	.041	1.30	1.04	---	2.34	**3.07**	1.81	**2.34**
1/4" x 4' x 8'	Inst	SF	CS	.041	1.54	1.04	---	2.58	**3.35**	2.00	**2.56**
1/4" x 4' x 10'	Inst	SF	CS	.041	1.72	1.04	---	2.76	**3.55**	2.17	**2.75**
1/2" x 4' x 10'	Inst	SF	CS	.041	1.96	1.04	---	3.00	**3.83**	2.38	**2.99**
3/4" x 4' x 8' vertical core	Inst	SF	CS	.041	2.13	1.04	---	3.17	**4.03**	2.53	**3.17**
3/4" x 4' x 8' lumber core	Inst	SF	CS	.041	3.04	1.04	---	4.08	**5.07**	3.32	**4.07**
Ash, V-grooved											
3/16" x 4' x 8'	Inst	SF	CS	.041	1.19	1.04	---	2.23	**2.94**	1.71	**2.22**
1/4" x 4' x 8'	Inst	SF	CS	.041	1.45	1.04	---	2.49	**3.24**	1.94	**2.49**
1/4" x 4' x 10'	Inst	SF	CS	.041	1.67	1.04	---	2.71	**3.50**	2.13	**2.71**

Description	Oper	Unit	Crew Size	Man-Hours Per Unit	Avg Mat'l Unit Cost	Avg Labor Unit Cost	Avg Equip Unit Cost	Avg Total Unit Cost	Avg Price Incl O&P	Avg Total Unit Cost	Avg Price Incl O&P
			Costs Based On Small Volume							Large Volume	
Birch, natural, "A" grade face											
Flush face											
1/8" x 4' x 8'	Inst	SF	CS	.041	.83	1.04	---	1.87	**2.53**	1.41	**1.88**
3/16" x 4' x 8'	Inst	SF	CS	.041	1.30	1.04	---	2.34	**3.07**	1.81	**2.34**
1/4" x 4' x 8'	Inst	SF	CS	.041	1.31	1.04	---	2.35	**3.08**	1.83	**2.36**
1/4" x 4' x 10'	Inst	SF	CS	.041	1.72	1.04	---	2.76	**3.55**	2.17	**2.75**
3/8" x 4' x 8'	Inst	SF	CS	.041	1.90	1.04	---	2.94	**3.76**	2.33	**2.94**
1/2" x 4' x 8'	Inst	SF	CS	.041	2.20	1.04	---	3.24	**4.11**	2.58	**3.22**
3/4" x 4' x 8' lumber core	Inst	SF	CS	.041	2.99	1.04	---	4.03	**5.01**	3.27	**4.02**
V-grooved											
1/4" x 4' x 8', mismatched	Inst	SF	CS	.041	1.64	1.04	---	2.68	**3.46**	2.10	**2.67**
1/4" x 4' x 8'	Inst	SF	CS	.041	1.65	1.04	---	2.69	**3.47**	2.11	**2.68**
1/4" x 4' x 10'	Inst	SF	CS	.041	1.87	1.04	---	2.91	**3.73**	2.31	**2.91**
Birch, select red											
1/4" x 4' x 8'	Inst	SF	CS	.041	1.65	1.04	---	2.69	**3.47**	2.11	**2.68**
Birch, select white											
1/4" x 4' x 8'	Inst	SF	CS	.041	1.65	1.04	---	2.69	**3.47**	2.11	**2.68**
Oak, flush face											
1/8" x 4' x 8'	Inst	SF	CS	.041	.82	1.04	---	1.86	**2.52**	1.39	**1.85**
1/4" x 4' x 8'	Inst	SF	CS	.041	1.04	1.04	---	2.08	**2.77**	1.58	**2.07**
1/2" x 4' x 8'	Inst	SF	CS	.041	2.03	1.04	---	3.07	**3.91**	2.44	**3.06**
Philippine mahogany											
Rotary cut											
1/8" x 4' x 8'	Inst	SF	CS	.041	.39	1.04	---	1.43	**2.02**	1.02	**1.43**
3/16" x 4' x 8'	Inst	SF	CS	.041	.58	1.04	---	1.62	**2.24**	1.18	**1.61**
1/4" x 4' x 8'	Inst	SF	CS	.041	.58	1.04	---	1.62	**2.24**	1.18	**1.61**
1/4" x 4' x 10'	Inst	SF	CS	.041	1.01	1.04	---	2.05	**2.74**	1.55	**2.04**
1/2" x 4' x 8'	Inst	SF	CS	.041	1.17	1.04	---	2.21	**2.92**	1.69	**2.20**
3/4" x 4' x 8' vert. core	Inst	SF	CS	.041	1.78	1.04	---	2.82	**3.62**	2.22	**2.81**
V-grooved											
3/16" x 4' x 8'	Inst	SF	CS	.041	.61	1.04	---	1.65	**2.28**	1.21	**1.65**
1/4" x 4' x 8'	Inst	SF	CS	.041	.68	1.04	---	1.72	**2.36**	1.28	**1.73**
1/4" x 4' x 10'	Inst	SF	CS	.041	1.08	1.04	---	2.12	**2.82**	1.62	**2.12**

1. **Dimensions (Lath)**

   a. Gypsum. Plain, perforated and insulating. Normally, each lath is 16" x 48" in $^{3}/_{8}$" or $^{1}/_{2}$" thicknesses; a five-piece bundle covers 3 SY.

   b. Wire. Only diamond and riblash are discussed here. Diamond lath is furnished in 27" x 96"-wide sheets covering 16 SY and 20 SY respectively.

   c. Wood. Can be fir, pine, redwood, spruce, etc. Bundles may consist of 50 or 100 pieces of $^{3}/_{8}$" x 1$^{1}/_{2}$" x 48" lath covering 3.4 SY and 6.8 SY respectively. There is usually a $^{3}/_{8}$" gap between lath.

2. **Dimensions (Plaster)**

Only two and three coat gypsum cement plaster are discussed here.

3. **Installation (Lath)**

   Laths are nailed. The types and size of nails vary with the type and thickness of lath. Quantity will vary with oc spacing of studs or joists. Only lath applied to wood will be considered here.

   a. Nails for gypsum lath. Common type is 13 gauge blued $^{19}/_{64}$" flathead, 1$^{1}/_{8}$" long, spaced approximately 4" oc, and 1$^{1}/_{4}$" long spaced approximately 5" oc for $^{3}/_{8}$" and $^{1}/_{2}$" lath respectively.

   b. Nails for wire lath. For ceiling, common is 1$^{1}/_{2}$" long 11 gauge barbed galvanized with a $^{7}/_{16}$" head diameter.

   c. Nails for wood lath. 3d fine common.

   d. Gypsum lath may be attached by the use of air-driven staples. Wire lath may be tied to support, usually with 18 gauge wire.

4. **Installation (Plaster)**

   Quantities of materials used to plaster vary with the type of lath and thickness of plaster.

   a. Two coat. Brown and finish coat.

   b. Three coat. Scratch, brown and finish coat.

   For types and quantities of material used, see **Notes on Material Pricing.**

5. **Notes on Labor**

   a. Lath. Average Manhour per Unit is based on what one lather can do in one day.

   b. Plaster. Average Manhour per Unit is based on what two plasterers and one laborer can do in one day.

Stucco

1. **Dimensions**

   a. 18 gauge wire

   b. 15 lb. felt paper

   c. 1" x 18 gauge galvanized netting

   d. Mortar of 1:3 mix

2. **Installation (Lathing)**

   a. 18 gauge wire stretched taut horizontally across studs at approximately 8" oc.

   b. 15 lb. felt paper placed over wire.

   c. 1" x 18 gauge galvanized netting placed over felt.

3. **Installation (Mortar)**

   a. The mortar mix used in this section is a 1:3 mix.

   b. One CY of mortar is comprised of 1 CY sand, 9 CF portland cement and 100 lbs. hydrated lime.

   c. For mortar requirements for 100 SY of stucco, see table below:

4. **Estimating Technique.** Determine area and deduct area of window and door openings. No waste has been included in the following figures unless otherwise noted. For waste, add 10% to total area.

	Cubic Yards Per CSY	
Stucco Thickness	**On Masonry**	**On Netting**
1/2"	1.50	1.75
5/8"	1.90	2.20
3/4"	2.20	2.60
1"	2.90	3.40

Description	Oper	Unit	Crew Size	Man-Hours Per Unit	Avg Mat'l Unit Cost	Avg Labor Unit Cost	Avg Equip Unit Cost	Avg Total Unit Cost	Avg Price Incl O&P	Large Volume Avg Total Unit Cost	Large Volume Avg Price Incl O&P
Plaster and stucco											
Remove both plaster or stucco and lath or netting to studs or sheathing											
Lath (wood or metal) and plaster, walls and ceilings											
2 coats	Demo	SY	LB	.188	---	4.13	---	4.13	**6.23**	2.70	**4.08**
3 coats	Demo	SY	LB	.205	---	4.50	---	4.50	**6.79**	2.92	**4.41**
Lath (only), nails included											
Gypsum lath, 16" x 48", applied with nails to ceilings or walls, 5% waste included, nails included (.067 lbs. per CY)											
3/8" T, perforated or plain	Inst	SY	LR	.154	5.23	4.12	---	9.35	**12.20**	6.87	**8.86**
1/2" T, perforated or plain	Inst	SY	LR	.154	5.55	4.12	---	9.67	**12.60**	7.12	**9.15**
3/8" T, insulating, aluminum foil back	Inst	SY	LR	.154	5.76	4.12	---	9.88	**12.90**	7.29	**9.35**
1/2" T, insulating, aluminum foil back	Inst	SY	LR	.154	6.06	4.12	---	10.18	**13.20**	7.53	**9.62**
For installation with staples, DEDUCT	Inst	SY	LR	-.038	---	1.02	---	1.02	**---**	.67	**---**
Metal lath, nailed to ceilings or walls, 5% waste and nails (0.067 lbs. per SY) included											
Diamond lath (junior mesh), 27" x 96" sheets, nailed to wood members @ 16" oc											
3.4 lb black painted	Inst	SY	LR	.103	3.08	2.76	---	5.84	**7.71**	4.26	**5.55**
3.4 lb galvanized	Inst	SY	LR	.103	3.87	2.76	---	6.63	**8.62**	4.89	**6.27**
Riblath, 3/8" high rib, 27" x 96" sheets, nailed to wood members @ 24" oc											
3.4 lb painted	Inst	SY	LR	.077	3.22	2.06	---	5.28	**6.82**	3.91	**4.98**
3.4 lb galvanized	Inst	SY	LR	.077	4.26	2.06	---	6.32	**8.01**	4.74	**5.93**
Wood lath, nailed to ceilings or walls, 5% waste and nails included, redwood, "A" grade and better											
3/8" x 1-1/2" x 48" @ 3/8" spacing	Inst	SY	LR	.154	5.39	4.12	---	9.51	**12.40**	6.99	**9.00**
Labor adjustments, lath											
For gypsum or wood lath above second floor ADD	Inst	SY	LR	.011	---	.29	---	.29	**.44**	.19	**.28**
For metal lath above second floor ADD	Inst	SY	LR	.026	---	.70	---	.70	**1.05**	.46	**.69**
Plaster (only), applied to ceilings and walls; 10% waste included											
Material price includes gypsum plaster, sand, hydrated lime and gauging plaster											
Two coats gypsum plaster											
On gypsum lath	Inst	SY	P3	.320	3.91	8.14	.48	12.53	**17.30**	8.72	**11.90**
On unit masonry, no lath	Inst	SY	P3	.329	4.26	8.37	.49	13.12	**18.00**	9.13	**12.40**

Description	Oper	Unit	Crew Size	Man-Hours Per Unit	Avg Mat'l Unit Cost	Avg Labor Unit Cost	Avg Equip Unit Cost	Avg Total Unit Cost	Avg Price Incl O&P	Large Volume Avg Total Unit Cost	Large Volume Avg Price Incl O&P
Three coats gypsum plaster											
On gypsum lath	Inst	SY	P3	.444	4.22	11.30	.66	16.18	**22.60**	11.12	**15.40**
On unit masonry, no lath	Inst	SY	P3	.444	4.62	11.30	.66	16.58	**23.00**	11.43	**15.70**
On wire lath	Inst	SY	P3	.462	6.88	11.80	.69	19.37	**26.40**	13.52	**18.20**
On wood lath	Inst	SY	P3	.462	4.52	11.80	.69	17.01	**23.60**	11.65	**16.10**
Labor adjustments, plaster											
For plaster above second floor ADD	Inst	SY	P3	.096	---	2.44	---	2.44	**3.69**	1.60	**2.42**
Thin coat plaster over sheetrock (in lieu of taping)	Inst	SY	P3	.296	.95	7.53	---	8.48	**12.50**	5.63	**8.24**
Stucco, exterior walls											
Netting, galvanized, 1" x 20 ga. x 36", with 18 ga. wire and 15 lb. felt	Inst	SY	LR	.205	8.10	5.49	---	13.59	**17.60**	10.03	**12.80**
Steel-Tex, 49" W x 11-1/2' L rolls, with felt backing	Inst	SY	LR	.154	7.58	4.12	---	11.70	**14.90**	8.76	**11.00**
1 coat work with float finish											
Over masonry	Inst	SY	P3	.320	.96	8.14	.48	9.58	**13.90**	6.40	**9.22**
2 coat work with float finish											
Over masonry	Inst	SY	P3	.615	2.71	15.70	.91	19.32	**27.70**	12.95	**18.40**
Over metal netting	Inst	SY	PE	.727	3.79	18.20	.65	22.64	**32.60**	15.24	**21.70**
3 coat work with float finish											
Over metal netting	Inst	SY	PE	.952	5.10	23.90	.85	29.85	**42.80**	20.02	**28.50**

Plumbing. See individual items

Range hoods

Metal finishes

Labor includes wiring and connection by electrician and installation by carpenter in stud-exposed structure only

Economy model, UL approved; mitered, welded construction; completely assembled and wired; includes fan, motor, washable aluminum filter, and light

Description	Oper	Unit	Crew Size	Man-Hours Per Unit	Avg Mat'l Unit Cost	Avg Labor Unit Cost	Avg Equip Unit Cost	Avg Total Unit Cost	Avg Price Incl O&P	Large Volume Avg Total Unit Cost	Large Volume Avg Price Incl O&P
24" wide											
3-1/4" x 10" duct, 160 CFM	Inst	Ea	ED	5.71	91.20	159.00	---	250.20	**341.00**	191.20	**256.00**
Non-ducted, 160 CFM	Inst	Ea	ED	5.71	99.20	159.00	---	258.20	**350.00**	198.10	**264.00**
30" wide											
7" round duct, 240 CFM	Inst	Ea	ED	5.71	108.00	159.00	---	267.00	**360.00**	206.10	**273.00**
3-1/4" x 10" duct, 160 CFM	Inst	Ea	ED	5.71	120.00	159.00	---	279.00	**373.00**	216.00	**285.00**
Non-ducted, 160 CFM	Inst	Ea	ED	5.71	143.00	159.00	---	302.00	**400.00**	236.00	**307.00**
36" wide											
7" round duct, 240 CFM	Inst	Ea	ED	5.71	160.00	159.00	---	319.00	**419.00**	251.00	**325.00**
3-1/4" x 10" duct, 160 CFM	Inst	Ea	ED	5.71	171.00	159.00	---	330.00	**432.00**	261.00	**336.00**
Non-ducted, 160 CFM	Inst	Ea	ED	5.71	200.00	159.00	---	359.00	**465.00**	285.00	**364.00**

Description	Oper	Unit	Crew Size	Man-Hours Per Unit	Avg Mat'l Unit Cost	Avg Labor Unit Cost	Avg Equip Unit Cost	Avg Total Unit Cost	Avg Price Incl O&P	Large Volume Avg Total Unit Cost	Large Volume Avg Price Incl O&P
Standard model, UL approved; deluxe mitered wrap-around styling; solid state fan control, infinite speed settings; removable filter; built-in damper and dual light assembly											
30" wide x 9" deep											
3-1/4" x 10" duct	Inst	Ea	ED	5.71	194.00	159.00	---	353.00	**459.00**	280.00	**359.00**
Non-ducted	Inst	Ea	ED	5.71	335.00	159.00	---	494.00	**621.00**	403.00	**500.00**
36" wide x 9" deep											
3-1/4" x 10" duct	Inst	Ea	ED	5.71	222.00	159.00	---	381.00	**491.00**	305.00	**387.00**
Non-ducted	Inst	Ea	ED	5.71	239.00	159.00	---	398.00	**511.00**	320.00	**404.00**
42" wide x 9" deep											
3-1/4" x 10" duct	Inst	Ea	ED	7.62	192.00	213.00	---	405.00	**535.00**	315.00	**411.00**
Non-ducted	Inst	Ea	ED	7.62	211.00	213.00	---	424.00	**557.00**	332.00	**431.00**
Decorator/designer model; solid state fan control, infinite speed settings; easy clean aluminum mesh grease filters; enclosed light assembly and switches											
Single faced hood-fan exhausts horizontally or vertically, 24" D canopy x 21" front to back											
30" wide	Inst	Ea	ED	7.62	365.00	213.00	---	578.00	**734.00**	466.00	**584.00**
36" wide	Inst	Ea	ED	7.62	393.00	213.00	---	606.00	**767.00**	491.00	**613.00**
Single faced, contemporary style, duct horizontally or vertically, 9" D canopy											
30" wide											
With 330 cfm power unit	Inst	Ea	ED	7.62	348.00	213.00	---	561.00	**714.00**	451.00	**567.00**
With 410 cfm power unit	Inst	Ea	ED	7.62	393.00	213.00	---	606.00	**767.00**	491.00	**613.00**
36" wide											
With 330 cfm power unit	Inst	Ea	ED	7.62	367.00	213.00	---	580.00	**737.00**	468.00	**587.00**
With 410 cfm power unit	Inst	Ea	ED	7.62	410.00	213.00	---	623.00	**787.00**	505.00	**630.00**
Material adjustments											
Stainless steel, ADD	Inst	%		---	33.0	---	---	---	**---**	---	**---**

Reinforcing steel. **See Concrete, page 60**

Resilient flooring

Sheet Products

Linoleum or vinyl sheet are normally installed either over a wood subfloor or a smooth concrete subfloor. When laid over wood, a layer of felt must first be laid in paste. This keeps irregularities in the wood from showing through. The amount of paste required to bond both felt and sheet is approximately 16 gallons (5% waste included) per 100 SY. When laid over smooth concrete subfloor, the sheet products can be bonded directly to the floor. However, the concrete subfloor should not be in direct contact with the ground because of excessive moisture. For bonding to concrete, 8 gallons of paste is required (5% waste included) per 100 SY. After laying the flooring over concrete or wood, wax is applied using 0.5 gallon per 100 SY. Paste, wax, felt (as needed), and 10% sheet waste are included in material costs.

Tile Products

All resilient tile can be bonded the same way as sheet products, either to smooth concrete or to felt over wood subfloor. When bonded to smooth concrete, a concrete primer (0.5 gal/100 SF) is first applied to seal the concrete. The tiles are then bonded to the sealed floor with resilient tile cement (0.6 gal/100 SF). On wood subfloors, felt is first laid in paste (0.9 gal/100 SF) and then the tiles are bonded to the felt with resilient tile emulsion (0.7 gal/100 SF). Bonding materials, felt (as needed), and 10% tile waste are included in material costs.

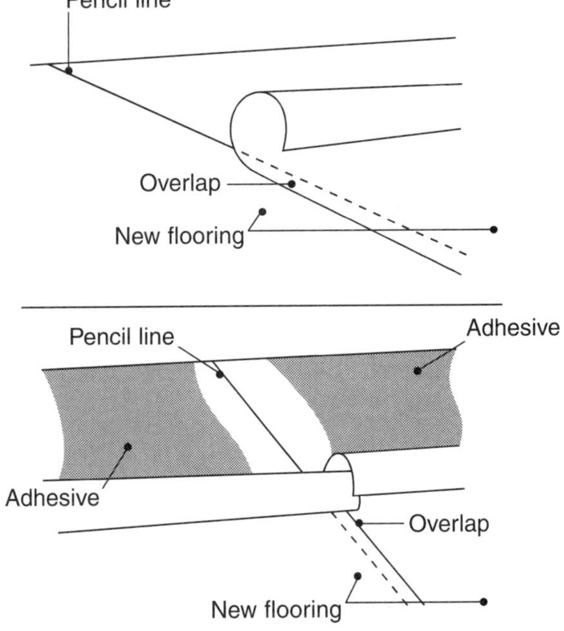

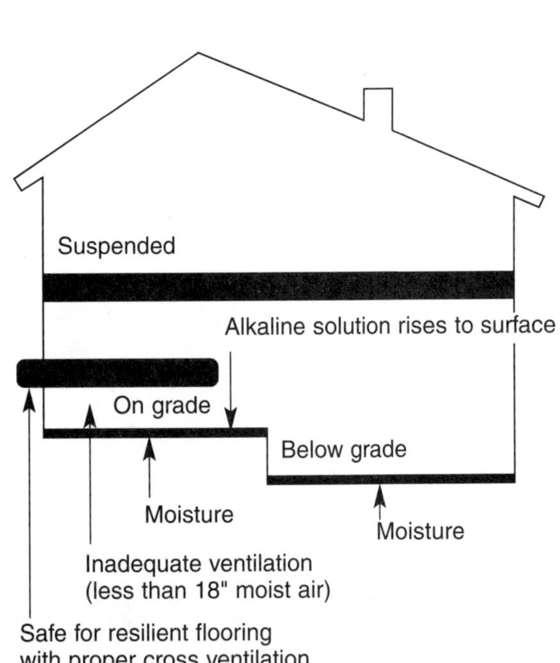

Description	Oper	Unit	Crew Size	Man-Hours Per Unit	Avg Mat'l Unit Cost	Avg Labor Unit Cost	Avg Equip Unit Cost	Avg Total Unit Cost	Avg Price Incl O&P	Avg Total Unit Cost	Avg Price Incl O&P
			Costs Based On Small Volume							Large Volume	
Resilient flooring											
Adhesive set sheet products	Demo	SY	LB	.143	---	3.14	---	3.14	**4.74**	2.20	**3.31**
Adhesive set tile products	Demo	SF	LB	.015	---	.33	---	.33	**.50**	.24	**.36**
Install over smooth concrete subfloor											
Tile											
Asphalt, 9" x 9", 1/8" thick, marbleized											
Group B colors, dark	Inst	SF	FB	.040	1.31	1.02	---	2.33	**3.03**	1.84	**2.36**
Group C colors, medium	Inst	SF	FB	.040	1.49	1.02	---	2.51	**3.24**	2.00	**2.54**
Group D colors, light	Inst	SF	FB	.040	1.49	1.02	---	2.51	**3.24**	2.00	**2.54**
Vinyl tile, including adhesive											
1/16" thick, 12" x 12"											
Vega II	Inst	SF	FB	.029	1.19	.74	---	1.93	**2.48**	1.53	**1.94**
.080" thick, 12" x 12"											
Designer slate	Inst	SF	FB	.029	2.55	.74	---	3.29	**4.04**	2.70	**3.28**
1/8" thick, 12" x 12"											
Embassy oak	Inst	SF	FB	.029	3.15	.74	---	3.89	**4.73**	3.21	**3.87**
Majestic slate	Inst	SF	FB	.029	2.55	.74	---	3.29	**4.04**	2.70	**3.28**
Mediterranean marble	Inst	SF	FB	.029	2.79	.74	---	3.53	**4.32**	2.90	**3.51**
Pecan	Inst	SF	FB	.029	3.27	.74	---	4.01	**4.87**	3.31	**3.98**
Plaza brick	Inst	SF	FB	.029	4.93	.74	---	5.67	**6.78**	4.73	**5.62**
Plymouth plank	Inst	SF	FB	.029	3.56	.74	---	4.30	**5.20**	3.56	**4.27**
Teak	Inst	SF	FB	.029	3.02	.74	---	3.76	**4.58**	3.10	**3.74**
Terrazzo	Inst	SF	FB	.029	2.43	.74	---	3.17	**3.90**	2.59	**3.16**
Vinyl tile, self stick											
.080" thick, 12" x 12"											
Elite	Inst	SF	FB	.035	1.90	.90	---	2.80	**3.52**	2.27	**2.83**
Stylglo	Inst	SF	FB	.035	1.42	.90	---	2.32	**2.97**	1.86	**2.36**
1/16" thick, 12" x 12"											
Decorator	Inst	SF	FB	.035	.95	.90	---	1.85	**2.43**	1.46	**1.90**
Proclaim	Inst	SF	FB	.035	1.24	.90	---	2.14	**2.76**	1.71	**2.18**
Composition vinyl tile											
3/32", 12" x 12"											
Designer colors	Inst	SF	FB	.035	1.06	.90	---	1.96	**2.55**	1.55	**2.00**
Standard colors (marbleized)	Inst	SF	FB	.035	1.37	.90	---	2.27	**2.91**	1.81	**2.30**
1/8", 12" x 12"											
Designer colors, pebbled	Inst	SF	FB	.035	2.02	.90	---	2.92	**3.66**	2.37	**2.94**
Standard colors (marbleized)	Inst	SF	FB	.035	1.19	.90	---	2.09	**2.70**	1.66	**2.13**
Solid black or white	Inst	SF	FB	.035	1.90	.90	---	2.80	**3.52**	2.27	**2.83**
Solid colors	Inst	SF	FB	.035	2.08	.90	---	2.98	**3.73**	2.43	**3.01**
Sheet vinyl											
6' wide, no wax, Armstrong											
Designer Solarian (.070")	Inst	SY	FB	.367	33.20	9.40	---	42.60	**52.20**	35.08	**42.60**
Designer Solarian II (.090")	Inst	SY	FB	.367	42.70	9.40	---	52.10	**63.20**	43.18	**51.90**
Imperial Accotone (.065")	Inst	SY	FB	.367	10.70	9.40	---	20.10	**26.30**	15.74	**20.30**
Solarian Supreme (.090")	Inst	SY	FB	.367	45.10	9.40	---	54.50	**65.90**	45.28	**54.30**
Sundial Solarian (.077")	Inst	SY	FB	.367	16.60	9.40	---	26.00	**33.10**	20.88	**26.20**

Description	Oper	Unit	Crew Size	Man-Hours Per Unit	Avg Mat'l Unit Cost	Avg Labor Unit Cost	Avg Equip Unit Cost	Avg Total Unit Cost	Avg Price Incl O&P	Avg Total Unit Cost (Large Volume)	Avg Price Incl O&P (Large Volume)
6' wide, Mannington											
Vega (.080")	Inst	SY	FB	.367	9.50	9.40	---	18.90	**24.90**	14.72	**19.20**
Vinyl Ease (.073")	Inst	SY	FB	.367	8.31	9.40	---	17.71	**23.60**	13.70	**18.00**
6' wide, Tarkett											
Preference (.065")	Inst	SY	FB	.367	7.72	9.40	---	17.12	**22.90**	13.19	**17.40**
Softred (.062")	Inst	SY	FB	.367	9.50	9.40	---	18.90	**24.90**	14.72	**19.20**
Install over wood subfloor											
Tile											
Asphalt, 9" x 9", 1/8" thick, marbleized											
Group B colors, dark	Inst	SF	FB	.046	1.31	1.18	---	2.49	**3.26**	1.94	**2.51**
Group C colors, medium	Inst	SF	FB	.046	1.49	1.18	---	2.67	**3.47**	2.10	**2.69**
Group D colors, light	Inst	SF	FB	.046	1.49	1.18	---	2.67	**3.47**	2.10	**2.69**
Vinyl tile, including adhesive											
1/16" thick, 12" x 12"											
Vega II	Inst	SF	FB	.034	1.19	.87	---	2.06	**2.67**	1.63	**2.09**
.080" thick, 12" x 12"											
Designer slate	Inst	SF	FB	.034	2.55	.87	---	3.42	**4.23**	2.80	**3.43**
1/8" thick, 12" x 12"											
Embassy oak	Inst	SF	FB	.034	3.15	.87	---	4.02	**4.92**	3.31	**4.02**
Majestic slate	Inst	SF	FB	.034	2.55	.87	---	3.42	**4.23**	2.80	**3.43**
Mediterranean marble	Inst	SF	FB	.034	2.79	.87	---	3.66	**4.51**	3.00	**3.66**
Pecan	Inst	SF	FB	.034	3.27	.87	---	4.14	**5.06**	3.41	**4.14**
Plaza brick	Inst	SF	FB	.034	4.93	.87	---	5.80	**6.97**	4.83	**5.77**
Plymouth plank	Inst	SF	FB	.034	3.56	.87	---	4.43	**5.39**	3.66	**4.42**
Teak	Inst	SF	FB	.034	3.02	.87	---	3.89	**4.77**	3.20	**3.89**
Terrazzo	Inst	SF	FB	.034	2.43	.87	---	3.30	**4.09**	2.69	**3.31**
Vinyl tile, self stick											
.080" thick, 12" x 12"											
Elite	Inst	SF	FB	.034	1.90	.87	---	2.77	**3.48**	2.24	**2.79**
Stylglo	Inst	SF	FB	.034	1.42	.87	---	2.29	**2.93**	1.83	**2.32**
1/16" thick, 12" x 12"											
Decorator	Inst	SF	FB	.034	.95	.87	---	1.82	**2.39**	1.43	**1.86**
Proclaim	Inst	SF	FB	.034	1.24	.87	---	2.11	**2.72**	1.68	**2.15**
Composition vinyl tile											
3/32", 12" x 12"											
Designer colors	Inst	SF	FB	.034	1.06	.87	---	1.93	**2.52**	1.52	**1.96**
Standard colors (marbleized)	Inst	SF	FB	.034	1.37	.87	---	2.24	**2.87**	1.78	**2.26**
1/8", 12" x 12"											
Designer colors, pebbled	Inst	SF	FB	.034	2.02	.87	---	2.89	**3.62**	2.34	**2.91**
Standard colors (marbleized)	Inst	SF	FB	.034	1.19	.87	---	2.06	**2.67**	1.63	**2.09**
Solid black or white	Inst	SF	FB	.034	1.90	.87	---	2.77	**3.48**	2.24	**2.79**
Solid colors	Inst	SF	FB	.034	2.08	.87	---	2.95	**3.69**	2.40	**2.97**
Sheet vinyl											
6' wide, no wax, Armstrong											
Designer Solarian (.070")	Inst	SY	FB	.429	33.20	11.00	---	44.20	**54.60**	36.18	**44.20**
Designer Solarian II (.090")	Inst	SY	FB	.429	42.70	11.00	---	53.70	**65.50**	44.28	**53.60**
Imperial Accotone (.065")	Inst	SY	FB	.429	10.70	11.00	---	21.70	**28.70**	16.84	**22.00**
Solarian Supreme (.090")	Inst	SY	FB	.429	45.10	11.00	---	56.10	**68.30**	46.38	**55.90**
Sundial Solarian (.077")	Inst	SY	FB	.429	16.60	11.00	---	27.60	**35.50**	21.98	**27.80**

			Costs Based On Small Volume							Large Volume	
Description	Oper	Unit	Crew Size	Man-Hours Per Unit	Avg Mat'l Unit Cost	Avg Labor Unit Cost	Avg Equip Unit Cost	Avg Total Unit Cost	Avg Price Incl O&P	Avg Total Unit Cost	Avg Price Incl O&P
6' wide, Mannington											
Vega (.080")	Inst	SY	FB	.429	9.50	11.00	---	20.50	**27.30**	15.82	**20.80**
Vinyl Ease (.073")	Inst	SY	FB	.429	8.31	11.00	---	19.31	**25.90**	14.80	**19.60**
6' wide, Tarkett											
Preference (.065")	Inst	SY	FB	.429	7.72	11.00	---	18.72	**25.30**	14.29	**19.10**
Softred (.062")	Inst	SY	FB	.429	9.50	11.00	---	20.50	**27.30**	15.82	**20.80**
Related materials and operations											
Top set base											
Vinyl											
2-1/2" H											
Colors	Inst	LF	FB	.051	.45	1.31	---	1.76	**2.46**	1.30	**1.81**
Wood grain	Inst	LF	FB	.051	.60	1.31	---	1.91	**2.64**	1.44	**1.97**
4" H											
Colors	Inst	LF	FB	.051	.53	1.31	---	1.84	**2.56**	1.37	**1.89**
Wood grain	Inst	LF	FB	.051	.72	1.31	---	2.03	**2.77**	1.53	**2.08**
Linoleum cove, 7/16" x 1-1/4"											
Softwood	Inst	LF	FB	.073	.53	1.87	---	2.40	**3.40**	1.76	**2.46**

Retaining walls, see Concrete, page 61, or Masonry, page 164

Roofing

Fiberglass Shingles, 225 lb., three tab strip. Three bundle/square; 20 year

1. **Dimensions.** Each shingle is 12" x 36". With a 5" exposure to the weather, 80 shingles are required to cover one square (100 SF).
2. **Installation**
   a. Over wood. After scatter-nailing one ply of 15 lb. felt, the shingles are installed. Four nails per shingle is customary.
   b. Over existing roofing. 15 lb. felt is not required. Shingles are installed the same as over wood, except longer nails are used, i.e., 1 1/4" in lieu of 1".
3. **Estimating Technique.** Determine roof area and add a percentage for starters, ridge, and valleys. The percent to be added varies, but generally:
   a. For plain gable and hip – add 10%.
   b. For gable or hip with dormers or intersecting roof(s) – add 15%.
   c. For gable or hip with dormers and intersecting roof(s) – add 20%.

When ridge or hip shingles are special ordered, reduce the above percentages by approximately 50%. Then apply unit costs to the area calculated (including the allowance above).

Mineral Surfaced Roll Roofing, 90 lb.

1. **Dimensions.** A roll is 36" wide x 36'-0" long, or 108 SF. It covers one square (100 SF), after including 8 SF for head and end laps.
2. **Installation.** Usually, lap cement and 7/8" galvanized nails are furnished with each roll.
   a. Over wood. Roll roofing is usually applied directly to sheathing with 7/8" galvanized nails. End and head laps are normally 6" and 2" or 3" respectively.
   b. Over existing roofing. Applied the same as over wood, except nails are not less than 1 1/4" long.
   c. Roll roofing may be installed by the exposed nail or the concealed nail method. In the exposed nail method, nails are visible at laps, and head laps are usually 2". In the concealed nail method, no nails are visible, the head lap is a minimum of 3", and there is a 9"-wide strip of roll installed along rakes and eaves.
3. **Estimating Technique.** Determine the area and add a percentage for hip and/or ridge. The percentage to be added varies, but generally:
   a. For plain gable or hip – add 5%.
   b. For cut-up roof – add 10%.

If metal ridge and/or hip are used, reduce the above percentages by approximately 50%. Then apply unit costs to the area calculated (including the allowance above).

Wood Shingles

1. **Dimensions.** Shingles are available in 16", 18", and 24" lengths and in uniform widths of 4", 5", or 6". But they are commonly furnished in random widths averaging 4" wide.
2. **Installation.** The normal exposure to weather for 16", 18" and 24" shingles on roofs with 1/4 or steeper pitch is 5", 5 1/2", and 7 1/2" respectively. Where the slope is less than 1/4, the exposure is usually reduced. Generally, 3d commons are used when shingles are applied to strip sheathing. 5d commons are used when shingles are applied over existing shingles. Two nails per shingle is the general rule, but on some roofs, one nail per shingle is used.
3. **Estimating Technique.** Determine the roof area and add a percentage for waste, starters, ridge, and hip shingles.
   a. Wood shingles. The amount of exposure determines the percentage of 1 square of shingles required to cover 100 SF of roof. Multiply the material cost per square of shingles by the appropriate percent (from the table on the next page) to determine the cost to cover 100 SF of roof with wood shingles. The table does not include cutting waste. NOTE: Nails and the exposure factors in this table have already been calculated into the Average Unit Material Costs.
   b. Nails. The weight of nails required per square varies with the size of the nail and with the shingle exposure. Multiply cost per pound of nails by the appropriate pounds per square. In the table on the next page, pounds of nails per square are based on two nails per shingle. For one nail per shingle, deduct 50%.
   c. The percentage to be added for starters, hip, and ridge shingles varies, but generally:
      1) Plain gable and hip – add 10%.
      2) Gable or hip with dormers or intersecting roof(s) – add 15%.
      3) Gable or hip with dormers and intersecting roof(s) – add 20%.

When ridge and/or hip shingles are special ordered, reduce these percentages by 50%. Then

Shingle Length	Exposure	% of Sq. Required to Cover 100 SF	Lbs of Nails per Square (2 Nails/Shingle) 3d	5d
24"	7½"	100	2.3	2.7
24"	7"	107	2.4	2.9
24"	6½"	115	2.6	3.2
24"	6"	125	2.8	3.4
24"	5¾"	130	2.9	3.6
18"	5½"	100	3.1	3.7
18"	5"	110	3.4	4.1
18"	4½"	122	3.8	4.6
16"	5"	100	3.4	4.1
16"	4½"	111	3.8	4.6
16"	4"	125	4.2	5.1
16"	3¾"	133	4.5	5.5

apply unit costs to the area calculated (including the allowance above).

Built-Up Roofing

1. **Dimensions**

   a. 15 lb. felt. Rolls are 36" wide x 144'-0" long or 432 SF – 4 squares per roll. 32 SF (or 8 SF/sq) is for laps.

   b. 30 lb. felt. Rolls are 36" wide x 72'-0" long or 216 SF – 2 squares per roll. 16 SF (or 8 SF/sq) is for laps.

   c. Asphalt. Available in 100 lb. cartons. Average asphalt usage is:

      1) Top coat without gravel – 35 lbs. per square.

      2) Top coat with gravel – 60 lbs. per square.

      3) Each coat (except top coat) – 25 lbs. per square.

   d. Tar. Available in 550 lb. kegs. Average tar usage is:

      1) Top coat without gravel – 40 lbs. per square.

      2) Top coat without gravel – 75 lbs. per square.

      3) Each coat (except top coat) – 30 lbs. per square.

   e. Gravel or slag. Normally, 400 lbs. of gravel or 275 lbs. of slag are used to cover one square.

2. **Installation**

   a. Over wood. Normally, one or two plies of felt are applied by scatter nailing before hot mopping is commenced. Subsequent plies of felt and gravel or slag are imbedded in hot asphalt or tar.

   b. Over concrete. Every ply of felt and gravel or slag is imbedded in hot asphalt or tar.

3. **Estimating Technique**

   a. For buildings with parapet walls, use the outside dimensions of the building to determine the area. This area is usually sufficient to include the flashing.

   b. For buildings without parapet walls, determine the area.

   c. Don't deduct any opening less than 10'-0" x 10'-0" (100 SF). Deduct 50% of the opening for openings larger than 100 SF but smaller than 300 SF, and 100% of the openings exceeding 300 SF.

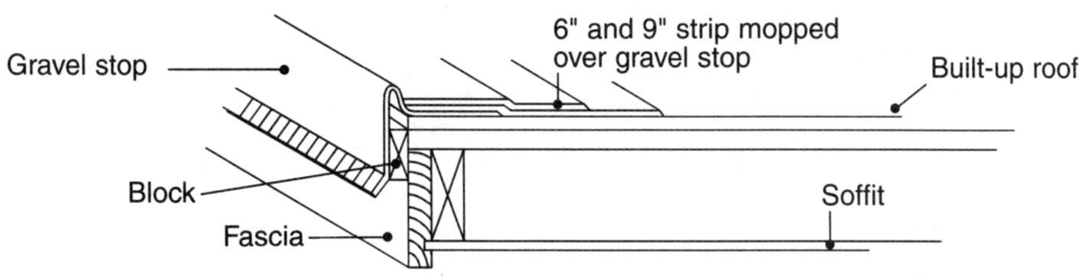

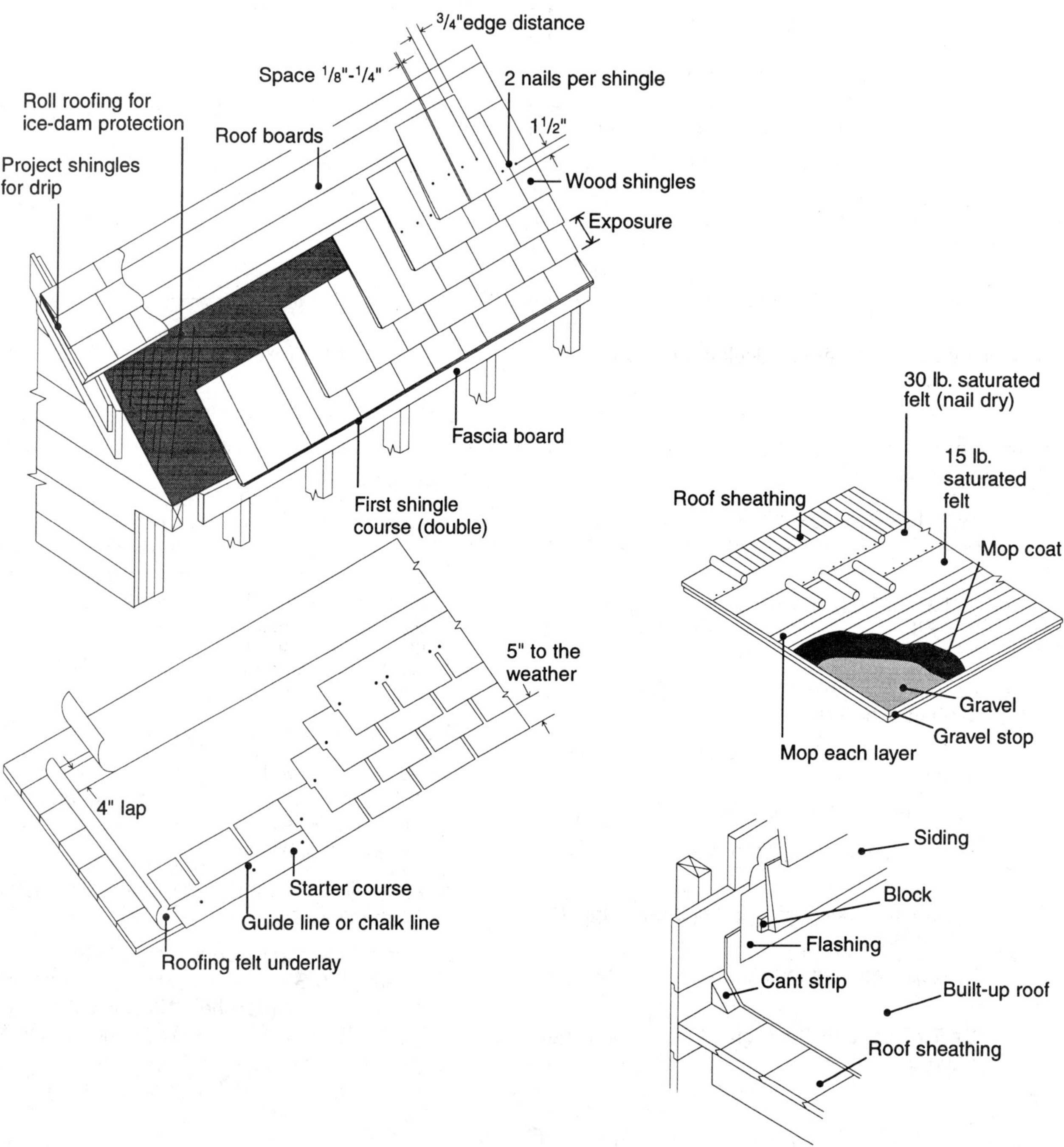
3/4"edge distance
Space 1/8"-1/4"
2 nails per shingle
Roll roofing for
ice-dam protection
Roof boards
1 1/2"
Project shingles
for drip
Wood shingles
Exposure
Fascia board
First shingle
course (double)
30 lb. saturated
felt (nail dry)
15 lb.
saturated
felt
Roof sheathing
Mop coat
5" to the
weather
Gravel
Gravel stop
Mop each layer
4" lap
Starter course
Guide line or chalk line
Roofing felt underlay
Siding
Block
Flashing
Cant strip
Built-up roof
Roof sheathing

Description	Oper	Unit	Crew Size	Man-Hours Per Unit	Avg Mat'l Unit Cost	Avg Labor Unit Cost	Avg Equip Unit Cost	Avg Total Unit Cost	Avg Price Incl O&P	Large Volume Avg Total Unit Cost	Large Volume Avg Price Incl O&P
Roofing											
Aluminum, nailed to wood											
Corrugated (2-1/2"), 26" W, with 3-3/4" side lap and 6" end lap	Demo	Sq	LB	2.00	---	43.90	---	43.90	**66.30**	35.10	**53.00**
Corrugated (2-1/2"), 26" W, with 3-3/4" side lap and 6" end lap											
0.0175" thick											
Natural	Inst	Sq	UC	2.67	144.00	69.10	---	213.10	**270.00**	178.10	**225.00**
Painted	Inst	Sq	UC	2.67	162.00	69.10	---	231.10	**290.00**	194.10	**242.00**
0.019" thick											
Natural	Inst	Sq	UC	2.67	144.00	69.10	---	213.10	**270.00**	178.10	**225.00**
Painted	Inst	Sq	UC	2.67	162.00	69.10	---	231.10	**290.00**	194.10	**242.00**
Composition, asphalt or fiberglass											
Shingle, 240#/Sq, strip, 3 tab, 5" exposure	Demo	Sq	LB	1.25	---	27.40	---	27.40	**41.40**	22.00	**33.10**
Shingles nailed; 3 tab strips, 12" x 36" with 5" exp.; 240#/Sq, seal down type											
Over existing roofing											
Gable, plain	Inst	Sq	RL	1.67	42.50	47.90	---	90.40	**124.00**	74.60	**102.00**
Gable with dormers	Inst	Sq	RL	1.85	42.90	53.10	---	96.00	**133.00**	79.20	**109.00**
Gable with intersecting roofs	Inst	Sq	RL	1.85	43.30	53.10	---	96.40	**133.00**	79.50	**109.00**
Gable w/dormers & intersections	Inst	Sq	RL	2.08	43.70	59.70	---	103.40	**144.00**	85.20	**118.00**
Hip, plain	Inst	Sq	RL	1.75	42.90	50.20	---	93.10	**128.00**	76.90	**105.00**
Hip with dormers	Inst	Sq	RL	1.95	43.30	56.00	---	99.30	**138.00**	81.80	**113.00**
Hip with intersecting roofs	Inst	Sq	RL	1.95	43.70	56.00	---	99.70	**138.00**	82.10	**113.00**
Hip with dormers & intersections	Inst	Sq	RL	2.21	44.10	63.40	---	107.50	**150.00**	88.50	**123.00**
Over wood decks, includes felt											
Gable, plain	Inst	Sq	RL	1.79	46.00	51.40	---	97.40	**134.00**	80.30	**110.00**
Gable with dormers	Inst	Sq	RL	2.00	46.40	57.40	---	103.80	**143.00**	85.50	**118.00**
Gable with intersecting roofs	Inst	Sq	RL	2.00	46.80	57.40	---	104.20	**144.00**	85.90	**118.00**
Gable w/dormers & intersections	Inst	Sq	RL	2.27	47.10	65.10	---	112.20	**156.00**	92.50	**128.00**
Hip, plain	Inst	Sq	RL	1.88	46.40	53.90	---	100.30	**138.00**	82.60	**113.00**
Hip with dormers	Inst	Sq	RL	2.12	46.80	60.80	---	107.60	**149.00**	88.50	**122.00**
Hip with intersecting roofs	Inst	Sq	RL	2.12	47.10	60.80	---	107.90	**150.00**	88.80	**122.00**
Hip with dormers & intersections	Inst	Sq	RL	2.43	47.50	69.70	---	117.20	**164.00**	96.30	**134.00**
Related materials and operations											
Ridge or hip roll, 90# mineral surfaced	Inst	LF	2R	.025	---	.76	---	.76	**1.19**	.61	**.95**
Valley roll, 90# mineral surfaced	Inst	LF	2R	.025	.31	.76	---	1.07	**1.55**	.87	**1.25**
Ridge or hip shingles, 9" x 12" with 5" exp.; 80 pieces/bundle @ 33.3 LF/bundle											
Over existing roofing	Inst	LF	2R	.029	.78	.88	---	1.66	**2.28**	1.37	**1.87**
Over wood decks	Inst	LF	2R	.030	.78	.91	---	1.69	**2.33**	1.40	**1.91**
Built-up or membrane roofing											
Over existing roofing											
One mop coat over smooth surface											
With asphalt	Inst	Sq	RS	.444	10.40	12.70	---	23.10	**31.80**	19.20	**26.30**
With tar	Inst	Sq	RS	.444	17.10	12.70	---	29.80	**39.60**	25.00	**33.00**

Description	Oper	Unit	Crew Size	Man-Hours Per Unit	Avg Mat'l Unit Cost	Avg Labor Unit Cost	Avg Equip Unit Cost	Avg Total Unit Cost	Avg Price Incl O&P	Large Volume Avg Total Unit Cost	Large Volume Avg Price Incl O&P
Remove gravel, mop one coat, redistribute old gravel											
With asphalt	Inst	Sq	RS	1.13	16.10	32.20	---	48.30	**69.10**	39.90	**56.70**
With tar	Inst	Sq	RS	1.13	26.40	32.20	---	58.60	**81.00**	48.80	**66.90**
Mop in one 30# cap sheet, plus smooth top mop coat											
With asphalt	Inst	Sq	RS	1.05	23.80	30.00	---	53.80	**74.40**	44.60	**61.40**
With tar	Inst	Sq	RS	1.05	34.40	30.00	---	64.40	**86.60**	53.70	**71.90**
Remove gravel, mop in one 30# cap sheet, mop in and redistribute old gravel											
With asphalt	Inst	Sq	RS	1.82	29.50	51.90	---	81.40	**115.00**	67.00	**94.30**
With tar	Inst	Sq	RS	1.82	43.70	51.90	---	95.60	**132.00**	79.20	**108.00**
Mop in two 15# felt plies, plus smooth top mop coat											
With asphalt	Inst	Sq	RS	1.67	29.50	47.70	---	77.20	**109.00**	63.50	**89.00**
With tar	Inst	Sq	RS	1.67	44.00	47.70	---	91.70	**125.00**	75.90	**103.00**
Remove gravel, mop in two 15# felt plies, mop in and redistribute old gravel											
With asphalt	Inst	Sq	RS	2.40	35.30	68.50	---	103.80	**148.00**	85.40	**121.00**
With tar	Inst	Sq	RS	2.40	53.30	68.50	---	121.80	**169.00**	100.90	**139.00**
Over smooth wood or concrete deck											
3 ply											
With gravel	Demo	Sq	LB	1.82	---	40.00	---	40.00	**60.30**	31.80	**48.10**
Without gravel	Demo	Sq	LB	1.43	---	31.40	---	31.40	**47.40**	25.00	**37.80**
5 ply											
With gravel	Demo	Sq	LB	2.00	---	43.90	---	43.90	**66.30**	35.10	**53.00**
Without gravel	Demo	Sq	LB	1.54	---	33.80	---	33.80	**51.00**	27.00	**40.80**
Installed over smooth wood decks											
Nail one 15# felt ply, plus mop in one 90# mineral surfaced ply											
With asphalt	Inst	Sq	RS	.750	39.90	21.40	---	61.30	**79.40**	51.30	**66.20**
With tar	Inst	Sq	RS	.750	44.40	21.40	---	65.80	**84.60**	55.20	**70.60**
Nail one and mop in two 15# felt plies, plus smooth top mop coat											
With asphalt	Inst	Sq	RS	1.82	34.90	51.90	---	86.80	**122.00**	71.50	**99.60**
With tar	Inst	Sq	RS	1.82	49.30	51.90	---	101.20	**138.00**	84.00	**114.00**
Nail one and mop in two 15# felt plies, plus mop in and distribute gravel											
With asphalt	Inst	Sq	RS	2.40	44.60	68.50	---	113.10	**159.00**	93.90	**131.00**
With tar	Inst	Sq	RS	2.40	62.60	68.50	---	131.10	**180.00**	109.50	**149.00**
Nail one and mop in three 15# felt plies, plus mop in and distribute gravel											
With asphalt	Inst	Sq	RS	3.00	54.20	85.60	---	139.80	**197.00**	115.90	**162.00**
With tar	Inst	Sq	RS	3.00	76.10	85.60	---	161.70	**222.00**	134.80	**184.00**
Nail one and mop in four 15# plies, plus mop in and distribute gravel											
With asphalt	Inst	Sq	RS	3.75	63.80	107.00	---	170.80	**241.00**	141.30	**198.00**
With tar	Inst	Sq	RS	3.75	89.50	107.00	---	196.50	**271.00**	163.50	**224.00**
Installed over smooth concrete decks											
Nail one 15# felt ply, plus mop in one 90# mineral surfaced ply											
With asphalt	Inst	Sq	RS	.968	45.60	27.60	---	73.20	**95.80**	61.30	**79.70**
With tar	Inst	Sq	RS	.968	53.70	27.60	---	81.30	**105.00**	68.20	**87.70**
Nail one and mop in two 15# felt plies, plus smooth top mop coat											
With asphalt	Inst	Sq	RS	2.22	39.10	63.30	---	102.40	**144.00**	84.60	**119.00**
With tar	Inst	Sq	RS	2.22	57.40	63.30	---	120.70	**165.00**	100.40	**137.00**
Nail one and mop in two 15# felt plies, plus mop in and distribute gravel											
With asphalt	Inst	Sq	RS	2.86	48.90	81.60	---	130.50	**184.00**	108.10	**152.00**
With tar	Inst	Sq	RS	2.86	70.70	81.60	---	152.30	**209.00**	127.00	**173.00**

Description	Oper	Unit	Crew Size	Man-Hours Per Unit	Avg Mat'l Unit Cost	Avg Labor Unit Cost	Avg Equip Unit Cost	Avg Total Unit Cost	Avg Price Incl O&P	Large Volume Avg Total Unit Cost	Large Volume Avg Price Incl O&P
Nail one and mop in three 15# felt plies, plus mop in and distribute gravel											
With asphalt	Inst	Sq	RS	3.53	58.40	101.00	---	159.40	**225.00**	131.60	**185.00**
With tar	Inst	Sq	RS	3.53	84.10	101.00	---	185.10	**255.00**	153.80	**211.00**
Nail one and mop in four 15# plies, plus mop in and distribute gravel											
With asphalt	Inst	Sq	RS	4.00	68.00	114.00	---	182.00	**257.00**	150.70	**212.00**
With tar	Inst	Sq	RS	4.00	97.60	114.00	---	211.60	**291.00**	176.20	**241.00**
Clay tile											
2 piece interlocking	Demo	Sq	LB	2.00	---	43.90	---	43.90	**66.30**	35.10	**53.00**
1 piece	Demo	Sq	LB	1.82	---	40.00	---	40.00	**60.30**	31.80	**48.10**
Clay tile; over wood; includes felt											
Interlocking tile shingles											
Early American	Inst	Sq	RG	5.56	411.00	153.00	---	564.00	**713.00**	473.00	**596.00**
Lanai	Inst	Sq	RG	5.56	455.00	153.00	---	608.00	**764.00**	511.00	**639.00**
Tile shingles											
Mission	Inst	Sq	RG	10.0	588.00	276.00	---	864.00	**1110.00**	722.00	**924.00**
Spanish	Inst	Sq	RG	10.0	446.00	276.00	---	722.00	**946.00**	601.00	**785.00**
Mineral surfaced roll											
Single coverage 90#/Sq roll, with 6" end lap and 2" head lap	Demo	Sq	LB	.625	---	13.70	---	13.70	**20.70**	11.00	**16.60**
Single coverage roll on											
Plain gable over											
Existing roofing with											
Nails concealed	Inst	Sq	2R	.870	33.60	26.40	---	60.00	**80.20**	49.90	**66.30**
Nails exposed	Inst	Sq	2R	.800	30.60	24.30	---	54.90	**73.30**	45.50	**60.60**
Wood deck with											
Nails concealed	Inst	Sq	2R	.909	34.60	27.60	---	62.20	**83.10**	51.70	**68.70**
Nails exposed	Inst	Sq	2R	.833	31.50	25.30	---	56.80	**75.90**	47.20	**62.70**
Plain hip over											
Existing roofing with											
Nails concealed	Inst	Sq	2R	.909	34.60	27.60	---	62.20	**83.10**	51.70	**68.70**
Nails exposed	Inst	Sq	2R	.833	31.50	25.30	---	56.80	**75.90**	47.20	**62.70**
Wood deck with											
Nails concealed	Inst	Sq	2R	1.00	35.20	30.40	---	65.60	**88.20**	54.40	**72.80**
Nails exposed	Inst	Sq	2R	.870	32.00	26.40	---	58.40	**78.30**	48.50	**64.70**
Double coverage selvage roll, with 19" lap and 17" exposure	Demo	Sq	LB	.909	---	20.00	---	20.00	**30.10**	16.00	**24.10**
Double coverage selvage roll, with 19" lap and 17" exposure (2 rolls/Sq)											
Plain gable over											
Wood deck with nails exposed	Inst	Sq	2R	1.43	70.30	43.40	---	113.70	**149.00**	94.70	**123.00**
Related materials and operations											
Starter strips (36' L rolls) along eaves and up rakes and gable ends on new roofing over wood decks											
9" W	Inst	LF	---	---	.37	---	---	.37	**.43**	.31	**.36**
12" W	Inst	LF	---	---	.46	---	---	.46	**.53**	.39	**.45**
18" W	Inst	LF	---	---	.60	---	---	.60	**.69**	.51	**.59**
24" W	Inst	LF	---	---	.74	---	---	.74	**.85**	.63	**.72**

Description	Oper	Unit	Crew Size	Man-Hours Per Unit	Avg Mat'l Unit Cost	Avg Labor Unit Cost	Avg Equip Unit Cost	Avg Total Unit Cost	Avg Price Incl O&P	Avg Total Unit Cost	Avg Price Incl O&P
			Costs Based On Small Volume							Large Volume	
Wood											
Shakes											
24" L with 10" exposure											
1/2" to 3/4" T	Demo	Sq	LB	.800	---	17.60	---	17.60	**26.50**	14.10	**21.20**
3/4" to 5/4" T	Demo	Sq	LB	.889	---	19.50	---	19.50	**29.50**	15.60	**23.60**
Shakes, over wood deck; 2 nails/shake, 24" L with 10" exp., 6" W (avg.), red cedar, sawn one side											
Gable											
1/2" to 3/4" T	Inst	Sq	RQ	2.28	180.00	65.80	---	245.80	**311.00**	206.80	**260.00**
3/4" to 5/4" T	Inst	Sq	RQ	2.54	225.00	73.30	---	298.30	**374.00**	250.90	**314.00**
Gable with dormers											
1/2" to 3/4" T	Inst	Sq	RQ	2.40	184.00	69.30	---	253.30	**320.00**	212.40	**267.00**
3/4" to 5/4" T	Inst	Sq	RQ	2.69	229.00	77.60	---	306.60	**386.00**	258.10	**323.00**
Gable with valleys											
1/2" to 3/4" T	Inst	Sq	RQ	2.40	184.00	69.30	---	253.30	**320.00**	212.40	**267.00**
3/4" to 5/4" T	Inst	Sq	RQ	2.69	229.00	77.60	---	306.60	**386.00**	258.10	**323.00**
Hip											
1/2" to 3/4" T	Inst	Sq	RQ	2.40	184.00	69.30	---	253.30	**320.00**	212.40	**267.00**
3/4" to 5/4" T	Inst	Sq	RQ	2.69	229.00	77.60	---	306.60	**386.00**	258.10	**323.00**
Hip with valleys											
1/2" to 3/4" T	Inst	Sq	RQ	2.54	187.00	73.30	---	260.30	**330.00**	218.90	**276.00**
3/4" to 5/4" T	Inst	Sq	RQ	2.82	234.00	81.40	---	315.40	**396.00**	264.90	**331.00**
Related materials and operations											
Ridge/hip units, 10" W with 10" exp., 20 pcs/bundle	Inst	LF	RJ	.022	1.59	.70	---	2.29	**2.93**	1.93	**2.47**
Roll valley, galvanized, unpainted, 28 gauge, 50' L rolls											
14" W	Inst	LF	UA	.083	1.18	2.47	---	3.65	**5.06**	3.00	**4.15**
20" W	Inst	LF	UA	.083	1.86	2.47	---	4.33	**5.84**	3.58	**4.82**
Rosin sized sheathing paper 36" W, 500 SF/roll; nailed											
Over open sheathing	Inst	Sq	RJ	.211	4.40	6.73	---	11.13	**15.60**	9.12	**12.70**
Over solid sheathing	Inst	Sq	RJ	.143	4.40	4.56	---	8.96	**12.20**	7.40	**10.00**
Shingles											
16" L with 5" exposure	Demo	Sq	LB	1.67	---	36.70	---	36.70	**55.30**	29.20	**44.10**
18" L with 5-1/2" exposure	Demo	Sq	LB	1.59	---	34.90	---	34.90	**52.70**	27.90	**42.10**
24" L with 7-1/2" exposure	Demo	Sq	LB	1.11	---	24.40	---	24.40	**36.80**	19.50	**29.50**
Shingles, red cedar, No. 1 perfect, 4" W (avg.), 2 nails per shingle											
Over existing roofing materials on											
Gable, plain											
16" L with 5" exposure	Inst	Sq	RM	4.46	220.00	133.00	---	353.00	**462.00**	295.00	**384.00**
18" L with 5-1/2" exposure	Inst	Sq	RM	4.03	228.00	121.00	---	349.00	**451.00**	290.60	**375.00**
24" L with 7-1/2" exposure	Inst	Sq	RM	2.91	216.00	87.00	---	303.00	**385.00**	254.70	**322.00**
Gable with dormers											
16" L with 5" exposure	Inst	Sq	RM	4.63	222.00	138.00	---	360.00	**472.00**	300.00	**392.00**
18" L with 5-1/2" exposure	Inst	Sq	RM	4.17	230.00	125.00	---	355.00	**460.00**	295.60	**382.00**
24" L with 7-1/2" exposure	Inst	Sq	RM	3.97	218.00	119.00	---	337.00	**437.00**	280.80	**363.00**

Description	Oper	Unit	Crew Size	Man-Hours Per Unit	Avg Mat'l Unit Cost	Avg Labor Unit Cost	Avg Equip Unit Cost	Avg Total Unit Cost	Avg Price Incl O&P	Large Volume Avg Total Unit Cost	Large Volume Avg Price Incl O&P
Gable with intersecting roofs											
16" L with 5" exposure	Inst	Sq	RM	4.63	222.00	138.00	---	360.00	**472.00**	300.00	**392.00**
18" L with 5-1/2" exposure	Inst	Sq	RM	4.17	230.00	125.00	---	355.00	**460.00**	295.60	**382.00**
24" L with 7-1/2" exposure	Inst	Sq	RM	3.97	218.00	119.00	---	337.00	**437.00**	280.80	**363.00**
Gable with dormers & intersecting roofs											
16" L with 5" exposure	Inst	Sq	RM	4.81	225.00	144.00	---	369.00	**485.00**	308.00	**402.00**
18" L with 5-1/2" exposure	Inst	Sq	RM	4.31	234.00	129.00	---	363.00	**471.00**	303.00	**391.00**
24" L with 7-1/2" exposure	Inst	Sq	RM	3.09	222.00	92.40	---	314.40	**400.00**	262.90	**334.00**
Hip, plain											
16" L with 5" exposure	Inst	Sq	RM	4.63	222.00	138.00	---	360.00	**472.00**	300.00	**392.00**
18" L with 5-1/2" exposure	Inst	Sq	RM	4.17	230.00	125.00	---	355.00	**460.00**	295.60	**382.00**
24" L with 7-1/2" exposure	Inst	Sq	RM	3.01	218.00	90.00	---	308.00	**392.00**	258.10	**327.00**
Hip with dormers											
16" L with 5" exposure	Inst	Sq	RM	4.81	224.00	144.00	---	368.00	**483.00**	306.00	**400.00**
18" L with 5-1/2" exposure	Inst	Sq	RM	4.39	232.00	131.00	---	363.00	**472.00**	303.00	**392.00**
24" L with 7-1/2" exposure	Inst	Sq	RM	3.16	220.00	94.50	---	314.50	**401.00**	263.70	**335.00**
Hip with intersecting roofs											
16" L with 5" exposure	Inst	Sq	RM	4.81	224.00	144.00	---	368.00	**483.00**	306.00	**400.00**
18" L with 5-1/2" exposure	Inst	Sq	RM	4.39	232.00	131.00	---	363.00	**472.00**	303.00	**392.00**
24" L with 7-1/2" exposure	Inst	Sq	RM	3.16	220.00	94.50	---	314.50	**401.00**	263.70	**335.00**
Hip with dormers & intersecting roofs											
16" L with 5" exposure	Inst	Sq	RM	5.00	227.00	150.00	---	377.00	**496.00**	314.00	**411.00**
18" L with 5-1/2" exposure	Inst	Sq	RM	4.55	236.00	136.00	---	372.00	**484.00**	310.00	**402.00**
24" L with 7-1/2" exposure	Inst	Sq	RM	3.25	224.00	97.20	---	321.20	**410.00**	268.70	**342.00**
Over wood decks on											
Gable, plain											
16" L with 5" exposure	Inst	Sq	RM	4.17	215.00	125.00	---	340.00	**443.00**	283.60	**368.00**
18" L with 5-1/2" exposure	Inst	Sq	RM	3.79	223.00	113.00	---	336.00	**435.00**	281.60	**362.00**
24" L with 7-1/2" exposure	Inst	Sq	RM	2.78	212.00	83.10	---	295.10	**374.00**	247.40	**313.00**
Gable with dormers											
16" L with 5" exposure	Inst	Sq	RM	4.31	217.00	129.00	---	346.00	**452.00**	288.00	**375.00**
18" L with 5-1/2" exposure	Inst	Sq	RM	3.91	225.00	117.00	---	342.00	**443.00**	286.60	**368.00**
24" L with 7-1/2" exposure	Inst	Sq	RM	2.87	214.00	85.80	---	299.80	**381.00**	251.80	**318.00**
Gable with intersecting roofs											
16" L with 5" exposure	Inst	Sq	RM	4.31	217.00	129.00	---	346.00	**452.00**	288.00	**375.00**
18" L with 5-1/2" exposure	Inst	Sq	RM	3.91	225.00	117.00	---	342.00	**443.00**	286.60	**368.00**
24" L with 7-1/2" exposure	Inst	Sq	RM	2.87	214.00	85.80	---	299.80	**381.00**	251.80	**318.00**
Gable with dormers & intersecting roofs											
16" L with 5" exposure	Inst	Sq	RM	4.46	221.00	133.00	---	354.00	**463.00**	296.00	**385.00**
18" L with 5-1/2" exposure	Inst	Sq	RM	4.03	229.00	121.00	---	350.00	**453.00**	292.60	**377.00**
24" L with 7-1/2" exposure	Inst	Sq	RM	2.94	218.00	87.90	---	305.90	**389.00**	256.30	**324.00**
Hip, plain											
16" L with 5" exposure	Inst	Sq	RM	4.31	217.00	129.00	---	346.00	**452.00**	288.00	**375.00**
18" L with 5-1/2" exposure	Inst	Sq	RM	3.91	225.00	117.00	---	342.00	**443.00**	286.60	**368.00**
24" L with 7-1/2" exposure	Inst	Sq	RM	2.87	214.00	85.80	---	299.80	**381.00**	251.80	**318.00**
Hip with dormers											
16" L with 5" exposure	Inst	Sq	RM	4.46	219.00	133.00	---	352.00	**461.00**	294.00	**383.00**
18" L with 5-1/2" exposure	Inst	Sq	RM	4.10	227.00	123.00	---	350.00	**454.00**	292.10	**377.00**
24" L with 7-1/2" exposure	Inst	Sq	RM	3.01	216.00	90.00	---	306.00	**390.00**	257.10	**325.00**

Description	Oper	Unit	Crew Size	Man-Hours Per Unit	Avg Mat'l Unit Cost	Avg Labor Unit Cost	Avg Equip Unit Cost	Avg Total Unit Cost	Avg Price Incl O&P	Avg Total Unit Cost	Avg Price Incl O&P
			Costs Based On Small Volume							Large Volume	
Hip with intersecting roofs											
16" L with 5" exposure	Inst	Sq	RM	4.46	219.00	133.00	---	352.00	**461.00**	294.00	**383.00**
18" L with 5-1/2" exposure	Inst	Sq	RM	4.10	227.00	123.00	---	350.00	**454.00**	292.10	**377.00**
24" L with 7-1/2" exposure	Inst	Sq	RM	3.01	216.00	90.00	---	306.00	**390.00**	257.10	**325.00**
Hip with dormers & intersecting roofs											
16" L with 5" exposure	Inst	Sq	RM	4.63	223.00	138.00	---	361.00	**473.00**	301.00	**393.00**
18" L with 5-1/2" exposure	Inst	Sq	RM	4.24	231.00	127.00	---	358.00	**465.00**	299.00	**386.00**
24" L with 7-1/2" exposure	Inst	Sq	RM	3.09	220.00	92.40	---	312.40	**398.00**	261.90	**332.00**
Related materials and operations											
Ridge/hip units, 40 pcs./bundle											
Over existing roofing											
5" exposure	Inst	LF	RJ	.036	1.55	1.15	---	2.70	**3.58**	2.24	**2.97**
5-1/2" exposure	Inst	LF	RJ	.034	1.55	1.08	---	2.63	**3.48**	2.18	**2.87**
7-1/2" exposure	Inst	LF	RJ	.029	1.31	.92	---	2.23	**2.96**	1.85	**2.44**
Over wood decks											
5" exposure	Inst	LF	RJ	.033	1.55	1.05	---	2.60	**3.43**	2.18	**2.87**
5-1/2" exposure	Inst	LF	RJ	.032	1.55	1.02	---	2.57	**3.38**	2.12	**2.77**
7-1/2" exposure	Inst	LF	RJ	.027	1.31	.86	---	2.17	**2.86**	1.79	**2.34**
Roll valley, galvanized, 28 gauge 50' L rolls, unpainted											
14" W	Inst	LF	UA	.083	1.13	2.47	---	3.60	**5.00**	2.95	**4.09**
20" W	Inst	LF	UA	.083	1.77	2.47	---	4.24	**5.74**	3.50	**4.73**
Rosin sized sheathing paper, 36" W, 500 SF/roll; nailed											
Over open sheathing	Inst	Sq	RJ	.211	4.35	6.73	---	11.08	**15.60**	9.08	**12.70**
Over solid sheathing	Inst	Sq	RJ	.143	4.35	4.56	---	8.91	**12.20**	7.36	**9.98**
For No. 2 (red label) grade											
DEDUCT	Inst	%		---	-10.0	---	---	---	**---**	---	**---**

Sheathing. See Framing, page 103

Sheet metal

Flashing, general; galvanized

Description	Oper	Unit	Crew Size	Man-Hours Per Unit	Avg Mat'l Unit Cost	Avg Labor Unit Cost	Avg Equip Unit Cost	Avg Total Unit Cost	Avg Price Incl O&P	Avg Total Unit Cost	Avg Price Incl O&P
Vertical chimney flashing	Inst	LF	UA	.084	.42	2.50	---	2.92	**4.23**	2.13	**3.06**
"Z" bar flashing											
Standard	Inst	LF	UA	.027	.42	.80	---	1.22	**1.69**	.94	**1.27**
Old style	Inst	LF	UA	.027	.42	.80	---	1.22	**1.69**	.94	**1.27**
For plywood siding	Inst	LF	UA	.027	.31	.80	---	1.11	**1.56**	.84	**1.16**
Rain diverter, 1" x 3"											
4' long	Inst	Ea	UA	.229	2.39	6.82	---	9.21	**13.00**	6.85	**9.55**
5' long	Inst	Ea	UA	.229	2.69	6.82	---	9.51	**13.30**	7.12	**9.86**
10' long	Inst	Ea	UA	.320	4.53	9.52	---	14.05	**19.50**	10.78	**14.80**
Roof flashing											
Nosing, roof edging at 90 degree angle or open at 105 degree; galvanized											
3/4" x 3/4"	Inst	LF	UA	.023	.19	.68	---	.87	**1.25**	.65	**.91**
1" x 1"	Inst	LF	UA	.023	.19	.68	---	.87	**1.25**	.65	**.91**
1" x 2"	Inst	LF	UA	.023	.23	.68	---	.91	**1.29**	.68	**.94**
1-1/2" x 1-1/2"	Inst	LF	UA	.023	.23	.68	---	.91	**1.29**	.68	**.94**
2" x 2"	Inst	LF	UA	.023	.29	.68	---	.97	**1.36**	.73	**1.00**
2" x 3"	Inst	LF	UA	.023	.35	.68	---	1.03	**1.43**	.79	**1.07**
2" x 4"	Inst	LF	UA	.023	.49	.68	---	1.17	**1.59**	.91	**1.21**

Description	Oper	Unit	Crew Size	Man-Hours Per Unit	Avg Mat'l Unit Cost	Avg Labor Unit Cost	Avg Equip Unit Cost	Avg Total Unit Cost	Avg Price Incl O&P	Large Volume Avg Total Unit Cost	Large Volume Avg Price Incl O&P
3" x 3"	Inst	LF	UA	.029	.49	.86	---	1.35	**1.86**	1.03	**1.39**
3" x 4"	Inst	LF	UA	.029	.57	.86	---	1.43	**1.95**	1.10	**1.47**
3" x 5"	Inst	LF	UA	.038	.68	1.13	---	1.81	**2.48**	1.40	**1.90**
4" x 4"	Inst	LF	UA	.038	.68	1.13	---	1.81	**2.48**	1.40	**1.90**
4" x 6"	Inst	LF	UA	.038	.86	1.13	---	1.99	**2.69**	1.55	**2.07**
5" x 5"	Inst	LF	UA	.038	.86	1.13	---	1.99	**2.69**	1.55	**2.07**
6" x 6"	Inst	LF	UA	.038	1.01	1.13	---	2.14	**2.86**	1.68	**2.22**
Roll valley, 50' rolls											
Galvanized, 28 gauge, unpainted											
8" wide	Inst	LF	UA	.038	.56	1.13	---	1.69	**2.34**	1.29	**1.77**
10" wide	Inst	LF	UA	.038	.91	1.13	---	2.04	**2.74**	1.60	**2.13**
20" wide	Inst	LF	UA	.038	1.28	1.13	---	2.41	**3.17**	1.92	**2.49**
Tin seamless, painted											
8" wide	Inst	LF	UA	.038	.44	1.13	---	1.57	**2.20**	1.18	**1.64**
10" wide	Inst	LF	UA	.038	.68	1.13	---	1.81	**2.48**	1.40	**1.90**
20" wide	Inst	LF	UA	.038	.95	1.13	---	2.08	**2.79**	1.63	**2.16**
Aluminum seamless, .016 gauge											
8" wide	Inst	LF	UA	.038	.52	1.13	---	1.65	**2.29**	1.25	**1.72**
10" wide	Inst	LF	UA	.038	.87	1.13	---	2.00	**2.70**	1.56	**2.08**
20" wide	Inst	LF	UA	.038	1.24	1.13	---	2.37	**3.12**	1.88	**2.45**
"W" valley											
Galvanized hemmed tin											
18" girth	Inst	LF	UA	.038	1.35	1.13	---	2.48	**3.25**	1.98	**2.56**
24" girth	Inst	LF	UA	.038	1.78	1.13	---	2.91	**3.74**	2.36	**3.00**
Gravel stop											
4-1/2" galvanized	Inst	LF	UA	.029	.31	.86	---	1.17	**1.65**	.87	**1.20**
6" galvanized	Inst	LF	UA	.033	.41	.98	---	1.39	**1.94**	1.04	**1.44**
7-1/4" galvanized	Inst	LF	UA	.038	.45	1.13	---	1.58	**2.21**	1.19	**1.65**
4-1/2" bonderized	Inst	LF	UA	.029	.39	.86	---	1.25	**1.74**	.95	**1.30**
6" bonderized	Inst	LF	UA	.033	.53	.98	---	1.51	**2.08**	1.14	**1.56**
7-1/4" bonderized	Inst	LF	UA	.038	.58	1.13	---	1.71	**2.36**	1.31	**1.79**
Gutters and downspouts. See page 144											
Roof edging (drip edge) with 1/4" kickout, galvanized											
1" x 1-1/2"	Inst	LF	UA	.023	.24	.68	---	.92	**1.30**	.69	**.96**
1-1/2" x 1-1/2"	Inst	LF	UA	.023	.24	.68	---	.92	**1.30**	.69	**.96**
2" x 2"	Inst	LF	UA	.023	.30	.68	---	.98	**1.37**	.74	**1.01**
Vents											
Clothes dryer vent set, aluminum, hood, duct, inside plate											
3" diameter	Inst	Set	UA	.727	3.33	21.60	---	24.93	**36.30**	17.82	**25.70**
4" diameter	Inst	Set	UA	.727	3.67	21.60	---	25.27	**36.70**	18.11	**26.00**
Dormer louvers, half round, 1/8" or 1/4" mesh											
18" x 9"											
Galvanized, 5-12 pitch	Inst	Ea	UA	1.00	20.70	29.80	---	50.50	**68.50**	38.10	**50.70**
Painted, 3-12 pitch	Inst	Ea	UA	1.00	24.50	29.80	---	54.30	**72.80**	41.30	**54.40**
24" x 12"											
Galvanized, 5-12 pitch	Inst	Ea	UA	1.00	26.00	29.80	---	55.80	**74.60**	42.70	**56.00**
Painted, 3-12 pitch	Inst	Ea	UA	1.00	29.50	29.80	---	59.30	**78.60**	45.70	**59.50**

			Costs Based On Small Volume							Large Volume	
Description	Oper	Unit	Crew Size	Man-Hours Per Unit	Avg Mat'l Unit Cost	Avg Labor Unit Cost	Avg Equip Unit Cost	Avg Total Unit Cost	Avg Price Incl O&P	Avg Total Unit Cost	Avg Price Incl O&P
Foundation vents, galvanized, 1/4" mesh											
6" x 14", stucco	Inst	Ea	UA	.471	1.59	14.00	---	15.59	**22.90**	11.30	**16.50**
8" x 14", stucco	Inst	Ea	UA	.471	1.90	14.00	---	15.90	**23.20**	11.58	**16.80**
6" x 14", two-way	Inst	Ea	UA	.471	1.90	14.00	---	15.90	**23.20**	11.58	**16.80**
6" x 14", for siding (flat type)	Inst	Ea	UA	.471	1.59	14.00	---	15.59	**22.90**	11.30	**16.50**
6" x 14", louver type	Inst	Ea	UA	.471	1.84	14.00	---	15.84	**23.10**	11.52	**16.70**
6" x 14", foundation insert	Inst	Ea	UA	.471	1.59	14.00	---	15.59	**22.90**	11.30	**16.50**
Louver vents, round, 1/8" or 1/4" mesh, galvanized											
12" diameter	Inst	Ea	UA	.727	21.20	21.60	---	42.80	**56.90**	33.50	**43.70**
14" diameter	Inst	Ea	UA	.727	22.50	21.60	---	44.10	**58.40**	34.60	**45.00**
16" diameter	Inst	Ea	UA	.727	24.40	21.60	---	46.00	**60.60**	36.30	**46.90**
18" diameter	Inst	Ea	UA	.727	28.00	21.60	---	49.60	**64.70**	39.40	**50.50**
24" diameter	Inst	Ea	UA	.727	47.60	21.60	---	69.20	**87.20**	56.60	**70.20**
Access doors											
Multi-purpose, galvanized											
24" x 18", nail-on type	Inst	Ea	UA	.727	13.50	21.60	---	35.10	**47.90**	26.70	**35.90**
24" x 24", nail-on type	Inst	Ea	UA	.727	15.20	21.60	---	36.80	**49.90**	28.20	**37.60**
24" x 18" x 6" deep box type	Inst	Ea	UA	.727	18.40	21.60	---	40.00	**53.60**	31.00	**40.80**
Attic access doors, galvanized											
22" x 22"	Inst	Ea	UA	1.00	19.00	29.80	---	48.80	**66.50**	36.60	**48.90**
22" x 30"	Inst	Ea	UA	1.00	22.00	29.80	---	51.80	**70.00**	39.20	**52.00**
30" x 30"	Inst	Ea	UA	1.00	23.30	29.80	---	53.10	**71.40**	40.30	**53.20**
Tub access doors, galvanized											
14" x 12"	Inst	Ea	UA	.727	9.45	21.60	---	31.05	**43.30**	23.17	**31.80**
14" x 15"	Inst	Ea	UA	.727	9.79	21.60	---	31.39	**43.70**	23.47	**32.20**
Utility louvers, aluminum or galvanized											
4" x 6"	Inst	Ea	UA	.286	1.13	8.51	---	9.64	**14.10**	6.94	**10.10**
8" x 6"	Inst	Ea	UA	.286	1.59	8.51	---	10.10	**14.60**	7.34	**10.50**
8" x 8"	Inst	Ea	UA	.286	1.84	8.51	---	10.35	**14.90**	7.56	**10.80**
8" x 12"	Inst	Ea	UA	.286	2.15	8.51	---	10.66	**15.20**	7.83	**11.10**
8" x 18"	Inst	Ea	UA	.286	2.57	8.51	---	11.08	**15.70**	8.20	**11.50**
12" x 12"	Inst	Ea	UA	.286	2.57	8.51	---	11.08	**15.70**	8.20	**11.50**
14" x 14"	Inst	Ea	UA	.286	3.25	8.51	---	11.76	**16.50**	8.79	**12.20**

Shower and tub doors

Shower doors

Hinged shower doors, anodized aluminum frame, tempered hammered glass, hardware included

Description	Oper	Unit	Crew Size	Man-Hours Per Unit	Avg Mat'l Unit Cost	Avg Labor Unit Cost	Avg Equip Unit Cost	Avg Total Unit Cost	Avg Price Incl O&P	Avg Total Unit Cost	Avg Price Incl O&P
64" H door, adjustable											
35"-37"											
Obscure glass											
Brass plated	Inst	Ea	CA	1.78	293.00	48.00	---	341.00	**410.00**	254.90	**306.00**
Silver plated	Inst	Ea	CA	1.78	265.00	48.00	---	313.00	**378.00**	233.90	**282.00**
Designer glass											
Brass plated	Inst	Ea	CA	1.78	366.00	48.00	---	414.00	**494.00**	309.90	**370.00**
Silver plated	Inst	Ea	CA	1.78	338.00	48.00	---	386.00	**462.00**	288.90	**346.00**
Clear glass											
Brass plated	Inst	Ea	CA	1.78	330.00	48.00	---	378.00	**453.00**	282.90	**338.00**
Silver plated	Inst	Ea	CA	1.78	302.00	48.00	---	350.00	**420.00**	261.90	**314.00**

Description	Oper	Unit	Crew Size	Man-Hours Per Unit	Avg Mat'l Unit Cost	Avg Labor Unit Cost	Avg Equip Unit Cost	Avg Total Unit Cost	Avg Price Incl O&P	Large Volume Avg Total Unit Cost	Large Volume Avg Price Incl O&P
			Costs Based On Small Volume							**Large Volume**	
33"-35"											
Obscure glass											
Brass plated	Inst	Ea	CA	1.78	293.00	48.00	---	341.00	**410.00**	254.90	**306.00**
Silver plated	Inst	Ea	CA	1.78	265.00	48.00	---	313.00	**378.00**	233.90	**282.00**
Designer glass											
Brass plated	Inst	Ea	CA	1.78	366.00	48.00	---	414.00	**494.00**	309.90	**370.00**
Silver plated	Inst	Ea	CA	1.78	338.00	48.00	---	386.00	**462.00**	288.90	**346.00**
Clear glass											
Brass plated	Inst	Ea	CA	1.78	330.00	48.00	---	378.00	**453.00**	282.90	**338.00**
Silver plated	Inst	Ea	CA	1.78	302.00	48.00	---	350.00	**420.00**	261.90	**314.00**
31"-33"											
Obscure glass											
Brass plated	Inst	Ea	CA	1.78	285.00	48.00	---	333.00	**400.00**	248.90	**299.00**
Silver plated	Inst	Ea	CA	1.78	257.00	48.00	---	305.00	**368.00**	227.90	**275.00**
Designer glass											
Brass plated	Inst	Ea	CA	1.78	358.00	48.00	---	406.00	**485.00**	303.90	**362.00**
Silver plated	Inst	Ea	CA	1.78	330.00	48.00	---	378.00	**453.00**	282.90	**338.00**
Clear glass											
Brass plated	Inst	Ea	CA	1.78	322.00	48.00	---	370.00	**443.00**	276.90	**331.00**
Silver plated	Inst	Ea	CA	1.78	294.00	48.00	---	342.00	**411.00**	255.90	**307.00**
27"-29"											
Obscure glass											
Brass plated	Inst	Ea	CA	1.78	278.00	48.00	---	326.00	**393.00**	243.90	**294.00**
Silver plated	Inst	Ea	CA	1.78	250.00	48.00	---	298.00	**360.00**	222.90	**270.00**
Designer glass											
Brass plated	Inst	Ea	CA	1.78	352.00	48.00	---	400.00	**477.00**	298.90	**357.00**
Silver plated	Inst	Ea	CA	1.78	324.00	48.00	---	372.00	**445.00**	277.90	**333.00**
Clear glass											
Brass plated	Inst	Ea	CA	1.78	315.00	48.00	---	363.00	**435.00**	271.90	**326.00**
Silver plated	Inst	Ea	CA	1.78	287.00	48.00	---	335.00	**403.00**	250.90	**302.00**
25"-27"											
Obscure glass											
Brass plated	Inst	Ea	CA	1.78	279.00	48.00	---	327.00	**394.00**	244.90	**294.00**
Silver plated	Inst	Ea	CA	1.78	251.00	48.00	---	299.00	**362.00**	223.90	**270.00**
Designer glass											
Brass plated	Inst	Ea	CA	1.78	353.00	48.00	---	401.00	**478.00**	299.90	**358.00**
Silver plated	Inst	Ea	CA	1.78	325.00	48.00	---	373.00	**446.00**	278.90	**334.00**
Clear glass											
Brass plated	Inst	Ea	CA	1.78	316.00	48.00	---	364.00	**437.00**	271.90	**326.00**
Silver plated	Inst	Ea	CA	1.78	288.00	48.00	---	336.00	**404.00**	251.90	**302.00**
Sliding shower doors, 2 bypassing panels, anodized aluminum frame, tempered hammered glass, hardware included											
70" H doors											
41" to 43" wide opening											
Obscure glass											
Brass plated	Inst	Ea	CA	1.94	288.00	52.30	---	340.30	**411.00**	255.10	**307.00**
Silver plated	Inst	Ea	CA	1.94	260.00	52.30	---	312.30	**378.00**	233.10	**283.00**
Designer glass											
Brass plated	Inst	Ea	CA	1.94	362.00	52.30	---	414.30	**496.00**	310.10	**371.00**
Silver plated	Inst	Ea	CA	1.94	334.00	52.30	---	386.30	**463.00**	289.10	**346.00**

Description	Oper	Unit	Crew Size	Man-Hours Per Unit	Avg Mat'l Unit Cost	Avg Labor Unit Cost	Avg Equip Unit Cost	Avg Total Unit Cost	Avg Price Incl O&P	Large Volume Avg Total Unit Cost	Large Volume Avg Price Incl O&P
Clear glass											
Brass plated	Inst	Ea	CA	1.94	325.00	52.30	---	377.30	**454.00**	282.10	**339.00**
Silver plated	Inst	Ea	CA	1.94	297.00	52.30	---	349.30	**422.00**	261.10	**315.00**
57.5 to 59.5 wide opening											
Obscure glass											
Brass plated	Inst	Ea	CA	1.94	325.00	52.30	---	377.30	**454.00**	282.10	**339.00**
Silver plated	Inst	Ea	CA	1.94	297.00	52.30	---	349.30	**422.00**	261.10	**315.00**
Designer glass											
Brass plated	Inst	Ea	CA	1.94	399.00	52.30	---	451.30	**538.00**	337.10	**403.00**
Silver plated	Inst	Ea	CA	1.94	371.00	52.30	---	423.30	**506.00**	316.10	**378.00**
Clear glass											
Brass plated	Inst	Ea	CA	1.94	363.00	52.30	---	415.30	**497.00**	310.10	**371.00**
Silver plated	Inst	Ea	CA	1.94	335.00	52.30	---	387.30	**464.00**	289.10	**347.00**
66" H doors											
47" to 49" wide opening											
Obscure glass											
Brass plated	Inst	Ea	CA	1.94	272.00	52.30	---	324.30	**392.00**	242.10	**293.00**
Silver plated	Inst	Ea	CA	1.94	244.00	52.30	---	296.30	**360.00**	221.10	**269.00**
Designer glass											
Brass plated	Inst	Ea	CA	1.94	345.00	52.30	---	397.30	**476.00**	297.10	**356.00**
Silver plated	Inst	Ea	CA	1.94	317.00	52.30	---	369.30	**444.00**	276.10	**332.00**
Clear glass											
Brass plated	Inst	Ea	CA	1.94	309.00	52.30	---	361.30	**435.00**	270.10	**325.00**
Silver plated	Inst	Ea	CA	1.94	281.00	52.30	---	333.30	**402.00**	249.10	**301.00**
45" to 47" wide opening											
Obscure glass											
Brass plated	Inst	Ea	CA	1.94	270.00	52.30	---	322.30	**391.00**	241.10	**292.00**
Silver plated	Inst	Ea	CA	1.94	243.00	52.30	---	295.30	**358.00**	220.10	**268.00**
Designer glass											
Brass plated	Inst	Ea	CA	1.94	344.00	52.30	---	396.30	**475.00**	296.10	**355.00**
Silver plated	Inst	Ea	CA	1.94	316.00	52.30	---	368.30	**443.00**	275.10	**331.00**
Clear glass											
Brass plated	Inst	Ea	CA	1.94	308.00	52.30	---	360.30	**433.00**	269.10	**324.00**
Silver plated	Inst	Ea	CA	1.94	280.00	52.30	---	332.30	**401.00**	248.10	**300.00**
43" to 45" wide opening											
Obscure glass											
Brass plated	Inst	Ea	CA	1.94	274.00	52.30	---	326.30	**395.00**	244.10	**295.00**
Silver plated	Inst	Ea	CA	1.94	246.00	52.30	---	298.30	**363.00**	223.10	**271.00**
Designer glass											
Brass plated	Inst	Ea	CA	1.94	348.00	52.30	---	400.30	**480.00**	299.10	**359.00**
Silver plated	Inst	Ea	CA	1.94	320.00	52.30	---	372.30	**447.00**	278.10	**335.00**
Clear glass											
Brass plated	Inst	Ea	CA	1.94	312.00	52.30	---	364.30	**438.00**	272.10	**327.00**
Silver plated	Inst	Ea	CA	1.94	284.00	52.30	---	336.30	**406.00**	251.10	**303.00**
41" to 43" wide opening											
Obscure glass											
Brass plated	Inst	Ea	CA	1.94	261.00	52.30	---	313.30	**380.00**	234.10	**284.00**
Silver plated	Inst	Ea	CA	1.94	233.00	52.30	---	285.30	**348.00**	213.10	**260.00**

Description	Oper	Unit	Crew Size	Man-Hours Per Unit	Avg Mat'l Unit Cost	Avg Labor Unit Cost	Avg Equip Unit Cost	Avg Total Unit Cost	Avg Price Incl O&P	Large Volume Avg Total Unit Cost	Large Volume Avg Price Incl O&P
Designer glass											
Brass plated	Inst	Ea	CA	1.94	335.00	52.30	---	387.30	**465.00**	289.10	**347.00**
Silver plated	Inst	Ea	CA	1.94	307.00	52.30	---	359.30	**432.00**	268.10	**323.00**
Clear glass											
Brass plated	Inst	Ea	CA	1.94	299.00	52.30	---	351.30	**423.00**	262.10	**316.00**
Silver plated	Inst	Ea	CA	1.94	271.00	52.30	---	323.30	**391.00**	241.10	**292.00**
Tub doors											
Sliding tub doors, 2 bypassing panels, adjustable width x 57" H											
69" to 71.5" wide opening											
Obscure glass											
Brass plated	Inst	Ea	CA	1.94	349.00	52.30	---	401.30	**480.00**	300.10	**359.00**
Silver plated	Inst	Ea	CA	1.94	321.00	52.30	---	373.30	**448.00**	279.10	**335.00**
Designer glass											
Brass plated	Inst	Ea	CA	1.94	422.00	52.30	---	474.30	**565.00**	355.10	**423.00**
Silver plated	Inst	Ea	CA	1.94	394.00	52.30	---	446.30	**533.00**	334.10	**398.00**
Clear glass											
Brass plated	Inst	Ea	CA	1.94	386.00	52.30	---	438.30	**523.00**	328.10	**391.00**
Silver plated	Inst	Ea	CA	1.94	358.00	52.30	---	410.30	**491.00**	307.10	**367.00**
Obscure mirror glass											
Brass plated	Inst	Ea	CA	1.94	397.00	52.30	---	449.30	**536.00**	336.10	**401.00**
Silver plated	Inst	Ea	CA	1.94	369.00	52.30	---	421.30	**504.00**	315.10	**377.00**
57" to 59" wide opening											
Obscure glass											
Brass plated	Inst	Ea	CA	1.94	275.00	52.30	---	327.30	**396.00**	245.10	**296.00**
Silver plated	Inst	Ea	CA	1.94	247.00	52.30	---	299.30	**364.00**	224.10	**272.00**
Designer glass											
Brass plated	Inst	Ea	CA	1.94	349.00	52.30	---	401.30	**481.00**	300.10	**359.00**
Silver plated	Inst	Ea	CA	1.94	321.00	52.30	---	373.30	**448.00**	279.10	**335.00**
Clear glass											
Brass plated	Inst	Ea	CA	1.94	284.00	52.30	---	336.30	**407.00**	252.10	**304.00**
Silver plated	Inst	Ea	CA	1.94	349.00	52.30	---	401.30	**481.00**	300.10	**359.00**
Obscure mirror glass											
Brass plated	Inst	Ea	CA	1.94	324.00	52.30	---	376.30	**452.00**	281.10	**338.00**
Silver plated	Inst	Ea	CA	1.94	296.00	52.30	---	348.30	**419.00**	260.10	**314.00**
55" to 57" wide opening											
Obscure glass											
Brass plated	Inst	Ea	CA	1.94	276.00	52.30	---	328.30	**397.00**	246.10	**297.00**
Silver plated	Inst	Ea	CA	1.94	248.00	52.30	---	300.30	**365.00**	225.10	**273.00**
Designer glass											
Brass plated	Inst	Ea	CA	1.94	350.00	52.30	---	402.30	**482.00**	301.10	**360.00**
Silver plated	Inst	Ea	CA	1.94	322.00	52.30	---	374.30	**450.00**	280.10	**336.00**
Clear glass											
Brass plated	Inst	Ea	CA	1.94	313.00	52.30	---	365.30	**440.00**	273.10	**329.00**
Silver plated	Inst	Ea	CA	1.94	285.00	52.30	---	337.30	**408.00**	252.10	**305.00**
Obscure mirror glass											
Brass plated	Inst	Ea	CA	1.94	325.00	52.30	---	377.30	**453.00**	282.10	**339.00**
Silver plated	Inst	Ea	CA	1.94	297.00	52.30	---	349.30	**421.00**	261.10	**314.00**
54" to 56" wide opening											
Obscure glass											
Brass plated	Inst	Ea	CA	1.94	278.00	52.30	---	330.30	**399.00**	247.10	**299.00**
Silver plated	Inst	Ea	CA	1.94	250.00	52.30	---	302.30	**367.00**	226.10	**274.00**

Description	Oper	Unit	Crew Size	Man-Hours Per Unit	Avg Mat'l Unit Cost	Avg Labor Unit Cost	Avg Equip Unit Cost	Avg Total Unit Cost	Avg Price Incl O&P	Large Volume Avg Total Unit Cost	Large Volume Avg Price Incl O&P
Designer glass											
Brass plated	Inst	Ea	CA	1.94	352.00	52.30	---	404.30	**484.00**	302.10	**362.00**
Silver plated	Inst	Ea	CA	1.94	324.00	52.30	---	376.30	**452.00**	281.10	**338.00**
Clear glass											
Brass plated	Inst	Ea	CA	1.94	315.00	52.30	---	367.30	**442.00**	275.10	**331.00**
Silver plated	Inst	Ea	CA	1.94	287.00	52.30	---	339.30	**410.00**	254.10	**306.00**
Obscure mirror glass											
Brass plated	Inst	Ea	CA	1.94	326.00	52.30	---	378.30	**455.00**	283.10	**340.00**
Silver plated	Inst	Ea	CA	1.94	298.00	52.30	---	350.30	**423.00**	262.10	**316.00**
52" to 54" wide opening											
Obscure glass											
Brass plated	Inst	Ea	CA	1.94	271.00	52.30	---	323.30	**392.00**	242.10	**293.00**
Silver plated	Inst	Ea	CA	1.94	244.00	52.30	---	296.30	**360.00**	221.10	**269.00**
Designer glass											
Brass plated	Inst	Ea	CA	1.94	345.00	52.30	---	397.30	**476.00**	297.10	**356.00**
Silver plated	Inst	Ea	CA	1.94	317.00	52.30	---	369.30	**444.00**	276.10	**332.00**
Clear glass											
Brass plated	Inst	Ea	CA	1.94	309.00	52.30	---	361.30	**434.00**	270.10	**325.00**
Silver plated	Inst	Ea	CA	1.94	281.00	52.30	---	333.30	**402.00**	249.10	**301.00**
Obscure mirror glass											
Brass plated	Inst	Ea	CA	1.94	320.00	52.30	---	372.30	**447.00**	278.10	**334.00**
Silver plated	Inst	Ea	CA	1.94	292.00	52.30	---	344.30	**415.00**	257.10	**311.00**
Swing door -- dual side panels for neo-angle showers											
72" H doors											
38" x 36"											
Obscure glass											
Brass plated	Inst	Ea	CA	2.37	627.00	63.90	---	690.90	**819.00**	517.00	**612.00**
Silver plated	Inst	Ea	CA	2.37	599.00	63.90	---	662.90	**786.00**	496.00	**588.00**
Designer glass											
Brass plated	Inst	Ea	CA	2.37	701.00	63.90	---	764.90	**903.00**	572.00	**676.00**
Silver plated	Inst	Ea	CA	2.37	673.00	63.90	---	736.90	**871.00**	552.00	**652.00**
Clear glass											
Brass plated	Inst	Ea	CA	2.37	665.00	63.90	---	728.90	**861.00**	545.00	**644.00**
Silver plated	Inst	Ea	CA	2.37	637.00	63.90	---	700.90	**829.00**	524.00	**620.00**
50" x 50"											
Obscure glass											
Brass plated	Inst	Ea	CA	2.37	763.00	63.90	---	826.90	**974.00**	619.00	**729.00**
Silver plated	Inst	Ea	CA	2.37	735.00	63.90	---	798.90	**942.00**	598.00	**705.00**
Designer glass											
Brass plated	Inst	Ea	CA	2.37	837.00	63.90	---	900.90	**1060.00**	674.00	**792.00**
Silver plated	Inst	Ea	CA	2.37	809.00	63.90	---	872.90	**1030.00**	653.00	**768.00**
Clear glass											
Brass plated	Inst	Ea	CA	2.37	800.00	63.90	---	863.90	**1020.00**	646.00	**761.00**
Silver plated	Inst	Ea	CA	2.37	772.00	63.90	---	835.90	**985.00**	625.00	**737.00**

Description	Oper	Unit	Crew Size	Man-Hours Per Unit	Avg Mat'l Unit Cost	Avg Labor Unit Cost	Avg Equip Unit Cost	Avg Total Unit Cost	Avg Price Incl O&P	Large Volume Avg Total Unit Cost	Large Volume Avg Price Incl O&P
			Costs Based On Small Volume							Large Volume	

Shower stalls

Shower stall units with slip-resistant fiberglass floors, and reinforced plastic integral wall surrounds (Aqua Glass), available in white or color, with good quality fittings, single control faucets, and shower head sprayer

Shower stall, no door, with plastic drain, one integral piece

Three wall model

Description	Oper	Unit	Crew Size	Man-Hours Per Unit	Avg Mat'l Unit Cost	Avg Labor Unit Cost	Avg Equip Unit Cost	Avg Total Unit Cost	Avg Price Incl O&P	Large Volume Avg Total Unit Cost	Large Volume Avg Price Incl O&P
Kohler Puebla shower module											
32" x 36" x 77" H											
Regular color	Inst	Ea	SB	12.7	523.00	333.00	---	856.00	**1100.00**	667.00	**847.00**
White	Inst	Ea	SB	12.7	507.00	333.00	---	840.00	**1080.00**	653.00	**831.00**
Kohler Trinidad shower module											
36" x 36" x 75" H											
Premium color	Inst	Ea	SB	12.7	802.00	333.00	---	1135.00	**1420.00**	899.00	**1110.00**
Regular color	Inst	Ea	SB	12.7	772.00	333.00	---	1105.00	**1380.00**	874.00	**1080.00**
White	Inst	Ea	SB	12.7	744.00	333.00	---	1077.00	**1350.00**	850.00	**1060.00**
Kohler Vallerta shower module											
36" x 36" x 77" H											
Regular color	Inst	Ea	SB	12.7	535.00	333.00	---	868.00	**1110.00**	677.00	**858.00**
White	Inst	Ea	SB	12.7	518.00	333.00	---	851.00	**1090.00**	663.00	**842.00**
Lasco compact shower stalls											
32" x 32" x 72" H											
Premium color	Inst	Ea	SB	12.7	519.00	333.00	---	852.00	**1090.00**	664.00	**843.00**
Standard color	Inst	Ea	SB	12.7	495.00	333.00	---	828.00	**1070.00**	644.00	**820.00**
Stock color	Inst	Ea	SB	12.7	478.00	333.00	---	811.00	**1050.00**	630.00	**804.00**
34" x 34" x 72" H											
Premium color	Inst	Ea	SB	12.7	546.00	333.00	---	879.00	**1120.00**	686.00	**869.00**
Standard color	Inst	Ea	SB	12.7	522.00	333.00	---	855.00	**1100.00**	667.00	**846.00**
Stock color	Inst	Ea	SB	12.7	506.00	333.00	---	839.00	**1080.00**	653.00	**830.00**
36" x 36" x 72" H											
Premium color	Inst	Ea	SB	12.7	548.00	333.00	---	881.00	**1130.00**	688.00	**870.00**
Standard color	Inst	Ea	SB	12.7	524.00	333.00	---	857.00	**1100.00**	668.00	**847.00**
Stock color	Inst	Ea	SB	12.7	507.00	333.00	---	840.00	**1080.00**	654.00	**831.00**
42" x 34" x 72" H											
Premium color	Inst	Ea	SB	12.7	573.00	333.00	---	906.00	**1160.00**	709.00	**894.00**
Standard color	Inst	Ea	SB	12.7	550.00	333.00	---	883.00	**1130.00**	689.00	**872.00**
Stock color	Inst	Ea	SB	12.7	533.00	333.00	---	866.00	**1110.00**	675.00	**856.00**
Lasco deluxe showers with molded seats											
48" x 34" x 72" H											
Premium color	Inst	Ea	SB	12.7	611.00	333.00	---	944.00	**1200.00**	741.00	**931.00**
Standard color	Inst	Ea	SB	12.7	588.00	333.00	---	921.00	**1170.00**	721.00	**908.00**
Stock color	Inst	Ea	SB	12.7	571.00	333.00	---	904.00	**1150.00**	707.00	**892.00**
54" x 35" x 72" H											
Premium color	Inst	Ea	SB	12.7	752.00	333.00	---	1085.00	**1360.00**	858.00	**1070.00**
Standard color	Inst	Ea	SB	12.7	721.00	333.00	---	1054.00	**1320.00**	831.00	**1040.00**
Stock color	Inst	Ea	SB	12.7	698.00	333.00	---	1031.00	**1300.00**	813.00	**1010.00**

Description	Oper	Unit	Crew Size	Man-Hours Per Unit	Avg Mat'l Unit Cost	Avg Labor Unit Cost	Avg Equip Unit Cost	Avg Total Unit Cost	Avg Price Incl O&P	Large Volume Avg Total Unit Cost	Large Volume Avg Price Incl O&P
Two wall model, one piece											
Lasco space saver shower module											
38" x 38" x 72" H, neo angle											
Premium color	Inst	Ea	SB	12.7	553.00	333.00	---	886.00	**1130.00**	692.00	**875.00**
Standard color	Inst	Ea	SB	12.7	529.00	333.00	---	862.00	**1100.00**	672.00	**852.00**
Stock color	Inst	Ea	SB	12.7	512.00	333.00	---	845.00	**1090.00**	658.00	**836.00**
36" x 36" x 72" H											
Premium color	Inst	Ea	SB	12.7	578.00	333.00	---	911.00	**1160.00**	713.00	**899.00**
Standard color	Inst	Ea	SB	12.7	555.00	333.00	---	888.00	**1130.00**	693.00	**877.00**
Stock color	Inst	Ea	SB	12.7	538.00	333.00	---	871.00	**1110.00**	679.00	**860.00**
Shower stall, nonintegral, nonassembled units with drain											
Shower floors, slip resistant fiberglass											
To remove and replace fiberglass shower stall receptor (floors)											
32" x 32"	R&R	Ea	SB	7.14	113.00	187.00	---	300.00	**409.00**	224.70	**303.00**
32" x 48"	R&R	Ea	SB	7.14	159.00	187.00	---	346.00	**462.00**	263.00	**347.00**
34" x 48"	R&R	Ea	SB	7.14	161.00	187.00	---	348.00	**464.00**	264.00	**349.00**
34" x 60"	R&R	Ea	SB	7.14	207.00	187.00	---	394.00	**517.00**	303.00	**393.00**
36" x 36" neo angle	R&R	Ea	SB	7.14	136.00	187.00	---	323.00	**435.00**	244.00	**325.00**
38" x 38" neo angle	R&R	Ea	SB	7.14	153.00	187.00	---	340.00	**455.00**	258.00	**342.00**
Shower wall surrounds with adjustable walls											
For square shower base floors											
30" - 36" wide x 30" - 36" deep	Inst	Ea	SB	12.7	404.00	333.00	---	737.00	**961.00**	568.00	**733.00**
For rectangular shower base floors											
36" - 60" wide x 30" - 34" deep	Inst	Ea	SB	12.7	464.00	333.00	---	797.00	**1030.00**	618.00	**790.00**
For neo-angle & single threshold											
36"	Inst	Ea	SB	12.7	514.00	333.00	---	847.00	**1090.00**	659.00	**837.00**
For neo-angle & corner threshold											
36"	Inst	Ea	SB	12.7	473.00	333.00	---	806.00	**1040.00**	626.00	**799.00**
38"	Inst	Ea	SB	12.7	499.00	333.00	---	832.00	**1070.00**	647.00	**824.00**
Adjustments											
Shower stall											
Remove only	Demo	Ea	SB	3.52	---	92.30	---	92.30	**137.00**	64.50	**96.10**
Remove and reset	Reset	Ea	SB	7.14	---	187.00	---	187.00	**279.00**	131.00	**195.00**
Install rough-in, ADD	Inst	Ea	SB	19.0	85.00	498.00	---	583.00	**840.00**	419.60	**601.00**

Description	Oper	Unit	Crew Size	Man-Hours Per Unit	Avg Mat'l Unit Cost	Avg Labor Unit Cost	Avg Equip Unit Cost	Avg Total Unit Cost	Avg Price Incl O&P	Avg Total Unit Cost	Avg Price Incl O&P
			Costs Based On Small Volume							Large Volume	

Shower & tub combinations

Shower/tub three-wall stall combinations are fiberglass reinforced plastic with integral bath/shower and wall surrounds; includes good quality fittings, single control faucets, and shower head sprayer.

One piece combination, no door

Description	Oper	Unit	Crew Size	Man-Hours Per Unit	Avg Mat'l Unit Cost	Avg Labor Unit Cost	Avg Equip Unit Cost	Avg Total Unit Cost	Avg Price Incl O&P	Avg Total Unit Cost (Large Volume)	Avg Price Incl O&P (Large Volume)
Kohler Barbados shower/tub module											
60" x 35" x 75" H											
Premium color	Inst	Ea	SB	12.3	888.00	322.00	---	1210.00	**1500.00**	970.00	**1190.00**
Regular color	Inst	Ea	SB	12.3	865.00	322.00	---	1187.00	**1480.00**	951.00	**1170.00**
White	Inst	Ea	SB	12.3	815.00	322.00	---	1137.00	**1420.00**	909.00	**1120.00**
Kohler Cancun II series with seat											
48" x 36" x 77" H											
Regular color	Inst	Ea	SB	12.3	668.00	322.00	---	990.00	**1250.00**	787.00	**984.00**
White	Inst	Ea	SB	12.3	691.00	322.00	---	1013.00	**1280.00**	807.00	**1010.00**
Kohler Veracruz II series with seat											
60" x 32" x 77" H											
Regular color	Inst	Ea	SB	12.3	679.00	322.00	---	1001.00	**1260.00**	797.00	**996.00**
White	Inst	Ea	SB	12.3	704.00	322.00	---	1026.00	**1290.00**	817.00	**1020.00**

Multiple piece combination, no door

Description	Oper	Unit	Crew Size	Man-Hours Per Unit	Avg Mat'l Unit Cost	Avg Labor Unit Cost	Avg Equip Unit Cost	Avg Total Unit Cost	Avg Price Incl O&P	Avg Total Unit Cost (Large Volume)	Avg Price Incl O&P (Large Volume)
Lasco Danube I shower/tub module											
36" x 36" x 84" H											
Premium color	Inst	Ea	SB	12.3	1060.00	322.00	---	1382.00	**1690.00**	1109.00	**1350.00**
Regular color	Inst	Ea	SB	12.3	1030.00	322.00	---	1352.00	**1670.00**	1090.00	**1330.00**
White	Inst	Ea	SB	12.3	1020.00	322.00	---	1342.00	**1650.00**	1076.00	**1320.00**
Lasco Danube II shower/tub module											
48" x 36" x 84" H											
Premium color	Inst	Ea	SB	12.3	1070.00	322.00	---	1392.00	**1710.00**	1117.00	**1360.00**
Regular color	Inst	Ea	SB	12.3	1040.00	322.00	---	1362.00	**1680.00**	1097.00	**1340.00**
White	Inst	Ea	SB	12.3	1020.00	322.00	---	1342.00	**1660.00**	1083.00	**1320.00**
Lasco Dalila shower/tub module											
60" x 34" x 84" H											
Premium color	Inst	Ea	SB	12.3	1090.00	322.00	---	1412.00	**1730.00**	1134.00	**1380.00**
Regular color	Inst	Ea	SB	12.3	1060.00	322.00	---	1382.00	**1700.00**	1114.00	**1360.00**
White	Inst	Ea	SB	12.3	1040.00	322.00	---	1362.00	**1680.00**	1100.00	**1340.00**
Lasco Mali shower/tub module											
60" x 30" x 72" H											
Premium color	Inst	Ea	SB	12.3	635.00	322.00	---	957.00	**1210.00**	760.00	**954.00**
Regular color	Inst	Ea	SB	12.3	612.00	322.00	---	934.00	**1180.00**	741.00	**931.00**
White	Inst	Ea	SB	12.3	595.00	322.00	---	917.00	**1160.00**	727.00	**915.00**

Adjustments

To remove only

Description	Oper	Unit	Crew Size	Man-Hours Per Unit	Avg Mat'l Unit Cost	Avg Labor Unit Cost	Avg Equip Unit Cost	Avg Total Unit Cost	Avg Price Incl O&P	Avg Total Unit Cost (Large Volume)	Avg Price Incl O&P (Large Volume)
Shower stall	Demo	Ea	SB	3.48	---	91.20	---	91.20	**136.00**	64.50	**96.10**
Shower/tub	Demo	Ea	SB	3.81	---	99.90	---	99.90	**149.00**	70.00	**104.00**
Tub	Demo	Ea	SB	3.48	---	91.20	---	91.20	**136.00**	64.50	**96.10**

Description	Oper	Unit	Crew Size	Man-Hours Per Unit	Avg Mat'l Unit Cost	Avg Labor Unit Cost	Avg Equip Unit Cost	Avg Total Unit Cost	Avg Price Incl O&P	Large Volume Avg Total Unit Cost	Large Volume Avg Price Incl O&P
To remove and reset											
Shower stall	Reset	Ea	SB	7.27	---	191.00	---	191.00	**284.00**	131.00	**195.00**
Shower/tub	Reset	Ea	SB	7.62	---	200.00	---	200.00	**298.00**	140.00	**208.00**
Tub	Reset	Ea	SB	7.27	---	191.00	---	191.00	**284.00**	131.00	**195.00**
To install rough-in, ADD											
Shower stall	Inst	Ea	SB	20.0	85.00	524.00	---	609.00	**879.00**	419.60	**601.00**
Shower/tub	Inst	Ea	SB	20.0	125.00	524.00	---	649.00	**925.00**	453.00	**639.00**
Tub	Inst	Ea	SB	14.5	100.00	380.00	---	480.00	**681.00**	345.00	**486.00**

Shutters

Exterior

Description	Oper	Unit	Crew Size	Man-Hours Per Unit	Avg Mat'l Unit Cost	Avg Labor Unit Cost	Avg Equip Unit Cost	Avg Total Unit Cost	Avg Price Incl O&P	Large Volume Avg Total Unit Cost	Large Volume Avg Price Incl O&P
Aluminum, louvered, 14" W											
3'-0" long	Inst	Pair	CA	1.33	49.10	35.90	---	85.00	**111.00**	66.20	**86.00**
6'-8" long	Inst	Pair	CA	1.33	81.80	35.90	---	117.70	**149.00**	92.30	**116.00**
Pine or fir, provincial raised panel, primed, in stock, 1-1/16" T											
12" W											
2'-1" long	Inst	Pair	CA	1.33	30.40	35.90	---	66.30	**89.50**	51.30	**68.90**
3'-1" long	Inst	Pair	CA	1.33	32.20	35.90	---	68.10	**91.50**	52.70	**70.50**
4'-1" long	Inst	Pair	CA	1.33	36.90	35.90	---	72.80	**96.90**	56.50	**74.90**
5'-1" long	Inst	Pair	CA	1.33	44.70	35.90	---	80.60	**106.00**	62.70	**82.00**
6'-1" long	Inst	Pair	CA	1.33	52.80	35.90	---	88.70	**115.00**	69.20	**89.50**
18" W											
2'-1" long	Inst	Pair	CA	1.33	36.10	35.90	---	72.00	**96.10**	55.80	**74.20**
3'-1" long	Inst	Pair	CA	1.33	38.00	35.90	---	73.90	**98.30**	57.40	**75.90**
4'-1" long	Inst	Pair	CA	1.33	45.20	35.90	---	81.10	**107.00**	63.10	**82.50**
5'-1" long	Inst	Pair	CA	1.33	53.10	35.90	---	89.00	**116.00**	69.40	**89.70**
6'-1" long	Inst	Pair	CA	1.33	63.70	35.90	---	99.60	**128.00**	77.90	**99.50**
Door blinds, 6'-9" long											
1'-3" wide	Inst	Pair	CA	1.33	65.40	35.90	---	101.30	**130.00**	79.20	**101.00**
1'-6" wide	Inst	Pair	CA	1.33	81.80	35.90	---	117.70	**149.00**	92.30	**116.00**
Birch, special order stationary slat blinds 1-1/16" T											
12" W											
2'-1" long	Inst	Pair	CA	1.33	59.20	35.90	---	95.10	**123.00**	74.30	**95.40**
3'-1" long	Inst	Pair	CA	1.33	70.50	35.90	---	106.40	**136.00**	83.30	**106.00**
4'-1" long	Inst	Pair	CA	1.33	82.60	35.90	---	118.50	**149.00**	92.90	**117.00**
5'-1" long	Inst	Pair	CA	1.33	95.10	35.90	---	131.00	**164.00**	102.90	**128.00**
6'-1" long	Inst	Pair	CA	1.33	109.00	35.90	---	144.90	**180.00**	114.10	**141.00**
18" W											
2'-1" long	Inst	Pair	CA	1.33	64.40	35.90	---	100.30	**129.00**	78.40	**100.00**
3'-1" long	Inst	Pair	CA	1.33	79.30	35.90	---	115.20	**146.00**	90.30	**114.00**
4'-1" long	Inst	Pair	CA	1.33	94.00	35.90	---	129.90	**163.00**	102.00	**127.00**
5'-1" long	Inst	Pair	CA	1.33	111.00	35.90	---	146.90	**182.00**	115.50	**143.00**
6'-1" long	Inst	Pair	CA	1.33	125.00	35.90	---	160.90	**198.00**	126.70	**156.00**

Description	Oper	Unit	Crew Size	Man-Hours Per Unit	Avg Mat'l Unit Cost	Avg Labor Unit Cost	Avg Equip Unit Cost	Avg Total Unit Cost	Avg Price Incl O&P	Large Volume Avg Total Unit Cost	Large Volume Avg Price Incl O&P
			Costs Based On Small Volume							Large Volume	
Western hemlock, colonial style, primed, 1-1/8" T											
14" wide											
5'-7" long	Inst	Pair	CA	1.33	57.80	35.90	---	93.70	**121.00**	73.10	**94.00**
16" wide											
3'-0" long	Inst	Pair	CA	1.33	36.00	35.90	---	71.90	**95.90**	55.70	**74.00**
4'-3" long	Inst	Pair	CA	1.33	43.60	35.90	---	79.50	**105.00**	61.80	**81.00**
5'-0" long	Inst	Pair	CA	1.33	50.10	35.90	---	86.00	**112.00**	67.00	**87.00**
6'-0" long	Inst	Pair	CA	1.33	55.60	35.90	---	91.50	**118.00**	71.40	**92.00**
Door blinds, 6'-9" long											
1'-3" wide	Inst	Pair	CA	1.33	71.90	35.90	---	107.80	**137.00**	84.40	**107.00**
1'-6" wide	Inst	Pair	CA	1.33	74.10	35.90	---	110.00	**140.00**	86.20	**109.00**
Cellwood shutters, molded structural foam polystyrene											
Prefinished louver, 16" wide											
3'-3" long	Inst	Pair	CA	1.33	35.30	35.90	---	71.20	**95.10**	55.10	**73.30**
4'-7" long	Inst	Pair	CA	1.33	44.50	35.90	---	80.40	**106.00**	62.50	**81.80**
5'-3" long	Inst	Pair	CA	1.33	50.30	35.90	---	86.20	**112.00**	67.10	**87.10**
6'-3" long	Inst	Pair	CA	1.33	68.00	35.90	---	103.90	**133.00**	81.30	**103.00**
Door blinds											
16" wide x 6'-9" long	Inst	Pair	CA	1.33	70.30	35.90	---	106.20	**135.00**	83.10	**105.00**
Prefinished panel, 16" wide											
3'-3" long	Inst	Pair	CA	1.33	36.60	35.90	---	72.50	**96.60**	56.20	**74.60**
4'-7" long	Inst	Pair	CA	1.33	46.80	35.90	---	82.70	**108.00**	64.30	**83.90**
5'-3" long	Inst	Pair	CA	1.33	52.10	35.90	---	88.00	**114.00**	68.60	**88.80**
6'-3" long	Inst	Pair	CA	1.33	68.00	35.90	---	103.90	**133.00**	81.30	**103.00**
Door blinds											
16" wide x 6'-9" long	Inst	Pair	CA	1.33	74.10	35.90	---	110.00	**140.00**	86.10	**109.00**

Siding

Aluminum Siding

1. **Dimensions and Descriptions.** The types of aluminum siding discussed are: (1) painted (2) available in various colors, (3) either 0.024" gauge or 0.019" gauge. The types of aluminum siding discussed in this section are:

   a. Clapboard: 12'-6" long; 8" exposure; 5/8" butt; 0.024" gauge.

   b. Board and batten: 10'-0" long; 10" exposure; 1/2" butt; 0.024" gauge.

   c. "Double-five-inch" clapboard: 12'-0" long; 10" exposure; 1/2" butt; 0.024" gauge.

   d. Sculpture clapboard: 12'-6" long; 8" exposure; 1/2" butt; 0.024" gauge.

   e. Shingle siding: 24" long; 12" exposure; 1 1/8" butt; 0.019" gauge.

Most types of aluminum siding may be available in a smooth or
rough texture.

2. **Installation.** Applied horizontally with nails. See following chart:
3. **Estimating Technique.** Determine area and deduct area of window and door openings then add for cutting and fitting waste. In the figures on aluminum siding, no waste has been included.

Type	Exposure	Butt	Normal Range of Cut & Fit Waste*
Clapboard	8"	5/8"	6% - 10%
Board & Batten	10"	1/2"	4% - 8%
Double-Five-Inch	10"	1/2"	6% - 10%
Sculptured	8"	1/2"	6% - 10%
Shingle Siding	12"	1 1/8"	4% - 6%

*The amount of waste will vary with the methods of installing siding and can be affected by dimension of the area to be covered.

Wood Siding

The types of wood exterior finishes covered in this section are:

a. Bevel siding.

b. Drop siding.

c. Vertical siding.

e. Batten siding.

f. Plywood siding with battens.

g. Wood shingle siding.

Bevel Siding

1. **Dimension.** Boards are 4", 6", 8", or 12" wide with a thickness of the top edge of 3/16". The thickness of the bottom edge is 9/16" for 4" and 6" widths and 11/16" for 8", 10", 12" widths.
2. **Installation.** Applied horizontally with 6d or 8d common nails. See exposure and waste table below:
3. **Estimating Technique.** Determine area and deduct area of window and door openings, then add for lap, exposure, and cutting and fitting waste. In the figures on bevel siding, waste has been included, unless otherwise noted.

Board Size	Actual Width	Lap	Exposure	Milling & Lap Waste	Normal Range of Cut & Fit Waste*
1/2" x 4"	3 1/4"	1/2"	2 3/4"	31.25%	19% - 23%
1/2" x 6"	5 1/4"	1/2"	4 3/4"	21.00%	10% - 13%
1/2" x 8"	7 1/4"	1/2"	6 3/4"	15.75%	7% - 10%
5/8" x 4"	3 1/4"	1/2"	2 3/4"	31.25%	19% - 23%
5/8" x 6"	5 1/4"	1/2"	4 3/4"	21.00%	10% - 13%
5/8" x 8"	7 1/4"	1/2"	6 3/4"	15.75%	7% - 10%
5/8" x 10"	9 1/4"	1/2"	8 3/4"	12.50%	6% - 10%
3/4" x 6"	5 1/4"	1/2"	4 3/4"	21.00%	10% - 13%
3/4" x 8"	7 1/4"	1/2"	6 3/4"	15.75%	7% - 10%
3/4" x 10"	9 1/4"	1/2"	8 3/4"	12.50%	6% - 10%
3/4" x 12"	11 1/4"	1/2"	10 3/4"	10.50%	6% - 10%

*The amount of waste will vary with the methods of installing siding and can be affected by dimension of the area to be covered.

Drop Siding, Tongue and Grooved

1. **Dimension.** The boards are 4", 6", 8", 10", or 12" wide, and all boards are 3/4" thick at both edges.
2. **Installation.** Applied horizontally, usually with 8d common nails. See exposure and waste table below:

Board Size	Actual Width	Lap +1/16"	Exposure	Milling & Lap Waste	Normal Range of Cut & Waste*
1" x 4"	3 1/4"	3/16"	3 1/16"	23.50%	7% - 10%
1" x 6"	5 1/4"	3/16"	5 1/16"	15.75%	6% - 8%
1" x 8"	7 1/4"	3/16"	7 1/16"	11.75%	5% - 6%
1" x 10"	9 1/4"	3/16"	9 1/16"	9.50%	4% - 5%
1" x 12"	11 1/4"	3/16"	11 1/16"	8.00%	4% - 5%

*The amount of waste will vary with the methods of installing siding and can be affected by dimensions of the area to be covered.

3. **Estimating Technique.** Determine area and deduct area of window and door openings; then add for lap, exposure, butting and fitting waste. In the figures on drop siding, waste has been included, unless otherwise noted.

Vertical Siding, Tongue and Grooved

1. **Dimensions.** Boards are 8", 10", or 12" wide, and all boards are 3/4" thick at both top and bottom edges.
2. **Installation.** Applied vertically over horizontal furring strips usually with 8d common nails. See exposure and waste table below:

Board Size	Actual Width	Lap 1/16"	Exposure	Milling & Lap Waste	Normal Range of Cut & Fit Waste
1" x 8"	7 1/4"	3/16"	7 1/16"	11.75%	5% - 7%
1" x 10"	9 1/4"	3/16"	9 1/16"	9.50%	4% - 5%
1" x 12"	11 1/4"	3/16"	11 1/16"	8.00%	4% - 5%

3. **Estimating Technique.** Determine area and deduct area of window and door openings; then add for lap, exposure, cutting and fitting waste. In the figures on vertical siding, waste has been included, unless otherwise noted.

Batten Siding

1. **Dimension.** The boards are 8", 10", or 12" wide. If the boards are rough, they are 1" thick; if the boards are dressed, then they are 25/32" thick. The battens are 1" thick and 2" wide.
2. **Installation.** The 8", 10", or 12"-wide boards are installed vertically over 1" x 3" or 1" x 4" horizontal furring strips. The battens are then installed over the seams of the vertical boards. See exposure and waste table below:

Board Size & Type		Actual Width & Exposure	Milling Waste	Normal Range of Cut & Fit Waste*
1" x 8"	rough sawn	8"	—	3% - 5%
	dressed	7 1/2"	6.25%	3% - 5%
1" x 10"	rough sawn	10"	—	3% - 5%
	dressed	9 1/2"	5.00%	3% - 5%
1" x 12"	rough sawn	12"	—	3% - 5%
	dressed	11 1/2"	4.25%	3% - 5%

*The amount of waste will vary with the methods of installing siding and can be affected by dimension of the area to be covered.

3. **Estimating Technique.** Determine area and deduct area of window and door openings; then add for exposure, cutting and fitting waste. In the figures on batten siding, waste has been included, unless otherwise noted.

Quantity of Battens - No Cut and Fit Waste Included

Board Size and Type		LF per SF or BF
1" x 6"	rough sawn	2.00
	dressed	2.25
1" x 8"	rough sawn	1.55
	dressed	1.60
1" x 10"	rough sawn	1.20
	dressed	1.30
1" x 12"	rough sawn	1.00
	dressed	1.05

Plywood Siding with Battens

1. **Dimensions.** The plywood panels are 4' wide and 8', 9', or 10' long. The panels are 3/8", 1/2" or 5/8" thick. The battens are 1" thick and 2" wide.
2. **Installation.** The panels are applied directly to the studs with the 4'-0" width parallel to the plate. Nails used are 6d commons. The waste will normally range from 3% to 5%.

Quantity of Battens - No Cut and Fit Waste Included

Spacing	LF per SF or BF
16" oc	.75
24" oc	.50

3. **Estimating Technique.** Determine area and deduct area of window and door openings; then add for cutting and fitting waste. In the figures on plywood siding with battens, waste has been included, unless otherwise noted.

Wood Shingle Siding

1. **Dimension.** The wood shingles have an average width of 4" and length of 16", 18" or 24". The thickness may vary in all cases.
2. **Installation.** On roofs with wood shingles, there should be three thicknesses of wood shingle at every point on the roof. But on exterior walls with wood shingles, only two thicknesses of wood siding shingles are needed. Nail to 1" x 3" or 1" x 4" furring strips using 3d rust-resistant nails. Use 5d rust-resistant nails when applying new shingles over old shingles. See table below:

Shingle Length	Exposure	Shingles per Square, No Waste Included	Lbs. of Nails at Two Nails per Shingle 3d	Lbs. of Nails at Two Nails per Shingle 5d
16"	$7^1/_2$"	480	2.3	2.7
18"	$8^1/_2$"	424	2.0	2.4
24"	$11^1/_2$"	313	1.5	1.8

3. **Estimating Technique.** Determine the area and deduct the area of window and door openings. Then add approximately 10% for waste. In the figures on wood shingle siding, waste has been included, unless otherwise noted. In the above table, pounds of nails per square are based on two nails per shingle. For one nail per shingle, deduct 50%.

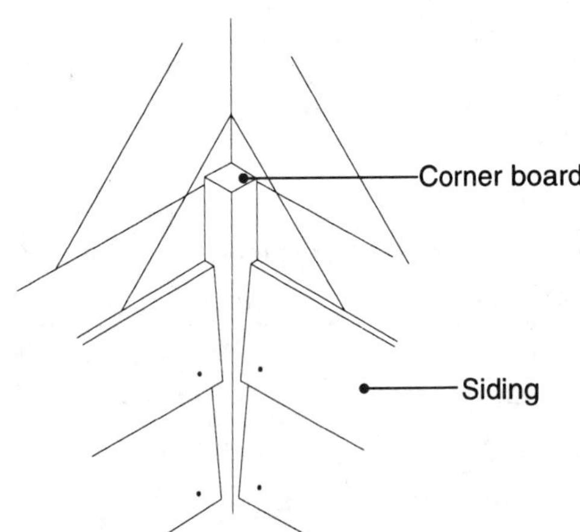

Corner board for application of horizontal siding at interior corner

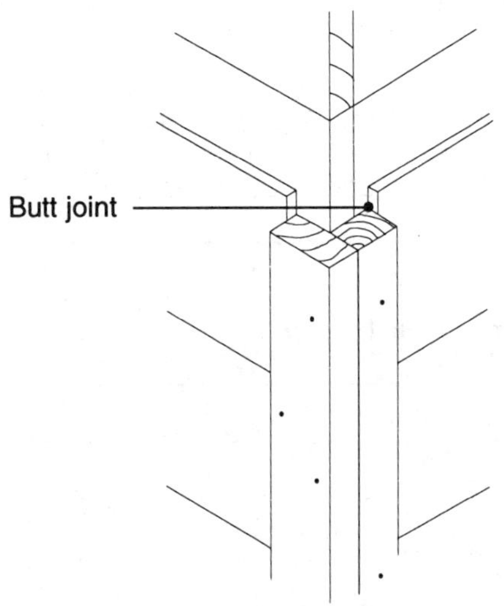

Corner boards for application of horizontal siding at exterior corner

Description	Oper	Unit	Crew Size	Man-Hours Per Unit	Avg Mat'l Unit Cost	Avg Labor Unit Cost	Avg Equip Unit Cost	Avg Total Unit Cost	Avg Price Incl O&P	Large Volume Avg Total Unit Cost	Large Volume Avg Price Incl O&P
Siding											
Aluminum, nailed to wood											
Clapboard (i.e. lap, drop)											
8" exposure	Demo	SF	LB	.022	---	.48	---	.48	**.73**	.35	**.53**
10" exposure	Demo	SF	LB	.018	---	.40	---	.40	**.60**	.29	**.43**
Clapboard, horizontal; 1/2" butt											
Straight wall											
8" exposure	Inst	SF	2C	.037	2.24	1.00	---	3.24	**4.09**	2.49	**3.12**
10" exposure	Inst	SF	2C	.030	1.86	.81	---	2.67	**3.37**	2.06	**2.57**
Cut-up wall											
8" exposure	Inst	SF	2C	.044	2.95	1.19	---	4.14	**5.20**	3.20	**3.98**
10" exposure	Inst	SF	2C	.036	2.56	.97	---	3.53	**4.42**	2.72	**3.38**
Corrugated (2-1/2"), 26" W, with 2-1/2" side lap and 4" end lap											
	Demo	SF	LB	.019	---	.42	---	.42	**.63**	.29	**.43**
0.0175" thick											
Natural	Inst	SF	UC	.028	2.58	.72	---	3.30	**4.05**	2.55	**3.11**
Painted	Inst	SF	UC	.028	2.80	.72	---	3.52	**4.31**	2.73	**3.31**
0.019" thick											
Natural	Inst	SF	UC	.028	2.62	.72	---	3.34	**4.10**	2.59	**3.15**
Painted	Inst	SF	UC	.028	2.85	.72	---	3.57	**4.36**	2.77	**3.36**
Panels, 4' x 8' board and batten (12" oc)	Demo	SF	LB	.010	---	.22	---	.22	**.33**	.15	**.23**
Straight wall											
Smooth	Inst	SF	2C	.019	2.31	.51	---	2.82	**3.44**	2.19	**2.65**
Rough sawn	Inst	SF	2C	.019	2.53	.51	---	3.04	**3.69**	2.37	**2.86**
Cut-up wall											
Smooth	Inst	SF	2C	.022	3.01	.59	---	3.60	**4.36**	2.84	**3.43**
Rough sawn	Inst	SF	2C	.022	3.25	.59	---	3.84	**4.64**	3.02	**3.63**
Shingle, 24" L with 12" exposure											
	Demo	SF	LB	.016	---	.35	---	.35	**.53**	.24	**.36**
Straight wall	Inst	SF	2C	.026	.05	.70	---	.75	**1.12**	.53	**.78**
Cut-up wall	Inst	SF	2C	.032	.05	.86	---	.91	**1.37**	.66	**.99**
Hardboard siding, Masonite											
Lap, 1/2" T x 12" W x 16' L, with 11" exposure											
	Demo	SF	LB	.017	---	.37	---	.37	**.56**	.26	**.40**
Straight wall											
Smooth finish	Inst	SF	2C	.033	1.46	.89	---	2.35	**3.03**	1.79	**2.29**
Textured finish	Inst	SF	2C	.033	1.73	.89	---	2.62	**3.34**	2.00	**2.53**
Wood-like finish	Inst	SF	2C	.033	1.84	.89	---	2.73	**3.47**	2.09	**2.63**
Cut-up wall											
Smooth finish	Inst	SF	2C	.039	1.50	1.05	---	2.55	**3.32**	1.93	**2.49**
Textured finish	Inst	SF	2C	.039	1.77	1.05	---	2.82	**3.63**	2.15	**2.74**
Wood-like finish	Inst	SF	2C	.039	1.89	1.05	---	2.94	**3.77**	2.24	**2.84**

Description	Oper	Unit	Crew Size	Man-Hours Per Unit	Avg Mat'l Unit Cost	Avg Labor Unit Cost	Avg Equip Unit Cost	Avg Total Unit Cost	Avg Price Incl O&P	Large Volume Avg Total Unit Cost	Large Volume Avg Price Incl O&P
Panels, 7/16" T x 4' W x 8' H	Demo	SF	LB	.009	---	.20	---	.20	**.30**	.13	**.20**
Grooved/planked panels, textured or wood-like finish											
Primecote	Inst	SF	2C	.019	1.38	.51	---	1.89	**2.37**	1.46	**1.81**
Unprimed	Inst	SF	2C	.019	1.49	.51	---	2.00	**2.49**	1.54	**1.90**
Panel with batten preattached											
Unprimed	Inst	SF	2C	.019	1.57	.51	---	2.08	**2.58**	1.61	**1.98**
Stucco textured panel											
Primecote	Inst	SF	2C	.019	1.62	.51	---	2.13	**2.64**	1.64	**2.02**
Vinyl											
Solid vinyl panels .024 gauge, woodgrained or colors											
10" or double 5" panels	Demo	SF	LB	.017	---	.37	---	.37	**.56**	.26	**.40**
10" or double 5" panels, standard	Inst	SF	2C	.046	.78	1.24	---	2.02	**2.78**	1.48	**2.02**
10" or double 5" panels, insulated	Inst	SF	2C	.046	1.09	1.24	---	2.33	**3.14**	1.73	**2.31**
Starter strip	Inst	LF	2C	.057	.28	1.54	---	1.82	**2.66**	1.31	**1.90**
J channel for corners	Inst	LF	2C	.051	.66	1.37	---	2.03	**2.85**	1.49	**2.07**
Outside corner and corner post	Inst	LF	2C	.057	.97	1.54	---	2.51	**3.45**	1.85	**2.52**
Casing and trim	Inst	LF	2C	.065	.51	1.75	---	2.26	**3.25**	1.64	**2.35**
Soffit and facia	Inst	LF	2C	.057	.72	1.54	---	2.26	**3.16**	1.65	**2.29**
Wood											
Bevel											
1/2" x 8" with 6-3/4" exp.	Demo	SF	LB	.025	---	.55	---	.55	**.83**	.40	**.60**
5/8" x 10" with 8-3/4" exp.	Demo	SF	LB	.019	---	.42	---	.42	**.63**	.31	**.46**
3/4" x 12" with 10-3/4" exp.	Demo	SF	LB	.016	---	.35	---	.35	**.53**	.24	**.36**
Bevel (horizontal), 1/2" lap, 10' L											
Redwood, kiln-dried, clear VG, pattern											
1/2" x 8", with 6-3/4" exp.											
Rough ends	Inst	SF	2C	.052	2.74	1.40	---	4.14	**5.28**	3.19	**4.03**
Fitted ends	Inst	SF	2C	.068	2.80	1.83	---	4.63	**6.01**	3.53	**4.54**
Mitered ends	Inst	SF	2C	.080	2.86	2.16	---	5.02	**6.57**	3.80	**4.93**
5/8" x 10", with 8-3/4" exp.											
Rough ends	Inst	SF	2C	.041	2.65	1.11	---	3.76	**4.73**	2.90	**3.63**
Fitted ends	Inst	SF	2C	.053	2.72	1.43	---	4.15	**5.30**	3.18	**4.02**
Mitered ends	Inst	SF	2C	.062	2.78	1.67	---	4.45	**5.74**	3.39	**4.33**
3/4" x 12", with 10-3/4" exp.											
Rough ends	Inst	SF	2C	.033	4.62	.89	---	5.51	**6.67**	4.32	**5.20**
Fitted ends	Inst	SF	2C	.043	4.74	1.16	---	5.90	**7.21**	4.60	**5.59**
Mitered ends	Inst	SF	2C	.050	4.85	1.35	---	6.20	**7.63**	4.82	**5.90**
Red cedar, clear, kiln-dried											
1/2" x 8", with 6-3/4" exp.											
Rough ends	Inst	SF	2C	.052	2.42	1.40	---	3.82	**4.91**	2.93	**3.74**
Fitted ends	Inst	SF	2C	.068	2.47	1.83	---	4.30	**5.63**	3.27	**4.24**
Mitered ends	Inst	SF	2C	.080	2.53	2.16	---	4.69	**6.19**	3.53	**4.62**
5/8" x 10", with 8-3/4" exp.											
Rough ends	Inst	SF	2C	.041	2.53	1.11	---	3.64	**4.59**	2.81	**3.52**
Fitted ends	Inst	SF	2C	.053	2.59	1.43	---	4.02	**5.15**	3.08	**3.91**
Mitered ends	Inst	SF	2C	.062	2.65	1.67	---	4.32	**5.59**	3.29	**4.21**
3/4" x 12", with 10-3/4" exp.											
Rough ends	Inst	SF	2C	.033	4.18	.89	---	5.07	**6.16**	3.96	**4.78**

Description	Oper	Unit	Costs Based On Small Volume							Large Volume	
			Crew Size	Man-Hours Per Unit	Avg Mat'l Unit Cost	Avg Labor Unit Cost	Avg Equip Unit Cost	Avg Total Unit Cost	Avg Price Incl O&P	Avg Total Unit Cost	Avg Price Incl O&P
Fitted ends	Inst	SF	2C	.043	4.29	1.16	---	5.45	**6.70**	4.24	**5.17**
Mitered ends	Inst	SF	2C	.050	4.39	1.35	---	5.74	**7.10**	4.45	**5.47**
Drop (horizontal) T&G, 1/4" lap											
1" x 8" with 7" exp.	Demo	SF	LB	.024	---	.53	---	.53	**.80**	.37	**.56**
1" x 10" with 9" exp.	Demo	SF	LB	.019	---	.42	---	.42	**.63**	.29	**.43**
Drop (horizontal) T&G, 1/4" lap, 10' L (avg.)											
Redwood, kiln-dried, clear VG, patterns											
1" x 8", with 7" exp.											
Rough ends	Inst	SF	2C	.049	4.47	1.32	---	5.79	**7.15**	4.49	**5.50**
Fitted ends	Inst	SF	2C	.063	4.58	1.70	---	6.28	**7.85**	4.85	**6.01**
Mitered ends	Inst	SF	2C	.072	4.69	1.94	---	6.63	**8.34**	5.12	**6.40**
1" x 10", with 9" exp.											
Rough ends	Inst	SF	2C	.038	4.69	1.02	---	5.71	**6.95**	4.49	**5.43**
Fitted ends	Inst	SF	2C	.049	4.81	1.32	---	6.13	**7.54**	4.77	**5.82**
Mitered ends	Inst	SF	2C	.056	4.93	1.51	---	6.44	**7.96**	5.02	**6.17**
1" x 8", with 7-3/4" exp.											
Rough ends	Inst	SF	2C	.044	4.06	1.19	---	5.25	**6.47**	4.09	**5.01**
Fitted ends	Inst	SF	2C	.057	4.17	1.54	---	5.71	**7.13**	4.41	**5.47**
Mitered ends	Inst	SF	2C	.065	4.28	1.75	---	6.03	**7.59**	4.66	**5.82**
1" x 10", with 9-3/4" exp.											
Rough ends	Inst	SF	2C	.035	4.36	.94	---	5.30	**6.45**	4.16	**5.04**
Fitted ends	Inst	SF	2C	.045	4.48	1.21	---	5.69	**7.00**	4.44	**5.43**
Mitered ends	Inst	SF	2C	.052	4.59	1.40	---	5.99	**7.41**	4.65	**5.71**
Red cedar, clear, kiln-dried											
1" x 8", with 7" exp.											
Rough ends	Inst	SF	2C	.049	4.08	1.32	---	5.40	**6.70**	4.19	**5.15**
Fitted ends	Inst	SF	2C	.063	4.18	1.70	---	5.88	**7.39**	4.54	**5.66**
Mitered ends	Inst	SF	2C	.072	4.28	1.94	---	6.22	**7.87**	4.79	**6.02**
1" x 10", with 9" exp.											
Rough ends	Inst	SF	2C	.038	4.23	1.02	---	5.25	**6.42**	4.12	**5.00**
Fitted ends	Inst	SF	2C	.049	4.34	1.32	---	5.66	**7.00**	4.39	**5.38**
Mitered ends	Inst	SF	2C	.056	4.44	1.51	---	5.95	**7.40**	4.63	**5.72**
Pine, kiln-dried, select											
Rough sawn											
1" x 8", with 7-3/4" exp.											
Rough ends	Inst	SF	2C	.044	2.85	1.19	---	4.04	**5.08**	3.12	**3.89**
Fitted ends	Inst	SF	2C	.057	2.93	1.54	---	4.47	**5.71**	3.42	**4.33**
Mitered ends	Inst	SF	2C	.065	3.00	1.75	---	4.75	**6.11**	3.64	**4.65**
1" x 10", with 9-3/4" exp.											
Rough ends	Inst	SF	2C	.035	3.60	.94	---	4.54	**5.57**	3.55	**4.34**
Fitted ends	Inst	SF	2C	.045	3.70	1.21	---	4.91	**6.10**	3.82	**4.72**
Mitered ends	Inst	SF	2C	.052	3.79	1.40	---	5.19	**6.49**	4.01	**4.97**
Dressed/milled											
1" x 8", with 7" exp.											
Rough ends	Inst	SF	2C	.049	3.40	1.32	---	4.72	**5.92**	3.65	**4.53**
Fitted ends	Inst	SF	2C	.063	3.49	1.70	---	5.19	**6.60**	3.98	**5.01**
Mitered ends	Inst	SF	2C	.072	3.57	1.94	---	5.51	**7.06**	4.23	**5.38**
1" x 10", with 9" exp.											
Rough ends	Inst	SF	2C	.038	4.25	1.02	---	5.27	**6.44**	4.13	**5.02**

Description	Oper	Unit	Crew Size	Man-Hours Per Unit	Avg Mat'l Unit Cost	Avg Labor Unit Cost	Avg Equip Unit Cost	Avg Total Unit Cost	Avg Price Incl O&P	Large Volume Avg Total Unit Cost	Large Volume Avg Price Incl O&P
Fitted ends	Inst	SF	2C	.049	4.36	1.32	---	5.68	**7.02**	4.40	**5.40**
Mitered ends	Inst	SF	2C	.056	4.47	1.51	---	5.98	**7.44**	4.65	**5.74**
Board (1" x 12") and batten (1" x 2", 12" oc)											
Horizontal	Demo	SF	LB	.018	---	.40	---	.40	**.60**	.29	**.43**
Vertical											
Standard	Demo	SF	LB	.022	---	.48	---	.48	**.73**	.35	**.53**
Reverse	Demo	SF	LB	.021	---	.46	---	.46	**.70**	.33	**.50**
Board (1" x 12") and batten (1" x 2", 12" oc); rough sawn, common, std. and better, green											
Horizontal application, @ 10' L											
Redwood	Inst	SF	2C	.041	2.11	1.11	---	3.22	**4.11**	2.46	**3.12**
Red cedar	Inst	SF	2C	.041	1.70	1.11	---	2.81	**3.64**	2.15	**2.76**
Pine	Inst	SF	2C	.041	2.08	1.11	---	3.19	**4.07**	2.45	**3.11**
Vertical, std. (batten on board), @ 8' L											
Redwood	Inst	SF	2C	.051	2.11	1.37	---	3.48	**4.52**	2.65	**3.41**
Red cedar	Inst	SF	2C	.051	1.70	1.37	---	3.07	**4.04**	2.34	**3.05**
Pine	Inst	SF	2C	.051	2.08	1.37	---	3.45	**4.48**	2.64	**3.40**
Vertical, reverse (board on batten), @ 8' L											
Redwood	Inst	SF	2C	.052	2.11	1.40	---	3.51	**4.56**	2.65	**3.41**
Red cedar	Inst	SF	2C	.052	1.70	1.40	---	3.10	**4.09**	2.34	**3.05**
Pine	Inst	SF	2C	.052	2.08	1.40	---	3.48	**4.52**	2.64	**3.40**
Board on board (1" x 12" with 1-1/2" overlap), vertical	Demo	SF	LB	.024	---	.53	---	.53	**.80**	.37	**.56**
Board on board (1" x 12" with 1-1/2" overlap) rough sawn common, std. & better green, 8' L; vertical											
Redwood	Inst	SF	2C	.039	2.03	1.05	---	3.08	**3.93**	2.35	**2.97**
Red cedar	Inst	SF	2C	.039	1.69	1.05	---	2.74	**3.54**	2.11	**2.69**
Pine	Inst	SF	2C	.039	2.12	1.05	---	3.17	**4.04**	2.43	**3.06**
Plywood (1/2" T) with battens (1" x 2")											
16" oc battens	Demo	SF	LB	.010	---	.22	---	.22	**.33**	.15	**.23**
24" oc battens	Demo	SF	LB	.010	---	.22	---	.22	**.33**	.15	**.23**
Plywood (1/2" T) sanded exterior, AB grade with battens (1" x 2") S4S, common, std. & better green											
16" oc											
Fir	Inst	SF	2C	.031	1.52	.84	---	2.36	**3.02**	1.80	**2.29**
Pine	Inst	SF	2C	.031	1.56	.84	---	2.40	**3.06**	1.83	**2.33**
24" oc											
Fir	Inst	SF	2C	.026	1.47	.70	---	2.17	**2.76**	1.66	**2.08**
Pine	Inst	SF	2C	.026	1.50	.70	---	2.20	**2.79**	1.68	**2.11**
Shakes only, no sheathing included											
24" L with 11-1/2" exposure											
1/2" to 3/4" T	Demo	SF	LB	.014	---	.31	---	.31	**.46**	.22	**.33**
3/4" to 5/4" T	Demo	SF	LB	.016	---	.35	---	.35	**.53**	.24	**.36**
Shakes only, over sheathing, 2 nails/shake											
24" L with 11-1/2" exp., @ 6" W, red cedar, sawn one side											
Straight wall											
1/2" to 3/4" T	Inst	SF	CN	.030	1.59	.78	---	2.37	**3.01**	1.82	**2.29**
3/4" to 5/4" T	Inst	SF	CN	.034	2.10	.88	---	2.98	**3.76**	2.30	**2.88**

Description	Oper	Unit	Crew Size	Man-Hours Per Unit	Avg Mat'l Unit Cost	Avg Labor Unit Cost	Avg Equip Unit Cost	Avg Total Unit Cost	Avg Price Incl O&P	Large Volume Avg Total Unit Cost	Large Volume Avg Price Incl O&P
			Costs Based On Small Volume							Large Volume	
Cut-up wall											
1/2" to 3/4" T	Inst	SF	CN	.038	1.62	.99	---	2.61	**3.36**	1.99	**2.55**
3/4" to 5/4" T	Inst	SF	CN	.042	2.14	1.09	---	3.23	**4.12**	2.46	**3.11**
Related materials and operations											
Rosin sized sheathing paper, 36" W 500 SF/roll; nailed											
Over solid sheathing	Inst	SF	2C	.002	---	.05	---	.05	**.08**	.03	**.04**
Shingles only, no sheathing included											
16" L with 7-1/2" exposure	Demo	SF	LB	.023	---	.50	---	.50	**.76**	.35	**.53**
18" L with 8-1/2" exposure	Demo	SF	LB	.020	---	.44	---	.44	**.66**	.31	**.46**
24" L with 11-1/2" exposure	Demo	SF	LB	.015	---	.33	---	.33	**.50**	.22	**.33**
Shingles only, red cedar, No 1. perfect, @ 4" W, 2 nails/shake, over sheathing											
Straight wall											
16" L with 7-1/2" exposure	Inst	SF	2C	.039	1.85	1.05	---	2.90	**3.73**	2.21	**2.81**
18" L with 8-1/2" exposure	Inst	SF	2C	.035	1.42	.94	---	2.36	**3.07**	1.79	**2.29**
24" L with 11-1/2" exposure	Inst	SF	2C	.026	1.82	.70	---	2.52	**3.16**	1.95	**2.42**
Cut-up wall											
16" L with 7-1/2" exposure	Inst	SF	2C	.049	1.89	1.32	---	3.21	**4.18**	2.43	**3.13**
18" L with 8-1/2" exposure	Inst	SF	2C	.043	1.45	1.16	---	2.61	**3.43**	1.97	**2.56**
24" L with 11-1/2" exposure	Inst	SF	2C	.032	1.85	.86	---	2.71	**3.44**	2.07	**2.60**
Related materials and operations											
Rosin sized sheathing paper, 36" W 500 SF/roll; nailed											
Over solid sheathing	Inst	SF	2C	.002	---	.05	---	.05	**.08**	.03	**.04**

Sinks

Bathroom

Lavatories (bathroom sinks), with good quality fittings and faucets, concealed overflow

Vitreous china

Countertop units, with self rim (American Standard Products)

Description	Oper	Unit	Crew Size	Man-Hours Per Unit	Avg Mat'l Unit Cost	Avg Labor Unit Cost	Avg Equip Unit Cost	Avg Total Unit Cost	Avg Price Incl O&P	Large Volume Avg Total Unit Cost	Large Volume Avg Price Incl O&P
Antiquity											
24-1/2" x 18" with 8" or 4" cc											
Color	Inst	Ea	SA	3.81	368.00	116.00	---	484.00	**596.00**	386.30	**472.00**
White	Inst	Ea	SA	3.81	320.00	116.00	---	436.00	**540.00**	346.30	**426.00**
Aqualyn											
20" x 17" with 8" or 4" cc											
Color	Inst	Ea	SA	3.81	319.00	116.00	---	435.00	**539.00**	345.30	**425.00**
White	Inst	Ea	SA	3.81	285.00	116.00	---	401.00	**500.00**	317.30	**393.00**
Elisse Petite											
24" x 19" with 8" or 4" cc											
Color	Inst	Ea	SA	3.81	438.00	116.00	---	554.00	**677.00**	445.30	**539.00**
White	Inst	Ea	SA	3.81	370.00	116.00	---	486.00	**599.00**	388.30	**475.00**
Heritage											
26" x 20" with 8" or 4" cc											
Color	Inst	Ea	SA	3.81	728.00	116.00	---	844.00	**1010.00**	685.30	**816.00**
White	Inst	Ea	SA	3.81	577.00	116.00	---	693.00	**837.00**	560.30	**672.00**
Hexalyn											
22" x 19" with 8" or 4" cc											
Color	Inst	Ea	SA	3.81	399.00	116.00	---	515.00	**632.00**	412.30	**502.00**
White	Inst	Ea	SA	3.81	342.00	116.00	---	458.00	**567.00**	365.30	**448.00**

Description	Oper	Unit	Crew Size	Man-Hours Per Unit	Avg Mat'l Unit Cost	Avg Labor Unit Cost	Avg Equip Unit Cost	Avg Total Unit Cost	Avg Price Incl O&P	Large Volume Avg Total Unit Cost	Large Volume Avg Price Incl O&P
Lexington											
22" x 19" with 8" or 4" cc											
Color	Inst	Ea	SA	3.81	494.00	116.00	---	610.00	**741.00**	491.30	**593.00**
White	Inst	Ea	SA	3.81	411.00	116.00	---	527.00	**646.00**	422.30	**514.00**
Pallas Petite											
24" x 18" with 8" or 4" cc											
Color	Inst	Ea	SA	3.81	525.00	116.00	---	641.00	**776.00**	516.30	**622.00**
White	Inst	Ea	SA	3.81	432.00	116.00	---	548.00	**670.00**	440.30	**534.00**
Renaissance											
19" diameter with 8" or 4" cc											
Color	Inst	Ea	SA	3.81	319.00	116.00	---	435.00	**539.00**	345.30	**425.00**
White	Inst	Ea	SA	3.81	285.00	116.00	---	401.00	**500.00**	317.30	**393.00**
Roma											
26" x 20" with 8" or 4" cc w/ sealant											
Color	Inst	Ea	SA	3.81	701.00	116.00	---	817.00	**979.00**	663.30	**790.00**
White	Inst	Ea	SA	3.81	558.00	116.00	---	674.00	**814.00**	544.30	**653.00**
Rondalyn											
19" diameter with 8" or 4" cc											
Color	Inst	Ea	SA	3.81	307.00	116.00	---	423.00	**526.00**	336.30	**414.00**
White	Inst	Ea	SA	3.81	276.00	116.00	---	392.00	**491.00**	310.30	**385.00**
Countertop units (Kohler Products, Artists Edition)											
Escapade											
Vintage countertop 17" x 14"											
White	Inst	Ea	SA	3.81	1020.00	116.00	---	1136.00	**1350.00**	932.30	**1100.00**
Vintage undercounter 17" x 14"											
White	Inst	Ea	SA	3.81	898.00	116.00	---	1014.00	**1210.00**	826.30	**978.00**
Fables and Flowers											
Vintage countertop 17" x 14"											
White	Inst	Ea	SA	3.81	1310.00	116.00	---	1426.00	**1680.00**	1171.30	**1370.00**
Vintage undercounter 17" x 14"											
White	Inst	Ea	SA	3.81	1010.00	116.00	---	1126.00	**1330.00**	915.30	**1080.00**
Graphite											
Vintage countertop 17" x 14"											
White	Inst	Ea	SA	3.81	1020.00	116.00	---	1136.00	**1350.00**	932.30	**1100.00**
Vintage undercounter 17" x 14"											
White	Inst	Ea	SA	3.81	983.00	116.00	---	1099.00	**1300.00**	896.30	**1060.00**
Isis											
Vintage countertop 17" x 14"											
White	Inst	Ea	SA	3.81	1020.00	116.00	---	1136.00	**1350.00**	932.30	**1100.00**
Vintage undercounter 17" x 14"											
White	Inst	Ea	SA	3.81	898.00	116.00	---	1014.00	**1210.00**	826.30	**978.00**
Oriental Patchwork											
Chenoile countertop											
Color	Inst	Ea	SA	3.81	1350.00	116.00	---	1466.00	**1730.00**	1201.30	**1410.00**
Provencial											
Vintage countertop 17" x 14"											
White	Inst	Ea	SA	3.81	353.00	116.00	---	469.00	**579.00**	374.30	**459.00**
Vintage undercounter 17" x 14"											
White	Inst	Ea	SA	3.81	314.00	116.00	---	430.00	**534.00**	341.30	**420.00**

Description	Oper	Unit	Crew Size	Man-Hours Per Unit	Avg Mat'l Unit Cost	Avg Labor Unit Cost	Avg Equip Unit Cost	Avg Total Unit Cost	Avg Price Incl O&P	Avg Total Unit Cost	Avg Price Incl O&P
			Costs Based On Small Volume							Large Volume	
Pedestal sinks (American Standard Products)											
Antiquity w/ pedestal											
24-1/2" x 19" with 8" or 4" cc											
Color	Inst	Ea	SA	3.81	297.00	116.00	---	413.00	**514.00**	327.30	**404.00**
White	Inst	Ea	SA	3.81	413.00	116.00	---	529.00	**648.00**	424.30	**515.00**
Lavatory slab only											
Color	Inst	Ea	SA	2.00	178.00	60.90	---	238.90	**295.00**	196.70	**243.00**
White	Inst	Ea	SA	2.00	128.00	60.90	---	188.90	**238.00**	154.70	**195.00**
Pedestal only											
Color	Inst	Ea	SA	2.00	119.00	60.90	---	179.90	**227.00**	147.20	**186.00**
White	Inst	Ea	SA	2.00	84.80	60.90	---	145.70	**188.00**	119.00	**154.00**
Cadet w/ pedestal											
24" x 20" with 8" or 4" cc											
Color	Inst	Ea	SA	3.81	476.00	116.00	---	592.00	**721.00**	476.30	**576.00**
White	Inst	Ea	SA	3.81	404.00	116.00	---	520.00	**638.00**	417.30	**507.00**
Lavatory slab only											
Color	Inst	Ea	SA	2.00	197.00	60.90	---	257.90	**317.00**	211.70	**260.00**
White	Inst	Ea	SA	2.00	145.00	60.90	---	205.90	**257.00**	168.70	**211.00**
Pedestal only											
Color	Inst	Ea	SA	2.00	79.70	60.90	---	140.60	**182.00**	114.80	**149.00**
White	Inst	Ea	SA	2.00	59.30	60.90	---	120.20	**159.00**	97.90	**129.00**
Elisse Petite w/ pedestal											
25" x 21" with 8" or 4" cc											
Color	Inst	Ea	SA	3.81	759.00	116.00	---	875.00	**1050.00**	711.30	**846.00**
White	Inst	Ea	SA	3.81	603.00	116.00	---	719.00	**866.00**	581.30	**696.00**
Lavatory slab only											
Color	Inst	Ea	SA	2.00	334.00	60.90	---	394.90	**475.00**	325.70	**391.00**
White	Inst	Ea	SA	2.00	241.00	60.90	---	301.90	**368.00**	248.70	**302.00**
Pedestal only											
Color	Inst	Ea	SA	2.00	225.00	60.90	---	285.90	**350.00**	235.70	**288.00**
White	Inst	Ea	SA	2.00	162.00	60.90	---	222.90	**277.00**	182.70	**227.00**
Lexington w/ pedestal											
24" x 18" with 8" or 4" cc											
Color	Inst	Ea	SA	3.81	649.00	116.00	---	765.00	**919.00**	620.30	**741.00**
White	Inst	Ea	SA	3.81	522.00	116.00	---	638.00	**773.00**	514.30	**619.00**
Lavatory slab only											
Color	Inst	Ea	SA	2.00	308.00	60.90	---	368.90	**446.00**	304.70	**367.00**
White	Inst	Ea	SA	2.00	224.00	60.90	---	284.90	**348.00**	234.70	**286.00**
Pedestal only											
Color	Inst	Ea	SA	2.00	141.00	60.90	---	201.90	**253.00**	165.70	**207.00**
White	Inst	Ea	SA	2.00	98.30	60.90	---	159.20	**204.00**	130.30	**166.00**
Renaissance w/ pedestal											
23" x 18-1/4" with 8" or 4" cc											
Color	Inst	Ea	SA	3.81	385.00	116.00	---	501.00	**615.00**	400.30	**488.00**
White	Inst	Ea	SA	3.81	337.00	116.00	---	453.00	**561.00**	361.30	**443.00**
Lavatory slab only											
Color	Inst	Ea	SA	2.00	112.00	60.90	---	172.90	**219.00**	141.60	**179.00**
White	Inst	Ea	SA	2.00	82.20	60.90	---	143.10	**185.00**	116.90	**151.00**
Pedestal only											
Color	Inst	Ea	SA	2.00	72.90	60.90	---	133.80	**175.00**	109.20	**142.00**
White	Inst	Ea	SA	2.00	55.10	60.90	---	116.00	**154.00**	94.40	**125.00**

			Costs Based On Small Volume						Large Volume	
Description	Oper	Unit	Crew Size	Man-Hours Per Unit	Avg Mat'l Unit Cost	Avg Labor Unit Cost	Avg Equip Unit Cost	Avg Total Unit Cost	Avg Price Incl O&P	Avg Total Unit Cost	Avg Price Incl O&P
Pedestal sinks (Kohler Products)											
Anatole											
21" x 18" with 8" or 4" cc											
Color	Inst	Ea	SA	3.81	776.00	116.00	---	892.00	**1060.00**	725.30	**862.00**
Premium color	Inst	Ea	SA	3.81	862.00	116.00	---	978.00	**1160.00**	796.30	**944.00**
White	Inst	Ea	SA	3.81	643.00	116.00	---	759.00	**912.00**	615.30	**735.00**
Arabesque											
32" x 20" x 33" with 8" or 4" cc											
Color	Inst	Ea	SA	3.81	1230.00	116.00	---	1346.00	**1580.00**	1101.30	**1290.00**
Premium color	Inst	Ea	SA	3.81	1380.00	116.00	---	1496.00	**1760.00**	1231.30	**1440.00**
White	Inst	Ea	SA	3.81	990.00	116.00	---	1106.00	**1310.00**	902.30	**1070.00**
Required towel bar	Inst	Ea	SA	1.33	530.00	40.50	---	570.50	**670.00**	470.50	**551.00**
Cabernet											
22" x 18" x 33" with 8" or 4" cc											
Color	Inst	Ea	SA	3.81	594.00	116.00	---	710.00	**856.00**	574.30	**688.00**
Premium color	Inst	Ea	SA	3.81	653.00	116.00	---	769.00	**923.00**	623.30	**744.00**
White	Inst	Ea	SA	3.81	503.00	116.00	---	619.00	**751.00**	498.30	**601.00**
27" x 20" x 33" with 8" or 4" cc											
Color	Inst	Ea	SA	3.81	822.00	116.00	---	938.00	**1120.00**	763.30	**905.00**
Premium color	Inst	Ea	SA	3.81	915.00	116.00	---	1031.00	**1220.00**	840.30	**994.00**
White	Inst	Ea	SA	3.81	678.00	116.00	---	794.00	**953.00**	644.30	**768.00**
Chablis											
24" x 19" with 8" or 4" cc											
Color	Inst	Ea	SA	3.81	647.00	116.00	---	763.00	**917.00**	618.30	**738.00**
Premium color	Inst	Ea	SA	3.81	714.00	116.00	---	830.00	**994.00**	673.30	**802.00**
White	Inst	Ea	SA	3.81	544.00	116.00	---	660.00	**798.00**	532.30	**640.00**
Chardonnay											
28" x 21-1/2" with 8" or 4" cc											
Color	Inst	Ea	SA	3.81	1070.00	116.00	---	1186.00	**1400.00**	969.30	**1140.00**
Premium color	Inst	Ea	SA	3.81	1200.00	116.00	---	1316.00	**1550.00**	1078.30	**1270.00**
White	Inst	Ea	SA	3.81	869.00	116.00	---	985.00	**1170.00**	803.30	**951.00**
Figurante											
32" x 20-1/4" with 8" cc											
Color	Inst	Ea	SA	3.81	1230.00	116.00	---	1346.00	**1580.00**	1101.30	**1290.00**
Premium color	Inst	Ea	SA	3.81	1380.00	116.00	---	1496.00	**1760.00**	1231.30	**1440.00**
White	Inst	Ea	SA	3.81	990.00	116.00	---	1106.00	**1310.00**	902.30	**1070.00**
Fleur											
28" x 21-1/2" with 8" or 4" cc											
Color	Inst	Ea	SA	3.81	979.00	116.00	---	1095.00	**1300.00**	893.30	**1060.00**
White	Inst	Ea	SA	3.81	799.00	116.00	---	915.00	**1090.00**	744.30	**884.00**
Pillow Talk											
27-1/2" x 20-1/2" x 33" with 8" or 4" cc											
Color	Inst	Ea	SA	3.81	1270.00	116.00	---	1386.00	**1630.00**	1131.30	**1330.00**
Premium color	Inst	Ea	SA	3.81	1430.00	116.00	---	1546.00	**1810.00**	1261.30	**1480.00**
White	Inst	Ea	SA	3.81	1020.00	116.00	---	1136.00	**1350.00**	928.30	**1100.00**
Wall hung units (American Standard Products)											
Corner Minette											
11" x 16" with 8" or 4" cc											
Color	Inst	Ea	SA	2.86	464.00	87.10	---	551.10	**663.00**	445.90	**533.00**
White	Inst	Ea	SA	2.86	403.00	87.10	---	490.10	**593.00**	394.90	**475.00**

Description	Oper	Unit	Crew Size	Man-Hours Per Unit	Avg Mat'l Unit Cost	Avg Labor Unit Cost	Avg Equip Unit Cost	Avg Total Unit Cost	Avg Price Incl O&P	Avg Total Unit Cost	Avg Price Incl O&P
			Costs Based On Small Volume							Large Volume	
Declyn											
19" x 17" with 8" or 4" cc											
Color	Inst	Ea	SA	3.81	296.00	116.00	---	412.00	**513.00**	326.30	**404.00**
White	Inst	Ea	SA	3.81	280.00	116.00	---	396.00	**495.00**	313.30	**388.00**
Penlyn											
18" x 16" with 8" or 4" cc											
White	Inst	Ea	SA	3.81	379.00	116.00	---	495.00	**609.00**	395.30	**483.00**
Roxalyn											
20" x 18" with 8" or 4" cc											
Color	Inst	Ea	SA	3.81	425.00	116.00	---	541.00	**661.00**	433.30	**526.00**
White	Inst	Ea	SA	3.81	368.00	116.00	---	484.00	**597.00**	387.30	**473.00**
Wheelchair, meets ADA requirements											
27" x 20" with 8" or 4" cc											
Color	Inst	Ea	SA	3.81	728.00	116.00	---	844.00	**1010.00**	685.30	**816.00**
White	Inst	Ea	SA	3.81	555.00	116.00	---	671.00	**811.00**	542.30	**651.00**
Wall hung units (Kohler Products)											
Chesapeake											
19" x 17" with 4" cc											
Color	Inst	Ea	SA	3.81	357.00	116.00	---	473.00	**583.00**	377.30	**461.00**
White	Inst	Ea	SA	3.81	320.00	116.00	---	436.00	**541.00**	347.30	**427.00**
20" x 18" with 4" cc											
Color	Inst	Ea	SA	3.81	397.00	116.00	---	513.00	**630.00**	411.30	**500.00**
White	Inst	Ea	SA	3.81	352.00	116.00	---	468.00	**577.00**	373.30	**457.00**
Fleur											
19-7/8" x 14-3/8" with single hole											
Color	Inst	Ea	SA	3.81	754.00	116.00	---	870.00	**1040.00**	707.30	**841.00**
White	Inst	Ea	SA	3.81	626.00	116.00	---	742.00	**893.00**	601.30	**719.00**
Greenwich											
24" x 20" with 8" or 4" cc											
White	Inst	Ea	SA	3.81	452.00	116.00	---	568.00	**692.00**	456.30	**552.00**
20" x 18" with 8 or 4" cc											
Color	Inst	Ea	SA	3.81	402.00	116.00	---	518.00	**636.00**	415.30	**505.00**
White	Inst	Ea	SA	3.81	336.00	116.00	---	452.00	**559.00**	360.30	**442.00**
18" x 15" with 8 or 4" cc											
White	Inst	Ea	SA	3.81	363.00	116.00	---	479.00	**590.00**	382.30	**468.00**
Jamestown											
20" x 18" with 8" or 4" cc											
White	Inst	Ea	SA	3.81	371.00	116.00	---	487.00	**599.00**	389.30	**475.00**
Juneau											
19" x 17"											
White	Inst	Ea	SA	3.81	590.00	116.00	---	706.00	**851.00**	570.30	**684.00**
Kingston											
21" x 18" with 8" or 4" cc											
Color	Inst	Ea	SA	3.81	385.00	116.00	---	501.00	**616.00**	401.30	**489.00**
White	Inst	Ea	SA	3.81	342.00	116.00	---	458.00	**567.00**	365.30	**448.00**
Portrait											
16-1/2" x 12-1/2" bowl w/ cistern											
White	Inst	Ea	SA	3.81	840.00	116.00	---	956.00	**1140.00**	779.30	**923.00**
27" x 19-3/8" with 8" cc											
Color	Inst	Ea	SA	3.81	913.00	116.00	---	1029.00	**1220.00**	839.30	**993.00**
White	Inst	Ea	SA	3.81	748.00	116.00	---	864.00	**1030.00**	702.30	**836.00**

Description	Oper	Unit	Crew Size	Man-Hours Per Unit	Avg Mat'l Unit Cost	Avg Labor Unit Cost	Avg Equip Unit Cost	Avg Total Unit Cost	Avg Price Incl O&P	Avg Total Unit Cost	Avg Price Incl O&P
			Costs Based On Small Volume							Large Volume	
Adjustments											
To only remove and reset lavatory											
	Reset	Ea	SA	2.86	---	87.10	---	87.10	**130.00**	60.90	**90.80**
To install rough-in	Inst	Ea	SA	10.0	---	305.00	---	305.00	**454.00**	203.00	**303.00**
For colors, ADD	Inst	%		---	30.0	---	---	---	**---**	---	**---**
For two 15" L towel bars, ADD	Inst	Pr	---	---	35.60	---	---	35.60	**41.00**	35.60	**41.00**
Enameled cast iron											
Countertop units (Kohler Products)											
Farmington											
19" x 16" with 8" or 4" cc											
Color	Inst	Ea	SA	3.81	354.00	116.00	---	470.00	**580.00**	375.30	**459.00**
White	Inst	Ea	SA	3.81	319.00	116.00	---	435.00	**539.00**	345.30	**425.00**
Radiant											
19" diameter with 4" cc											
Color	Inst	Ea	SA	3.81	356.00	116.00	---	472.00	**582.00**	376.30	**461.00**
White	Inst	Ea	SA	3.81	320.00	116.00	---	436.00	**541.00**	346.30	**426.00**
Tahoe											
20" x 19" with 8" or 4" cc											
Color	Inst	Ea	SA	3.81	378.00	116.00	---	494.00	**608.00**	395.30	**482.00**
White	Inst	Ea	SA	3.81	337.00	116.00	---	453.00	**561.00**	361.30	**443.00**
Countertop units, with self rim (Kohler Products)											
Castelle											
21" x 19" with 8" or 4" cc											
Color	Inst	Ea	SA	3.81	477.00	116.00	---	593.00	**721.00**	477.30	**576.00**
White	Inst	Ea	SA	3.81	385.00	116.00	---	501.00	**616.00**	401.30	**489.00**
Ellipse											
33" x 19" with 8" cc											
Color	Inst	Ea	SA	3.81	724.00	116.00	---	840.00	**1010.00**	682.30	**812.00**
White	Inst	Ea	SA	3.81	550.00	116.00	---	666.00	**806.00**	538.30	**646.00**
Farmington											
19-1/4" x 16-1/4" with 8" or 4" cc											
Color	Inst	Ea	SA	3.81	395.00	116.00	---	511.00	**627.00**	409.30	**498.00**
White	Inst	Ea	SA	3.81	331.00	116.00	---	447.00	**553.00**	355.30	**437.00**
Hexsign											
22-1/2" x 19" with 8" or 4" cc											
Color	Inst	Ea	SA	3.81	417.00	116.00	---	533.00	**653.00**	427.30	**519.00**
White	Inst	Ea	SA	3.81	345.00	116.00	---	461.00	**570.00**	368.30	**451.00**
Radiant											
19" with 8" or 4" cc											
Color	Inst	Ea	SA	3.81	356.00	116.00	---	472.00	**582.00**	376.30	**461.00**
White	Inst	Ea	SA	3.81	320.00	116.00	---	436.00	**541.00**	346.30	**426.00**
Terragon											
19-3/4" x 15-3/4"											
Color	Inst	Ea	SA	3.81	498.00	116.00	---	614.00	**746.00**	495.30	**597.00**
White	Inst	Ea	SA	3.81	400.00	116.00	---	516.00	**633.00**	413.30	**503.00**
Wall hung units, with self rim (American Standard Products)											
Regalyn											
20" x 18" with 8" or 4" cc											
Color	Inst	Ea	SA	3.81	415.00	116.00	---	531.00	**651.00**	426.30	**518.00**
White	Inst	Ea	SA	3.81	380.00	116.00	---	496.00	**610.00**	396.30	**484.00**

Description	Oper	Unit	Crew Size	Man-Hours Per Unit	Avg Mat'l Unit Cost	Avg Labor Unit Cost	Avg Equip Unit Cost	Avg Total Unit Cost	Avg Price Incl O&P	Large Volume Avg Total Unit Cost	Large Volume Avg Price Incl O&P
19" x 17" with 8" or 4" cc											
Color	Inst	Ea	SA	3.81	404.00	116.00	---	520.00	**638.00**	417.30	**507.00**
White	Inst	Ea	SA	3.81	370.00	116.00	---	486.00	**598.00**	388.30	**474.00**
Wall hung units, with shelf back (Kohler Products)											
Hampton											
22" x 19"											
Color	Inst	Ea	SA	3.81	439.00	116.00	---	555.00	**678.00**	445.30	**540.00**
White	Inst	Ea	SA	3.81	384.00	116.00	---	500.00	**615.00**	400.30	**488.00**
19" x 17"											
White	Inst	Ea	SA	3.81	419.00	116.00	---	535.00	**655.00**	429.30	**521.00**
Marston, corner unit											
16" x 16"											
Color	Inst	Ea	SA	3.81	655.00	116.00	---	771.00	**926.00**	625.30	**746.00**
White	Inst	Ea	SA	3.81	578.00	116.00	---	694.00	**838.00**	561.30	**673.00**
Taunton											
16" x 14"											
Color	Inst	Ea	SA	3.81	630.00	116.00	---	746.00	**897.00**	604.30	**722.00**
White	Inst	Ea	SA	3.81	559.00	116.00	---	675.00	**816.00**	545.30	**655.00**
Trailer											
13" x 13"											
White	Inst	Ea	SA	4.00	380.00	122.00	---	502.00	**618.00**	375.90	**453.00**
Adjustments											
To only remove and reset lavatory	Reset	Ea	SA	2.86	---	87.10	---	87.10	**130.00**	60.90	**90.80**
To install rough-in	Inst	Ea	SA	10.0	---	305.00	---	305.00	**454.00**	203.00	**303.00**
For colors, ADD	Inst	%		---	25.0	---	---	---	**---**	---	**---**
For two 15" L towel bars, ADD	Inst	Pr	---	---	35.60	---	---	35.60	**41.00**	35.60	**41.00**
Enameled steel											
Countertop units, with self rim (American Standard Products)											
Ovation											
19" diameter with 4" cc											
Color	Inst	Ea	SA	3.81	178.00	116.00	---	294.00	**377.00**	228.30	**291.00**
White	Inst	Ea	SA	3.81	172.00	116.00	---	288.00	**371.00**	224.30	**286.00**
Vanette											
19" diameter with 4" cc											
Color	Inst	Ea	SA	3.81	176.00	116.00	---	292.00	**375.00**	227.30	**289.00**
White	Inst	Ea	SA	3.81	170.00	116.00	---	286.00	**368.00**	222.30	**283.00**
Adjustments											
To only remove and reset lavatory	Reset	Ea	SA	2.86	---	87.10	---	87.10	**130.00**	60.90	**90.80**
To install rough-in	Inst	Ea	SA	10.0	---	305.00	---	305.00	**454.00**	203.00	**303.00**
For colors, ADD	Inst	%		---	15.0	---	---	---	**---**	---	**---**
Americast - Porcelain enameled finish over steel composite material											
Countertop units, with self rim (American Standard Products)											
Acclivity											
19" diameter with 8" or 4" cc											
Color	Inst	Ea	SA	3.81	296.00	116.00	---	412.00	**514.00**	327.30	**404.00**
White	Inst	Ea	SA	3.81	252.00	116.00	---	368.00	**463.00**	290.30	**362.00**
Affinity											
20" x 17" with 8" or 4" cc											
Color	Inst	Ea	SA	3.81	317.00	116.00	---	433.00	**538.00**	344.30	**424.00**
White	Inst	Ea	SA	3.81	267.00	116.00	---	383.00	**480.00**	303.30	**376.00**

Description	Oper	Unit	Crew Size	Man-Hours Per Unit	Avg Mat'l Unit Cost	Avg Labor Unit Cost	Avg Equip Unit Cost	Avg Total Unit Cost	Avg Price Incl O&P	Avg Total Unit Cost	Avg Price Incl O&P
			Costs Based On Small Volume							Large Volume	
Adjustments											
To only remove and reset lavatory	Reset	Ea	SA	2.86	---	87.10	---	87.10	**130.00**	60.90	**90.80**
To install rough-in	Inst	Ea	SA	10.0	---	305.00	---	305.00	**454.00**	203.00	**303.00**
For colors, ADD	Inst	%		---	15.0	---	---	---	**---**	---	**---**

Kitchen

(Kitchen and utility), with good quality fittings, faucets, and sprayer

Americast - Porcelain enameled finish over steel composite material (American Standard)

Description	Oper	Unit	Crew Size	Man-Hours Per Unit	Avg Mat'l Unit Cost	Avg Labor Unit Cost	Avg Equip Unit Cost	Avg Total Unit Cost	Avg Price Incl O&P	Avg Total Unit Cost	Avg Price Incl O&P
Silhouette, double bowl with self rim											
33" x 22"											
Color	Inst	Ea	SA	7.27	466.00	221.00	---	687.00	**866.00**	539.00	**672.00**
Premium color	Inst	Ea	SA	7.27	507.00	221.00	---	728.00	**913.00**	573.00	**711.00**
White	Inst	Ea	SA	7.27	414.00	221.00	---	635.00	**806.00**	495.00	**622.00**
Silhouette, double bowl with tile edge											
33" x 22"											
Color	Inst	Ea	SA	7.27	484.00	221.00	---	705.00	**887.00**	554.00	**689.00**
Premium color	Inst	Ea	SA	7.27	526.00	221.00	---	747.00	**935.00**	589.00	**729.00**
White	Inst	Ea	SA	7.27	427.00	221.00	---	648.00	**821.00**	507.00	**635.00**
Silhouette, double bowl with color matched mounting rim											
33" x 22"											
Color	Inst	Ea	SA	7.27	483.00	221.00	---	704.00	**886.00**	553.00	**688.00**
White	Inst	Ea	SA	7.27	428.00	221.00	---	649.00	**822.00**	507.00	**636.00**
Silhouette, dual level double bowl with self rim											
38" x 22"											
Color	Inst	Ea	SA	7.27	591.00	221.00	---	812.00	**1010.00**	642.00	**791.00**
Premium color	Inst	Ea	SA	7.27	647.00	221.00	---	868.00	**1070.00**	689.00	**844.00**
White	Inst	Ea	SA	7.27	517.00	221.00	---	738.00	**925.00**	581.00	**720.00**
Silhouette, dual level double bowl with tile edge or color matched rim											
38" x 22"											
Color	Inst	Ea	SA	7.27	608.00	221.00	---	829.00	**1030.00**	656.00	**807.00**
Premium color	Inst	Ea	SA	7.27	666.00	221.00	---	887.00	**1100.00**	705.00	**863.00**
White	Inst	Ea	SA	7.27	531.00	221.00	---	752.00	**940.00**	592.00	**733.00**
Silhouette, dual level double bowl with self rim											
33" x 22"											
Color	Inst	Ea	SA	7.27	487.00	221.00	---	708.00	**890.00**	556.00	**691.00**
Premium color	Inst	Ea	SA	7.27	529.00	221.00	---	750.00	**938.00**	591.00	**732.00**
White	Inst	Ea	SA	7.27	430.00	221.00	---	651.00	**824.00**	509.00	**637.00**
Silhouette, dual level double bowl with tile edge or color matched rim											
33" x 22"											
Color	Inst	Ea	SA	7.27	503.00	221.00	---	724.00	**909.00**	570.00	**708.00**
Premium color	Inst	Ea	SA	7.27	549.00	221.00	---	770.00	**962.00**	608.00	**751.00**
White	Inst	Ea	SA	7.27	444.00	221.00	---	665.00	**841.00**	521.00	**651.00**
Silhouette, single bowl with self rim											
25" x 22"											
Color	Inst	Ea	SA	5.71	370.00	174.00	---	544.00	**684.00**	429.00	**534.00**
Premium color	Inst	Ea	SA	5.71	398.00	174.00	---	572.00	**716.00**	452.00	**561.00**
White	Inst	Ea	SA	5.71	332.00	174.00	---	506.00	**641.00**	398.00	**499.00**
Silhouette, single bowl with tile edge											
25" x 22"											
Color	Inst	Ea	SA	5.71	397.00	174.00	---	571.00	**715.00**	451.00	**560.00**
Premium color	Inst	Ea	SA	5.71	418.00	174.00	---	592.00	**740.00**	469.00	**580.00**
White	Inst	Ea	SA	5.71	348.00	174.00	---	522.00	**659.00**	410.00	**513.00**

Description	Oper	Unit	Crew Size	Man-Hours Per Unit	Avg Mat'l Unit Cost	Avg Labor Unit Cost	Avg Equip Unit Cost	Avg Total Unit Cost	Avg Price Incl O&P	Large Volume Avg Total Unit Cost	Large Volume Avg Price Incl O&P
Silhouette, single bowl with self rim											
18" x 18"											
Color	Inst	Ea	SA	3.81	487.00	116.00	---	603.00	**732.00**	485.30	**586.00**
Premium color	Inst	Ea	SA	3.81	529.00	116.00	---	645.00	**781.00**	520.30	**626.00**
White	Inst	Ea	SA	3.81	430.00	116.00	---	546.00	**667.00**	438.30	**531.00**
Silhouette, single bowl with tile edge											
18" x 18"											
Color	Inst	Ea	SA	3.81	503.00	116.00	---	619.00	**752.00**	499.30	**602.00**
Premium color	Inst	Ea	SA	3.81	549.00	116.00	---	665.00	**805.00**	537.30	**645.00**
White	Inst	Ea	SA	3.81	444.00	116.00	---	560.00	**684.00**	450.30	**545.00**
Enameled cast iron (Kohler Products)											
Bakersfield, single bowl											
31" x 22"											
Color	Inst	Ea	SA	7.27	442.00	221.00	---	663.00	**838.00**	519.00	**649.00**
Premium color	Inst	Ea	SA	7.27	486.00	221.00	---	707.00	**889.00**	556.00	**691.00**
White	Inst	Ea	SA	7.27	383.00	221.00	---	604.00	**770.00**	469.00	**592.00**
Bon Vivant, triple bowl, 2 large, 1 small with tile or self rim											
48" x 22"											
Color	Inst	Ea	SA	7.27	855.00	221.00	---	1076.00	**1310.00**	861.00	**1040.00**
Premium color	Inst	Ea	SA	7.27	961.00	221.00	---	1182.00	**1440.00**	950.00	**1140.00**
White	Inst	Ea	SA	7.27	713.00	221.00	---	934.00	**1150.00**	744.00	**907.00**
Cantina, double with tile or self rim, 2 large											
43" x 22"											
Color	Inst	Ea	SA	7.27	639.00	221.00	---	860.00	**1060.00**	682.00	**837.00**
Premium color	Inst	Ea	SA	7.27	713.00	221.00	---	934.00	**1150.00**	744.00	**908.00**
White	Inst	Ea	SA	7.27	540.00	221.00	---	761.00	**951.00**	600.00	**743.00**
Ecocycle, double bowl with regular rim											
43" x 22"											
Color	Inst	Ea	SA	7.27	1480.00	221.00	---	1701.00	**2040.00**	1382.00	**1640.00**
Premium color	Inst	Ea	SA	7.27	1680.00	221.00	---	1901.00	**2270.00**	1552.00	**1830.00**
White	Inst	Ea	SA	7.27	1220.00	221.00	---	1441.00	**1730.00**	1162.00	**1390.00**
Epicurean, double bowl with cutting board and drainboard, 1 large, 1 small											
43" x 22"											
Color	Inst	Ea	SA	7.27	754.00	221.00	---	975.00	**1200.00**	778.00	**946.00**
White	Inst	Ea	SA	7.27	632.00	221.00	---	853.00	**1060.00**	677.00	**830.00**
Executive Chef, double bowl with tile or self rim, 1 large, 1 medium											
33" x 22"											
Color	Inst	Ea	SA	7.27	582.00	221.00	---	803.00	**999.00**	635.00	**782.00**
Premium color	Inst	Ea	SA	7.27	647.00	221.00	---	868.00	**1070.00**	689.00	**845.00**
White	Inst	Ea	SA	7.27	494.00	221.00	---	715.00	**899.00**	562.00	**699.00**
Julienne, single bowl with self rim											
38" x 22"											
Color	Inst	Ea	SA	7.27	848.00	221.00	---	1069.00	**1300.00**	856.00	**1040.00**
Premium color	Inst	Ea	SA	7.27	953.00	221.00	---	1174.00	**1430.00**	943.00	**1140.00**
White	Inst	Ea	SA	7.27	707.00	221.00	---	928.00	**1140.00**	739.00	**902.00**
Lakefield double bowl with tile or self rim, 1 large, 1 small											
33" x 22"											
Color	Inst	Ea	SA	7.27	490.00	221.00	---	711.00	**893.00**	559.00	**694.00**
Premium color	Inst	Ea	SA	7.27	542.00	221.00	---	763.00	**953.00**	601.00	**744.00**
White	Inst	Ea	SA	7.27	421.00	221.00	---	642.00	**814.00**	501.00	**629.00**
Marsala Hi-Low with self or tile rim											
33" x 22"											
Color	Inst	Ea	SA	7.27	582.00	221.00	---	803.00	**999.00**	635.00	**782.00**

Description	Oper	Unit	Crew Size	Man-Hours Per Unit	Avg Mat'l Unit Cost	Avg Labor Unit Cost	Avg Equip Unit Cost	Avg Total Unit Cost	Avg Price Incl O&P	Avg Total Unit Cost	Avg Price Incl O&P
			Costs Based On Small Volume							Large Volume	
Premium color	Inst	Ea	SA	7.27	647.00	221.00	---	868.00	**1070.00**	689.00	**845.00**
White	Inst	Ea	SA	7.27	494.00	221.00	---	715.00	**899.00**	562.00	**699.00**
Mayfield, single for frame mount with self or tile rim											
24" x 21"											
Color	Inst	Ea	SA	5.71	338.00	174.00	---	512.00	**648.00**	403.00	**505.00**
White	Inst	Ea	SA	5.71	300.00	174.00	---	474.00	**604.00**	371.00	**468.00**
Porcelain on steel											
Double bowl, with self rim											
33" x 22"											
Color	Inst	Ea	SA	7.27	237.00	221.00	---	458.00	**602.00**	348.00	**453.00**
White	Inst	Ea	SA	7.27	228.00	221.00	---	449.00	**592.00**	341.00	**445.00**
Double bowl, dual level with self rim											
33" x 22"											
Color	Inst	Ea	SA	7.27	263.00	221.00	---	484.00	**632.00**	370.00	**478.00**
White	Inst	Ea	SA	7.27	252.00	221.00	---	473.00	**620.00**	361.00	**468.00**
Single bowl											
25" x 22"											
Color	Inst	Ea	SA	5.71	228.00	174.00	---	402.00	**521.00**	311.00	**399.00**
White	Inst	Ea	SA	5.71	220.00	174.00	---	394.00	**512.00**	304.00	**391.00**
Stainless steel (Kohler)											
Triple bowl with self rim											
Ravinia											
43" x 22"	Inst	Ea	SA	7.27	1060.00	221.00	---	1281.00	**1550.00**	1034.00	**1240.00**
Double bowl with self rim											
Ravinia											
42" x 22"	Inst	Ea	SA	7.27	801.00	221.00	---	1022.00	**1250.00**	817.00	**992.00**
33" x 22"	Inst	Ea	SA	7.27	657.00	221.00	---	878.00	**1090.00**	697.00	**854.00**
Lyric											
33" x 22"	Inst	Ea	SA	7.27	344.00	221.00	---	565.00	**726.00**	438.00	**555.00**
Single bowl with self rim											
Ravinia											
25" x 22"	Inst	Ea	SA	5.71	477.00	174.00	---	651.00	**808.00**	518.00	**637.00**
Lyric											
33" x 22"	Inst	Ea	SA	5.71	346.00	174.00	---	520.00	**657.00**	409.00	**512.00**
33" x 21"	Inst	Ea	SA	5.71	310.00	174.00	---	484.00	**616.00**	380.00	**478.00**
25" x 22"	Inst	Ea	SA	5.71	263.00	174.00	---	437.00	**562.00**	340.00	**433.00**
25" x 21"	Inst	Ea	SA	5.71	263.00	174.00	---	437.00	**562.00**	340.00	**433.00**
Bar											
Enameled cast iron (Kohler)											
Addison											
13" x 13"											
Color	Inst	Ea	SA	3.81	303.00	116.00	---	419.00	**521.00**	332.30	**410.00**
Premium color	Inst	Ea	SA	3.81	326.00	116.00	---	442.00	**548.00**	352.30	**433.00**
White	Inst	Ea	SA	3.81	271.00	116.00	---	387.00	**485.00**	306.30	**380.00**

Description	Oper	Unit	Crew Size	Man-Hours Per Unit	Avg Mat'l Unit Cost	Avg Labor Unit Cost	Avg Equip Unit Cost	Avg Total Unit Cost	Avg Price Incl O&P	Large Volume Avg Total Unit Cost	Large Volume Avg Price Incl O&P
			Costs Based On Small Volume							Large Volume	
Apertif with tile or self rim											
16" x19"											
Color	Inst	Ea	SA	3.81	353.00	116.00	---	469.00	**579.00**	374.30	**458.00**
Premium color	Inst	Ea	SA	3.81	385.00	116.00	---	501.00	**615.00**	400.30	**488.00**
White	Inst	Ea	SA	3.81	312.00	116.00	---	428.00	**531.00**	340.30	**419.00**
Entertainer with semi circular self rim or undercounter self rim											
22" x18"											
Color	Inst	Ea	SA	5.71	376.00	174.00	---	550.00	**691.00**	434.00	**540.00**
Premium color	Inst	Ea	SA	5.71	410.00	174.00	---	584.00	**731.00**	462.00	**573.00**
White	Inst	Ea	SA	5.71	330.00	174.00	---	504.00	**638.00**	395.00	**496.00**
Sorbet with tile or self or undercounter rim											
15" x 15"											
Color	Inst	Ea	SA	3.81	327.00	116.00	---	443.00	**549.00**	352.30	**433.00**
Premium color	Inst	Ea	SA	3.81	354.00	116.00	---	470.00	**580.00**	375.30	**459.00**
White	Inst	Ea	SA	3.81	290.00	116.00	---	406.00	**507.00**	322.30	**398.00**
Stainless steel (Kohler)											
Ravinia											
22" x 15"	Inst	Ea	SA	3.81	523.00	116.00	---	639.00	**774.00**	515.30	**620.00**
Lyric											
15" x 15"	Inst	Ea	SA	3.81	234.00	116.00	---	350.00	**442.00**	275.30	**344.00**
Utility/service											
Enameled cast iron (Kohler)											
Sutton high back											
24" x 18", 8" High back	Inst	Ea	SA	5.71	496.00	174.00	---	670.00	**829.00**	533.00	**655.00**
Tech											
24" x 18", Bubbler mound	Inst	Ea	SA	5.71	337.00	174.00	---	511.00	**647.00**	402.00	**504.00**
Whitby											
28" x 28" Corner	Inst	Ea	SA	5.71	632.00	174.00	---	806.00	**986.00**	646.00	**785.00**
Glen Falls, laundry sink											
25" x 22"											
Color	Inst	Ea	SA	5.71	494.00	174.00	---	668.00	**827.00**	532.00	**653.00**
White	Inst	Ea	SA	5.71	420.00	174.00	---	594.00	**742.00**	471.00	**583.00**
River Falls, laundry sink											
25" x 22"											
Color	Inst	Ea	SA	5.71	539.00	174.00	---	713.00	**878.00**	569.00	**696.00**
White	Inst	Ea	SA	5.71	456.00	174.00	---	630.00	**783.00**	500.00	**617.00**
Westover, double deep, laundry sink											
42" x 21"											
Color	Inst	Ea	SA	7.27	768.00	221.00	---	989.00	**1210.00**	790.00	**960.00**
White	Inst	Ea	SA	7.27	640.00	221.00	---	861.00	**1070.00**	683.00	**837.00**
Vitreous china (Kohler)											
Hollister											
28" x 22	Inst	Ea	SA	5.71	900.00	174.00	---	1074.00	**1290.00**	869.00	**1040.00**
Sudbury											
22" x 20	Inst	Ea	SA	5.71	554.00	174.00	---	728.00	**896.00**	582.00	**710.00**

Description	Oper	Unit	Crew Size	Man-Hours Per Unit	Avg Mat'l Unit Cost	Avg Labor Unit Cost	Avg Equip Unit Cost	Avg Total Unit Cost	Avg Price Incl O&P	Large Volume Avg Total Unit Cost	Large Volume Avg Price Incl O&P
Tyrrell											
20" x 20 w/ siphon jet	Inst	Ea	SA	7.27	959.00	221.00	---	1180.00	**1430.00**	948.00	**1140.00**
Adjustments											
To only remove and reset sink											
Single bowl	Reset	Ea	SA	2.86	---	87.10	---	87.10	**130.00**	60.90	**90.80**
Double bowl	Reset	Ea	SA	2.96	---	90.20	---	90.20	**134.00**	64.30	**95.80**
Triple bowl	Reset	Ea	SA	3.20	---	97.50	---	97.50	**145.00**	69.80	**104.00**
To install rough-in	Inst	Ea	SA	10.0	---	305.00	---	305.00	**454.00**	203.00	**303.00**
For 18 gauge steel, ADD	Inst	%		---	40.0	---	---	---	**---**	---	**---**

Skylights

Labor costs are for installation of skylight, flashing and roller shade only.

No carpentry or roofing work included.

Polycarbonate dome, clear transparent or tinted, roof opening sizes, curb or flush mount, self-flashing

Description	Oper	Unit	Crew Size	Man-Hours Per Unit	Avg Mat'l Unit Cost	Avg Labor Unit Cost	Avg Equip Unit Cost	Avg Total Unit Cost	Avg Price Incl O&P	Large Volume Avg Total Unit Cost	Large Volume Avg Price Incl O&P
Single dome											
22" x 22"	Inst	Ea	CA	2.61	47.60	70.40	---	118.00	**162.00**	81.90	**110.00**
30" x 30"	Inst	Ea	CA	4.30	85.70	116.00	---	201.70	**275.00**	140.90	**188.00**
46" x 46"	Inst	Ea	CA	5.80	252.00	156.00	---	408.00	**528.00**	303.80	**384.00**
22" x 46"	Inst	Ea	CA	4.30	102.00	116.00	---	218.00	**294.00**	154.70	**204.00**
30" x 46"	Inst	Ea	CA	4.94	150.00	133.00	---	283.00	**375.00**	204.80	**265.00**
Double dome											
22" x 22"	Inst	Ea	CA	2.61	67.80	70.40	---	138.20	**185.00**	98.70	**129.00**
30" x 30"	Inst	Ea	CA	4.30	131.00	116.00	---	247.00	**327.00**	178.60	**231.00**
46" x 46"	Inst	Ea	CA	5.80	313.00	156.00	---	469.00	**598.00**	353.80	**442.00**
22" x 46"	Inst	Ea	CA	4.30	146.00	116.00	---	262.00	**345.00**	191.60	**246.00**
30" x 46"	Inst	Ea	CA	4.94	190.00	133.00	---	323.00	**421.00**	237.80	**303.00**
Triple dome											
22" x 22"	Inst	Ea	CA	2.61	88.10	70.40	---	158.50	**208.00**	115.60	**149.00**
30" x 30"	Inst	Ea	CA	4.30	167.00	116.00	---	283.00	**368.00**	208.60	**265.00**
46" x 46"	Inst	Ea	CA	5.80	390.00	156.00	---	546.00	**687.00**	418.80	**516.00**
22" x 46"	Inst	Ea	CA	4.30	205.00	116.00	---	321.00	**412.00**	239.60	**302.00**
30" x 46"	Inst	Ea	CA	4.94	256.00	133.00	---	389.00	**497.00**	292.80	**366.00**
Operable skylight, Velux model VS											
21-9/16" x 27-1/2"	Inst	Ea	CA	5.71	333.00	154.00	---	487.00	**617.00**	369.20	**459.00**
21-9/16" x 38-1/2"	Inst	Ea	CA	5.71	357.00	154.00	---	511.00	**645.00**	389.20	**482.00**
21-9/16" x 46-3/8"	Inst	Ea	CA	5.71	393.00	154.00	---	547.00	**686.00**	419.20	**516.00**
21-9/16" x 55"	Inst	Ea	CA	5.71	417.00	154.00	---	571.00	**713.00**	439.20	**539.00**
30-5/8" x 38-1/2"	Inst	Ea	CA	5.71	417.00	154.00	---	571.00	**713.00**	439.20	**539.00**
30-5/8" x 55"	Inst	Ea	CA	5.93	488.00	160.00	---	648.00	**804.00**	502.00	**613.00**
44-3/4" x 27-1/2"	Inst	Ea	CA	5.93	452.00	160.00	---	612.00	**763.00**	472.00	**579.00**
44-3/4" x 46-1/2"	Inst	Ea	CA	5.93	547.00	160.00	---	707.00	**873.00**	551.00	**670.00**
Fixed skylight, Velux model FS											
21-9/16" x 27-1/2"	Inst	Ea	CA	5.71	226.00	154.00	---	380.00	**494.00**	280.20	**356.00**
21-9/16" x 38-1/2"	Inst	Ea	CA	5.71	256.00	154.00	---	410.00	**528.00**	305.20	**385.00**
21-9/16" x 46-3/8"	Inst	Ea	CA	5.71	286.00	154.00	---	440.00	**562.00**	330.20	**413.00**
21-9/16" x 55"	Inst	Ea	CA	5.71	298.00	154.00	---	452.00	**576.00**	340.20	**425.00**
21-9/16" x 70-7/8"	Inst	Ea	CA	5.71	345.00	154.00	---	499.00	**631.00**	379.20	**470.00**
30-5/8" x 38-1/2"	Inst	Ea	CA	5.71	303.00	154.00	---	457.00	**583.00**	344.20	**430.00**
30-5/8" x 55"	Inst	Ea	CA	5.93	357.00	160.00	---	517.00	**654.00**	393.00	**487.00**
44-3/4" x 27-1/2"	Inst	Ea	CA	5.93	333.00	160.00	---	493.00	**626.00**	373.00	**465.00**
44-3/4" x 46-1/2"	Inst	Ea	CA	5.93	405.00	160.00	---	565.00	**708.00**	433.00	**533.00**

Description	Oper	Unit	Crew Size	Man-Hours Per Unit	Avg Mat'l Unit Cost	Avg Labor Unit Cost	Avg Equip Unit Cost	Avg Total Unit Cost	Avg Price Incl O&P	Large Volume Avg Total Unit Cost	Large Volume Avg Price Incl O&P
			Costs Based On Small Volume							Large Volume	
Fixed ventilation with removable filter on ventilation flap, Velux model FSF											
21-9/16" x 27-1/2"	Inst	Ea	CA	5.23	284.00	141.00	---	425.00	**541.00**	320.70	**400.00**
21-9/16" x 38-1/2"	Inst	Ea	CA	5.23	326.00	141.00	---	467.00	**589.00**	355.70	**440.00**
21-9/16" x 46-3/8"	Inst	Ea	CA	5.23	343.00	141.00	---	484.00	**609.00**	370.70	**457.00**
21-9/16" x 55"	Inst	Ea	CA	5.71	470.00	154.00	---	624.00	**775.00**	483.20	**590.00**
21-9/16" x 70-7/8"	Inst	Ea	CA	5.63	380.00	152.00	---	532.00	**668.00**	407.40	**503.00**
30-5/8" x 38-1/2"	Inst	Ea	CA	5.23	379.00	141.00	---	520.00	**650.00**	399.70	**491.00**
30-5/8" x 55"	Inst	Ea	CA	5.63	440.00	152.00	---	592.00	**736.00**	457.40	**560.00**
44-3/4" x 27-1/2"	Inst	Ea	CA	5.59	403.00	151.00	---	554.00	**693.00**	425.60	**523.00**
44-3/4" x 46-1/2"	Inst	Ea	CA	5.63	487.00	152.00	---	639.00	**791.00**	496.40	**605.00**
Low profile, acrylic double domes with molded edge. Preassembled units include plywood curb-liner, 16 oz. copper flashing for pitched or flat roof, screen, and all hardware. Ventarama.											
Ventilating type											
30" x 72"	Inst	Ea	CA	4.30	530.00	116.00	---	646.00	**785.00**	510.60	**612.00**
45-1/2" x 45-1/2"	Inst	Ea	CA	4.30	476.00	116.00	---	592.00	**724.00**	465.60	**561.00**
45-1/2" x 30"	Inst	Ea	CA	4.30	393.00	116.00	---	509.00	**628.00**	396.60	**481.00**
30" x 45-1/2"	Inst	Ea	CA	4.30	393.00	116.00	---	509.00	**628.00**	396.60	**481.00**
30" x 30"	Inst	Ea	CA	3.45	321.00	93.00	---	414.00	**511.00**	322.80	**392.00**
30" x 22"	Inst	Ea	CA	3.45	292.00	93.00	---	385.00	**477.00**	298.80	**364.00**
22" x 30"	Inst	Ea	CA	3.45	292.00	93.00	---	385.00	**477.00**	298.80	**364.00**
22" x 45-1/2"	Inst	Ea	CA	3.45	351.00	93.00	---	444.00	**545.00**	347.80	**421.00**
Add for insulated glass	Inst	%		---	30.0	---	---	---	**---**	---	**---**
Add for bronze tinted outer dome	Inst	Ea	---	---	41.70	---	---	---	**47.90**	---	**39.90**
Add for white inner dome	Inst	Ea	---	---	20.20	---	---	---	**23.30**	---	**19.40**
Add for roll shade											
22" width	Inst	Ea	---	---	119.00	---	---	---	**137.00**	---	**114.00**
30" width	Inst	Ea	---	---	129.00	---	---	---	**148.00**	---	**123.00**
45-1/2" width	Inst	Ea	---	---	137.00	---	---	---	**157.00**	---	**131.00**
Add for storm panel with weatherstripping											
30" x 30"	Inst	Ea	---	---	59.50	---	---	---	**68.40**	---	**56.90**
45-1/2" x 45-1/2"	Inst	Ea	---	---	82.10	---	---	---	**94.40**	---	**78.60**
Add for motorization	Inst	Ea	CA	5.37	238.00	145.00	---	383.00	**494.00**	285.10	**360.00**
Fixed type											
30" x 72"	Inst	Ea	CA	4.30	417.00	116.00	---	533.00	**655.00**	416.60	**504.00**
45-1/2" x 45-1/2"	Inst	Ea	CA	4.30	357.00	116.00	---	473.00	**587.00**	366.60	**447.00**
45-1/2" x 30"	Inst	Ea	CA	4.30	315.00	116.00	---	431.00	**539.00**	331.60	**407.00**
30" x 45-1/2"	Inst	Ea	CA	4.30	315.00	116.00	---	431.00	**539.00**	331.60	**407.00**
30" x 30"	Inst	Ea	CA	3.45	256.00	93.00	---	349.00	**436.00**	268.80	**330.00**
30" x 22"	Inst	Ea	CA	3.45	226.00	93.00	---	319.00	**401.00**	243.80	**301.00**
22" x 30"	Inst	Ea	CA	3.45	226.00	93.00	---	319.00	**401.00**	243.80	**301.00**
22" x 45-1/2"	Inst	Ea	CA	3.45	286.00	93.00	---	379.00	**470.00**	293.80	**358.00**
Add for insulated laminated glass	Inst	%		---	30.0	---	---	---	**---**	---	**---**

Description	Oper	Unit	Crew Size	Man-Hours Per Unit	Avg Mat'l Unit Cost	Avg Labor Unit Cost	Avg Equip Unit Cost	Avg Total Unit Cost	Avg Price Incl O&P	Large Volume Avg Total Unit Cost	Large Volume Avg Price Incl O&P
Skywindows											
Labor costs are for installation of skywindows only.											
No carpentry or roofing work included.											
Ultraseal self-flashing units. Clear insulated glass, flexible flange, no mastic or step flashing required. Other glazings available.											
Skywindow E-Class											
22-1/2" x 22-1/2" fixed	Inst	Ea	CA	1.33	238.00	35.90	---	273.90	**328.00**	219.60	**260.00**
22-1/2" x 30-1/2" fixed	Inst	Ea	CA	1.33	274.00	35.90	---	309.90	**369.00**	249.60	**295.00**
22-1/2" x 46-1/2" fixed	Inst	Ea	CA	1.33	321.00	35.90	---	356.90	**424.00**	288.60	**340.00**
22-1/2" x 70-1/2" fixed	Inst	Ea	CA	1.33	440.00	35.90	---	475.90	**561.00**	387.60	**454.00**
30-1/2" x 30-1/2" fixed	Inst	Ea	CA	1.33	321.00	35.90	---	356.90	**424.00**	288.60	**340.00**
30-1/2" x 46-1/2" fixed	Inst	Ea	CA	1.33	393.00	35.90	---	428.90	**506.00**	348.60	**408.00**
46-1/2" x 46-1/2" fixed	Inst	Ea	CA	1.33	440.00	35.90	---	475.90	**561.00**	387.60	**454.00**
22-1/2" x 22-1/2" venting	Inst	Ea	CA	1.33	452.00	35.90	---	487.90	**575.00**	397.60	**465.00**
22-1/2" x 30-1/2" venting	Inst	Ea	CA	1.33	488.00	35.90	---	523.90	**616.00**	427.60	**500.00**
22-1/2" x 46-1/2" venting	Inst	Ea	CA	1.33	553.00	35.90	---	588.90	**691.00**	481.60	**562.00**
30-1/2" x 30-1/2" venting	Inst	Ea	CA	1.33	524.00	35.90	---	559.90	**657.00**	457.60	**534.00**
30-1/2" x 46-1/2" venting	Inst	Ea	CA	1.33	601.00	35.90	---	636.90	**746.00**	521.60	**608.00**
Step flash-pan flashed units. Insulated clear tempered glass. Deck mounted step flash unit. Other glazings available.											
Skywindow Excel-10											
22-1/2" x 22-1/2" fixed	Inst	Ea	CA	4.30	220.00	116.00	---	336.00	**429.00**	252.60	**316.00**
22-1/2" x 30-1/2" fixed	Inst	Ea	CA	4.30	232.00	116.00	---	348.00	**443.00**	262.60	**328.00**
22-1/2" x 46-1/2" fixed	Inst	Ea	CA	4.30	244.00	116.00	---	360.00	**457.00**	272.60	**339.00**
22-1/2" x 70-1/2" fixed	Inst	Ea	CA	4.30	381.00	116.00	---	497.00	**614.00**	386.60	**470.00**
30-1/2" x 30-1/2" fixed	Inst	Ea	CA	4.30	262.00	116.00	---	378.00	**477.00**	287.60	**356.00**
30-1/2" x 46-1/2" fixed	Inst	Ea	CA	4.30	315.00	116.00	---	431.00	**539.00**	331.60	**407.00**
22-1/2" x 22-1/2" vented	Inst	Ea	CA	4.30	369.00	116.00	---	485.00	**600.00**	376.60	**459.00**
22-1/2" x 30-1/2" vented	Inst	Ea	CA	4.30	399.00	116.00	---	515.00	**635.00**	401.60	**487.00**
22-1/2" x 46-1/2" vented	Inst	Ea	CA	4.30	476.00	116.00	---	592.00	**724.00**	465.60	**561.00**
30-1/2" x 30-1/2" vented	Inst	Ea	CA	4.30	458.00	116.00	---	574.00	**703.00**	450.60	**544.00**
30-1/2" x 46-1/2" vented	Inst	Ea	CA	4.30	512.00	116.00	---	628.00	**765.00**	495.60	**595.00**
Add for flashing kit	Inst	Ea	---	---	60.00	---	---	---	**60.00**	---	**60.00**
Low profile insulated glass skywindow. Deck mount, clear tempered glass, self-flashing models. Other glazings available.											
Skywindow Genra-1											
22-1/2" x 22-1/2" fixed	Inst	Ea	CA	4.30	196.00	116.00	---	312.00	**402.00**	232.60	**294.00**
22-1/2" x 30-1/2" fixed	Inst	Ea	CA	4.30	232.00	116.00	---	348.00	**443.00**	262.60	**328.00**
22-1/2" x 46-1/2" fixed	Inst	Ea	CA	4.30	286.00	116.00	---	402.00	**505.00**	307.60	**379.00**
22-1/2" x 70-1/2" fixed	Inst	Ea	CA	4.30	381.00	116.00	---	497.00	**614.00**	386.60	**470.00**
30-1/2" x 30-1/2" fixed	Inst	Ea	CA	4.30	268.00	116.00	---	384.00	**484.00**	292.60	**362.00**
30-1/2" x 46-1/2" fixed	Inst	Ea	CA	4.30	327.00	116.00	---	443.00	**553.00**	341.60	**419.00**
46-1/2" x 46-1/2" fixed	Inst	Ea	CA	4.30	417.00	116.00	---	533.00	**655.00**	416.60	**504.00**
22-1/2" x 22-1/2" vented	Inst	Ea	CA	4.30	405.00	116.00	---	521.00	**642.00**	406.60	**493.00**
22-1/2" x 30-1/2" vented	Inst	Ea	CA	4.30	434.00	116.00	---	550.00	**676.00**	430.60	**521.00**
22-1/2" x 46-1/2" vented	Inst	Ea	CA	4.30	512.00	116.00	---	628.00	**765.00**	495.60	**595.00**
30-1/2" x 30-1/2" vented	Inst	Ea	CA	4.30	476.00	116.00	---	592.00	**724.00**	465.60	**561.00**
30-1/2" x 46-1/2" vented	Inst	Ea	CA	4.30	559.00	116.00	---	675.00	**819.00**	534.60	**641.00**
46-1/2" x 46-1/2" vented	Inst	Ea	CA	4.30	690.00	116.00	---	806.00	**970.00**	643.60	**766.00**

Description	Oper	Unit	Crew Size	Man-Hours Per Unit	Avg Mat'l Unit Cost	Avg Labor Unit Cost	Avg Equip Unit Cost	Avg Total Unit Cost	Avg Price Incl O&P	Large Volume Avg Total Unit Cost	Large Volume Avg Price Incl O&P
Roof windows											
Includes prefabricated flashing, exterior awning, interior rollerblind and insect screen. Sash rotates 180 degrees. Labor costs are for installation on roofs are for installation on roofs with 10 to 85 degree slope and include installation of roof window, flashing and sun screen accessories. Labor costs do not include carpentry or roofing work other than curb. Add for interior trim. Listed by actual dimensions, top hung.											
Aluminum-clad wood frame, double-insulated tempered glass.											
22-1/2" x 46-1/2" fixed	Inst	Ea	CA	5.80	470.00	156.00	---	626.00	**778.00**	484.80	**592.00**
22-1/2" x 70-1/2" fixed	Inst	Ea	CA	5.93	506.00	160.00	---	666.00	**825.00**	517.00	**630.00**
30-1/2" x 30-1/2" fixed	Inst	Ea	CA	5.80	524.00	156.00	---	680.00	**840.00**	529.80	**644.00**
30-1/2" x 46-1/2" fixed	Inst	Ea	CA	5.93	595.00	160.00	---	755.00	**927.00**	591.00	**715.00**
46-1/2" x 46-1/2" fixed	Inst	Ea	CA	5.93	655.00	160.00	---	815.00	**996.00**	641.00	**772.00**
Spas											
Spas, with good quality fittings and faucets											
Whirlpool baths, with color matched jet trim & suction cover (Jacuzzi)											
Builders Group											
Nova model											
72" x 42" x 20"											
Color	Inst	Ea	SB	26.7	1770.00	700.00	---	2470.00	**3080.00**	2014.00	**2490.00**
White	Inst	Ea	SB	26.7	1630.00	700.00	---	2330.00	**2920.00**	1894.00	**2360.00**
Riva model											
62" x 43" x 18"											
Color	Inst	Ea	SB	26.7	1490.00	700.00	---	2190.00	**2750.00**	1774.00	**2220.00**
White	Inst	Ea	SB	26.7	1400.00	700.00	---	2100.00	**2650.00**	1694.00	**2130.00**
Tara model											
60" x 60" x 20"											
Color	Inst	Ea	SB	26.7	1990.00	700.00	---	2690.00	**3330.00**	2194.00	**2700.00**
White	Inst	Ea	SB	26.7	1900.00	700.00	---	2600.00	**3220.00**	2114.00	**2610.00**
Torino model											
66" x 42" x 20"											
Color	Inst	Ea	SB	26.7	1750.00	700.00	---	2450.00	**3060.00**	1994.00	**2480.00**
White	Inst	Ea	SB	26.7	1660.00	700.00	---	2360.00	**2950.00**	1924.00	**2390.00**
Designer Group											
Amea model											
70" x 36" x 20"											
Color	Inst	Ea	SB	26.7	3040.00	700.00	---	3740.00	**4540.00**	3074.00	**3720.00**
White	Inst	Ea	SB	26.7	2950.00	700.00	---	3650.00	**4430.00**	2994.00	**3630.00**
Aura model											
72" x 60" x 20"											
Color	Inst	Ea	SB	26.7	5040.00	700.00	---	5740.00	**6840.00**	4754.00	**5650.00**
White	Inst	Ea	SB	26.7	4950.00	700.00	---	5650.00	**6730.00**	4684.00	**5560.00**
Ciprea model											
72" x 48" x 20"											
Color	Inst	Ea	SB	26.7	4710.00	700.00	---	5410.00	**6450.00**	4474.00	**5330.00**
White	Inst	Ea	SB	26.7	4610.00	700.00	---	5310.00	**6350.00**	4404.00	**5240.00**

Description	Oper	Unit	Crew Size	Man-Hours Per Unit	Avg Mat'l Unit Cost	Avg Labor Unit Cost	Avg Equip Unit Cost	Avg Total Unit Cost	Avg Price Incl O&P	Large Volume Avg Total Unit Cost	Large Volume Avg Price Incl O&P
Fiore model											
66" x 66" x 21-3/4"											
Color	Inst	Ea	SB	26.7	6060.00	700.00	---	6760.00	**8020.00**	5614.00	**6640.00**
White	Inst	Ea	SB	26.7	5970.00	700.00	---	6670.00	**7910.00**	5544.00	**6550.00**
Fontana model											
72" x 54" x 28"											
Color	Inst	Ea	SB	26.7	5320.00	700.00	---	6020.00	**7160.00**	4994.00	**5920.00**
White	Inst	Ea	SB	26.7	5230.00	700.00	---	5930.00	**7060.00**	4914.00	**5830.00**
Options for whirlpool baths and spas											
4 hydrojet fiberglass/acrylic, ADD	Inst	Ea	---	---	950.00	---	---	950.00	**950.00**	798.00	**798.00**
4 mini jet fiberglass/acrylic, ADD	Inst	Ea	---	---	680.00	---	---	680.00	**680.00**	572.00	**572.00**
Jet covers, ADD	Inst	Ea	---	---	92.60	---	---	92.60	**92.60**	77.80	**77.80**
In-line heater, ADD	Inst	Ea	---	---	259.00	---	---	259.00	**259.00**	218.00	**218.00**
Lamp package, ADD	Inst	Ea	---	---	173.00	---	---	173.00	**173.00**	145.00	**145.00**
Two speed pump, ADD	Inst	Ea	---	---	346.00	---	---	346.00	**346.00**	290.00	**290.00**
One speed pump, ADD	Inst	Ea	---	---	216.00	---	---	216.00	**216.00**	181.00	**181.00**
Timer kit, ADD	Inst	Ea	---	---	56.90	---	---	56.90	**56.90**	47.80	**47.80**
Trim kit, ADD	Inst	Ea	---	---	115.00	---	---	115.00	**115.00**	96.40	**96.40**
Trip lever drain, ADD	Inst	Ea	---	---	107.00	---	---	107.00	**107.00**	90.20	**90.20**
Water rainbow spout kit, ADD	Inst	Ea	---	---	198.00	---	---	198.00	**198.00**	167.00	**167.00**
Adjustments/related operations											
To remove and reset											
Whirlpool bath	Reset	Ea	SB	9.41	---	247.00	---	247.00	**367.00**	175.00	**260.00**
Whirlpool spa	Reset	Ea	SB	11.4	---	299.00	---	299.00	**445.00**	210.00	**312.00**
To rough-in											
Whirlpool bath	Inst	Ea	SB	14.5	189.00	380.00	---	569.00	**784.00**	421.00	**573.00**
Whirlpool spa	Inst	Ea	SB	14.5	189.00	380.00	---	569.00	**784.00**	421.00	**573.00**

Stairs

Stair parts

Description	Oper	Unit	Crew Size	Man-Hours Per Unit	Avg Mat'l Unit Cost	Avg Labor Unit Cost	Avg Equip Unit Cost	Avg Total Unit Cost	Avg Price Incl O&P	Large Volume Avg Total Unit Cost	Large Volume Avg Price Incl O&P
Balusters, stock pine											
1 -1/16" x 1-1/16"	Inst	LF	CA	.067	2.50	1.81	---	4.31	**5.62**	3.29	**4.18**
1-5/8" x 1-5/8"	Inst	LF	CA	.083	2.31	2.24	---	4.55	**6.06**	3.39	**4.40**
Balusters, turned											
30" high											
Pine	Inst	Ea	CA	.556	9.40	15.00	---	24.40	**33.60**	17.27	**23.20**
Birch	Inst	Ea	CA	.556	11.00	15.00	---	26.00	**35.40**	18.67	**24.80**
42" high											
Pine	Inst	Ea	CA	.667	11.00	18.00	---	29.00	**40.00**	20.49	**27.50**
Birch	Inst	Ea	CA	.667	14.20	18.00	---	32.20	**43.60**	23.30	**30.80**
Newels, 3-1/4" wide											
Starting	Inst	Ea	CA	2.22	152.00	59.90	---	211.90	**266.00**	169.90	**209.00**
Landing	Inst	Ea	CA	3.33	216.00	89.80	---	305.80	**385.00**	243.90	**301.00**
Railings, built-up											
Oak	Inst	LF	CA	.267	15.20	7.20	---	22.40	**28.50**	17.71	**22.00**

Description	Oper	Unit	Crew Size	Man-Hours Per Unit	Avg Mat'l Unit Cost	Avg Labor Unit Cost	Avg Equip Unit Cost	Avg Total Unit Cost	Avg Price Incl O&P	Large Volume Avg Total Unit Cost	Large Volume Avg Price Incl O&P
			Costs Based On Small Volume							Large Volume	
Railings, subrail											
Oak	Inst	LF	CA	.133	20.30	3.59	---	23.89	**28.80**	20.06	**23.90**
Risers, 3/4" x 7-1/2" high											
Beech	Inst	LF	CA	.222	6.60	5.99	---	12.59	**16.70**	9.41	**12.10**
Fir	Inst	LF	CA	.222	1.91	5.99	---	7.90	**11.30**	5.27	**7.38**
Oak	Inst	LF	CA	.222	6.03	5.99	---	12.02	**16.00**	8.91	**11.60**
Pine	Inst	LF	CA	.222	1.91	5.99	---	7.90	**11.30**	5.27	**7.38**
Skirt board, pine											
1" x 10"	Inst	LF	CA	.267	2.10	7.20	---	9.30	**13.40**	6.16	**8.68**
1" x 12"	Inst	LF	CA	.296	2.54	7.98	---	10.52	**15.10**	7.04	**9.87**
Treads, oak											
1-16" x 9-1/2" wide											
3' long	Inst	Ea	CA	.833	27.90	22.50	---	50.40	**66.30**	38.10	**48.80**
4' long	Inst	Ea	CA	.889	36.80	24.00	---	60.80	**78.80**	46.90	**59.20**
1-1/16" x 11-1/2" wide											
3' long	Inst	Ea	CA	.833	29.90	22.50	---	52.40	**68.50**	39.80	**50.80**
6' long	Inst	Ea	CA	1.11	71.10	29.90	---	101.00	**127.00**	80.70	**99.50**
For beech treads, ADD	Inst	%		---	40.0	---	---	---	**---**	---	**---**
For mitered return nosings, ADD	Inst	LF	CA	.222	10.70	5.99	---	16.69	**21.40**	13.00	**16.30**
Stairs, shop fabricated, per flight											
Basement stairs, soft wood, open risers											
3' wide, 8' high	Inst	Flt	2C	6.67	476.00	180.00	---	656.00	**821.00**	528.00	**647.00**
Box stairs, no handrails, 3' wide											
Oak treads											
2' high	Inst	Flt	2C	5.33	189.00	144.00	---	333.00	**436.00**	253.30	**323.00**
4' high	Inst	Flt	2C	6.67	406.00	180.00	---	586.00	**741.00**	466.00	**576.00**
6' high	Inst	Flt	2C	7.62	660.00	205.00	---	865.00	**1070.00**	705.00	**857.00**
8' high	Inst	Flt	2C	8.89	826.00	240.00	---	1066.00	**1310.00**	872.00	**1060.00**
Pine treads for carpet											
2' high	Inst	Flt	2C	5.33	150.00	144.00	---	294.00	**391.00**	218.30	**283.00**
4' high	Inst	Flt	2C	6.67	255.00	180.00	---	435.00	**567.00**	333.00	**423.00**
6' high	Inst	Flt	2C	7.62	387.00	205.00	---	592.00	**758.00**	465.00	**580.00**
8' high	Inst	Flt	2C	8.89	483.00	240.00	---	723.00	**919.00**	570.00	**708.00**
Stair rail with balusters, 5 risers	Inst	Ea	2C	1.78	222.00	48.00	---	270.00	**329.00**	224.90	**269.00**
For 4' wide stairs, ADD	Inst	%	2C	---	25.0	---	5.0	---	**---**	---	**---**
Open stairs, prefinished, metal stringers, 3'-6" wide treads, no railings											
3' high	Inst	Flt	2C	5.33	552.00	144.00	---	696.00	**854.00**	573.30	**691.00**
4' high	Inst	Flt	2C	6.67	692.00	180.00	---	872.00	**1070.00**	718.00	**866.00**
6' high	Inst	Flt	2C	7.62	1210.00	205.00	---	1415.00	**1700.00**	1183.00	**1410.00**
8' high	Inst	Flt	2C	8.89	1940.00	240.00	---	2180.00	**2590.00**	1854.00	**2180.00**

Description	Oper	Unit	Crew Size	Man-Hours Per Unit	Avg Mat'l Unit Cost	Avg Labor Unit Cost	Avg Equip Unit Cost	Avg Total Unit Cost	Avg Price Incl O&P	Large Volume Avg Total Unit Cost	Large Volume Avg Price Incl O&P
Adjustments											
For 3 piece wood railings and balusters, ADD											
3' high	Inst	Ea	2C	1.78	189.00	48.00	---	237.00	**291.00**	195.90	**236.00**
4' high	Inst	Ea	2C	1.90	243.00	51.20	---	294.20	**357.00**	244.70	**293.00**
6' high	Inst	Ea	2C	2.05	373.00	55.30	---	428.30	**513.00**	362.20	**429.00**
8' high	Inst	Ea	2C	2.22	464.00	59.90	---	523.90	**624.00**	444.90	**525.00**
For 3'-6" x 3'-6" platform, ADD	Inst	Ea	2C	6.67	216.00	180.00	---	396.00	**522.00**	298.00	**383.00**
Curved stairways, oak, unfinished, curved balustrade											
Open one side											
9' high	Inst	Flt	2C	38.1	8100.00	1030.00	---	9130.00	**10900**	7757.00	**9150.00**
10' high	Inst	Flt	2C	38.1	9140.00	1030.00	---	10170	**12100**	8677.00	**10200**
Open both sides											
9' high	Inst	Flt	2C	53.3	12700	1440.00	---	14140	**16800**	12063	**14200**
10' high	Inst	Flt	2C	53.3	13700	1440.00	---	15140	**18000**	12963	**15200**

Steel, reinforcing. See Concrete, page 60

Stucco. See Plaster, page 198

Studs. See Framing, page 119

Subflooring. See Framing, page 126

Suspended Ceilings

Suspended ceilings consist of a grid of small metal hangers which are hung from the ceiling framing with wire or strap, and drop-in panels sized to fit the grid system. The main advantage to this type of ceiling finish is that it can be adjusted to any height desired. It can cover a lot of flaws and obstructions (uneven plaster, pipes, wiring, and ductwork). With a high ceiling, the suspended ceiling reduces the sound transfer from the floor above and increases the ceiling's insulating value. One great advantage of suspended ceilings is that the area directly above the tiles is readily accessible for repair and maintenance work; the tiles merely have to be lifted out from the grid. This system also eliminates the need for any other ceiling finish.

1. **Installation Procedure**
   a. Decide on ceiling height. It must clear the lowest obstacles but remain above windows and doors. Snap a chalk line around walls at the new height. Check it with a carpenter's level.

   b. Install wall molding/angle at the marked line; this supports the tiles around the room's perimeter. Check with a level. Use nails (6d) or screws to fasten molding to studs (16" or 24" oc); on masonry, use concrete nails 24" oc. The molding should be flush along the chalk line with the bottom leg of the L-angle facing into the room.

   c. Fit wall molding at inside corners by straight cutting (90 degrees) and overlapping; for outside corners, miter cut corner (45 degrees) and butt together. Cut the molding to required lengths with tin snips or a hacksaw.

   d. Mark center line of room above the wall molding. Decide where the first main runner should be to get the desired border width. Position main runners by running taut string lines across room. Repeat for the first row of cross tees closest to one wall. Then attach hanger wires with screw eyes to the joists at 4' intervals.

   e. Trim main runners at wall to line up the slot with cross tee string. Rest trimmed end on wall molding; insert wire through holes; recheck with level; twist wire several times.

   f. Insert cross tee and tabs into main runner's slots; push down to lock. Check with level. Repeat.

   g. Drop 2' x 4' panels into place.

2. **Estimating Materials**
   a. Measure the ceiling and plot it on graph paper. Mark the direction of ceiling joists. On the ceiling itself, snap chalk lines along the joist lines.

   b. Draw ceiling layout for grid system onto graph paper. Plan the ceiling layout, figuring full panels across the main ceiling and evenly trimmed partial panels at the edges. To calculate the width of border panels in each direction, determine the width of the gap left after full panels are placed all across the dimension; then divide by 2.

   c. For purposes of ordering material, determine the room size in even number of feet. If the room length or width is not divisible by 2 feet, increase the dimension to the next larger unit divisible by 2. For example, a room that measured 11'-6" x 14'-4" would be considered a 12' x 16' room. This allows for waste and/or breakage. In this example, you would order 192 SF of material.

   d. Main runners and tees must run perpendicular to ceiling joists. For a 2' x 2' grid, cross and main tees are 2' apart. For a 2' x 4' grid, main tees are 4' apart and 4' cross tees connect the main tees; then 2' cross tees can be used to connect the 4' cross tees. The long panels of the grid are set parallel to the ceiling joists, so the T-shaped main runner is attached perpendicular to joists at 4' oc.

   e. Wall molding/angle is manufactured in 10' lengths. Main runners and tees are manufactured in 12' lengths. Cross tees are either 2' long or 4' long. Hanger wire is 12 gauge, and attached to joists with either screw eyes or hook-and-nail. Drop-in panels are either 8 SF (2' x 4') or 4 SF (2' x 2').

   f. To find the number of drop-in panels required, divide the nominal room size in square feet (e.g., 192 SF in above example) by square feet of panel to be used.

   g. The quantity of 12 gauge wire depends on the "drop" distance of the ceiling. Figure one suspension wire and screw eye for each 4' of main runner; if any main run is longer than 12' then splice plates are needed, with a wire and screw eye on each side of the splice. The length of each wire should be at least 2" longer than the "drop" distance, to allow for a twist after passing through runner or tee.

Description	Oper	Unit	Crew Size	Man-Hours Per Unit	Avg Mat'l Unit Cost	Avg Labor Unit Cost	Avg Equip Unit Cost	Avg Total Unit Cost	Avg Price Incl O&P	Large Volume Avg Total Unit Cost	Large Volume Avg Price Incl O&P
Suspended ceiling systems											
Suspended ceiling panels and grid system											
	Demo	SF	LB	.013	---	.29	---	.29	**.43**	.20	**.30**
Complete 2' x 4' slide lock grid system (baked enamel)											
Labor includes installing grid and laying in panels											
Luminous panels											
Prismatic panels											
Acrylic	Inst	SF	CA	.030	1.86	.81	---	2.67	**3.37**	2.24	**2.78**
Polystyrene	Inst	SF	CA	.030	1.14	.81	---	1.95	**2.54**	1.60	**2.05**
Ribbed panels											
Acrylic	Inst	SF	CA	.030	1.98	.81	---	2.79	**3.51**	2.35	**2.91**
Polystyrene	Inst	SF	CA	.030	1.26	.81	---	2.07	**2.68**	1.70	**2.16**
Fiberglass panels											
1/2" T											
Plain white	Inst	SF	CA	.028	1.02	.75	---	1.77	**2.32**	1.46	**1.88**
Sculptured white	Inst	SF	CA	.028	1.14	.75	---	1.89	**2.46**	1.57	**2.00**
5/8" T											
Fissured	Inst	SF	CA	.028	1.20	.75	---	1.95	**2.53**	1.62	**2.06**
Fissured, fire rated	Inst	SF	CA	.028	1.32	.75	---	2.07	**2.67**	1.73	**2.19**
Mineral fiber panels											
1/2" T	Inst	SF	CA	.028	.84	.75	---	1.59	**2.11**	1.30	**1.69**
9/16" T	Inst	SF	CA	.028	1.36	.75	---	2.11	**2.71**	1.76	**2.22**
For wood grained grid components											
ADD	Inst	%		---	7.0	---	---	---	**---**	---	**---**
Panels only (labor to lay panels in grid system)											
Luminous panels											
Prismatic panels											
Acrylic	Inst	SF	CA	.011	1.55	.30	---	1.85	**2.23**	1.61	**1.93**
Polystyrene	Inst	SF	CA	.011	.83	.30	---	1.13	**1.41**	.97	**1.19**
Ribbed panels											
Acrylic	Inst	SF	CA	.011	1.67	.30	---	1.97	**2.37**	1.72	**2.05**
Polystyrene	Inst	SF	CA	.011	.95	.30	---	1.25	**1.54**	1.08	**1.32**
Fiberglass panels											
1/2" T											
Plain white	Inst	SF	CA	.009	.71	.24	---	.95	**1.19**	.83	**1.02**
Sculptured white	Inst	SF	CA	.009	.83	.24	---	1.07	**1.32**	.94	**1.15**
5/8" T											
Fissured	Inst	SF	CA	.009	.89	.24	---	1.13	**1.39**	.99	**1.21**
Fissured, fire rated	Inst	SF	CA	.009	1.01	.24	---	1.25	**1.53**	1.10	**1.33**
Mineral fiber panels											
5/8" T	Inst	SF	CA	.009	.54	.24	---	.78	**.99**	.67	**.84**
3/4" T	Inst	SF	CA	.009	1.05	.24	---	1.29	**1.58**	1.13	**1.37**

Description	Oper	Unit	Crew Size	Man-Hours Per Unit	Avg Mat'l Unit Cost	Avg Labor Unit Cost	Avg Equip Unit Cost	Avg Total Unit Cost	Avg Price Incl O&P	Avg Total Unit Cost	Avg Price Incl O&P
			Costs Based On Small Volume							Large Volume	
Grid system components											
Main runner, 12' L pieces											
Enamel	Inst	Ea	---	---	5.30	---	---	5.30	**5.30**	4.76	**4.76**
Wood grain	Inst	Ea	---	---	6.07	---	---	6.07	**6.07**	5.46	**5.46**
Cross tees											
2' L pieces											
Enamel	Inst	Ea	---	---	.95	---	---	.95	**.95**	.86	**.86**
Wood grain	Inst	Ea	---	---	1.07	---	---	1.07	**1.07**	.96	**.96**
4' L pieces											
Enamel	Inst	Ea	---	---	1.79	---	---	1.79	**1.79**	1.61	**1.61**
Wood grain	Inst	Ea	---	---	2.02	---	---	2.02	**2.02**	1.82	**1.82**
Wall molding, 10' L pieces											
Enamel	Inst	Ea	---	---	3.39	---	---	3.39	**3.39**	3.05	**3.05**
Wood grain	Inst	Ea	---	---	4.46	---	---	4.46	**4.46**	4.01	**4.01**
Main runner hold-down clips											
1/2" T panels (1000/carton)	Inst	Ctn	---	---	9.52	---	---	9.52	**9.52**	8.56	**8.56**
5/8" T panels (1000/carton)	Inst	Ctn	---	---	13.10	---	---	13.10	**13.10**	11.80	**11.80**

Telephone prewiring. See Electrical, page 95

Television antenna. See Electrical, page 95

Thermostat. See Electrical, page 95

Tile, ceiling. See Acoustical treatment, page 22

Tile, ceramic. See Ceramic tile, page 52

Tile, quarry. See Masonry, page 168

Tile, vinyl or asphalt. See Resilient flooring, page 201

Toilets, bidets

Toilets (water closets), vitreous china, white or color, 1.6 gpf, includes seat, shut-off valve, connectors, flanges, water supply valve (American Standard)

Description	Oper	Unit	Crew Size	Man-Hours Per Unit	Avg Mat'l Unit Cost	Avg Labor Unit Cost	Avg Equip Unit Cost	Avg Total Unit Cost	Avg Price Incl O&P	Avg Total Unit Cost	Avg Price Incl O&P
One piece, floor mounted											
Ellise, elongated model											
Color	Inst	Ea	SA	6.15	999.00	187.00	---	1186.00	**1430.00**	1017.00	**1220.00**
Premium color	Inst	Ea	SA	6.15	1120.00	187.00	---	1307.00	**1560.00**	1120.00	**1330.00**
White	Inst	Ea	SA	6.15	799.00	187.00	---	986.00	**1200.00**	841.00	**1010.00**
Fontaine, pressure assisted											
Elongated w/ seat											
Color	Inst	Ea	SA	6.15	895.00	187.00	---	1082.00	**1310.00**	925.00	**1110.00**
Premium color	Inst	Ea	SA	6.15	1000.00	187.00	---	1187.00	**1430.00**	1021.00	**1220.00**
White	Inst	Ea	SA	6.15	720.00	187.00	---	907.00	**1110.00**	771.00	**933.00**

Description	Oper	Unit	Crew Size	Man-Hours Per Unit	Avg Mat'l Unit Cost	Avg Labor Unit Cost	Avg Equip Unit Cost	Avg Total Unit Cost	Avg Price Incl O&P	Large Volume Avg Total Unit Cost	Large Volume Avg Price Incl O&P
Hamilton, space saving model											
Elongated w/ seat											
Color	Inst	Ea	SA	6.15	375.00	187.00	---	562.00	**710.00**	466.00	**582.00**
Premium color	Inst	Ea	SA	6.15	425.00	187.00	---	612.00	**767.00**	510.00	**633.00**
White	Inst	Ea	SA	6.15	291.00	187.00	---	478.00	**614.00**	392.00	**497.00**
High Hamilton EL, space saving model											
Elongated w/ seat											
Color	Inst	Ea	SA	6.15	512.00	187.00	---	699.00	**868.00**	587.00	**721.00**
Premium color	Inst	Ea	SA	6.15	570.00	187.00	---	757.00	**935.00**	638.00	**781.00**
White	Inst	Ea	SA	6.15	408.00	187.00	---	595.00	**748.00**	495.00	**616.00**
Heritage, pressure assisted model											
Color	Inst	Ea	SA	6.15	882.00	187.00	---	1069.00	**1290.00**	914.00	**1100.00**
Premium color	Inst	Ea	SA	6.15	991.00	187.00	---	1178.00	**1420.00**	1010.00	**1210.00**
White	Inst	Ea	SA	6.15	708.00	187.00	---	895.00	**1090.00**	760.00	**920.00**
Lexington model											
Elongated seat											
Color	Inst	Ea	SA	6.15	645.00	187.00	---	832.00	**1020.00**	705.00	**857.00**
Premium color	Inst	Ea	SA	6.15	724.00	187.00	---	911.00	**1110.00**	774.00	**937.00**
White	Inst	Ea	SA	6.15	516.00	187.00	---	703.00	**873.00**	591.00	**726.00**
Two piece, close coupled, floor mounted, water saver models, seat not included											
Antiquity											
Elongated											
White	Inst	Ea	SA	6.15	279.00	187.00	---	466.00	**600.00**	381.00	**485.00**
Round											
White	Inst	Ea	SA	6.15	237.00	187.00	---	424.00	**552.00**	344.00	**442.00**
Cadet, pressure assisted											
Round											
Color	Inst	Ea	SA	6.15	375.00	187.00	---	562.00	**710.00**	466.00	**582.00**
Premium color	Inst	Ea	SA	6.15	420.00	187.00	---	607.00	**762.00**	505.00	**628.00**
White	Inst	Ea	SA	6.15	312.00	187.00	---	499.00	**638.00**	411.00	**519.00**
Elongated											
Color	Inst	Ea	SA	6.15	415.00	187.00	---	602.00	**756.00**	501.00	**622.00**
Premium color	Inst	Ea	SA	6.15	465.00	187.00	---	652.00	**813.00**	545.00	**673.00**
White	Inst	Ea	SA	6.15	345.00	187.00	---	532.00	**676.00**	440.00	**552.00**
Cadet elongated, 17" rim height											
Color	Inst	Ea	SA	6.15	509.00	187.00	---	696.00	**865.00**	585.00	**719.00**
Premium color	Inst	Ea	SA	6.15	570.00	187.00	---	757.00	**935.00**	638.00	**781.00**
White	Inst	Ea	SA	6.15	425.00	187.00	---	612.00	**767.00**	510.00	**633.00**
Heritage											
Elongated seat											
Color	Inst	Ea	SA	6.15	683.00	187.00	---	870.00	**1060.00**	738.00	**895.00**
Premium color	Inst	Ea	SA	6.15	624.00	187.00	---	811.00	**997.00**	686.00	**835.00**
White	Inst	Ea	SA	6.15	874.00	187.00	---	1061.00	**1280.00**	907.00	**1090.00**
Renaissance											
Round											
Color	Inst	Ea	SA	6.15	162.00	187.00	---	349.00	**466.00**	278.00	**366.00**
Premium color	Inst	Ea	SA	6.15	183.00	187.00	---	370.00	**490.00**	297.00	**387.00**
White	Inst	Ea	SA	6.15	135.00	187.00	---	322.00	**434.00**	254.00	**338.00**

Description	Oper	Unit	Crew Size	Man-Hours Per Unit	Avg Mat'l Unit Cost	Avg Labor Unit Cost	Avg Equip Unit Cost	Avg Total Unit Cost	Avg Price Incl O&P	Avg Total Unit Cost	Avg Price Incl O&P
			Costs Based On Small Volume							Large Volume	
Elongated											
Color	Inst	Ea	SA	6.15	201.00	187.00	---	388.00	**511.00**	313.00	**406.00**
Premium color	Inst	Ea	SA	6.15	225.00	187.00	---	412.00	**538.00**	333.00	**430.00**
White	Inst	Ea	SA	6.15	167.00	187.00	---	354.00	**472.00**	283.00	**371.00**
Venice II											
Round											
Color	Inst	Ea	SA	6.15	138.00	187.00	---	325.00	**438.00**	257.00	**342.00**
White	Inst	Ea	SA	6.15	115.00	187.00	---	302.00	**411.00**	236.00	**318.00**
Bidets, deck mounted											
Ellisse model											
Color	Inst	Ea	SA	5.33	483.00	162.00	---	645.00	**797.00**	561.00	**692.00**
White	Inst	Ea	SA	5.33	345.00	162.00	---	507.00	**639.00**	440.00	**552.00**
Heritage model											
Color	Inst	Ea	SA	5.33	641.00	162.00	---	803.00	**979.00**	701.00	**852.00**
White	Inst	Ea	SA	5.33	458.00	162.00	---	620.00	**768.00**	539.00	**666.00**
Lexington model											
Color	Inst	Ea	SA	5.33	437.00	162.00	---	599.00	**745.00**	521.00	**645.00**
White	Inst	Ea	SA	5.33	312.00	162.00	---	474.00	**601.00**	411.00	**519.00**
Roma model											
Color	Inst	Ea	SA	5.33	616.00	162.00	---	778.00	**950.00**	679.00	**827.00**
White	Inst	Ea	SA	5.33	437.00	162.00	---	599.00	**745.00**	521.00	**645.00**
Urinals, wall mounted											
Innsbrook model											
w/ integral proximity flush											
Color	Inst	Ea	SA	5.33	980.00	162.00	---	1142.00	**1370.00**	976.00	**1160.00**
White	Inst	Ea	SA	5.33	725.00	162.00	---	887.00	**1080.00**	751.00	**901.00**
w/ push button integral flush											
Color	Inst	Ea	SA	5.33	957.00	162.00	---	1119.00	**1340.00**	956.00	**1140.00**
White	Inst	Ea	SA	5.33	709.00	162.00	---	871.00	**1060.00**	737.00	**885.00**
Maybrook model											
Color	Inst	Ea	SA	5.33	209.00	162.00	---	371.00	**482.00**	295.00	**377.00**
White	Inst	Ea	SA	5.33	155.00	162.00	---	317.00	**420.00**	248.00	**322.00**
Washbrook model											
Color	Inst	Ea	SA	5.33	473.00	162.00	---	635.00	**786.00**	528.00	**645.00**
White	Inst	Ea	SA	5.33	350.00	162.00	---	512.00	**644.00**	420.00	**520.00**
Adjustments											
To remove and reset toilet or bidet	Reset	Ea	SA	2.86	---	87.10	---	87.10	**130.00**	60.90	**90.80**
To install rough-in											
Toilet	Inst	Ea	SA	13.3	167.00	405.00	---	572.00	**795.00**	424.00	**582.00**
Bidet	Inst	Ea	SA	11.4	139.00	347.00	---	486.00	**677.00**	367.00	**504.00**

Description	Oper	Unit	Crew Size	Man-Hours Per Unit	Avg Mat'l Unit Cost	Avg Labor Unit Cost	Avg Equip Unit Cost	Avg Total Unit Cost	Avg Price Incl O&P	Large Volume Avg Total Unit Cost	Large Volume Avg Price Incl O&P
Trash compactors											
Includes wiring, connection, and installation in a pre-cutout area											
White	Inst	Ea	EA	2.67	741.00	77.00	---	818.00	**966.00**	716.70	**843.00**
Colors	Inst	Ea	EA	2.67	836.00	77.00	---	913.00	**1080.00**	800.70	**940.00**
To remove and reset	Reset	Ea	EA	.964	---	27.80	---	27.80	**41.10**	21.00	**31.00**
Material adjustments											
Stainless steel trim kit	R&R	LS	---	---	28.10	---	---	28.10	**28.10**	25.00	**25.00**
Stainless steel panel and trim kit	R&R	LS	---	---	69.10	---	---	69.10	**69.10**	61.40	**61.40**
"Black Glas" acrylic front panels and trim kit	R&R	LS	---	---	69.10	---	---	69.10	**69.10**	61.40	**61.40**
Hardwood top	R&R	LS	---	---	54.00	---	---	54.00	**54.00**	48.00	**48.00**

Trim. See Molding, page 169

Trusses. See Framing, page 126

Vanity units. See Cabinetry, page 46

Ventilation

Flue piping or vent chimney

Prefabricated metal, UL listed

Description	Oper	Unit	Crew Size	Man-Hours Per Unit	Avg Mat'l Unit Cost	Avg Labor Unit Cost	Avg Equip Unit Cost	Avg Total Unit Cost	Avg Price Incl O&P	Large Volume Avg Total Unit Cost	Large Volume Avg Price Incl O&P
Gas, double wall, galv. steel											
3" diameter	Inst	VLF	UB	.327	3.85	9.73	---	13.58	**19.00**	10.17	**14.10**
4" diameter	Inst	VLF	UB	.348	4.72	10.40	---	15.12	**21.00**	11.30	**15.50**
5" diameter	Inst	VLF	UB	.372	5.60	11.10	---	16.70	**23.10**	12.54	**17.10**
6" diameter	Inst	VLF	UB	.390	6.50	11.60	---	18.10	**24.90**	13.85	**18.80**
7" diameter	Inst	VLF	UB	.421	9.58	12.50	---	22.08	**29.80**	17.13	**22.80**
8" diameter	Inst	VLF	UB	.457	10.70	13.60	---	24.30	**32.70**	18.78	**24.90**
10" diameter	Inst	VLF	UB	.500	22.50	14.90	---	37.40	**48.20**	30.00	**38.00**
12" diameter	Inst	VLF	UB	.552	30.20	16.40	---	46.60	**59.40**	37.50	**47.20**
16" diameter	Inst	VLF	UB	.593	68.40	17.70	---	86.10	**105.00**	71.90	**87.10**
20" diameter	Inst	VLF	UB	.667	105.00	19.90	---	124.90	**150.00**	105.10	**126.00**
24" diameter	Inst	VLF	UB	.762	163.00	22.70	---	185.70	**221.00**	157.90	**187.00**
Vent damper, bi-metal, 6" flue											
Gas, auto., electric	Inst	Ea	UB	4.00	148.00	119.00	---	267.00	**349.00**	208.50	**267.00**
Oil, auto., electric	Inst	Ea	UB	4.00	178.00	119.00	---	297.00	**383.00**	233.50	**297.00**
All fuel, double wall, stainless steel											
6" diameter	Inst	VLF	UB	.381	31.90	11.30	---	43.20	**53.70**	35.65	**43.80**
7" diameter	Inst	VLF	UB	.410	41.00	12.20	---	53.20	**65.50**	44.11	**53.80**
8" diameter	Inst	VLF	UB	.444	47.90	13.20	---	61.10	**74.90**	50.77	**61.60**
10" diameter	Inst	VLF	UB	.471	69.50	14.00	---	83.50	**101.00**	70.31	**84.30**
12" diameter	Inst	VLF	UB	.516	93.50	15.40	---	108.90	**131.00**	92.00	**110.00**
14" diameter	Inst	VLF	UB	.571	123.00	17.00	---	140.00	**167.00**	118.90	**141.00**

Description	Oper	Unit	Crew Size	Man-Hours Per Unit	Avg Mat'l Unit Cost	Avg Labor Unit Cost	Avg Equip Unit Cost	Avg Total Unit Cost	Avg Price Incl O&P	Large Volume Avg Total Unit Cost	Large Volume Avg Price Incl O&P
				Costs Based On Small Volume						Large Volume	
All fuel, double wall, stainless steel fittings											
Roof support											
6" diameter	Inst	Ea	UB	.762	81.50	22.70	---	104.20	**128.00**	86.70	**105.00**
7" diameter	Inst	Ea	UB	.800	91.80	23.80	---	115.60	**141.00**	96.70	**117.00**
8" diameter	Inst	Ea	UB	.889	99.80	26.50	---	126.30	**154.00**	104.90	**127.00**
10" diameter	Inst	Ea	UB	.941	131.00	28.00	---	159.00	**193.00**	133.90	**161.00**
12" diameter	Inst	Ea	UB	1.07	158.00	31.80	---	189.80	**230.00**	159.60	**191.00**
14" diameter	Inst	Ea	UB	1.14	201.00	33.90	---	234.90	**282.00**	197.80	**236.00**
Elbow 15 degree											
6" diameter	Inst	Ea	UB	.762	71.30	22.70	---	94.00	**116.00**	77.80	**95.00**
7" diameter	Inst	Ea	UB	.800	79.80	23.80	---	103.60	**127.00**	86.30	**105.00**
8" diameter	Inst	Ea	UB	.889	91.20	26.50	---	117.70	**145.00**	97.50	**119.00**
10" diameter	Inst	Ea	UB	.941	119.00	28.00	---	147.00	**178.00**	122.90	**148.00**
12" diameter	Inst	Ea	UB	1.07	145.00	31.80	---	176.80	**214.00**	147.60	**177.00**
14" diameter	Inst	Ea	UB	1.14	173.00	33.90	---	206.90	**250.00**	173.80	**209.00**
Insulated tee with insulated tee cap											
6" diameter	Inst	Ea	UB	.762	135.00	22.70	---	157.70	**189.00**	132.90	**158.00**
7" diameter	Inst	Ea	UB	.800	176.00	23.80	---	199.80	**238.00**	169.00	**201.00**
8" diameter	Inst	Ea	UB	.889	198.00	26.50	---	224.50	**268.00**	190.30	**226.00**
10" diameter	Inst	Ea	UB	.941	279.00	28.00	---	307.00	**363.00**	262.90	**309.00**
12" diameter	Inst	Ea	UB	1.07	393.00	31.80	---	424.80	**500.00**	363.60	**425.00**
14" diameter	Inst	Ea	UB	1.14	513.00	33.90	---	546.90	**641.00**	469.80	**548.00**
Joist shield											
6" diameter	Inst	Ea	UB	.762	41.00	22.70	---	63.70	**81.20**	51.50	**64.80**
7" diameter	Inst	Ea	UB	.800	44.50	23.80	---	68.30	**86.80**	55.60	**69.90**
8" diameter	Inst	Ea	UB	.889	54.70	26.50	---	81.20	**103.00**	65.80	**82.10**
10" diameter	Inst	Ea	UB	.941	62.10	28.00	---	90.10	**113.00**	73.90	**91.80**
12" diameter	Inst	Ea	UB	1.07	91.80	31.80	---	123.60	**153.00**	101.30	**124.00**
14" diameter	Inst	Ea	UB	1.14	114.00	33.90	---	147.90	**182.00**	122.80	**150.00**
Round top											
6" diameter	Inst	Ea	UB	.762	45.60	22.70	---	68.30	**86.50**	55.50	**69.30**
7" diameter	Inst	Ea	UB	.800	62.10	23.80	---	85.90	**107.00**	71.00	**87.50**
8" diameter	Inst	Ea	UB	.889	83.20	26.50	---	109.70	**135.00**	90.60	**111.00**
10" diameter	Inst	Ea	UB	.941	149.00	28.00	---	177.00	**214.00**	149.90	**179.00**
12" diameter	Inst	Ea	UB	1.07	213.00	31.80	---	244.80	**293.00**	206.60	**245.00**
14" diameter	Inst	Ea	UB	1.14	280.00	33.90	---	313.90	**373.00**	267.80	**316.00**
Adjustable roof flashing											
6" diameter	Inst	Ea	UB	.762	54.20	22.70	---	76.90	**96.30**	62.90	**77.90**
7" diameter	Inst	Ea	UB	.800	61.60	23.80	---	85.40	**107.00**	70.50	**87.00**
8" diameter	Inst	Ea	UB	.889	67.30	26.50	---	93.80	**117.00**	76.70	**94.60**
10" diameter	Inst	Ea	UB	.941	86.10	28.00	---	114.10	**141.00**	94.70	**116.00**
12" diameter	Inst	Ea	UB	1.07	111.00	31.80	---	142.80	**176.00**	118.10	**143.00**
14" diameter	Inst	Ea	UB	1.14	139.00	33.90	---	172.90	**211.00**	144.80	**175.00**
Minimum Job Charge	Inst	Job	UB	6.67	---	199.00	---	199.00	**298.00**	159.00	**238.00**

Wallpaper

Most wall coverings today are really not wallpapers in the technical sense. Wallpaper is a paper material (which may or not be coated with washable plastic). Most of today's products, though still called wallpaper, are actually composed of a vinyl on a fabric, not a paper, backing. Some vinyl/fabric coverings come prepasted.

Vinyl coverings with non-woven fabric, woven fabric or synthetic fiber backing are fire, mildew and fade resistant, and they can be stripped from plaster walls. Coverings with paper backing must be steamed or scraped from the walls. Vinyl coverings may also be stripped from gypsum wallboard, but unless the vinyl covering has a synthetic fiber backing, it's likely to damage the wallboard paper surface.

One gallon of paste (approximately two-thirds pound of dry paste and water) should adequately cover 12 rolls with paper backing and 6 rolls with fabric backing. The rougher the texture of the surface to be pasted, the greater the quantity of wet paste needed.

1. **Dimensions**
   a. Single roll is 36 SF. The paper is 18" wide x 24'-0" long.
   b. Double roll is 72 SF. The paper is 18" wide x 48'-0" long.
2. **Installation.** New paper may be applied over existing paper if the existing paper has butt joints, a smooth surface, and is tight to the wall. When new paper is applied direct to plaster or drywall, the wall should receive a coat of glue size before the paper is applied.
3. **Estimating Technique.** Determine the gross area, deduct openings and other areas not to be papered, and add 20% to the net area for waste. To find the number of rolls needed, divide the net area plus 20% by the number of SF per roll.

The steps in wallpapering

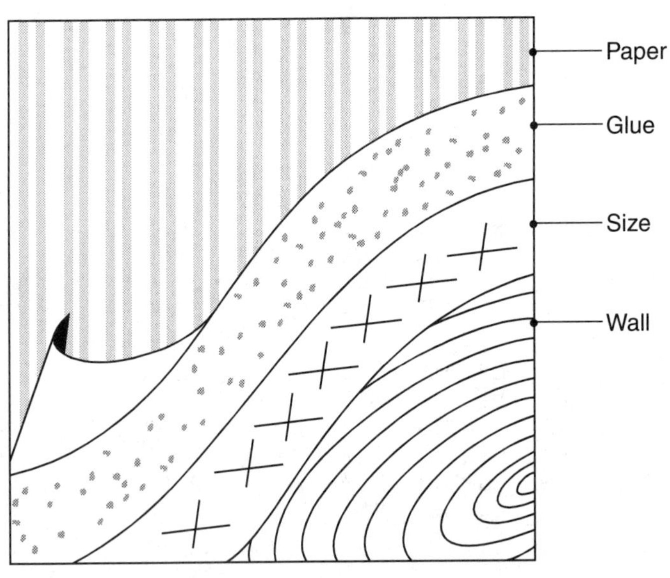

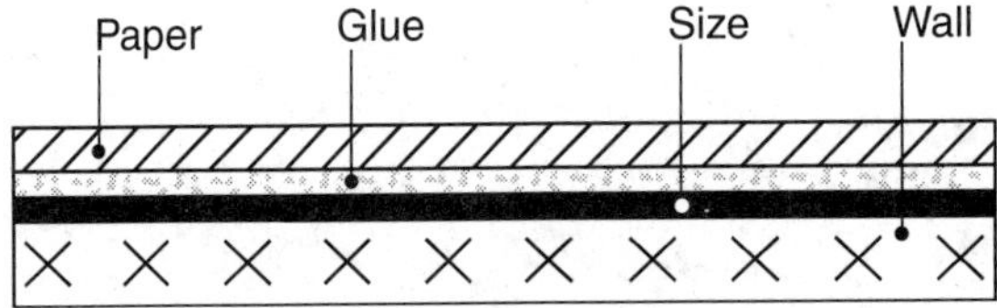

Description	Oper	Unit	Crew Size	Man-Hours Per Unit	Avg Mat'l Unit Cost	Avg Labor Unit Cost	Avg Equip Unit Cost	Avg Total Unit Cost	Avg Price Incl O&P	Large Volume Avg Total Unit Cost	Large Volume Avg Price Incl O&P
			Costs Based On Small Volume							Large Volume	
Wallpaper											
Butt joint											
Includes application of glue sizing. Material prices are not given because they vary radically											
Paper, single rolls, 36 SF/Roll	Inst	Roll	QA	.549	---	16.20	6.36	22.56	**30.70**	14.64	**19.90**
Vinyl, with fabric backing (prepasted), single rolls, 36 SF/Roll	Inst	Roll	QA	.615	---	18.10	7.13	25.23	**34.30**	16.43	**22.30**
Vinyl, with fabric backing (not prepasted), single rolls, 36 SF/Roll	Inst	Roll	QA	.669	---	19.70	7.75	27.45	**37.40**	17.84	**24.30**
Labor adjustments											
When ceiling is 9'-0" high or more ADD	Inst	%	QA	20.0	---	---	---	---	**---**	---	**---**
In kitchens, baths, and in other rooms, when floor area is less than 50.0 SF ADD	Inst	%	QA	32.0	---	---	---	---	**---**	---	**---**
Water closets. See Toilets, page 251											
Water heaters											
(A.O. Smith/National Products)											
Electric, glass lined, 2 element heater, baked enamel, installation and connection only											
Standard model, tall											
5 year Energy Saver, foam insulated											
30 gal round	Inst	Ea	SA	6.15	293.00	187.00	---	480.00	**616.00**	389.00	**493.00**
40 gal round	Inst	Ea	SA	6.15	316.00	187.00	---	503.00	**643.00**	409.00	**517.00**
50 gal round	Inst	Ea	SA	6.67	364.00	203.00	---	567.00	**721.00**	459.00	**577.00**
66 gal round	Inst	Ea	SA	6.67	498.00	203.00	---	701.00	**876.00**	575.00	**711.00**
80 gal round	Inst	Ea	SA	6.67	568.00	203.00	---	771.00	**956.00**	636.00	**780.00**
120 gal round	Inst	Ea	SA	7.27	837.00	221.00	---	1058.00	**1290.00**	878.00	**1060.00**
Standard model, low boy, top connection											
5 year Energy Saver, foam insulated											
6 gal round	Inst	Ea	SA	5.33	223.00	162.00	---	385.00	**498.00**	315.00	**404.00**
15 gal round	Inst	Ea	SA	6.15	242.00	187.00	---	429.00	**558.00**	345.00	**443.00**
30 gal round	Inst	Ea	SA	6.15	312.00	187.00	---	499.00	**638.00**	405.00	**512.00**
40 gal round	Inst	Ea	SA	6.67	337.00	203.00	---	540.00	**690.00**	435.00	**550.00**
50 gal round	Inst	Ea	SA	6.67	388.00	203.00	---	591.00	**748.00**	479.00	**600.00**
Standard model, low boy, side connection											
5 year Energy Saver, foam insulated											
10 gal round	Inst	Ea	SA	5.71	231.00	174.00	---	405.00	**525.00**	328.00	**421.00**
15 gal round	Inst	Ea	SA	6.15	252.00	187.00	---	439.00	**568.00**	353.00	**452.00**
30 gal round	Inst	Ea	SA	6.15	302.00	187.00	---	489.00	**627.00**	397.00	**503.00**
40 gal round	Inst	Ea	SA	6.67	327.00	203.00	---	530.00	**679.00**	427.00	**540.00**

Description	Oper	Unit	Crew Size	Man-Hours Per Unit	Avg Mat'l Unit Cost	Avg Labor Unit Cost	Avg Equip Unit Cost	Avg Total Unit Cost	Avg Price Incl O&P	Large Volume Avg Total Unit Cost	Large Volume Avg Price Incl O&P
Deluxe model											
10 year Series II											
30 gal round	Inst	Ea	SA	6.15	386.00	187.00	---	573.00	**723.00**	470.00	**587.00**
40 gal round	Inst	Ea	SA	6.15	410.00	187.00	---	597.00	**750.00**	490.00	**610.00**
50 gal round	Inst	Ea	SA	6.67	454.00	203.00	---	657.00	**825.00**	537.00	**667.00**
66 gal round	Inst	Ea	SA	6.67	582.00	203.00	---	785.00	**972.00**	648.00	**794.00**
80 gal round	Inst	Ea	SA	6.67	649.00	203.00	---	852.00	**1050.00**	706.00	**861.00**
120 gal round	Inst	Ea	SA	7.27	987.00	221.00	---	1208.00	**1470.00**	1008.00	**1210.00**
Gas, 3 stage automatic flame control, baked enamel, installation and connection only											
Standard model, tall											
5 year Energy Saver											
30 gal round, 40,000 btu	Inst	Ea	SA	6.15	359.00	187.00	---	546.00	**692.00**	446.00	**560.00**
40 gal round, 40,000 btu	Inst	Ea	SA	6.15	383.00	187.00	---	570.00	**719.00**	467.00	**583.00**
50 gal round, 40,000 btu	Inst	Ea	SA	6.67	430.00	203.00	---	633.00	**798.00**	516.00	**643.00**
75 gal round, 75,000 btu	Inst	Ea	SA	6.67	763.00	203.00	---	966.00	**1180.00**	804.00	**974.00**
Deluxe model, tall											
10 year Energy Saver											
30 gal round, 32,000 btu	Inst	Ea	SA	6.15	403.00	187.00	---	590.00	**743.00**	485.00	**604.00**
38 gal round, 50,000 btu	Inst	Ea	SA	6.15	459.00	187.00	---	646.00	**807.00**	533.00	**659.00**
40 gal round, 40,000 btu	Inst	Ea	SA	6.15	427.00	187.00	---	614.00	**770.00**	505.00	**628.00**
48 gal round, 60,000 btu	Inst	Ea	SA	6.15	475.00	187.00	---	662.00	**825.00**	547.00	**675.00**
50 gal round, 50,000 btu	Inst	Ea	SA	6.67	438.00	203.00	---	641.00	**807.00**	523.00	**651.00**
Deluxe model, low boy											
10 year Energy Saver											
30 gal round, 40,000 btu	Inst	Ea	SA	6.15	388.00	187.00	---	575.00	**725.00**	471.00	**588.00**
40 gal round, 40,000 btu	Inst	Ea	SA	6.15	411.00	187.00	---	598.00	**752.00**	492.00	**612.00**
Standard model, tall											
5 year, Indirect Water Heater											
30 gal round, 32,000 btu	Inst	Ea	SA	6.15	324.00	187.00	---	511.00	**652.00**	416.00	**525.00**
40 gal round, 32,500 btu	Inst	Ea	SA	6.15	348.00	187.00	---	535.00	**679.00**	437.00	**549.00**
50 gal round, 40,000 btu	Inst	Ea	SA	6.67	396.00	203.00	---	599.00	**758.00**	486.00	**608.00**
75 gal round, 75,100 btu	Inst	Ea	SA	6.67	712.00	203.00	---	915.00	**1120.00**	760.00	**924.00**
100 gal round, 80,000 btu	Inst	Ea	SA	6.67	1170.00	203.00	---	1373.00	**1650.00**	1163.00	**1380.00**
Standard model, low boy											
5 year, Indirect Water Heater											
30 gal round, 32,000 btu	Inst	Ea	SA	6.15	324.00	187.00	---	511.00	**652.00**	416.00	**525.00**
40 gal round, 40,000 btu	Inst	Ea	SA	6.15	348.00	187.00	---	535.00	**679.00**	437.00	**549.00**
Deluxe model, tall											
10 year, Indirect Water Heater											
30 gal round	Inst	Ea	SA	6.15	570.00	187.00	---	757.00	**934.00**	629.00	**770.00**
50 gal round	Inst	Ea	SA	6.15	715.00	187.00	---	902.00	**1100.00**	755.00	**915.00**
80 gal round	Inst	Ea	SA	6.67	1040.00	203.00	---	1243.00	**1500.00**	1044.00	**1250.00**

Description	Oper	Unit	Crew Size	Man-Hours Per Unit	Avg Mat'l Unit Cost	Avg Labor Unit Cost	Avg Equip Unit Cost	Avg Total Unit Cost	Avg Price Incl O&P	Large Volume Avg Total Unit Cost	Large Volume Avg Price Incl O&P
Optional accessories											
Water heater cabinets											
24" x 24" x 72"											
heaters up to 40 gallons	---	Ea	---	---	82.50	---	---	82.50	**94.90**	71.60	**82.30**
30" x 30" x 72"											
heaters up to 75 gallons	---	Ea	---	---	102.00	---	---	102.00	**118.00**	88.60	**104.00**
36" x 36" x 83"											
heaters up to 100 gallons	---	Ea	---	---	205.00	---	---	205.00	**236.00**	178.00	**209.00**
Water heater stands											
22" x 22", heaters up to 52 gal	---	Ea	---	---	45.90	---	---	45.90	**52.80**	39.80	**46.70**
26" x 26", heaters up to 75 gal	---	Ea	---	---	62.90	---	---	62.90	**72.30**	54.50	**64.00**
Free-standing restraint system											
75 gal capacity	---	Ea	---	---	160.00	---	---	160.00	**184.00**	138.00	**162.00**
100 gal capacity	---	Ea	---	---	171.00	---	---	171.00	**196.00**	148.00	**174.00**
Wall mount platform restraint system											
75 gal capacity	---	Ea	---	---	151.00	---	---	151.00	**174.00**	131.00	**154.00**
100 gal capacity	---	Ea	---	---	168.00	---	---	168.00	**193.00**	146.00	**171.00**

Solar water heating systems

Complete closed loop solar system with solar electric water heater with exchanger, differential control, heating element, circulator, collector and tank and panel sensors. The material costs given include the subcontractor's overhead and profit.

Description	Oper	Unit	Crew Size	Man-Hours Per Unit	Avg Mat'l Unit Cost	Avg Labor Unit Cost	Avg Equip Unit Cost	Avg Total Unit Cost	Avg Price Incl O&P	Large Volume Avg Total Unit Cost	Large Volume Avg Price Incl O&P
82 gallon capacity collector											
One deluxe collector	---	LS	---	---	2880.00	---	---	2880.00	---	2500.00	---
Two deluxe collectors											
Standard collectors	---	LS	---	---	3570.00	---	---	3570.00	---	3090.00	---
Deluxe collectors	---	LS	---	---	3670.00	---	---	3670.00	---	3190.00	---
Three collectors											
Economy collectors	---	LS	---	---	4190.00	---	---	4190.00	---	3630.00	---
Standard collectors	---	LS	---	---	4290.00	---	---	4290.00	---	3720.00	---
Four collectors											
Economy collectors	---	LS	---	---	4520.00	---	---	4520.00	---	3920.00	---
Standard collectors	---	LS	---	---	4630.00	---	---	4630.00	---	4020.00	---
120 gallon capacity system											
Three collectors											
Economy collectors	---	LS	---	---	5190.00	---	---	5190.00	---	4500.00	---
Standard collectors	---	LS	---	---	5310.00	---	---	5310.00	---	4610.00	---
Four standard collectors	---	LS	---	---	6220.00	---	---	6220.00	---	5390.00	---
Five collectors											
Economy collectors	---	LS	---	---	6780.00	---	---	6780.00	---	5880.00	---
Standard collectors	---	LS	---	---	6890.00	---	---	6890.00	---	5980.00	---
Six collectors											
Economy collectors	---	LS	---	---	7350.00	---	---	7350.00	---	6370.00	---
Standard collectors	---	LS	---	---	7460.00	---	---	7460.00	---	6470.00	---

Description	Oper	Unit	Crew Size	Man-Hours Per Unit	Avg Mat'l Unit Cost	Avg Labor Unit Cost	Avg Equip Unit Cost	Avg Total Unit Cost	Avg Price Incl O&P	Large Volume Avg Total Unit Cost	Large Volume Avg Price Incl O&P
			Costs Based On Small Volume							Large Volume	
Material adjustments											
Collector mounting kits, one required per collector panel											
Adjustable position hinge	---	Ea	---	---	64.40	---	---	64.40	---	55.90	---
Integral flange	---	Ea	---	---	5.65	---	---	5.65	---	4.90	---
Additional panels											
Economy collectors	---	Ea	---	---	565.00	---	---	565.00	---	490.00	---
Standard collectors	---	Ea	---	---	678.00	---	---	678.00	---	588.00	---
Deluxe collectors	---	Ea	---	---	791.00	---	---	791.00	---	686.00	---
Individual components											
Solar storage tanks, glass lined with fiberglass insulation											
66 gallon	---	Ea	---	---	672.00	---	---	672.00	---	583.00	---
82 gallon	---	Ea	---	---	763.00	---	---	763.00	---	662.00	---
120 gallon	---	Ea	---	---	1160.00	---	---	1160.00	---	1000.00	---
Solar electric water heaters, glass lined with fiberglass insulation, heating element with thermostat											
66 gallon, 4.5 KW	---	Ea	---	---	671.00	---	---	671.00	---	582.00	---
82 gallon, 4.5 KW	---	Ea	---	---	772.00	---	---	772.00	---	669.00	---
120 gallon, 4.5 KW	---	Ea	---	---	920.00	---	---	920.00	---	798.00	---
Additional element	---	Ea	---	---	37.30	---	---	37.30	---	32.30	---
Solar electric water heaters with heat exchangers, glass lined with fiberglass insulation, two copper exchangers, powered circulator, differential control, adjustable thermostat											
82 gallon, 4.5 KW, 1/20 HP	---	Ea	---	---	1000.00	---	---	1000.00	---	870.00	---
120 gallon, 4.5 KW, 1/20 HP	---	Ea	---	---	1320.00	---	---	1320.00	---	1140.00	---

Water softeners

(Bruner/Calgon)

Automatic water softeners, complete with yoke with 3/4" I.P.S. supply

Description	Oper	Unit	Crew Size	Man-Hours Per Unit	Avg Mat'l Unit Cost	Avg Labor Unit Cost	Avg Equip Unit Cost	Avg Total Unit Cost	Avg Price Incl O&P	Large Volume Avg Total Unit Cost	Large Volume Avg Price Incl O&P
Single tank units											
8,000 grain exchange capacity, 160 lbs salt storage, 6 gpm	---	Ea	---	---	622.00	---	---	622.00	**715.00**	557.00	**641.00**
15,000 grain exchange capacity, 200 lbs salt storage, 8.8 gpm	---	Ea	---	---	685.00	---	---	685.00	**788.00**	614.00	**706.00**
25,000 grain exchange capacity, 175 lbs salt storage, 11.3 gpm	---	Ea	---	---	975.00	---	---	975.00	**1120.00**	873.00	**1000.00**
Two tank units, side-by-side											
5,000 grain exchange capacity, 200 lbs salt storage, 8.8 gpm	---	Ea	---	---	720.00	---	---	720.00	**828.00**	645.00	**742.00**
30,000 grain exchange capacity, 200 lbs salt storage, 9 gpm	---	Ea	---	---	850.00	---	---	850.00	**977.00**	761.00	**875.00**
45,000 grain exchange capacity, 200 lbs salt storage, 9.5 gpm	---	Ea	---	---	971.00	---	---	971.00	**1120.00**	869.00	**1000.00**
Softy 1; for 5 persons with up to 50 grains per gallon of hardness	---	Ea	---	---	524.00	---	---	524.00	**603.00**	470.00	**540.00**
Softy 2; above plus taste & odor removal capabilities	---	Ea	---	---	556.00	---	---	556.00	**639.00**	498.00	**573.00**
Softy 3; for 6 persons with up to 73 grains per gallon of hardness	---	Ea	---	---	599.00	---	---	599.00	**689.00**	537.00	**617.00**

Description	Oper	Unit	Crew Size	Man-Hours Per Unit	Avg Mat'l Unit Cost	Avg Labor Unit Cost	Avg Equip Unit Cost	Avg Total Unit Cost	Avg Price Incl O&P	Large Volume Avg Total Unit Cost	Large Volume Avg Price Incl O&P
Automatic water filters, complete with yoke and media with 3/4" I.P.S. supply											
Two tank units, side-by-side											
Eliminate rotten egg odor and rust	---	Ea	---	---	788.00	---	---	788.00	**906.00**	706.00	**811.00**
Eliminate chlorine taste	---	Ea	---	---	662.00	---	---	662.00	**762.00**	593.00	**682.00**
Clear up water discoloration	---	Ea	---	---	593.00	---	---	593.00	**682.00**	531.00	**611.00**
Clear up corroding pipes	---	Ea	---	---	616.00	---	---	616.00	**709.00**	552.00	**635.00**
Manual water filters, complete fiberglass tank with 3/4" pipe supplies with mineral packs											
Eliminate taste and odor	---	Ea	---	---	588.00	---	---	588.00	**676.00**	526.00	**605.00**
Eliminate acid water	---	Ea	---	---	500.00	---	---	500.00	**575.00**	448.00	**515.00**
Eliminate iron in solution	---	Ea	---	---	605.00	---	---	605.00	**696.00**	542.00	**623.00**
Eliminate sediment	---	Ea	---	---	500.00	---	---	500.00	**575.00**	448.00	**515.00**
Chemical feed pumps											
115 volt, 9 gal/day	---	Ea	---	---	348.00	---	---	348.00	**401.00**	312.00	**359.00**
230 volt, 9 gal/day	---	Ea	---	---	399.00	---	---	399.00	**459.00**	357.00	**411.00**

Weathervanes. See Cupolas, page 65

Windows

Windows, with related trim and frame

Description	Oper	Unit	Crew Size	Man-Hours Per Unit	Avg Mat'l Unit Cost	Avg Labor Unit Cost	Avg Equip Unit Cost	Avg Total Unit Cost	Avg Price Incl O&P	Large Volume Avg Total Unit Cost	Large Volume Avg Price Incl O&P
To 12 SF											
Aluminum	Demo	Ea	LB	1.17	---	25.70	---	25.70	**38.80**	16.70	**25.30**
Wood	Demo	Ea	LB	1.54	---	33.80	---	33.80	**51.00**	22.00	**33.10**
13 SF to 50 SF											
Aluminum	Demo	Ea	LB	1.88	---	41.30	---	41.30	**62.30**	27.00	**40.80**
Wood	Demo	Ea	LB	2.46	---	54.00	---	54.00	**81.50**	35.10	**53.00**

Aluminum

Vertical slide, satin anodized finish, includes screen and hardware

Description	Oper	Unit	Crew Size	Man-Hours Per Unit	Avg Mat'l Unit Cost	Avg Labor Unit Cost	Avg Equip Unit Cost	Avg Total Unit Cost	Avg Price Incl O&P	Large Volume Avg Total Unit Cost	Large Volume Avg Price Incl O&P
Single glazed											
1'-6" x 3'-0" H	Inst	Set	CA	1.74	73.90	46.90	---	120.80	**156.00**	94.10	**120.00**
2'-0" x 2'-0" H	Inst	Set	CA	1.74	67.20	46.90	---	114.10	**149.00**	88.30	**113.00**
2'-0" x 2'-6" H	Inst	Set	CA	1.74	76.20	46.90	---	123.10	**159.00**	96.00	**122.00**
2'-0" x 3'-0" H	Inst	Set	CA	1.90	81.80	51.20	---	133.00	**172.00**	103.30	**131.00**
2'-0" x 3'-6" H	Inst	Set	CA	1.90	87.40	51.20	---	138.60	**178.00**	108.10	**137.00**
2'-0" x 4'-0" H	Inst	Set	CA	1.90	91.80	51.20	---	143.00	**183.00**	111.90	**141.00**
2'-0" x 4'-6" H	Inst	Set	CA	1.90	98.60	51.20	---	149.80	**191.00**	117.70	**148.00**
2'-0" x 5'-0" H	Inst	Set	CA	2.05	103.00	55.30	---	158.30	**203.00**	124.20	**156.00**
2'-0" x 6'-0" H	Inst	Set	CA	2.05	113.00	55.30	---	168.30	**214.00**	132.90	**166.00**
2'-6" x 3'-0" H	Inst	Set	CA	1.90	89.60	51.20	---	140.80	**181.00**	110.00	**139.00**
2'-6" x 3'-6" H	Inst	Set	CA	1.90	95.20	51.20	---	146.40	**187.00**	114.80	**144.00**
2'-6" x 4'-0" H	Inst	Set	CA	1.90	101.00	51.20	---	152.20	**194.00**	119.60	**150.00**
2'-6" x 4'-6" H	Inst	Set	CA	1.90	109.00	51.20	---	160.20	**203.00**	126.30	**157.00**
2'-6" x 5'-0" H	Inst	Set	CA	2.05	113.00	55.30	---	168.30	**214.00**	132.90	**166.00**
2'-6" x 6'-0" H	Inst	Set	CA	2.05	124.00	55.30	---	179.30	**227.00**	142.90	**177.00**

			Costs Based On Small Volume							Large Volume	
Description	Oper	Unit	Crew Size	Man-Hours Per Unit	Avg Mat'l Unit Cost	Avg Labor Unit Cost	Avg Equip Unit Cost	Avg Total Unit Cost	Avg Price Incl O&P	Avg Total Unit Cost	Avg Price Incl O&P
3'-0" x 2'-0" H	Inst	Set	CA	1.90	86.20	51.20	---	137.40	**177.00**	107.10	**135.00**
3'-0" x 3'-0" H	Inst	Set	CA	1.90	96.30	51.20	---	147.50	**189.00**	115.80	**145.00**
3'-0" x 3'-6" H	Inst	Set	CA	1.90	103.00	51.20	---	154.20	**196.00**	121.50	**152.00**
3'-0" x 4'-0" H	Inst	Set	CA	1.90	110.00	51.20	---	161.20	**204.00**	127.30	**159.00**
3'-0" x 4'-6" H	Inst	Set	CA	1.90	119.00	51.20	---	170.20	**214.00**	135.20	**167.00**
3'-0" x 5'-0" H	Inst	Set	CJ	4.21	123.00	103.00	---	226.00	**298.00**	173.50	**224.00**
3'-0" x 6'-0" H	Inst	Set	CJ	4.71	136.00	115.00	---	251.00	**331.00**	191.30	**248.00**
3'-6" x 3'-6" H	Inst	Set	CA	1.90	111.00	51.20	---	162.20	**205.00**	128.20	**160.00**
3'-6" x 4'-0" H	Inst	Set	CA	2.22	119.00	59.90	---	178.90	**228.00**	141.10	**176.00**
3'-6" x 4'-6" H	Inst	Set	CA	2.76	130.00	74.40	---	204.40	**263.00**	159.00	**201.00**
3'-6" x 5'-0" H	Inst	Set	CJ	4.71	133.00	115.00	---	248.00	**328.00**	189.30	**246.00**
4'-0" x 3'-0" H	Inst	Set	CA	1.90	111.00	51.20	---	162.20	**205.00**	128.20	**160.00**
4'-0" x 3'-6" H	Inst	Set	CA	2.22	119.00	59.90	---	178.90	**228.00**	141.10	**176.00**
4'-0" x 4'-0" H	Inst	Set	CJ	4.21	127.00	103.00	---	230.00	**302.00**	175.50	**227.00**
4'-0" x 4'-6" H	Inst	Set	CJ	4.21	134.00	103.00	---	237.00	**311.00**	182.50	**235.00**
4'-0" x 5'-0" H	Inst	Set	CJ	4.71	143.00	115.00	---	258.00	**340.00**	198.30	**256.00**
Dual glazed											
1'-6" x 3'-0" H	Inst	Set	CA	1.82	99.70	49.10	---	148.80	**189.00**	117.20	**147.00**
2'-0" x 2'-6" H	Inst	Set	CA	1.82	103.00	49.10	---	152.10	**193.00**	120.10	**150.00**
2'-0" x 3'-0" H	Inst	Set	CA	1.95	110.00	52.60	---	162.60	**206.00**	128.30	**160.00**
2'-0" x 3'-6" H	Inst	Set	CA	1.95	118.00	52.60	---	170.60	**215.00**	135.20	**168.00**
2'-0" x 4'-0" H	Inst	Set	CA	1.95	124.00	52.60	---	176.60	**223.00**	141.20	**175.00**
2'-0" x 4'-6" H	Inst	Set	CA	1.95	137.00	52.60	---	189.60	**237.00**	151.20	**187.00**
2'-0" x 5'-0" H	Inst	Set	CA	2.11	147.00	56.90	---	203.90	**255.00**	163.20	**201.00**
2'-0" x 6'-0" H	Inst	Set	CA	2.11	161.00	56.90	---	217.90	**272.00**	175.20	**216.00**
2'-6" x 3'-0" H	Inst	Set	CA	1.95	121.00	52.60	---	173.60	**219.00**	138.20	**171.00**
2'-6" x 3'-6" H	Inst	Set	CA	1.95	129.00	52.60	---	181.60	**228.00**	144.20	**179.00**
2'-6" x 4'-0" H	Inst	Set	CA	1.95	138.00	52.60	---	190.60	**238.00**	152.20	**188.00**
2'-6" x 4'-6" H	Inst	Set	CA	1.95	150.00	52.60	---	202.60	**253.00**	163.20	**200.00**
2'-6" x 5'-0" H	Inst	Set	CA	2.11	161.00	56.90	---	217.90	**272.00**	175.20	**216.00**
2'-6" x 6'-0" H	Inst	Set	CA	2.11	178.00	56.90	---	234.90	**291.00**	190.20	**232.00**
3'-0" x 3'-0" H	Inst	Set	CA	1.95	131.00	52.60	---	183.60	**231.00**	146.20	**181.00**
3'-0" x 3'-6" H	Inst	Set	CA	1.95	141.00	52.60	---	193.60	**242.00**	155.20	**191.00**
3'-0" x 4'-0" H	Inst	Set	CA	1.95	150.00	52.60	---	202.60	**253.00**	163.20	**200.00**
3'-0" x 4'-6" H	Inst	Set	CA	1.95	165.00	52.60	---	217.60	**269.00**	175.20	**214.00**
3'-0" x 5'-0" H	Inst	Set	CJ	4.44	176.00	109.00	---	285.00	**367.00**	221.00	**280.00**
3'-0" x 6'-0" H	Inst	Set	CJ	4.85	194.00	119.00	---	313.00	**403.00**	244.30	**310.00**
3'-6" x 3'-6" H	Inst	Set	CA	1.95	158.00	52.60	---	210.60	**262.00**	169.20	**208.00**
3'-6" x 4'-0" H	Inst	Set	CA	2.35	164.00	63.40	---	227.40	**284.00**	180.70	**223.00**
3'-6" x 4'-6" H	Inst	Set	CA	2.86	179.00	77.10	---	256.10	**323.00**	204.20	**253.00**
3'-6" x 5'-0" H	Inst	Set	CJ	4.85	189.00	119.00	---	308.00	**398.00**	240.30	**306.00**
4'-0" x 3'-0" H	Inst	Set	CA	1.95	152.00	52.60	---	204.60	**255.00**	165.20	**202.00**
4'-0" x 3'-6" H	Inst	Set	CA	2.35	165.00	63.40	---	228.40	**286.00**	181.70	**224.00**
4'-0" x 4'-0" H	Inst	Set	CJ	4.44	176.00	109.00	---	285.00	**367.00**	221.00	**280.00**
4'-0" x 4'-6" H	Inst	Set	CJ	4.44	196.00	109.00	---	305.00	**390.00**	238.00	**300.00**
4'-0" x 5'-0" H	Inst	Set	CJ	4.85	213.00	119.00	---	332.00	**425.00**	260.30	**329.00**
For bronze finish ADD	Inst	%		---	12.0	---	---	---	**---**	---	**---**

Description	Oper	Unit	Crew Size	Man-Hours Per Unit	Avg Mat'l Unit Cost	Avg Labor Unit Cost	Avg Equip Unit Cost	Avg Total Unit Cost	Avg Price Incl O&P	Avg Total Unit Cost	Avg Price Incl O&P
			Costs Based On Small Volume							Large Volume	
Horizontal slide, satin anodized finish, includes screen and hardware											
1 sliding lite, 1 fixed lite											
Single glazed											
2'-0" x 2'-0" H	Inst	Set	CA	1.90	49.30	51.20	---	100.50	**135.00**	75.40	**99.00**
2'-0" x 3'-0" H	Inst	Set	CA	1.90	59.40	51.20	---	110.60	**146.00**	84.10	**109.00**
2'-6" x 3'-0" H	Inst	Set	CA	1.90	63.80	51.20	---	115.00	**151.00**	87.90	**113.00**
3'-0" x 1'-0" H	Inst	Set	CA	1.90	45.90	51.20	---	97.10	**131.00**	72.60	**95.70**
3'-0" x 1'-6" H	Inst	Set	CA	1.90	51.50	51.20	---	102.70	**137.00**	77.40	**101.00**
3'-0" x 2'-0" H	Inst	Set	CA	1.90	58.20	51.20	---	109.40	**145.00**	83.10	**108.00**
3'-0" x 2'-6" H	Inst	Set	CA	1.90	63.80	51.20	---	115.00	**151.00**	87.90	**113.00**
3'-0" x 3'-0" H	Inst	Set	CA	1.90	69.40	51.20	---	120.60	**158.00**	92.70	**119.00**
3'-0" x 3'-6" H	Inst	Set	CA	2.22	75.00	59.90	---	134.90	**177.00**	103.40	**133.00**
3'-0" x 4'-0" H	Inst	Set	CA	2.76	80.60	74.40	---	155.00	**206.00**	117.10	**152.00**
3'-0" x 5'-0" H	Inst	Set	CJ	6.40	93.00	157.00	---	250.00	**345.00**	179.70	**244.00**
3'-6" x 2'-0" H	Inst	Set	CA	1.90	61.60	51.20	---	112.80	**149.00**	86.00	**111.00**
3'-6" x 2'-6" H	Inst	Set	CA	1.90	68.30	51.20	---	119.50	**156.00**	91.80	**118.00**
3'-6" x 3'-0" H	Inst	Set	CA	2.22	73.90	59.90	---	133.80	**176.00**	102.50	**132.00**
3'-6" x 3'-6" H	Inst	Set	CA	2.22	80.60	59.90	---	140.50	**184.00**	108.20	**139.00**
3'-6" x 4'-0" H	Inst	Set	CA	2.76	86.20	74.40	---	160.60	**212.00**	121.90	**158.00**
4'-0" x 1'-0" H	Inst	Set	CA	2.22	52.60	59.90	---	112.50	**152.00**	84.20	**111.00**
4'-0" x 1'-6" H	Inst	Set	CA	2.22	58.20	59.90	---	118.10	**158.00**	89.00	**117.00**
4'-0" x 2'-0" H	Inst	Set	CA	2.76	66.10	74.40	---	140.50	**189.00**	104.60	**138.00**
4'-0" x 2'-6" H	Inst	Set	CA	2.76	72.80	74.40	---	147.20	**197.00**	110.40	**145.00**
4'-0" x 3'-0" H	Inst	Set	CA	2.76	79.50	74.40	---	153.90	**205.00**	116.20	**151.00**
4'-0" x 3'-6" H	Inst	Set	CA	2.96	86.20	79.80	---	166.00	**220.00**	125.10	**163.00**
4'-0" x 4'-0" H	Inst	Set	CA	2.96	93.00	79.80	---	172.80	**228.00**	130.90	**169.00**
4'-0" x 5'-0" H	Inst	Set	CJ	6.96	106.00	170.00	---	276.00	**381.00**	203.20	**275.00**
5'-0" x 2'-0" H	Inst	Set	CJ	5.93	73.90	145.00	---	218.90	**305.00**	156.60	**214.00**
5'-0" x 2'-6" H	Inst	Set	CJ	5.93	81.80	145.00	---	226.80	**314.00**	163.30	**222.00**
5'-0" x 3'-0" H	Inst	Set	CJ	6.40	88.50	157.00	---	245.50	**340.00**	175.80	**240.00**
5'-0" x 3'-6" H	Inst	Set	CJ	6.40	96.30	157.00	---	253.30	**349.00**	182.60	**247.00**
5'-0" x 4'-0" H	Inst	Set	CJ	6.96	104.00	170.00	---	274.00	**378.00**	201.30	**273.00**
5'-0" x 5'-0" H	Inst	Set	CJ	6.96	125.00	170.00	---	295.00	**403.00**	220.00	**294.00**
6'-0" x 2'-0" H	Inst	Set	CJ	5.93	81.80	145.00	---	226.80	**314.00**	165.50	**226.00**
6'-0" x 2'-6" H	Inst	Set	CJ	5.93	89.60	145.00	---	234.60	**323.00**	172.20	**233.00**
6'-0" x 3'-0" H	Inst	Set	CJ	6.40	98.60	157.00	---	255.60	**351.00**	184.50	**250.00**
8'-0" x 2'-0" H	Inst	Set	CJ	6.40	102.00	157.00	---	259.00	**355.00**	187.40	**253.00**
8'-0" x 2'-6" H	Inst	Set	CJ	6.40	113.00	157.00	---	270.00	**368.00**	197.00	**264.00**
8'-0" x 3'-0" H	Inst	Set	CJ	6.96	124.00	170.00	---	294.00	**402.00**	219.00	**292.00**
8'-0" x 4'-0" H	Inst	Set	CJ	7.62	157.00	186.00	---	343.00	**464.00**	256.00	**340.00**
8'-0" x 5'-0" H	Inst	Set	CJ	8.00	189.00	196.00	---	385.00	**515.00**	292.00	**385.00**
For bronze finish, ADD	Inst	%		---	20.0	---	---	---	**---**	---	**---**
Dual glazed											
2'-0" x 2'-0" H	Inst	Set	CA	1.95	79.50	52.60	---	132.10	**171.00**	102.40	**130.00**
2'-0" x 2'-6" H	Inst	Set	CA	1.95	87.40	52.60	---	140.00	**180.00**	109.10	**138.00**
2'-0" x 3'-0" H	Inst	Set	CA	1.95	95.20	52.60	---	147.80	**189.00**	115.80	**146.00**
2'-0" x 3'-6" H	Inst	Set	CA	1.95	103.00	52.60	---	155.60	**198.00**	122.50	**154.00**
2'-0" x 4'-0" H	Inst	Set	CA	1.95	111.00	52.60	---	163.60	**207.00**	129.20	**161.00**

Description	Oper	Unit	Crew Size	Man-Hours Per Unit	Avg Mat'l Unit Cost	Avg Labor Unit Cost	Avg Equip Unit Cost	Avg Total Unit Cost	Avg Price Incl O&P	Large Volume Avg Total Unit Cost	Large Volume Avg Price Incl O&P
2'-6" x 2'-0" H	Inst	Set	CA	1.95	85.10	52.60	---	137.70	**178.00**	107.20	**136.00**
2'-6" x 2'-6" H	Inst	Set	CA	1.95	95.20	52.60	---	147.80	**189.00**	115.80	**146.00**
2'-6" x 3'-0" H	Inst	Set	CA	2.00	103.00	53.90	---	156.90	**200.00**	123.60	**155.00**
2'-6" x 3'-6" H	Inst	Set	CA	2.00	112.00	53.90	---	165.90	**211.00**	131.30	**164.00**
2'-6" x 4'-0" H	Inst	Set	CA	2.11	121.00	56.90	---	177.90	**226.00**	141.20	**176.00**
3'-0" x 2'-0" H	Inst	Set	CA	1.95	91.80	52.60	---	144.40	**186.00**	112.90	**143.00**
3'-0" x 2'-6" H	Inst	Set	CA	1.95	102.00	52.60	---	154.60	**197.00**	121.60	**153.00**
3'-0" x 3'-0" H	Inst	Set	CA	1.95	111.00	52.60	---	163.60	**207.00**	129.20	**161.00**
3'-0" x 3'-6" H	Inst	Set	CA	2.35	121.00	63.40	---	184.40	**235.00**	144.70	**181.00**
3'-0" x 4'-0" H	Inst	Set	CA	2.86	131.00	77.10	---	208.10	**268.00**	162.20	**205.00**
3'-0" x 5'-0" H	Inst	Set	CA	3.33	151.00	89.80	---	240.80	**310.00**	188.20	**238.00**
3'-6" x 2'-0" H	Inst	Set	CA	1.95	98.60	52.60	---	151.20	**193.00**	118.70	**149.00**
3'-6" x 2'-6" H	Inst	Set	CA	1.95	109.00	52.60	---	161.60	**205.00**	127.30	**159.00**
3'-6" x 3'-0" H	Inst	Set	CA	2.35	120.00	63.40	---	183.40	**234.00**	143.70	**180.00**
3'-6" x 3'-6" H	Inst	Set	CA	2.35	130.00	63.40	---	193.40	**246.00**	151.70	**190.00**
3'-6" x 4'-0" H	Inst	Set	CA	2.86	141.00	77.10	---	218.10	**279.00**	171.20	**215.00**
3'-6" x 5'-0" H	Inst	Set	CA	2.86	165.00	77.10	---	242.10	**307.00**	191.20	**239.00**
4'-0" x 2'-0" H	Inst	Set	CA	2.86	104.00	77.10	---	181.10	**237.00**	139.50	**179.00**
4'-0" x 2'-6" H	Inst	Set	CA	2.86	116.00	77.10	---	193.10	**251.00**	150.00	**191.00**
4'-0" x 3'-0" H	Inst	Set	CA	2.86	128.00	77.10	---	205.10	**264.00**	159.20	**202.00**
4'-0" x 3'-6" H	Inst	Set	CA	3.08	139.00	83.00	---	222.00	**286.00**	172.90	**219.00**
4'-0" x 4'-0" H	Inst	Set	CA	3.08	151.00	83.00	---	234.00	**300.00**	183.90	**231.00**
4'-0" x 5'-0" H	Inst	Set	CA	3.81	176.00	103.00	---	279.00	**358.00**	216.20	**273.00**
5'-0" x 2'-0" H	Inst	Set	CA	3.08	118.00	83.00	---	201.00	**261.00**	154.90	**198.00**
5'-0" x 2'-6" H	Inst	Set	CA	3.08	131.00	83.00	---	214.00	**277.00**	165.90	**211.00**
5'-0" x 3'-0" H	Inst	Set	CJ	6.67	144.00	163.00	---	307.00	**414.00**	230.00	**303.00**
5'-0" x 3'-6" H	Inst	Set	CJ	6.67	158.00	163.00	---	321.00	**430.00**	241.00	**316.00**
5'-0" x 4'-0" H	Inst	Set	CJ	7.62	171.00	186.00	---	357.00	**480.00**	266.00	**349.00**
5'-0" x 5'-0" H	Inst	Set	CJ	7.62	214.00	186.00	---	400.00	**529.00**	302.00	**391.00**
6'-0" x 2'-0" H	Inst	Set	CJ	6.40	130.00	157.00	---	287.00	**387.00**	211.00	**280.00**
6'-0" x 2'-6" H	Inst	Set	CJ	6.40	146.00	157.00	---	303.00	**405.00**	225.00	**296.00**
6'-0" x 3'-0" H	Inst	Set	CJ	6.67	160.00	163.00	---	323.00	**432.00**	243.00	**318.00**
6'-0" x 3'-6" H	Inst	Set	CJ	7.62	174.00	186.00	---	360.00	**483.00**	268.00	**351.00**
6'-0" x 4'-0" H	Inst	Set	CJ	7.62	192.00	186.00	---	378.00	**503.00**	283.00	**369.00**
6'-0" x 5'-0" H	Inst	Set	CJ	7.62	242.00	186.00	---	428.00	**561.00**	326.00	**419.00**
For bronze finish, ADD	Inst	%		---	15.0	---	---	---	**---**	---	**---**
2 sliding lites, 1 fixed lite											
Single glazed											
6'-0" x 2'-0" H	Inst	Set	CJ	6.67	103.00	163.00	---	266.00	**366.00**	194.30	**262.00**
6'-0" x 2'-6" H	Inst	Set	CJ	6.67	113.00	163.00	---	276.00	**378.00**	203.00	**272.00**
6'-0" x 3'-0" H	Inst	Set	CJ	6.96	123.00	170.00	---	293.00	**400.00**	218.00	**291.00**
6'-0" x 4'-0" H	Inst	Set	CJ	6.96	147.00	170.00	---	317.00	**427.00**	238.00	**314.00**
6'-0" x 5'-0" H	Inst	Set	CJ	6.96	170.00	170.00	---	340.00	**454.00**	258.00	**338.00**
7'-0" x 2'-0" H	Inst	Set	CJ	6.67	109.00	163.00	---	272.00	**373.00**	199.10	**268.00**
7'-0" x 2'-6" H	Inst	Set	CJ	6.67	121.00	163.00	---	284.00	**387.00**	210.00	**280.00**
7'-0" x 3'-0" H	Inst	Set	CJ	6.96	133.00	170.00	---	303.00	**412.00**	226.00	**301.00**
7'-0" x 4'-0" H	Inst	Set	CJ	6.96	169.00	170.00	---	339.00	**453.00**	257.00	**337.00**
7'-0" x 5'-0" H	Inst	Set	CJ	8.89	214.00	217.00	---	431.00	**576.00**	323.00	**423.00**
8'-0" x 2'-0" H	Inst	Set	CJ	6.96	120.00	170.00	---	290.00	**397.00**	215.00	**288.00**
8'-0" x 2'-6" H	Inst	Set	CJ	6.96	132.00	170.00	---	302.00	**411.00**	225.00	**300.00**

Description	Oper	Unit	Crew Size	Man-Hours Per Unit	Avg Mat'l Unit Cost	Avg Labor Unit Cost	Avg Equip Unit Cost	Avg Total Unit Cost	Avg Price Incl O&P	Large Volume Avg Total Unit Cost	Large Volume Avg Price Incl O&P
			Costs Based On Small Volume							Large Volume	
8'-0" x 3'-0" H	Inst	Set	CJ	6.96	147.00	170.00	---	317.00	**427.00**	238.00	**314.00**
8'-0" x 3'-6" H	Inst	Set	CJ	6.96	159.00	170.00	---	329.00	**442.00**	248.00	**327.00**
8'-0" x 4'-0" H	Inst	Set	CJ	8.89	172.00	217.00	---	389.00	**529.00**	288.00	**382.00**
8'-0" x 5'-0" H	Inst	Set	CJ	8.89	216.00	217.00	---	433.00	**579.00**	325.00	**425.00**
10'-0" x 2'-0" H	Inst	Set	CJ	6.96	155.00	170.00	---	325.00	**436.00**	244.00	**322.00**
10'-0" x 3'-0" H	Inst	Set	CJ	6.96	177.00	170.00	---	347.00	**462.00**	264.00	**344.00**
10'-0" x 3'-6" H	Inst	Set	CJ	8.89	203.00	217.00	---	420.00	**564.00**	314.00	**412.00**
10'-0" x 4'-0" H	Inst	Set	CJ	8.89	213.00	217.00	---	430.00	**575.00**	322.00	**422.00**
10'-0" x 5'-0" H	Inst	Set	CJ	8.89	299.00	217.00	---	516.00	**674.00**	396.00	**507.00**
For bronze finish, ADD	Inst	%		---	13.5	---	---	---	**---**	---	**---**
Dual glazed											
6'-0" x 2'-0" H	Inst	Set	CJ	6.96	161.00	170.00	---	331.00	**444.00**	250.00	**329.00**
6'-0" x 2'-6" H	Inst	Set	CJ	6.96	179.00	170.00	---	349.00	**465.00**	266.00	**347.00**
6'-0" x 3'-0" H	Inst	Set	CJ	7.62	197.00	186.00	---	383.00	**510.00**	288.00	**375.00**
6'-0" x 3'-6" H	Inst	Set	CJ	7.62	215.00	186.00	---	401.00	**531.00**	303.00	**392.00**
6'-0" x 4'-0" H	Inst	Set	CJ	7.62	243.00	186.00	---	429.00	**563.00**	327.00	**420.00**
6'-0" x 5'-0" H	Inst	Set	CJ	7.62	281.00	186.00	---	467.00	**607.00**	360.00	**457.00**
7'-0" x 2'-0" H	Inst	Set	CJ	6.67	178.00	163.00	---	341.00	**453.00**	259.00	**336.00**
7'-0" x 2'-6" H	Inst	Set	CJ	6.67	196.00	163.00	---	359.00	**473.00**	274.00	**354.00**
7'-0" x 3'-0" H	Inst	Set	CJ	6.96	216.00	170.00	---	386.00	**507.00**	297.00	**383.00**
7'-0" x 3'-6" H	Inst	Set	CJ	7.62	248.00	186.00	---	434.00	**568.00**	331.00	**424.00**
7'-0" x 4'-0" H	Inst	Set	CJ	7.62	270.00	186.00	---	456.00	**594.00**	350.00	**446.00**
7'-0" x 5'-0" H	Inst	Set	CJ	9.41	332.00	230.00	---	562.00	**731.00**	434.00	**555.00**
8'-0" x 2'-0" H	Inst	Set	CJ	7.62	188.00	186.00	---	374.00	**500.00**	280.00	**366.00**
8'-0" x 2'-6" H	Inst	Set	CJ	7.62	209.00	186.00	---	395.00	**524.00**	299.00	**387.00**
8'-0" x 3'-0" H	Inst	Set	CJ	7.62	242.00	186.00	---	428.00	**561.00**	326.00	**419.00**
8'-0" x 3'-6" H	Inst	Set	CJ	7.62	263.00	186.00	---	449.00	**586.00**	345.00	**440.00**
8'-0" x 4'-0" H	Inst	Set	CJ	9.41	286.00	230.00	---	516.00	**678.00**	395.00	**510.00**
8'-0" x 5'-0" H	Inst	Set	CJ	9.41	339.00	230.00	---	569.00	**740.00**	441.00	**563.00**
10'-0" x 2'-0" H	Inst	Set	CJ	6.96	264.00	170.00	---	434.00	**563.00**	339.00	**430.00**
10'-0" x 3'-0" H	Inst	Set	CJ	6.96	325.00	170.00	---	495.00	**632.00**	390.00	**490.00**
10'-0" x 3'-6" H	Inst	Set	CJ	8.89	370.00	217.00	---	587.00	**755.00**	457.00	**577.00**
10'-0" x 4'-0" H	Inst	Set	CJ	8.89	400.00	217.00	---	617.00	**790.00**	483.00	**606.00**
10'-0" x 5'-0" H	Inst	Set	CJ	8.89	485.00	217.00	---	702.00	**888.00**	556.00	**690.00**
For bronze finish, ADD	Inst	%		---	15.0	---	---	---	**---**	---	**---**

Wood

Awning windows; pine frames, exterior treated and primed, interior natural finish; glazed 1/2" T insulating glass; unit includes weatherstripping and exterior trim

Unit is one ventilating lite wide

Description	Oper	Unit	Crew Size	Man-Hours Per Unit	Avg Mat'l Unit Cost	Avg Labor Unit Cost	Avg Equip Unit Cost	Avg Total Unit Cost	Avg Price Incl O&P	Large Volume Avg Total Unit Cost	Large Volume Avg Price Incl O&P
20", 24" x 32" W											
High quality workmanship	Inst	Set	CA	1.90	258.00	51.20	---	309.20	**374.00**	254.70	**305.00**
Good quality workmanship	Inst	Set	CA	1.54	230.00	41.50	---	271.50	**327.00**	224.00	**267.00**
Average quality workmanship	Inst	Set	CA	1.29	190.00	34.80	---	224.80	**272.00**	185.50	**222.00**
32", 36" x 32" W											
High quality workmanship	Inst	Set	CA	1.90	342.00	51.20	---	393.20	**471.00**	326.70	**388.00**
Good quality workmanship	Inst	Set	CA	1.54	302.00	41.50	---	343.50	**411.00**	286.00	**339.00**
Average quality workmanship	Inst	Set	CA	1.29	252.00	34.80	---	286.80	**343.00**	238.50	**283.00**

Description	Oper	Unit	Crew Size	Man-Hours Per Unit	Avg Mat'l Unit Cost	Avg Labor Unit Cost	Avg Equip Unit Cost	Avg Total Unit Cost	Avg Price Incl O&P	Large Volume Avg Total Unit Cost	Large Volume Avg Price Incl O&P
20", 24" x 40" W											
High quality workmanship	Inst	Set	CA	1.90	280.00	51.20	---	331.20	**400.00**	273.70	**327.00**
Good quality workmanship	Inst	Set	CA	1.54	269.00	41.50	---	310.50	**372.00**	257.00	**306.00**
Average quality workmanship	Inst	Set	CA	1.29	224.00	34.80	---	258.80	**310.00**	214.50	**255.00**
32", 36" x 40" W											
High quality workmanship	Inst	Set	CA	1.90	353.00	51.20	---	404.20	**484.00**	335.70	**399.00**
Good quality workmanship	Inst	Set	CA	1.54	297.00	41.50	---	338.50	**404.00**	281.00	**334.00**
Average quality workmanship	Inst	Set	CA	1.29	246.00	34.80	---	280.80	**336.00**	233.50	**277.00**
20", 24" x 48" W											
High quality workmanship	Inst	Set	CA	1.90	353.00	51.20	---	404.20	**484.00**	335.70	**399.00**
Good quality workmanship	Inst	Set	CA	1.54	291.00	41.50	---	332.50	**398.00**	277.00	**328.00**
Average quality workmanship	Inst	Set	CA	1.29	242.00	34.80	---	276.80	**331.00**	229.50	**273.00**
32", 36" x 48" W											
High quality workmanship	Inst	Set	CA	1.90	532.00	51.20	---	583.20	**690.00**	489.70	**576.00**
Good quality workmanship	Inst	Set	CA	1.54	330.00	41.50	---	371.50	**443.00**	310.00	**367.00**
Average quality workmanship	Inst	Set	CA	1.29	274.00	34.80	---	308.80	**368.00**	257.50	**305.00**
Unit is two ventilating lites wide											
20", 24" x 48" W											
High quality workmanship	Inst	Set	CJ	5.16	437.00	126.00	---	563.00	**694.00**	455.50	**554.00**
Good quality workmanship	Inst	Set	CJ	4.10	403.00	100.00	---	503.00	**616.00**	411.30	**497.00**
Average quality workmanship	Inst	Set	CJ	3.40	336.00	83.20	---	419.20	**513.00**	342.30	**414.00**
32", 36" x 48" W											
High quality workmanship	Inst	Set	CJ	5.16	482.00	126.00	---	608.00	**746.00**	494.50	**599.00**
Good quality workmanship	Inst	Set	CJ	4.10	459.00	100.00	---	559.00	**680.00**	459.30	**552.00**
Average quality workmanship	Inst	Set	CJ	3.40	382.00	83.20	---	465.20	**566.00**	381.30	**459.00**
20", 24" x 72" W											
High quality workmanship	Inst	Set	CJ	5.16	476.00	126.00	---	602.00	**739.00**	489.50	**593.00**
Good quality workmanship	Inst	Set	CJ	4.10	448.00	100.00	---	548.00	**668.00**	449.30	**541.00**
Average quality workmanship	Inst	Set	CJ	3.40	373.00	83.20	---	456.20	**555.00**	374.30	**450.00**
32", 36" x 72" W											
High quality workmanship	Inst	Set	CJ	5.16	560.00	126.00	---	686.00	**836.00**	561.50	**676.00**
Good quality workmanship	Inst	Set	CJ	4.10	532.00	100.00	---	632.00	**764.00**	521.30	**624.00**
Average quality workmanship	Inst	Set	CJ	3.40	442.00	83.20	---	525.20	**635.00**	433.30	**519.00**
20", 24" x 80" W											
High quality workmanship	Inst	Set	CJ	5.16	515.00	126.00	---	641.00	**784.00**	523.50	**632.00**
Good quality workmanship	Inst	Set	CJ	4.10	485.00	100.00	---	585.00	**710.00**	481.30	**577.00**
Average quality workmanship	Inst	Set	CJ	3.40	404.00	83.20	---	487.20	**591.00**	401.30	**481.00**
32", 36" x 80" W											
High quality workmanship	Inst	Set	CJ	5.16	594.00	126.00	---	720.00	**874.00**	590.50	**709.00**
Good quality workmanship	Inst	Set	CJ	4.10	560.00	100.00	---	660.00	**796.00**	545.30	**651.00**
Average quality workmanship	Inst	Set	CJ	3.40	466.00	83.20	---	549.20	**662.00**	453.30	**542.00**
20", 24" x 96" W											
High quality workmanship	Inst	Set	CJ	5.16	560.00	126.00	---	686.00	**836.00**	561.50	**676.00**
Good quality workmanship	Inst	Set	CJ	4.10	526.00	100.00	---	626.00	**758.00**	516.30	**618.00**
Average quality workmanship	Inst	Set	CJ	3.40	440.00	83.20	---	523.20	**633.00**	431.30	**516.00**
32", 36" x 96" W											
High quality workmanship	Inst	Set	CJ	5.16	661.00	126.00	---	787.00	**952.00**	647.50	**775.00**
Good quality workmanship	Inst	Set	CJ	4.10	622.00	100.00	---	722.00	**867.00**	598.30	**712.00**
Average quality workmanship	Inst	Set	CJ	3.40	519.00	83.20	---	602.20	**723.00**	498.30	**594.00**

Description	Oper	Unit	Crew Size	Man-Hours Per Unit	Avg Mat'l Unit Cost	Avg Labor Unit Cost	Avg Equip Unit Cost	Avg Total Unit Cost	Avg Price Incl O&P	Large Volume Avg Total Unit Cost	Large Volume Avg Price Incl O&P
Unit is three ventilating lites wide											
20", 24" x 72" W											
High quality workmanship	Inst	Set	CJ	6.96	661.00	170.00	---	831.00	**1020.00**	675.00	**816.00**
Good quality workmanship	Inst	Set	CJ	5.52	616.00	135.00	---	751.00	**914.00**	615.10	**740.00**
Average quality workmanship	Inst	Set	CJ	4.57	513.00	112.00	---	625.00	**760.00**	512.40	**616.00**
32", 36" x 72" W											
High quality workmanship	Inst	Set	CJ	6.96	728.00	170.00	---	898.00	**1100.00**	733.00	**883.00**
Good quality workmanship	Inst	Set	CJ	5.52	683.00	135.00	---	818.00	**991.00**	673.10	**806.00**
Average quality workmanship	Inst	Set	CJ	4.57	571.00	112.00	---	683.00	**827.00**	562.40	**673.00**
20", 24" x 108" W											
High quality workmanship	Inst	Set	CJ	6.96	717.00	170.00	---	887.00	**1080.00**	723.00	**872.00**
Good quality workmanship	Inst	Set	CJ	5.52	672.00	135.00	---	807.00	**978.00**	663.10	**795.00**
Average quality workmanship	Inst	Set	CJ	4.57	560.00	112.00	---	672.00	**814.00**	552.40	**662.00**
32", 36" x 108" W											
High quality workmanship	Inst	Set	CJ	6.96	851.00	170.00	---	1021.00	**1240.00**	839.00	**1000.00**
Good quality workmanship	Inst	Set	CJ	5.52	801.00	135.00	---	936.00	**1130.00**	773.10	**922.00**
Average quality workmanship	Inst	Set	CJ	4.57	672.00	112.00	---	784.00	**943.00**	648.40	**772.00**
20", 24" x 120" W											
High quality workmanship	Inst	Set	CJ	6.96	784.00	170.00	---	954.00	**1160.00**	781.00	**938.00**
Good quality workmanship	Inst	Set	CJ	5.52	739.00	135.00	---	874.00	**1060.00**	721.10	**861.00**
Average quality workmanship	Inst	Set	CJ	4.57	616.00	112.00	---	728.00	**878.00**	600.40	**717.00**
32", 36" x 120" W											
High quality workmanship	Inst	Set	CJ	6.96	902.00	170.00	---	1072.00	**1300.00**	882.00	**1050.00**
Good quality workmanship	Inst	Set	CJ	5.52	846.00	135.00	---	981.00	**1180.00**	812.10	**966.00**
Average quality workmanship	Inst	Set	CJ	4.57	706.00	112.00	---	818.00	**981.00**	677.40	**806.00**
20", 24" x 144" W											
High quality workmanship	Inst	Set	CJ	6.96	851.00	170.00	---	1021.00	**1240.00**	839.00	**1000.00**
Good quality workmanship	Inst	Set	CJ	5.52	801.00	135.00	---	936.00	**1130.00**	773.10	**922.00**
Average quality workmanship	Inst	Set	CJ	4.57	666.00	112.00	---	778.00	**936.00**	643.40	**767.00**
32", 36" x 144" W											
High quality workmanship	Inst	Set	CJ	6.96	997.00	170.00	---	1167.00	**1410.00**	963.00	**1150.00**
Good quality workmanship	Inst	Set	CJ	5.52	935.00	135.00	---	1070.00	**1280.00**	889.10	**1050.00**
Average quality workmanship	Inst	Set	CJ	4.57	784.00	112.00	---	896.00	**1070.00**	744.40	**883.00**
For aluminum clad (baked white enamel exterior)											
ADD	Inst	%		---	12.0	---	---	---	**---**	---	**---**
Casement windows; glazed 1/2" T insulating glass; unit includes: hardware, drip cap and weatherstripping; does not include screens; rough opening sizes											
1 ventilating lite, no fixed lites; 20", 24", 28" W											
36" H											
High quality workmanship	Inst	Set	CJ	3.08	258.00	75.30	---	333.30	**411.00**	269.90	**328.00**
Good quality workmanship	Inst	Set	CJ	2.46	237.00	60.20	---	297.20	**364.00**	243.10	**294.00**
Average quality workmanship	Inst	Set	CJ	2.05	198.00	50.10	---	248.10	**304.00**	202.50	**245.00**
41" H											
High quality workmanship	Inst	Set	CJ	3.27	274.00	80.00	---	354.00	**437.00**	286.60	**349.00**
Good quality workmanship	Inst	Set	CJ	2.58	254.00	63.10	---	317.10	**388.00**	259.10	**313.00**
Average quality workmanship	Inst	Set	CJ	2.16	212.00	52.80	---	264.80	**324.00**	215.20	**261.00**
48" H											
High quality workmanship	Inst	Set	CJ	3.40	314.00	83.20	---	397.20	**487.00**	323.30	**392.00**
Good quality workmanship	Inst	Set	CJ	2.71	290.00	66.30	---	356.30	**434.00**	292.50	**352.00**
Average quality workmanship	Inst	Set	CJ	2.29	242.00	56.00	---	298.00	**363.00**	243.20	**293.00**

Description	Oper	Unit	Crew Size	Man-Hours Per Unit	Avg Mat'l Unit Cost	Avg Labor Unit Cost	Avg Equip Unit Cost	Avg Total Unit Cost	Avg Price Incl O&P	Large Volume Avg Total Unit Cost	Large Volume Avg Price Incl O&P
53" H											
High quality workmanship	Inst	Set	CJ	3.64	342.00	89.00	---	431.00	**528.00**	350.50	**424.00**
Good quality workmanship	Inst	Set	CJ	2.91	316.00	71.20	---	387.20	**471.00**	317.00	**381.00**
Average quality workmanship	Inst	Set	CJ	2.42	263.00	59.20	---	322.20	**393.00**	264.40	**318.00**
60" H											
High quality workmanship	Inst	Set	CJ	3.81	375.00	93.20	---	468.20	**573.00**	383.20	**463.00**
Good quality workmanship	Inst	Set	CJ	3.08	347.00	75.30	---	422.30	**514.00**	346.90	**417.00**
Average quality workmanship	Inst	Set	CJ	2.58	289.00	63.10	---	352.10	**428.00**	288.90	**347.00**
66" H											
High quality workmanship	Inst	Set	CJ	3.81	403.00	93.20	---	496.20	**605.00**	407.20	**490.00**
Good quality workmanship	Inst	Set	CJ	3.08	373.00	75.30	---	448.30	**543.00**	368.90	**442.00**
Average quality workmanship	Inst	Set	CJ	2.58	310.00	63.10	---	373.10	**453.00**	306.90	**368.00**
72" H											
High quality workmanship	Inst	Set	CJ	4.10	431.00	100.00	---	531.00	**648.00**	435.30	**524.00**
Good quality workmanship	Inst	Set	CJ	3.27	399.00	80.00	---	479.00	**580.00**	394.10	**472.00**
Average quality workmanship	Inst	Set	CJ	2.71	333.00	66.30	---	399.30	**483.00**	328.50	**394.00**
2 ventilating lites, no fixed lites; 40" W											
36" H											
High quality workmanship	Inst	Set	CJ	3.81	426.00	93.20	---	519.20	**631.00**	426.20	**512.00**
Good quality workmanship	Inst	Set	CJ	3.08	393.00	75.30	---	468.30	**567.00**	385.90	**462.00**
Average quality workmanship	Inst	Set	CJ	2.58	328.00	63.10	---	391.10	**473.00**	321.90	**386.00**
41" H											
High quality workmanship	Inst	Set	CJ	4.10	448.00	100.00	---	548.00	**668.00**	449.30	**541.00**
Good quality workmanship	Inst	Set	CJ	3.27	414.00	80.00	---	494.00	**598.00**	407.10	**488.00**
Average quality workmanship	Inst	Set	CJ	2.71	345.00	66.30	---	411.30	**497.00**	339.50	**406.00**
48" H											
High quality workmanship	Inst	Set	CJ	4.44	515.00	109.00	---	624.00	**758.00**	512.00	**614.00**
Good quality workmanship	Inst	Set	CJ	3.48	476.00	85.10	---	561.10	**677.00**	464.00	**554.00**
Average quality workmanship	Inst	Set	CJ	2.91	396.00	71.20	---	467.20	**564.00**	386.50	**461.00**
53" H											
High quality workmanship	Inst	Set	CJ	4.71	560.00	115.00	---	675.00	**819.00**	555.30	**666.00**
Good quality workmanship	Inst	Set	CJ	3.81	519.00	93.20	---	612.20	**738.00**	504.20	**603.00**
Average quality workmanship	Inst	Set	CJ	3.14	431.00	76.80	---	507.80	**613.00**	420.10	**501.00**
60" H											
High quality workmanship	Inst	Set	CJ	5.16	622.00	126.00	---	748.00	**907.00**	614.50	**737.00**
Good quality workmanship	Inst	Set	CJ	4.10	575.00	100.00	---	675.00	**813.00**	557.30	**666.00**
Average quality workmanship	Inst	Set	CJ	3.40	479.00	83.20	---	562.20	**678.00**	465.30	**555.00**
66" H											
High quality workmanship	Inst	Set	CJ	5.16	650.00	126.00	---	776.00	**939.00**	638.50	**764.00**
Good quality workmanship	Inst	Set	CJ	4.10	601.00	100.00	---	701.00	**844.00**	581.30	**692.00**
Average quality workmanship	Inst	Set	CJ	3.40	501.00	83.20	---	584.20	**702.00**	483.30	**576.00**
72" H											
High quality workmanship	Inst	Set	CJ	5.52	678.00	135.00	---	813.00	**984.00**	670.00	**803.00**
Good quality workmanship	Inst	Set	CJ	4.44	627.00	109.00	---	736.00	**886.00**	609.20	**726.00**
Average quality workmanship	Inst	Set	CJ	3.72	522.00	91.00	---	613.00	**738.00**	506.20	**604.00**
2 ventilating lites, no fixed lites; 48" W											
36" H											
High quality workmanship	Inst	Set	CJ	3.81	459.00	93.20	---	552.20	**670.00**	455.20	**546.00**
Good quality workmanship	Inst	Set	CJ	3.08	426.00	75.30	---	501.30	**604.00**	413.90	**494.00**
Average quality workmanship	Inst	Set	CJ	2.58	354.00	63.10	---	417.10	**503.00**	343.90	**411.00**

Description	Oper	Unit	Crew Size	Man-Hours Per Unit	Avg Mat'l Unit Cost	Avg Labor Unit Cost	Avg Equip Unit Cost	Avg Total Unit Cost	Avg Price Incl O&P	Avg Total Unit Cost	Avg Price Incl O&P
			Costs Based On Small Volume							Large Volume	
41" H											
High quality workmanship	Inst	Set	CJ	4.10	498.00	100.00	---	598.00	**726.00**	492.30	**591.00**
Good quality workmanship	Inst	Set	CJ	3.27	461.00	80.00	---	541.00	**652.00**	448.10	**534.00**
Average quality workmanship	Inst	Set	CJ	2.71	384.00	66.30	---	450.30	**543.00**	372.50	**445.00**
48" H											
High quality workmanship	Inst	Set	CJ	5.16	549.00	126.00	---	675.00	**823.00**	551.50	**665.00**
Good quality workmanship	Inst	Set	CJ	4.10	507.00	100.00	---	607.00	**736.00**	500.30	**599.00**
Average quality workmanship	Inst	Set	CJ	3.40	423.00	83.20	---	506.20	**613.00**	417.30	**500.00**
53" H											
High quality workmanship	Inst	Set	CJ	4.71	599.00	115.00	---	714.00	**864.00**	589.30	**705.00**
Good quality workmanship	Inst	Set	CJ	3.81	554.00	93.20	---	647.20	**779.00**	535.20	**638.00**
Average quality workmanship	Inst	Set	CJ	3.14	461.00	76.80	---	537.80	**647.00**	446.10	**531.00**
60" H											
High quality workmanship	Inst	Set	CJ	5.16	633.00	126.00	---	759.00	**920.00**	623.50	**748.00**
Good quality workmanship	Inst	Set	CJ	4.10	586.00	100.00	---	686.00	**826.00**	567.30	**677.00**
Average quality workmanship	Inst	Set	CJ	3.40	487.00	83.20	---	570.20	**687.00**	472.30	**563.00**
66" H											
High quality workmanship	Inst	Set	CJ	5.16	678.00	126.00	---	804.00	**971.00**	662.50	**792.00**
Good quality workmanship	Inst	Set	CJ	4.10	627.00	100.00	---	727.00	**874.00**	603.30	**717.00**
Average quality workmanship	Inst	Set	CJ	3.40	522.00	83.20	---	605.20	**727.00**	501.30	**597.00**
72" H											
High quality workmanship	Inst	Set	CJ	5.52	722.00	135.00	---	857.00	**1040.00**	708.00	**847.00**
Good quality workmanship	Inst	Set	CJ	4.44	669.00	109.00	---	778.00	**934.00**	644.20	**767.00**
Average quality workmanship	Inst	Set	CJ	3.72	557.00	91.00	---	648.00	**778.00**	536.20	**639.00**
2 ventilating lites, no fixed lites; 56" W											
36" H											
High quality workmanship	Inst	Set	CJ	3.81	498.00	93.20	---	591.20	**715.00**	488.20	**584.00**
Good quality workmanship	Inst	Set	CJ	3.08	461.00	75.30	---	536.30	**645.00**	444.90	**529.00**
Average quality workmanship	Inst	Set	CJ	2.58	384.00	63.10	---	447.10	**538.00**	369.90	**441.00**
41" H											
High quality workmanship	Inst	Set	CJ	4.10	549.00	100.00	---	649.00	**784.00**	535.30	**640.00**
Good quality workmanship	Inst	Set	CJ	3.27	507.00	80.00	---	587.00	**705.00**	487.10	**579.00**
Average quality workmanship	Inst	Set	CJ	2.71	423.00	66.30	---	489.30	**588.00**	406.50	**483.00**
48" H											
High quality workmanship	Inst	Set	CJ	4.44	610.00	109.00	---	719.00	**867.00**	593.00	**708.00**
Good quality workmanship	Inst	Set	CJ	3.48	564.00	85.10	---	649.10	**779.00**	540.00	**642.00**
Average quality workmanship	Inst	Set	CJ	2.91	470.00	71.20	---	541.20	**649.00**	449.50	**534.00**
53" H											
High quality workmanship	Inst	Set	CJ	4.71	670.00	115.00	---	785.00	**945.00**	649.30	**775.00**
Good quality workmanship	Inst	Set	CJ	3.81	619.00	93.20	---	712.20	**854.00**	591.20	**702.00**
Average quality workmanship	Inst	Set	CJ	3.14	516.00	76.80	---	592.80	**710.00**	493.10	**585.00**
60" H											
High quality workmanship	Inst	Set	CJ	5.16	728.00	126.00	---	854.00	**1030.00**	705.50	**841.00**
Good quality workmanship	Inst	Set	CJ	4.10	674.00	100.00	---	774.00	**928.00**	643.30	**764.00**
Average quality workmanship	Inst	Set	CJ	3.40	561.00	83.20	---	644.20	**772.00**	535.30	**636.00**
66" H											
High quality workmanship	Inst	Set	CJ	5.16	778.00	126.00	---	904.00	**1090.00**	748.50	**891.00**
Good quality workmanship	Inst	Set	CJ	4.10	721.00	100.00	---	821.00	**982.00**	683.30	**810.00**
Average quality workmanship	Inst	Set	CJ	3.40	600.00	83.20	---	683.20	**817.00**	569.30	**674.00**

			Costs Based On Small Volume							Large Volume	
Description	Oper	Unit	Crew Size	Man-Hours Per Unit	Avg Mat'l Unit Cost	Avg Labor Unit Cost	Avg Equip Unit Cost	Avg Total Unit Cost	Avg Price Incl O&P	Avg Total Unit Cost	Avg Price Incl O&P
72" H											
High quality workmanship	Inst	Set	CJ	5.52	829.00	135.00	---	964.00	**1160.00**	799.00	**952.00**
Good quality workmanship	Inst	Set	CJ	4.44	767.00	109.00	---	876.00	**1050.00**	729.20	**864.00**
Average quality workmanship	Inst	Set	CJ	3.72	638.00	91.00	---	729.00	**872.00**	606.20	**719.00**
2 ventilating lites, no fixed lites; 72" W											
36" H											
High quality workmanship	Inst	Set	CJ	4.44	650.00	109.00	---	759.00	**912.00**	627.00	**747.00**
Good quality workmanship	Inst	Set	CJ	3.48	601.00	85.10	---	686.10	**821.00**	572.00	**678.00**
Average quality workmanship	Inst	Set	CJ	2.91	501.00	71.20	---	572.20	**684.00**	475.50	**564.00**
41" H											
High quality workmanship	Inst	Set	CJ	4.71	717.00	115.00	---	832.00	**999.00**	689.30	**821.00**
Good quality workmanship	Inst	Set	CJ	3.81	663.00	93.20	---	756.20	**904.00**	628.20	**745.00**
Average quality workmanship	Inst	Set	CJ	3.14	552.00	76.80	---	628.80	**752.00**	523.10	**620.00**
48" H											
High quality workmanship	Inst	Set	CJ	5.16	778.00	126.00	---	904.00	**1090.00**	748.50	**891.00**
Good quality workmanship	Inst	Set	CJ	4.10	720.00	100.00	---	820.00	**981.00**	682.30	**809.00**
Average quality workmanship	Inst	Set	CJ	3.40	600.00	83.20	---	683.20	**817.00**	569.30	**674.00**
60" H											
High quality workmanship	Inst	Set	CJ	5.16	885.00	126.00	---	1011.00	**1210.00**	839.50	**996.00**
Good quality workmanship	Inst	Set	CJ	4.10	819.00	100.00	---	919.00	**1090.00**	767.30	**906.00**
Average quality workmanship	Inst	Set	CJ	3.40	682.00	83.20	---	765.20	**911.00**	639.30	**755.00**
4 ventilating lites, no fixed lites, 68" W											
36" H											
High quality workmanship	Inst	Set	CJ	5.16	829.00	126.00	---	955.00	**1140.00**	791.50	**941.00**
Good quality workmanship	Inst	Set	CJ	4.10	767.00	100.00	---	867.00	**1030.00**	723.30	**855.00**
Average quality workmanship	Inst	Set	CJ	3.40	638.00	83.20	---	721.20	**861.00**	601.30	**712.00**
41" H											
High quality workmanship	Inst	Set	CJ	5.52	930.00	135.00	---	1065.00	**1270.00**	886.00	**1050.00**
Good quality workmanship	Inst	Set	CJ	4.44	860.00	109.00	---	969.00	**1150.00**	808.20	**956.00**
Average quality workmanship	Inst	Set	CJ	3.72	717.00	91.00	---	808.00	**963.00**	673.20	**797.00**
48" H											
High quality workmanship	Inst	Set	CJ	6.15	1060.00	150.00	---	1210.00	**1450.00**	1009.80	**1200.00**
Good quality workmanship	Inst	Set	CJ	4.85	986.00	119.00	---	1105.00	**1310.00**	923.30	**1090.00**
Average quality workmanship	Inst	Set	CJ	4.10	821.00	100.00	---	921.00	**1100.00**	769.30	**908.00**
60" H											
High quality workmanship	Inst	Set	CJ	6.96	1320.00	170.00	---	1490.00	**1770.00**	1239.00	**1460.00**
Good quality workmanship	Inst	Set	CJ	5.52	1220.00	135.00	---	1355.00	**1600.00**	1127.10	**1330.00**
Average quality workmanship	Inst	Set	CJ	4.57	1010.00	112.00	---	1122.00	**1340.00**	941.40	**1110.00**
72" H											
High quality workmanship	Inst	Set	CJ	7.62	1440.00	186.00	---	1626.00	**1940.00**	1362.00	**1610.00**
Good quality workmanship	Inst	Set	CJ	6.15	1340.00	150.00	---	1490.00	**1770.00**	1247.80	**1470.00**
Average quality workmanship	Inst	Set	CJ	5.16	1110.00	126.00	---	1236.00	**1470.00**	1036.50	**1220.00**
4 ventilating lites, no fixed lites, 82" W											
36" H											
High quality workmanship	Inst	Set	CJ	5.16	902.00	126.00	---	1028.00	**1230.00**	854.50	**1010.00**
Good quality workmanship	Inst	Set	CJ	4.10	834.00	100.00	---	934.00	**1110.00**	780.30	**922.00**
Average quality workmanship	Inst	Set	CJ	3.40	696.00	83.20	---	779.20	**926.00**	650.30	**768.00**

			Costs Based On Small Volume							Large Volume	
Description	Oper	Unit	Crew Size	Man-Hours Per Unit	Avg Mat'l Unit Cost	Avg Labor Unit Cost	Avg Equip Unit Cost	Avg Total Unit Cost	Avg Price Incl O&P	Avg Total Unit Cost	Avg Price Incl O&P
41" H											
High quality workmanship	Inst	Set	CJ	5.52	952.00	135.00	---	1087.00	**1300.00**	905.00	**1070.00**
Good quality workmanship	Inst	Set	CJ	4.44	881.00	109.00	---	990.00	**1180.00**	827.20	**977.00**
Average quality workmanship	Inst	Set	CJ	3.72	734.00	91.00	---	825.00	**982.00**	688.20	**813.00**
48" H											
High quality workmanship	Inst	Set	CJ	6.15	1090.00	150.00	---	1240.00	**1480.00**	1033.80	**1230.00**
Good quality workmanship	Inst	Set	CJ	4.85	1010.00	119.00	---	1129.00	**1340.00**	944.30	**1110.00**
Average quality workmanship	Inst	Set	CJ	4.10	842.00	100.00	---	942.00	**1120.00**	787.30	**929.00**
60" H											
High quality workmanship	Inst	Set	CJ	6.96	1330.00	170.00	---	1500.00	**1790.00**	1249.00	**1480.00**
Good quality workmanship	Inst	Set	CJ	5.52	1230.00	135.00	---	1365.00	**1620.00**	1147.10	**1350.00**
Average quality workmanship	Inst	Set	CJ	4.57	1030.00	112.00	---	1142.00	**1350.00**	953.40	**1120.00**
72" H											
High quality workmanship	Inst	Set	CJ	7.62	1530.00	186.00	---	1716.00	**2040.00**	1432.00	**1690.00**
Good quality workmanship	Inst	Set	CJ	6.15	1410.00	150.00	---	1560.00	**1860.00**	1307.80	**1540.00**
Average quality workmanship	Inst	Set	CJ	5.16	1180.00	126.00	---	1306.00	**1550.00**	1091.50	**1290.00**
4 ventilating lites, no fixed lites, 96" W											
36" H											
High quality workmanship	Inst	Set	CJ	5.16	963.00	126.00	---	1089.00	**1300.00**	907.50	**1070.00**
Good quality workmanship	Inst	Set	CJ	4.10	892.00	100.00	---	992.00	**1180.00**	829.30	**978.00**
Average quality workmanship	Inst	Set	CJ	3.40	743.00	83.20	---	826.20	**980.00**	690.30	**814.00**
41" H											
High quality workmanship	Inst	Set	CJ	5.52	1050.00	135.00	---	1185.00	**1420.00**	991.00	**1170.00**
Good quality workmanship	Inst	Set	CJ	4.44	974.00	109.00	---	1083.00	**1290.00**	906.20	**1070.00**
Average quality workmanship	Inst	Set	CJ	3.72	812.00	91.00	---	903.00	**1070.00**	755.20	**890.00**
48" H											
High quality workmanship	Inst	Set	CJ	6.15	1150.00	150.00	---	1300.00	**1560.00**	1086.80	**1290.00**
Good quality workmanship	Inst	Set	CJ	4.85	1070.00	119.00	---	1189.00	**1410.00**	994.30	**1170.00**
Average quality workmanship	Inst	Set	CJ	4.10	890.00	100.00	---	990.00	**1180.00**	828.30	**977.00**
60" H											
High quality workmanship	Inst	Set	CJ	6.96	1450.00	170.00	---	1620.00	**1930.00**	1349.00	**1590.00**
Good quality workmanship	Inst	Set	CJ	5.52	1340.00	135.00	---	1475.00	**1750.00**	1237.10	**1460.00**
Average quality workmanship	Inst	Set	CJ	4.57	1120.00	112.00	---	1232.00	**1460.00**	1031.40	**1210.00**
72" H											
High quality workmanship	Inst	Set	CJ	7.62	1600.00	186.00	---	1786.00	**2120.00**	1492.00	**1760.00**
Good quality workmanship	Inst	Set	CJ	6.15	1480.00	150.00	---	1630.00	**1930.00**	1367.80	**1600.00**
Average quality workmanship	Inst	Set	CJ	5.16	1230.00	126.00	---	1356.00	**1610.00**	1141.50	**1340.00**
4 ventilating lites, no fixed lites, 114" W											
36" H											
High quality workmanship	Inst	Set	CJ	5.16	1050.00	126.00	---	1176.00	**1400.00**	979.50	**1160.00**
Good quality workmanship	Inst	Set	CJ	4.10	970.00	100.00	---	1070.00	**1270.00**	896.30	**1060.00**
Average quality workmanship	Inst	Set	CJ	3.40	808.00	83.20	---	891.20	**1060.00**	746.30	**879.00**
41" H											
High quality workmanship	Inst	Set	CJ	5.52	1130.00	135.00	---	1265.00	**1500.00**	1054.00	**1240.00**
Good quality workmanship	Inst	Set	CJ	4.44	1040.00	109.00	---	1149.00	**1360.00**	964.20	**1130.00**
Average quality workmanship	Inst	Set	CJ	3.72	868.00	91.00	---	959.00	**1140.00**	803.20	**946.00**
48" H											
High quality workmanship	Inst	Set	CJ	6.15	1280.00	150.00	---	1430.00	**1700.00**	1197.80	**1410.00**
Good quality workmanship	Inst	Set	CJ	4.85	1190.00	119.00	---	1309.00	**1550.00**	1098.30	**1290.00**
Average quality workmanship	Inst	Set	CJ	4.10	989.00	100.00	---	1089.00	**1290.00**	913.30	**1070.00**

			Costs Based On Small Volume							Large Volume	
Description	Oper	Unit	Crew Size	Man-Hours Per Unit	Avg Mat'l Unit Cost	Avg Labor Unit Cost	Avg Equip Unit Cost	Avg Total Unit Cost	Avg Price Incl O&P	Avg Total Unit Cost	Avg Price Incl O&P
60" H											
High quality workmanship	Inst	Set	CJ	6.96	1530.00	170.00	---	1700.00	**2020.00**	1419.00	**1670.00**
Good quality workmanship	Inst	Set	CJ	5.52	1410.00	135.00	---	1545.00	**1830.00**	1297.10	**1530.00**
Average quality workmanship	Inst	Set	CJ	4.57	1180.00	112.00	---	1292.00	**1530.00**	1082.40	**1270.00**
72" H											
High quality workmanship	Inst	Set	CJ	7.62	1740.00	186.00	---	1926.00	**2290.00**	1612.00	**1900.00**
Good quality workmanship	Inst	Set	CJ	6.15	1610.00	150.00	---	1760.00	**2080.00**	1477.80	**1740.00**
Average quality workmanship	Inst	Set	CJ	5.16	1340.00	126.00	---	1466.00	**1740.00**	1231.50	**1450.00**
4 ventilating lites, 1 fixed lite, 86" W											
36" H											
High quality workmanship	Inst	Set	CJ	6.15	1050.00	150.00	---	1200.00	**1430.00**	995.80	**1180.00**
Good quality workmanship	Inst	Set	CJ	4.85	969.00	119.00	---	1088.00	**1290.00**	908.30	**1070.00**
Average quality workmanship	Inst	Set	CJ	4.10	808.00	100.00	---	908.00	**1080.00**	757.30	**895.00**
41" H											
High quality workmanship	Inst	Set	CJ	6.96	1180.00	170.00	---	1350.00	**1610.00**	1119.00	**1320.00**
Good quality workmanship	Inst	Set	CJ	5.52	1090.00	135.00	---	1225.00	**1460.00**	1020.10	**1210.00**
Average quality workmanship	Inst	Set	CJ	4.57	907.00	112.00	---	1019.00	**1210.00**	850.40	**1000.00**
48" H											
High quality workmanship	Inst	Set	CJ	7.62	1330.00	186.00	---	1516.00	**1820.00**	1262.00	**1500.00**
Good quality workmanship	Inst	Set	CJ	6.15	1030.00	150.00	---	1180.00	**1410.00**	978.80	**1160.00**
Average quality workmanship	Inst	Set	CJ	5.16	857.00	126.00	---	983.00	**1180.00**	815.50	**968.00**
64" H											
High quality workmanship	Inst	Set	CJ	8.89	1800.00	217.00	---	2017.00	**2400.00**	1680.00	**1980.00**
Good quality workmanship	Inst	Set	CJ	6.96	1660.00	170.00	---	1830.00	**2170.00**	1542.00	**1810.00**
Average quality workmanship	Inst	Set	CJ	5.93	1390.00	145.00	---	1535.00	**1810.00**	1283.20	**1510.00**
72" H											
High quality workmanship	Inst	Set	CJ	10.0	1980.00	245.00	---	2225.00	**2650.00**	1853.00	**2200.00**
Good quality workmanship	Inst	Set	CJ	8.00	1830.00	196.00	---	2026.00	**2400.00**	1700.00	**2000.00**
Average quality workmanship	Inst	Set	CJ	6.96	1520.00	170.00	---	1690.00	**2010.00**	1419.00	**1670.00**
4 ventilating lites, 1 fixed lite, 112" W											
36" H											
High quality workmanship	Inst	Set	CJ	6.15	1130.00	150.00	---	1280.00	**1520.00**	1062.80	**1260.00**
Good quality workmanship	Inst	Set	CJ	4.85	1040.00	119.00	---	1159.00	**1380.00**	971.30	**1150.00**
Average quality workmanship	Inst	Set	CJ	4.10	868.00	100.00	---	968.00	**1150.00**	809.30	**955.00**
41" H											
High quality workmanship	Inst	Set	CJ	6.96	1200.00	170.00	---	1370.00	**1640.00**	1139.00	**1350.00**
Good quality workmanship	Inst	Set	CJ	5.52	1110.00	135.00	---	1245.00	**1480.00**	1037.10	**1230.00**
Average quality workmanship	Inst	Set	CJ	4.57	924.00	112.00	---	1036.00	**1230.00**	864.40	**1020.00**
48" H											
High quality workmanship	Inst	Set	CJ	7.62	1360.00	186.00	---	1546.00	**1850.00**	1292.00	**1530.00**
Good quality workmanship	Inst	Set	CJ	6.15	1260.00	150.00	---	1410.00	**1680.00**	1177.80	**1390.00**
Average quality workmanship	Inst	Set	CJ	5.16	1050.00	126.00	---	1176.00	**1400.00**	981.50	**1160.00**
60" H											
High quality workmanship	Inst	Set	CJ	8.89	1650.00	217.00	---	1867.00	**2230.00**	1560.00	**1840.00**
Good quality workmanship	Inst	Set	CJ	6.96	1530.00	170.00	---	1700.00	**2020.00**	1422.00	**1680.00**
Average quality workmanship	Inst	Set	CJ	5.93	1270.00	145.00	---	1415.00	**1690.00**	1183.20	**1400.00**

Description	Oper	Unit	Crew Size	Man-Hours Per Unit	Avg Mat'l Unit Cost	Avg Labor Unit Cost	Avg Equip Unit Cost	Avg Total Unit Cost	Avg Price Incl O&P	Large Volume Avg Total Unit Cost	Large Volume Avg Price Incl O&P
72" H											
High quality workmanship	Inst	Set	CJ	10.0	1820.00	245.00	---	2065.00	**2460.00**	1723.00	**2040.00**
Good quality workmanship	Inst	Set	CJ	8.00	1680.00	196.00	---	1876.00	**2230.00**	1570.00	**1860.00**
Average quality workmanship	Inst	Set	CJ	6.96	1400.00	170.00	---	1570.00	**1870.00**	1309.00	**1550.00**
4 ventilating lites, 1 fixed lite, 120" W											
36" H											
High quality workmanship	Inst	Set	CJ	6.15	1200.00	150.00	---	1350.00	**1610.00**	1127.80	**1330.00**
Good quality workmanship	Inst	Set	CJ	4.85	1110.00	119.00	---	1229.00	**1460.00**	1029.30	**1210.00**
Average quality workmanship	Inst	Set	CJ	4.10	924.00	100.00	---	1024.00	**1220.00**	857.30	**1010.00**
41" H											
High quality workmanship	Inst	Set	CJ	6.96	1320.00	170.00	---	1490.00	**1780.00**	1239.00	**1470.00**
Good quality workmanship	Inst	Set	CJ	5.52	1220.00	135.00	---	1355.00	**1610.00**	1137.10	**1340.00**
Average quality workmanship	Inst	Set	CJ	4.57	1020.00	112.00	---	1132.00	**1340.00**	946.40	**1110.00**
48" H											
High quality workmanship	Inst	Set	CJ	7.62	1440.00	186.00	---	1626.00	**1940.00**	1362.00	**1610.00**
Good quality workmanship	Inst	Set	CJ	6.15	1340.00	150.00	---	1490.00	**1770.00**	1247.80	**1470.00**
Average quality workmanship	Inst	Set	CJ	5.16	1110.00	126.00	---	1236.00	**1470.00**	1036.50	**1220.00**
60" H											
High quality workmanship	Inst	Set	CJ	8.89	1660.00	217.00	---	1877.00	**2240.00**	1570.00	**1850.00**
Good quality workmanship	Inst	Set	CJ	6.96	1540.00	170.00	---	1710.00	**2030.00**	1432.00	**1690.00**
Average quality workmanship	Inst	Set	CJ	5.93	1280.00	145.00	---	1425.00	**1700.00**	1193.20	**1410.00**
72" H											
High quality workmanship	Inst	Set	CJ	10.00	1940.00	245.00	---	2185.00	**2610.00**	1833.00	**2160.00**
Good quality workmanship	Inst	Set	CJ	8.00	1800.00	196.00	---	1996.00	**2370.00**	1670.00	**1970.00**
Average quality workmanship	Inst	Set	CJ	6.96	1500.00	170.00	---	1670.00	**1980.00**	1389.00	**1640.00**
4 ventilating lites, 1 fixed lite, 142" W											
36" H											
High quality workmanship	Inst	Set	CJ	6.15	1310.00	150.00	---	1460.00	**1740.00**	1217.80	**1440.00**
Good quality workmanship	Inst	Set	CJ	4.85	1210.00	119.00	---	1329.00	**1580.00**	1118.30	**1310.00**
Average quality workmanship	Inst	Set	CJ	4.10	1010.00	100.00	---	1110.00	**1320.00**	932.30	**1100.00**
41" H											
High quality workmanship	Inst	Set	CJ	6.15	1410.00	150.00	---	1560.00	**1850.00**	1307.80	**1540.00**
Good quality workmanship	Inst	Set	CJ	4.85	1310.00	119.00	---	1429.00	**1680.00**	1198.30	**1410.00**
Average quality workmanship	Inst	Set	CJ	4.10	1090.00	100.00	---	1190.00	**1400.00**	998.30	**1170.00**
48" H											
High quality workmanship	Inst	Set	CJ	7.62	1610.00	186.00	---	1796.00	**2130.00**	1502.00	**1770.00**
Good quality workmanship	Inst	Set	CJ	6.15	1490.00	150.00	---	1640.00	**1940.00**	1367.80	**1610.00**
Average quality workmanship	Inst	Set	CJ	5.16	1240.00	126.00	---	1366.00	**1620.00**	1141.50	**1350.00**
60" H											
High quality workmanship	Inst	Set	CJ	8.89	1930.00	217.00	---	2147.00	**2550.00**	1790.00	**2110.00**
Good quality workmanship	Inst	Set	CJ	6.96	1780.00	170.00	---	1950.00	**2310.00**	1642.00	**1930.00**
Average quality workmanship	Inst	Set	CJ	5.93	1490.00	145.00	---	1635.00	**1930.00**	1363.20	**1610.00**
72" H											
High quality workmanship	Inst	Set	CJ	10.0	2210.00	245.00	---	2455.00	**2910.00**	2053.00	**2420.00**
Good quality workmanship	Inst	Set	CJ	8.00	2040.00	196.00	---	2236.00	**2650.00**	1880.00	**2210.00**
Average quality workmanship	Inst	Set	CJ	6.96	1700.00	170.00	---	1870.00	**2220.00**	1569.00	**1840.00**
For aluminum clad (baked white enamel) exterior											
ADD	Inst	%		---	20.0	---	---	---	**---**	---	**---**

Description	Oper	Unit	Crew Size	Man-Hours Per Unit	Avg Mat'l Unit Cost	Avg Labor Unit Cost	Avg Equip Unit Cost	Avg Total Unit Cost	Avg Price Incl O&P	Large Volume Avg Total Unit Cost	Large Volume Avg Price Incl O&P
Double hung windows; glazed 7/16" T insulating glass; two lites per unit; unit includes screens and weatherstripping; rough opening sizes											
1'-6" x 2'-0", 2'-6", 3'-0", 3'-6"											
High quality workmanship	Inst	Set	CA	3.81	269.00	103.00	---	372.00	**465.00**	297.40	**367.00**
Good quality workmanship	Inst	Set	CA	3.08	248.00	83.00	---	331.00	**411.00**	265.90	**326.00**
Average quality workmanship	Inst	Set	CA	2.58	220.00	69.60	---	289.60	**358.00**	233.00	**285.00**
1'-6" x 4'-0", 4'-6", 5'-0", 5'-6"											
High quality workmanship	Inst	Set	CA	3.81	385.00	103.00	---	488.00	**599.00**	397.40	**482.00**
Good quality workmanship	Inst	Set	CA	3.08	351.00	83.00	---	434.00	**529.00**	353.90	**428.00**
Average quality workmanship	Inst	Set	CA	2.58	305.00	69.60	---	374.60	**456.00**	306.00	**369.00**
2'-0" x 2'-0", 2'-6", 3'-0", 3'-6"											
High quality workmanship	Inst	Set	CA	3.81	278.00	103.00	---	381.00	**476.00**	305.40	**376.00**
Good quality workmanship	Inst	Set	CA	3.08	245.00	83.00	---	328.00	**408.00**	263.90	**324.00**
Average quality workmanship	Inst	Set	CA	2.58	230.00	69.60	---	299.60	**370.00**	242.00	**295.00**
2'-0" x 4'-0", 4'-6", 5'-0", 6'-0"											
High quality workmanship	Inst	Set	CA	3.81	365.00	103.00	---	468.00	**576.00**	380.40	**462.00**
Good quality workmanship	Inst	Set	CA	3.08	338.00	83.00	---	421.00	**515.00**	343.90	**415.00**
Average quality workmanship	Inst	Set	CA	2.58	314.00	69.60	---	383.60	**466.00**	314.00	**378.00**
2'-6" x 2'-0", 2'-6", 3'-0", 3'-6"											
High quality workmanship	Inst	Set	CA	3.81	289.00	103.00	---	392.00	**488.00**	315.40	**387.00**
Good quality workmanship	Inst	Set	CA	3.08	268.00	83.00	---	351.00	**434.00**	282.90	**346.00**
Average quality workmanship	Inst	Set	CA	2.58	237.00	69.60	---	306.60	**379.00**	249.00	**302.00**
2'-6" x 4'-0", 4'-6", 5'-0", 6'-0"											
High quality workmanship	Inst	Set	CA	3.81	408.00	103.00	---	511.00	**625.00**	416.40	**504.00**
Good quality workmanship	Inst	Set	CA	3.08	365.00	83.00	---	448.00	**546.00**	366.90	**442.00**
Average quality workmanship	Inst	Set	CA	2.58	342.00	69.60	---	411.60	**499.00**	338.00	**405.00**
3'-0" x 2'-0", 2'-6", 3'-0", 3'-6"											
High quality workmanship	Inst	Set	CA	3.81	335.00	103.00	---	438.00	**541.00**	354.40	**433.00**
Good quality workmanship	Inst	Set	CA	3.08	282.00	83.00	---	365.00	**451.00**	295.90	**360.00**
Average quality workmanship	Inst	Set	CA	2.58	258.00	69.60	---	327.60	**402.00**	266.00	**322.00**
3'-0" x 4'-0", 4'-6", 5'-0", 6'-0"											
High quality workmanship	Inst	Set	CA	3.81	449.00	103.00	---	552.00	**673.00**	452.40	**545.00**
Good quality workmanship	Inst	Set	CA	3.08	421.00	83.00	---	504.00	**611.00**	414.90	**497.00**
Average quality workmanship	Inst	Set	CA	2.58	372.00	69.60	---	441.60	**533.00**	364.00	**435.00**
3'-6" x 3'-0", 3'-6", 4'-0"											
High quality workmanship	Inst	Set	CA	3.81	401.00	103.00	---	504.00	**617.00**	411.40	**498.00**
Good quality workmanship	Inst	Set	CA	3.08	340.00	83.00	---	423.00	**518.00**	345.90	**418.00**
Average quality workmanship	Inst	Set	CA	2.58	301.00	69.60	---	370.60	**452.00**	303.00	**365.00**
3'-6" x 4'-6", 5'-0", 6'-0"											
High quality workmanship	Inst	Set	CA	3.81	551.00	103.00	---	654.00	**790.00**	539.40	**646.00**
Good quality workmanship	Inst	Set	CA	3.08	505.00	83.00	---	588.00	**707.00**	486.90	**580.00**
Average quality workmanship	Inst	Set	CA	2.58	451.00	69.60	---	520.60	**625.00**	432.00	**513.00**
4'-0" x 3'-0", 3'-6", 4'-0"											
High quality workmanship	Inst	Set	CA	3.81	585.00	103.00	---	688.00	**828.00**	568.40	**679.00**
Good quality workmanship	Inst	Set	CA	3.08	470.00	83.00	---	553.00	**667.00**	456.90	**546.00**
Average quality workmanship	Inst	Set	CA	2.58	352.00	69.60	---	421.60	**510.00**	346.00	**415.00**
4'-0" x 4'-6", 5'-0", 6'-0"											
High quality workmanship	Inst	Set	CA	3.81	616.00	103.00	---	719.00	**865.00**	595.40	**710.00**
Good quality workmanship	Inst	Set	CA	3.08	560.00	83.00	---	643.00	**770.00**	533.90	**634.00**
Average quality workmanship	Inst	Set	CA	2.58	492.00	69.60	---	561.60	**671.00**	466.00	**553.00**

Description	Oper	Unit	Crew Size	Man-Hours Per Unit	Avg Mat'l Unit Cost	Avg Labor Unit Cost	Avg Equip Unit Cost	Avg Total Unit Cost	Avg Price Incl O&P	Large Volume Avg Total Unit Cost	Large Volume Avg Price Incl O&P
4'-6" x 4'-0", 4'-6											
High quality workmanship	Inst	Set	CA	3.81	650.00	103.00	---	753.00	**903.00**	624.40	**743.00**
Good quality workmanship	Inst	Set	CA	3.08	504.00	83.00	---	587.00	**706.00**	485.90	**579.00**
Average quality workmanship	Inst	Set	CA	2.58	423.00	69.60	---	492.60	**593.00**	408.00	**486.00**
4'-6" x 5'-0", 6'-0"											
High quality workmanship	Inst	Set	CA	3.81	706.00	103.00	---	809.00	**968.00**	672.40	**798.00**
Good quality workmanship	Inst	Set	CA	3.08	582.00	83.00	---	665.00	**796.00**	552.90	**656.00**
Average quality workmanship	Inst	Set	CA	2.58	523.00	69.60	---	592.60	**707.00**	493.00	**584.00**
5'-0" x 4'-0", 4'-6											
High quality workmanship	Inst	Set	CA	3.81	762.00	103.00	---	865.00	**1030.00**	720.40	**853.00**
Good quality workmanship	Inst	Set	CA	3.08	661.00	83.00	---	744.00	**886.00**	619.90	**733.00**
Average quality workmanship	Inst	Set	CA	2.58	457.00	69.60	---	526.60	**631.00**	437.00	**519.00**
5'-0" x 5'-0", 6'-0"											
High quality workmanship	Inst	Set	CA	3.81	818.00	103.00	---	921.00	**1100.00**	768.40	**908.00**
Good quality workmanship	Inst	Set	CA	3.08	706.00	83.00	---	789.00	**938.00**	658.90	**777.00**
Average quality workmanship	Inst	Set	CA	2.58	557.00	69.60	---	626.60	**746.00**	522.00	**617.00**
Picture windows											
Unit with one fixed lite, 3/4" T insulating glass, frame sizes											
4'-0" W x 3'-6" H; 3'-6" W x 4'-0" H											
High quality workmanship	Inst	Ea	CJ	2.81	340.00	68.70	---	408.70	**496.00**	336.50	**403.00**
Good quality workmanship	Inst	Ea	CJ	2.22	314.00	54.30	---	368.30	**443.00**	304.50	**363.00**
Average quality workmanship	Inst	Ea	CJ	1.86	262.00	45.50	---	307.50	**371.00**	254.60	**303.00**
4'-8" W x 3'-6" H; 4'-0" W x 4'-0" H; 3'-6" W x 4'-8"H											
High quality workmanship	Inst	Ea	CJ	2.81	392.00	68.70	---	460.70	**555.00**	380.50	**454.00**
Good quality workmanship	Inst	Ea	CJ	2.22	363.00	54.30	---	417.30	**500.00**	346.50	**412.00**
Average quality workmanship	Inst	Ea	CJ	1.86	302.00	45.50	---	347.50	**417.00**	288.60	**343.00**
4'-8" W x 4'-0" H; 4'-0" W x 4'-10" H											
High quality workmanship	Inst	Ea	CJ	2.81	431.00	68.70	---	499.70	**600.00**	414.50	**493.00**
Good quality workmanship	Inst	Ea	CJ	2.22	399.00	54.30	---	453.30	**541.00**	377.50	**447.00**
Average quality workmanship	Inst	Ea	CJ	1.86	333.00	45.50	---	378.50	**452.00**	314.60	**373.00**
4'-8" W x 4'-8" H											
High quality workmanship	Inst	Ea	CJ	3.08	470.00	75.30	---	545.30	**655.00**	451.90	**538.00**
Good quality workmanship	Inst	Ea	CJ	2.46	435.00	60.20	---	495.20	**591.00**	411.10	**488.00**
Average quality workmanship	Inst	Ea	CJ	2.05	363.00	50.10	---	413.10	**494.00**	343.50	**407.00**
4'-8" W x 5'-6" H; 5'-4" W x 4'-8" H											
High quality workmanship	Inst	Ea	CJ	3.08	594.00	75.30	---	669.30	**797.00**	557.90	**659.00**
Good quality workmanship	Inst	Ea	CJ	2.46	549.00	60.20	---	609.20	**723.00**	509.10	**600.00**
Average quality workmanship	Inst	Ea	CJ	2.05	457.00	50.10	---	507.10	**602.00**	424.50	**500.00**
5'-4" W x 5'-6" H											
High quality workmanship	Inst	Ea	CJ	3.08	672.00	75.30	---	747.30	**887.00**	624.90	**737.00**
Good quality workmanship	Inst	Ea	CJ	2.46	622.00	60.20	---	682.20	**806.00**	572.10	**672.00**
Average quality workmanship	Inst	Ea	CJ	2.05	517.00	50.10	---	567.10	**671.00**	476.50	**559.00**
Unit with one fixed lite, glazed 3/4" T insulating glass and two glazed S.S.B. casement flankers, 15", 19" or 23" W; unit includes: hardware, casing and molding; rough opening sizes											
40" H											
88" W											
High quality workmanship	Inst	Set	CJ	3.81	913.00	93.20	---	1006.20	**1190.00**	843.20	**993.00**
Good quality workmanship	Inst	Set	CJ	3.08	844.00	75.30	---	919.30	**1090.00**	772.90	**907.00**
Average quality workmanship	Inst	Set	CJ	2.58	703.00	63.10	---	766.10	**905.00**	643.90	**755.00**

Description	Oper	Unit	Crew Size	Man-Hours Per Unit	Avg Mat'l Unit Cost	Avg Labor Unit Cost	Avg Equip Unit Cost	Avg Total Unit Cost	Avg Price Incl O&P	Avg Total Unit Cost	Avg Price Incl O&P
			Costs Based On Small Volume							Large Volume	
96" W											
High quality workmanship	Inst	Set	CJ	3.81	924.00	93.20	---	1017.20	**1200.00**	853.20	**1000.00**
Good quality workmanship	Inst	Set	CJ	3.08	855.00	75.30	---	930.30	**1100.00**	780.90	**917.00**
Average quality workmanship	Inst	Set	CJ	2.58	712.00	63.10	---	775.10	**915.00**	651.90	**764.00**
104" W											
High quality workmanship	Inst	Set	CJ	3.81	1050.00	93.20	---	1143.20	**1350.00**	963.20	**1130.00**
Good quality workmanship	Inst	Set	CJ	3.08	974.00	75.30	---	1049.30	**1240.00**	883.90	**1030.00**
Average quality workmanship	Inst	Set	CJ	2.58	812.00	63.10	---	875.10	**1030.00**	736.90	**862.00**
112" W											
High quality workmanship	Inst	Set	CJ	3.81	1060.00	93.20	---	1153.20	**1370.00**	973.20	**1140.00**
Good quality workmanship	Inst	Set	CJ	3.08	984.00	75.30	---	1059.30	**1250.00**	892.90	**1040.00**
Average quality workmanship	Inst	Set	CJ	2.58	818.00	63.10	---	881.10	**1040.00**	741.90	**868.00**
48" H											
88" W											
High quality workmanship	Inst	Set	CJ	3.81	1010.00	93.20	---	1103.20	**1310.00**	930.20	**1090.00**
Good quality workmanship	Inst	Set	CJ	3.08	937.00	75.30	---	1012.30	**1190.00**	852.90	**998.00**
Average quality workmanship	Inst	Set	CJ	2.58	782.00	63.10	---	845.10	**995.00**	710.90	**833.00**
96" W											
High quality workmanship	Inst	Set	CJ	4.44	1020.00	109.00	---	1129.00	**1340.00**	948.00	**1120.00**
Good quality workmanship	Inst	Set	CJ	3.48	949.00	85.10	---	1034.10	**1220.00**	869.00	**1020.00**
Average quality workmanship	Inst	Set	CJ	2.91	791.00	71.20	---	862.20	**1020.00**	724.50	**850.00**
104" W											
High quality workmanship	Inst	Set	CJ	4.44	1190.00	109.00	---	1299.00	**1540.00**	1090.00	**1280.00**
Good quality workmanship	Inst	Set	CJ	3.48	1100.00	85.10	---	1185.10	**1400.00**	1003.00	**1170.00**
Average quality workmanship	Inst	Set	CJ	2.91	920.00	71.20	---	991.20	**1170.00**	834.50	**977.00**
112" W											
High quality workmanship	Inst	Set	CJ	4.44	1200.00	109.00	---	1309.00	**1550.00**	1100.00	**1290.00**
Good quality workmanship	Inst	Set	CJ	3.48	1110.00	85.10	---	1195.10	**1410.00**	1011.00	**1180.00**
Average quality workmanship	Inst	Set	CJ	2.91	928.00	71.20	---	999.20	**1180.00**	842.50	**986.00**
60" H											
88" W											
High quality workmanship	Inst	Set	CJ	4.44	1200.00	109.00	---	1309.00	**1540.00**	1100.00	**1290.00**
Good quality workmanship	Inst	Set	CJ	3.48	1110.00	85.10	---	1195.10	**1400.00**	1006.00	**1180.00**
Average quality workmanship	Inst	Set	CJ	2.91	924.00	71.20	---	995.20	**1170.00**	838.50	**981.00**
96" W											
High quality workmanship	Inst	Set	CJ	4.44	1210.00	109.00	---	1319.00	**1560.00**	1110.00	**1300.00**
Good quality workmanship	Inst	Set	CJ	3.48	1120.00	85.10	---	1205.10	**1420.00**	1016.00	**1190.00**
Average quality workmanship	Inst	Set	CJ	2.91	933.00	71.20	---	1004.20	**1180.00**	846.50	**990.00**
104" W											
High quality workmanship	Inst	Set	CJ	5.16	1520.00	126.00	---	1646.00	**1940.00**	1381.50	**1620.00**
Good quality workmanship	Inst	Set	CJ	4.10	1400.00	100.00	---	1500.00	**1770.00**	1265.30	**1480.00**
Average quality workmanship	Inst	Set	CJ	3.40	1170.00	83.20	---	1253.20	**1470.00**	1054.30	**1240.00**
112" W											
High quality workmanship	Inst	Set	CJ	5.16	1530.00	126.00	---	1656.00	**1950.00**	1391.50	**1630.00**
Good quality workmanship	Inst	Set	CJ	4.10	1410.00	100.00	---	1510.00	**1780.00**	1275.30	**1490.00**
Average quality workmanship	Inst	Set	CJ	3.40	1180.00	83.20	---	1263.20	**1480.00**	1064.30	**1250.00**
72" H											
88" W											
High quality workmanship	Inst	Set	CJ	5.16	1410.00	126.00	---	1536.00	**1810.00**	1291.50	**1510.00**
Good quality workmanship	Inst	Set	CJ	4.10	1310.00	100.00	---	1410.00	**1650.00**	1185.30	**1390.00**
Average quality workmanship	Inst	Set	CJ	3.40	1090.00	83.20	---	1173.20	**1380.00**	987.30	**1160.00**

Description	Oper	Unit	Crew Size	Man-Hours Per Unit	Avg Mat'l Unit Cost	Avg Labor Unit Cost	Avg Equip Unit Cost	Avg Total Unit Cost	Avg Price Incl O&P	Large Volume Avg Total Unit Cost	Large Volume Avg Price Incl O&P
96" W											
High quality workmanship	Inst	Set	CJ	5.16	1420.00	126.00	---	1546.00	**1830.00**	1301.50	**1530.00**
Good quality workmanship	Inst	Set	CJ	4.10	1320.00	100.00	---	1420.00	**1670.00**	1195.30	**1400.00**
Average quality workmanship	Inst	Set	CJ	3.40	1100.00	83.20	---	1183.20	**1390.00**	994.30	**1160.00**
Material adjustments											
For glazed 1/2" insulated flankers											
ADD	Inst	%		---	15.0	---	---	---	**---**	---	**---**
For screen per flanker, ADD	Inst	Ea	---	---	5.60	---	---	5.60	**6.44**	4.80	**5.52**

Garden windows, unit extends out 13-1/4" from wall, bronze tempered glass in top panels of all units, wood shelves included

1 fixed lite

Description	Oper	Unit	Crew Size	Man-Hours Per Unit	Avg Mat'l Unit Cost	Avg Labor Unit Cost	Avg Equip Unit Cost	Avg Total Unit Cost	Avg Price Incl O&P	Large Volume Avg Total Unit Cost	Large Volume Avg Price Incl O&P
Single glazed											
3'-0" x 3'-0" H											
High quality workmanship	Inst	Ea	CJ	6.96	467.00	170.00	---	637.00	**796.00**	509.00	**625.00**
Good quality workmanship	Inst	Ea	CJ	5.52	428.00	135.00	---	563.00	**697.00**	454.10	**554.00**
Average quality workmanship	Inst	Ea	CJ	4.57	357.00	112.00	---	469.00	**581.00**	378.40	**462.00**
3'-0" x 4'-0" H											
High quality workmanship	Inst	Ea	CJ	6.96	506.00	170.00	---	676.00	**841.00**	543.00	**664.00**
Good quality workmanship	Inst	Ea	CJ	5.52	465.00	135.00	---	600.00	**740.00**	485.10	**590.00**
Average quality workmanship	Inst	Ea	CJ	4.57	389.00	112.00	---	501.00	**617.00**	405.40	**493.00**
4'-0" x 3'-0" H											
High quality workmanship	Inst	Ea	CJ	6.96	529.00	170.00	---	699.00	**867.00**	562.00	**686.00**
Good quality workmanship	Inst	Ea	CJ	5.52	485.00	135.00	---	620.00	**763.00**	503.10	**610.00**
Average quality workmanship	Inst	Ea	CJ	4.57	404.00	112.00	---	516.00	**635.00**	419.40	**509.00**
4'-0" x 4'-0" H											
High quality workmanship	Inst	Ea	CJ	7.62	542.00	186.00	---	728.00	**907.00**	587.00	**720.00**
Good quality workmanship	Inst	Ea	CJ	6.15	497.00	150.00	---	647.00	**800.00**	523.80	**639.00**
Average quality workmanship	Inst	Ea	CJ	5.16	414.00	126.00	---	540.00	**668.00**	436.50	**532.00**
4'-0" x 5'-0" H											
High quality workmanship	Inst	Ea	CJ	8.89	581.00	217.00	---	798.00	**999.00**	638.00	**785.00**
Good quality workmanship	Inst	Ea	CJ	6.96	533.00	170.00	---	703.00	**872.00**	569.00	**695.00**
Average quality workmanship	Inst	Ea	CJ	5.93	445.00	145.00	---	590.00	**732.00**	474.20	**580.00**
5'-0" x 3'-0" H											
High quality workmanship	Inst	Ea	CJ	7.62	540.00	186.00	---	726.00	**904.00**	585.00	**718.00**
Good quality workmanship	Inst	Ea	CJ	6.15	495.00	150.00	---	645.00	**798.00**	521.80	**637.00**
Average quality workmanship	Inst	Ea	CJ	5.16	413.00	126.00	---	539.00	**667.00**	435.50	**531.00**
5'-0" x 3'-8" H											
High quality workmanship	Inst	Ea	CJ	8.89	572.00	217.00	---	789.00	**989.00**	631.00	**776.00**
Good quality workmanship	Inst	Ea	CJ	6.96	525.00	170.00	---	695.00	**863.00**	562.00	**688.00**
Average quality workmanship	Inst	Ea	CJ	5.93	438.00	145.00	---	583.00	**724.00**	468.20	**573.00**
5'-0" x 4'-0" H											
High quality workmanship	Inst	Ea	CJ	8.89	581.00	217.00	---	798.00	**999.00**	638.00	**785.00**
Good quality workmanship	Inst	Ea	CJ	6.96	533.00	170.00	---	703.00	**872.00**	569.00	**695.00**
Average quality workmanship	Inst	Ea	CJ	5.93	445.00	145.00	---	590.00	**732.00**	474.20	**580.00**
5'-0" x 5'-0" H											
High quality workmanship	Inst	Ea	CJ	10.0	633.00	245.00	---	878.00	**1100.00**	705.00	**872.00**
Good quality workmanship	Inst	Ea	CJ	8.00	580.00	196.00	---	776.00	**965.00**	627.00	**770.00**
Average quality workmanship	Inst	Ea	CJ	6.96	484.00	170.00	---	654.00	**815.00**	524.00	**642.00**

Description	Oper	Unit	Crew Size	Man-Hours Per Unit	Avg Mat'l Unit Cost	Avg Labor Unit Cost	Avg Equip Unit Cost	Avg Total Unit Cost	Avg Price Incl O&P	Large Volume Avg Total Unit Cost	Large Volume Avg Price Incl O&P
6'-0" x 3'-0" H											
High quality workmanship	Inst	Ea	CJ	8.89	577.00	217.00	---	794.00	**994.00**	634.00	**781.00**
Good quality workmanship	Inst	Ea	CJ	6.96	529.00	170.00	---	699.00	**867.00**	565.00	**691.00**
Average quality workmanship	Inst	Ea	CJ	5.93	441.00	145.00	---	586.00	**728.00**	471.20	**577.00**
6'-0" x 3'-8" H											
High quality workmanship	Inst	Ea	CJ	9.41	603.00	230.00	---	833.00	**1040.00**	666.00	**823.00**
Good quality workmanship	Inst	Ea	CJ	7.62	553.00	186.00	---	739.00	**920.00**	596.00	**731.00**
Average quality workmanship	Inst	Ea	CJ	6.40	461.00	157.00	---	618.00	**769.00**	499.00	**611.00**
6'-0" x 4'-0" H											
High quality workmanship	Inst	Ea	CJ	10.0	628.00	245.00	---	873.00	**1090.00**	702.00	**867.00**
Good quality workmanship	Inst	Ea	CJ	8.00	576.00	196.00	---	772.00	**959.00**	623.00	**766.00**
Average quality workmanship	Inst	Ea	CJ	6.96	480.00	170.00	---	650.00	**811.00**	521.00	**639.00**
Dual glazed											
3'-0" x 3'-0" H											
High quality workmanship	Inst	Ea	CJ	6.96	640.00	170.00	---	810.00	**994.00**	657.00	**795.00**
Good quality workmanship	Inst	Ea	CJ	5.52	587.00	135.00	---	722.00	**880.00**	590.10	**711.00**
Average quality workmanship	Inst	Ea	CJ	4.57	489.00	112.00	---	601.00	**733.00**	492.40	**592.00**
3'-0" x 4'-0" H											
High quality workmanship	Inst	Ea	CJ	6.96	704.00	170.00	---	874.00	**1070.00**	713.00	**859.00**
Good quality workmanship	Inst	Ea	CJ	5.52	646.00	135.00	---	781.00	**948.00**	641.10	**769.00**
Average quality workmanship	Inst	Ea	CJ	4.57	539.00	112.00	---	651.00	**789.00**	534.40	**641.00**
4'-0" x 3'-0" H											
High quality workmanship	Inst	Ea	CJ	6.96	698.00	170.00	---	868.00	**1060.00**	707.00	**853.00**
Good quality workmanship	Inst	Ea	CJ	5.52	641.00	135.00	---	776.00	**942.00**	636.10	**764.00**
Average quality workmanship	Inst	Ea	CJ	4.57	534.00	112.00	---	646.00	**784.00**	530.40	**637.00**
4'-0" x 4'-0" H											
High quality workmanship	Inst	Ea	CJ	7.62	757.00	186.00	---	943.00	**1150.00**	771.00	**932.00**
Good quality workmanship	Inst	Ea	CJ	6.15	694.00	150.00	---	844.00	**1030.00**	692.80	**833.00**
Average quality workmanship	Inst	Ea	CJ	5.16	579.00	126.00	---	705.00	**858.00**	577.50	**695.00**
4'-0" x 5'-0" H											
High quality workmanship	Inst	Ea	CJ	8.89	831.00	217.00	---	1048.00	**1290.00**	852.00	**1030.00**
Good quality workmanship	Inst	Ea	CJ	6.96	763.00	170.00	---	933.00	**1140.00**	766.00	**922.00**
Average quality workmanship	Inst	Ea	CJ	5.93	636.00	145.00	---	781.00	**952.00**	638.20	**769.00**
5'-0" x 3'-0" H											
High quality workmanship	Inst	Ea	CJ	7.62	750.00	186.00	---	936.00	**1150.00**	765.00	**926.00**
Good quality workmanship	Inst	Ea	CJ	6.15	688.00	150.00	---	838.00	**1020.00**	686.80	**827.00**
Average quality workmanship	Inst	Ea	CJ	5.16	573.00	126.00	---	699.00	**851.00**	573.50	**689.00**
5'-0" x 3'-8" H											
High quality workmanship	Inst	Ea	CJ	8.89	801.00	217.00	---	1018.00	**1250.00**	826.00	**1000.00**
Good quality workmanship	Inst	Ea	CJ	6.96	735.00	170.00	---	905.00	**1100.00**	742.00	**894.00**
Average quality workmanship	Inst	Ea	CJ	5.93	613.00	145.00	---	758.00	**925.00**	618.20	**746.00**
5'-0" x 4'-0" H											
High quality workmanship	Inst	Ea	CJ	8.89	818.00	217.00	---	1035.00	**1270.00**	841.00	**1020.00**
Good quality workmanship	Inst	Ea	CJ	6.96	750.00	170.00	---	920.00	**1120.00**	755.00	**910.00**
Average quality workmanship	Inst	Ea	CJ	5.93	625.00	145.00	---	770.00	**939.00**	629.20	**758.00**
5'-0" x 5'-0" H											
High quality workmanship	Inst	Ea	CJ	10.0	909.00	245.00	---	1154.00	**1420.00**	943.00	**1140.00**
Good quality workmanship	Inst	Ea	CJ	8.00	834.00	196.00	---	1030.00	**1260.00**	845.00	**1020.00**
Average quality workmanship	Inst	Ea	CJ	6.96	696.00	170.00	---	866.00	**1060.00**	705.00	**851.00**

Description	Oper	Unit	Crew Size	Man-Hours Per Unit	Avg Mat'l Unit Cost	Avg Labor Unit Cost	Avg Equip Unit Cost	Avg Total Unit Cost	Avg Price Incl O&P	Large Volume Avg Total Unit Cost	Large Volume Avg Price Incl O&P
6'-0" x 3'-0" H											
High quality workmanship	Inst	Ea	CJ	8.89	818.00	217.00	---	1035.00	**1270.00**	841.00	**1020.00**
Good quality workmanship	Inst	Ea	CJ	6.96	750.00	170.00	---	920.00	**1120.00**	755.00	**910.00**
Average quality workmanship	Inst	Ea	CJ	5.93	625.00	145.00	---	770.00	**939.00**	629.20	**758.00**
6'-0" x 3'-8" H											
High quality workmanship	Inst	Ea	CJ	9.41	862.00	230.00	---	1092.00	**1340.00**	889.00	**1080.00**
Good quality workmanship	Inst	Ea	CJ	7.62	791.00	186.00	---	977.00	**1190.00**	800.00	**965.00**
Average quality workmanship	Inst	Ea	CJ	6.40	660.00	157.00	---	817.00	**997.00**	668.00	**807.00**
6'-0" x 4'-0" H											
High quality workmanship	Inst	Ea	CJ	10.0	907.00	245.00	---	1152.00	**1410.00**	941.00	**1140.00**
Good quality workmanship	Inst	Ea	CJ	8.00	832.00	196.00	---	1028.00	**1250.00**	843.00	**1020.00**
Average quality workmanship	Inst	Ea	CJ	6.96	693.00	170.00	---	863.00	**1060.00**	703.00	**848.00**
1 fixed lite, 1 sliding lite											
Single glazed											
3'-0" x 3'-0" H											
High quality workmanship	Inst	Ea	CJ	6.96	596.00	170.00	---	766.00	**944.00**	620.00	**752.00**
Good quality workmanship	Inst	Ea	CJ	5.52	547.00	135.00	---	682.00	**834.00**	555.10	**671.00**
Average quality workmanship	Inst	Ea	CJ	4.57	456.00	112.00	---	568.00	**694.00**	463.40	**559.00**
3'-0" x 4'-0" H											
High quality workmanship	Inst	Ea	CJ	6.96	655.00	170.00	---	825.00	**1010.00**	671.00	**811.00**
Good quality workmanship	Inst	Ea	CJ	5.52	600.00	135.00	---	735.00	**896.00**	602.10	**724.00**
Average quality workmanship	Inst	Ea	CJ	4.57	501.00	112.00	---	613.00	**746.00**	501.40	**604.00**
4'-0" x 3'-0" H											
High quality workmanship	Inst	Ea	CJ	6.96	641.00	170.00	---	811.00	**995.00**	658.00	**797.00**
Good quality workmanship	Inst	Ea	CJ	5.52	587.00	135.00	---	722.00	**880.00**	590.10	**711.00**
Average quality workmanship	Inst	Ea	CJ	4.57	489.00	112.00	---	601.00	**733.00**	492.40	**592.00**
4'-0" x 4'-0" H											
High quality workmanship	Inst	Ea	CJ	7.62	693.00	186.00	---	879.00	**1080.00**	716.00	**869.00**
Good quality workmanship	Inst	Ea	CJ	6.15	637.00	150.00	---	787.00	**961.00**	643.80	**777.00**
Average quality workmanship	Inst	Ea	CJ	5.16	531.00	126.00	---	657.00	**802.00**	536.50	**647.00**
4'-0" x 5'-0" H											
High quality workmanship	Inst	Ea	CJ	8.89	763.00	217.00	---	980.00	**1210.00**	794.00	**964.00**
Good quality workmanship	Inst	Ea	CJ	6.96	700.00	170.00	---	870.00	**1060.00**	712.00	**860.00**
Average quality workmanship	Inst	Ea	CJ	5.93	584.00	145.00	---	729.00	**891.00**	593.20	**717.00**
5'-0" x 3'-0" H											
High quality workmanship	Inst	Ea	CJ	7.62	684.00	186.00	---	870.00	**1070.00**	709.00	**860.00**
Good quality workmanship	Inst	Ea	CJ	6.15	627.00	150.00	---	777.00	**950.00**	635.80	**767.00**
Average quality workmanship	Inst	Ea	CJ	5.16	523.00	126.00	---	649.00	**793.00**	529.50	**639.00**
5'-0" x 3'-8" H											
High quality workmanship	Inst	Ea	CJ	8.89	734.00	217.00	---	951.00	**1170.00**	769.00	**935.00**
Good quality workmanship	Inst	Ea	CJ	6.96	673.00	170.00	---	843.00	**1030.00**	689.00	**833.00**
Average quality workmanship	Inst	Ea	CJ	5.93	561.00	145.00	---	706.00	**866.00**	574.20	**695.00**
5'-0" x 4'-0" H											
High quality workmanship	Inst	Ea	CJ	8.89	741.00	217.00	---	958.00	**1180.00**	776.00	**943.00**
Good quality workmanship	Inst	Ea	CJ	6.96	680.00	170.00	---	850.00	**1040.00**	695.00	**840.00**
Average quality workmanship	Inst	Ea	CJ	5.93	567.00	145.00	---	712.00	**872.00**	579.20	**700.00**

			Costs Based On Small Volume							Large Volume	
Description	**Oper**	**Unit**	**Crew Size**	**Man-Hours Per Unit**	**Avg Mat'l Unit Cost**	**Avg Labor Unit Cost**	**Avg Equip Unit Cost**	**Avg Total Unit Cost**	**Avg Price Incl O&P**	**Avg Total Unit Cost**	**Avg Price Incl O&P**
5'-0" x 5'-0" H											
High quality workmanship	Inst	Ea	CJ	10.0	833.00	245.00	---	1078.00	**1330.00**	877.00	**1070.00**
Good quality workmanship	Inst	Ea	CJ	8.00	765.00	196.00	---	961.00	**1180.00**	786.00	**952.00**
Average quality workmanship	Inst	Ea	CJ	6.96	618.00	170.00	---	788.00	**970.00**	639.00	**774.00**
6'-0" x 3'-0" H											
High quality workmanship	Inst	Ea	CJ	8.89	719.00	217.00	---	936.00	**1160.00**	756.00	**921.00**
Good quality workmanship	Inst	Ea	CJ	6.96	660.00	170.00	---	830.00	**1020.00**	677.00	**820.00**
Average quality workmanship	Inst	Ea	CJ	5.93	550.00	145.00	---	695.00	**853.00**	564.20	**684.00**
6'-0" x 3'-8" H											
High quality workmanship	Inst	Ea	CJ	10.0	763.00	245.00	---	1008.00	**1250.00**	817.00	**1000.00**
Good quality workmanship	Inst	Ea	CJ	8.00	700.00	196.00	---	896.00	**1100.00**	730.00	**888.00**
Average quality workmanship	Inst	Ea	CJ	6.96	584.00	170.00	---	754.00	**930.00**	609.00	**740.00**
6'-0" x 4'-0" H											
High quality workmanship	Inst	Ea	CJ	10.0	782.00	245.00	---	1027.00	**1270.00**	833.00	**1020.00**
Good quality workmanship	Inst	Ea	CJ	8.00	717.00	196.00	---	913.00	**1120.00**	744.00	**905.00**
Average quality workmanship	Inst	Ea	CJ	6.96	598.00	170.00	---	768.00	**947.00**	622.00	**755.00**

Wiring. See Electrical, page 92

Index

A

B

D

M

N

O

P

Q

R

T

U

V

W

XYZ

Practical References for Builders

Renovating & Restyling Older Homes

Any builder can turn a run-down old house into a showcase of perfection — if the customer has unlimited funds to spend. Unfortunately, most customers are on a tight budget. They usually want more improvements than they can afford — and they expect you to deliver. This book shows how to add economical improvements that can increase the property value by two, five or even ten times the cost of the remodel. Sound impossible? Here you'll find the secrets of a builder who has been putting these techniques to work on Victorian and Craftsman-style houses for twenty years. You'll see what to repair, what to replace and what to leave, so you can remodel or restyle older homes for the least amount of money and the greatest increase in value.
416 pages, 8½ x 11, $33.50

Basic Plumbing with Illustrations, Revised

This completely-revised edition brings this comprehensive manual fully up-to-date with all the latest plumbing codes. It is the journeyman's and apprentice's guide to installing plumbing, piping, and fixtures in residential and light commercial buildings: how to select the right materials, lay out the job and do professional-quality plumbing work, use essential tools and materials, make repairs, maintain plumbing systems, install fixtures, and add to existing systems. Includes extensive study questions at the end of each chapter, and a section with all the correct answers.
384 pages, 8½ x 11, $33.00

Blueprint Reading for the Building Trades

How to read and understand construction documents, blueprints, and schedules. Includes layouts of structural, mechanical, HVAC and electrical drawings. Shows how to interpret sectional views, follow diagrams and schematics, and covers common problems with construction specifications. **192 pages, 5½ x 8½, $14.75**

CD Estimator

If your computer has *Windows*™ and a CD-ROM drive, *CD Estimator* puts at your fingertips 85,000 construction costs for new construction, remodeling, renovation & insurance repair, electrical, plumbing, HVAC and painting. You'll also have the *National Estimator* program — a stand-alone estimating program for Windows that Remodeling magazine called a "computer wiz." Quarterly cost updates are available at no charge on the Internet. To help you create professional-looking estimates, the disk includes over 40 construction estimating and bidding forms in a format that's perfect for nearly any word processing or spreadsheet program for *Windows*™. And to top it off, a 70-minute interactive video teaches you how to use this CD-ROM to estimate construction costs. **CD Estimator is $68.50**

National Building Assembly Costs

Complete assembly costs for all construction across the 16 CSI divisions. These prices include all the individual line items included in a typical construction assembly, added together to generate an assembly "unit" price. With unit price estimates from the assemblies section, and subcontractor unit prices from the line item cost information, you can put together an estimate for just about any project in minutes. It's far more accurate than a square foot cost estimate, yet takes a fraction of the time a stick-by-stick estimate takes. Includes a CD-ROM with estimating data you can use, and a demo version of WinEst LT, a powerful estimating program for light construction that includes the cost data from Craftsman's National Construction Estimator. **320 pages, 8½ x 11, $44.75**

National Construction Estimator

Current building costs for residential, commercial, and industrial construction. Estimated prices for every common building material. Provides manhours, recommended crew, and gives the labor cost for installation. Includes a CD-ROM with an electronic version of the book with *National Estimator*, a stand-alone *Windows*™ estimating program, plus an interactive multimedia video that shows how to use the disk to compile construction cost estimates. **616 pages, 8½ x 11, $47.50. Revised annually**

Contractor's Guide to QuickBooks Pro 2000

This user-friendly manual walks you through QuickBooks Pro's detailed setup procedure and explains step-by-step how to create a first-rate accounting system. You'll learn in days, rather than weeks, how to use QuickBooks Pro to get your contracting business organized, with simple, fast accounting procedures. On the CD included with the book you'll find a QuickBooks Pro file preconfigured for a construction company (you drag it over onto your computer and plug in your own company's data). You'll also get a complete estimating program, including a database, and a job costing program that lets you export your estimates to QuickBooks Pro. It even includes many useful construction forms to use in your business.
304 pages, 8½ x 11, $44.50

Construction Forms & Contracts

125 forms you can copy and use — or load into your computer (from the FREE disk enclosed). Then you can customize the forms to fit your company, fill them out, and print. Loads into *Word* for *Windows*™, *Lotus 1-2-3*, *WordPerfect*, *Works*, or *Excel* programs. You'll find forms covering accounting, estimating, fieldwork, contracts, and general office. Each form comes with complete instructions on when to use it and how to fill it out. These forms were designed, tested and used by contractors, and will help keep your business organized, profitable and out of legal, accounting and collection troubles. Includes a CD-ROM for *Windows*™ and Macintosh. **400 pages, 8½ x 11, $41.75**

Profits in Buying & Renovating Homes

Step-by-step instructions for selecting, repairing, improving, and selling highly profitable "fixer-uppers." Shows which price ranges offer the highest profit-to-investment ratios, which neighborhoods offer the best return, practical directions for repairs, and tips on dealing with buyers, sellers, and real estate agents. Shows you how to determine your profit before you buy, what "bargains" to avoid, and how to make simple, profitable, inexpensive upgrades. **304 pages, 8½ x 11, $19.75**

Contracting in All 50 States

Every state has its own licensing requirements that you must meet to do business there. These are usually written exams, financial requirements, and letters of reference. This book shows how to get a building, mechanical or specialty contractor's license, qualify for DOT work, and register as an out-of-state corporation, for every state in the U.S. It lists addresses, phone numbers, application fees, requirements, where an exam is required, what's covered on the exam and how much weight each area of construction is given on the exam. You'll find just about everything you need to know in order to apply for your out-of-state license. **416 pages, 8½ x 11, $36.00**

Build Smarter with Alternative Materials

New building products are coming out almost every week. Some of them may become new standards, as sheetrock replaced lath and plaster some years ago. Others are little more than a gimmick. To write this manual, the author researched hundreds of products that have come on the market in recent years. The ones he describes in this book will do the job better, creating a superior, longer-lasting finished product, and in many cases also save you time and money. Some are made with recycled products — a good selling point with many customers. But most of all, they give you choices, so you can give your customers choices. In this book, you'll find materials for almost all areas of constructing a house, from the ground up. For each product described, you'll learn where you can get it, where to use it, what benefits it provides, any disadvantages, and how to install it — including tips from the author. And to help you price your jobs, each description ends with manhours — for both the first time you install it, and after you've done it a few times. **336 pages, 8½ x 11, $34.75**

Professional Kitchen Design

Remodeling kitchens requires a "special" touch — one that blends artistic flair with function to create a kitchen with charm and personality as well as one that is easy to work in. Here you'll find how to make the best use of the space available in any kitchen design job, as well as tips and lessons on how to design one-wall, two-wall, L-shaped, U-shaped, peninsula and island kitchens. Also includes what you need to know to run a profitable kitchen design business. **176 pages, 8½ x 11, $24.50**

Contractor's Guide to the Building Code Revised

This new edition was written in collaboration with the International Conference of Building Officials, writers of the code. It explains in plain English exactly what the latest edition of the *Uniform Building Code* requires. Based on the 1997 code, it explains the changes and what they mean for the builder. Also covers the *Uniform Mechanical Code* and the *Uniform Plumbing Code*. Shows how to design and construct residential and light commercial buildings that'll pass inspection the first time. Suggests how to work with an inspector to minimize construction costs, what common building shortcuts are likely to be cited, and where exceptions may be granted. **320 pages, 8½ x 11, $39.00**

How to Succeed With Your Own Construction Business

Everything you need to start your own construction business: setting up the paperwork, finding the work, advertising, using contracts, dealing with lenders, estimating, scheduling, finding and keeping good employees, keeping the books, and coping with success. If you're considering starting your own construction business, all the knowledge, tips, and blank forms you need are here. **336 pages, 8½ x 11, $28.50**

National Painting Cost Estimator

A complete guide to estimating painting costs for just about any type of residential, commercial, or industrial painting, whether by brush, spray, or roller. Shows typical costs and bid prices for fast, medium, and slow work, including material costs per gallon; square feet covered per gallon; square feet covered per manhour; labor, material, overhead, and taxes per 100 square feet; and how much to add for profit. Includes a CD-ROM with an electronic version of the book with *National Estimator*, a stand-alone *Windows*™ estimating program, plus an interactive multimedia video that shows how to use the disk to compile construction cost estimates.
440 pages, 8½ x 11, $48.00. Revised annually

Planning Drain, Waste & Vent Systems

How to design plumbing systems in residential, commercial, and industrial buildings. Covers designing systems that meet code requirements for homes, commercial buildings, private sewage disposal systems, and even mobile home parks. Includes relevant code sections and many illustrations to guide you through what the code requires in designing drainage, waste, and vent systems. **192 pages, 8½ x 11, $19.25**

Plumbing & HVAC Manhour Estimates

Hundreds of tested and proven manhours for installing just about any plumbing and HVAC component you're likely to use in residential, commercial, and industrial work. You'll find manhours for installing piping systems, specialties, fixtures and accessories, ducting systems, and HVAC equipment. If you estimate the price of plumbing, you shouldn't be without the reliable, proven manhours in this unique book.
224 pages, 5½ x 8½, $28.25

Basic Lumber Engineering for Builders

Beam and lumber requirements for many jobs aren't always clear, especially with changing building codes and lumber products. Most of the time you rely on your own "rules of thumb" when figuring spans or lumber engineering. This book can help you fill the gap between what you can find in the building code span tables and what you need to pay a certified engineer to do. With its large, clear illustrations and examples, this book shows you how to figure stresses for pre-engineered wood or wood structural members, how to calculate loads, and how to design your own girders, joists and beams. Included FREE with the book — an easy-to-use limited version of NorthBridge Software's *Wood Beam Sizing* program.
272 pages, 81/2 x 11, $38.00

National Renovation & Insurance Repair Estimator

Current prices in dollars and cents for hard-to-find items needed on most insurance, repair, remodeling, and renovation jobs. All price items include labor, material, and equipment breakouts, plus special charts that tell you exactly how these costs are calculated. Includes a CD-ROM with an electronic version of the book with *National Estimator*, a stand-alone *Windows*™ estimating program, plus an interactive multimedia video that shows how to use the disk to compile construction cost estimates.
568 pages, 8½ x 11, $49.50. Revised annually

Pipe & Excavation Contracting

Shows how to read plans and compute quantities for both trench and surface excavation, figure crew and equipment productivity rates, estimate unit costs, bid the work, and get the bonds you need. Explains what equipment will deliver maximum productivity for a job, how to lay all types of water and sewer pipe, and how to switch your business to excavation work when you don't have pipe contracts. Covers asphalt and rock removal, working on steep slopes or in high groundwater, and how to avoid the pitfalls that can wipe out your profits on any job.
400 pages, 5½ x 8½, $29.00

Plumber's Handbook Revised

This new edition shows what will and won't pass inspection in drainage, vent, and waste piping, septic tanks, water supply, graywater recycling systems, pools and spas, fire protection, and gas piping systems. All tables, standards, and specifications are completely up-to-date with recent plumbing code changes. Covers common layouts for residential work, how to size piping, select and hang fixtures, practical recommendations, and trade tips. It's the approved reference for the plumbing contractor's exam in many states. Includes an extensive set of multiple choice questions after each chapter, and in the back of the book, the answers and explanations. Also in the back of the book, a full sample plumber's exam.
352 pages, 8½ x 11, $32.00

Plumber's Exam Preparation Guide

Hundreds of questions and answers to help you pass the apprentice, journeyman, or master plumber's exam. Questions are in the style of the actual exam. Gives answers for both the Standard and Uniform plumbing codes. Includes tips on studying for the exam and the best way to prepare yourself for examination day. **320 pages, 8½ x 11, $29.00**

Troubleshooting Guide to Residential Construction

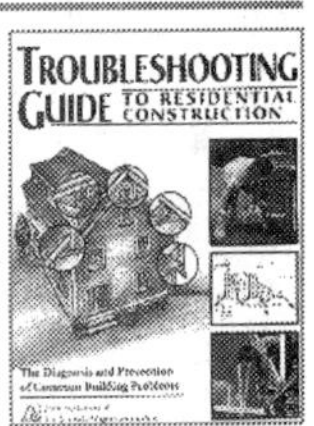

How to solve practically every construction problem — before it happens to you! With this book you'll learn from the mistakes other builders made as they faced 63 typical residential construction problems. Filled with clear photos and drawings that explain how to enhance your reputation as well as your bottom line by avoiding problems that plague most builders. Shows how to avoid, or fix, problems ranging from defective slabs, walls and ceilings, through roofing, plumbing & HVAC, to paint. **304 pages, 8½x 11, $32.50**

Basic Engineering for Builders

If you've ever been stumped by an engineering problem on the job, yet wanted to avoid the expense of hiring a qualified engineer, you should have this book. Here you'll find engineering principles explained in non-technical language and practical methods for applying them on the job. With the help of this book you'll be able to understand engineering functions in the plans and how to meet the requirements, how to get permits issued without the help of an engineer, and anticipate requirements for concrete, steel, wood and masonry. See why you sometimes have to hire an engineer and what you can undertake yourself: surveying, concrete, lumber loads and stresses, steel, masonry, plumbing, and HVAC systems. This book is designed to help the builder save money by understanding engineering principles that you can incorporate into the jobs you bid.
400 pages, 8½ x 11, $34.00

Concrete Construction & Estimating

Explains how to estimate the quantity of labor and materials needed, plan the job, erect fiberglass, steel, or prefabricated forms, install shores and scaffolding, handle the concrete into place, set joints, finish and cure the concrete. Full of practical reference data, cost estimates, and examples.
571 pages, 5½ x 8½, $25.00

BNI Public Works Costbook, 2000

This is the only book of its kind for public works construction. Here you'll find labor and material prices for most public works and infrastructure projects: roads and streets, utilities, street lighting, manholes, and much more. Includes manhour data and a 200-city geographic modifier chart. Includes FREE estimating software and data. **450 pages, 8½ x 11, $74.95**

Construction Estimating Reference Data

Provides the 300 most useful manhour tables for practically every item of construction. Labor requirements are listed for sitework, concrete work, masonry, steel, carpentry, thermal and moisture protection, doors and windows, finishes, mechanical and electrical. Each section details the work being estimated and gives appropriate crew size and equipment needed. Includes a CD-ROM with an electronic version of the book with *National Estimator*, a stand-alone *Windows*™ estimating program, plus an interactive multimedia video that shows how to use the disk to compile construction cost estimates.
432 pages, 11 x 8½, $39.50

Building Contractor's Exam Preparation Guide

Passing today's contractor's exams can be a major task. This book shows you how to study, how questions are likely to be worded, and the kinds of choices usually given for answers. Includes sample questions from actual state, county, and city examinations, plus a sample exam to practice on. This book isn't a substitute for the study material that your testing board recommends, but it will help prepare you for the types of questions — and their correct answers — that are likely to appear on the actual exam. Knowing how to answer these questions, as well as what to expect from the exam, can greatly increase your chances of passing.
320 pages, 8½ x 11, $35.00

Concrete & Formwork

This practical manual has all the information you need to select and pour the right mix for the job, lay out the structure, choose the right form materials, design and build the forms, and finish and cure the concrete, Nearly 100 pages of step-by-step instructions show how to construct and erect most types of site-fabricated wood forms used in residential construction.
176 pages, 8½ x 11, $17.75

Contractor's Growth and Profit Guide

Step-by-step instructions for planning growth and prosperity in a construction contracting or subcontracting company. Explains how to prepare a business plan: select reasonable goals, draft a market expansion plan, make income forecasts and expense budgets, and project cash flow. You'll learn everything that most lenders and investors require, as well as the best way to organize your business. **336 pages, 5½ x 8½, $19.00**

Contractor's Index to the 1997 *Uniform Building Code*

Finally, there's a common-sense index that helps you quickly and easily find the section you're looking for in the *UBC*. It lists topics under the names builders actually use in construction. Best of all, it gives the full section number and the actual page in the *UBC* where you'll find it. If you need to know the requirements for windows in exit access corridor walls, just look under *Windows*™. You'll find the requirements you need are in Section 1004.3.4.3.2.2 in the *UBC* — on page 115. This practical index was written by a former builder and building inspector who knows the *UBC* from both perspectives. If you hate to spend valuable time hunting through pages of fine print for the information you need, this is the book for you. **192 pages, 8½ x 11, paperback edition, $26.00**
192 pages, 8½ x 11, looseleaf edition, $29.00.

The Contractor's Legal Kit

Stop "eating" the costs of bad designs, hidden conditions, and job surprises. Set ground rules that assign those costs to the rightful party ahead of time. And it's all in plain English, not "legalese." For less than the cost of an hour with a lawyer you'll learn the exclusions to put in your agreements, why your insurance company may pay for your legal defense, how to avoid liability for injuries to your sub and his employees or damages they cause, how to collect on lawsuits you win, and much more. It also includes a FREE computer disk with contracts and forms you can customize for your own use. **352 pages, 8½ x 11, $59.95**

Contractor's Survival Manual

How to survive hard times and succeed during the up cycles. Shows what to do when the bills can't be paid, finding money and buying time, transferring debt, and all the alternatives to bankruptcy. Explains how to build profits, avoid problems in zoning and permits, taxes, time-keeping, and payroll. Unconventional advice on how to invest in inflation, get high appraisals, trade and postpone income, and stay hip-deep in profitable work. **160 pages, 8½ x 11, $22.25**

Craftsman's Illustrated Dictionary of Construction Terms

Almost everything you could possibly want to know about any word or technique in construction. Hundreds of up-to-date construction terms, materials, drawings and pictures with detailed, illustrated articles describing equipment and methods. Terms and techniques are explained or illustrated in vivid detail. Use this valuable reference to check spelling, find clear, concise definitions of construction terms used on plans and construction documents, or learn about little-known tools, equipment, tests and methods used in the building industry. It's all here.
416 pages, 8½ x 11, $36.00

Estimating Electrical Construction

Like taking a class in how to estimate materials and labor for residential and commercial electrical construction. Written by an A.S.P.E. National Estimator of the Year, it teaches you how to use labor units, the plan takeoff, and the bid summary to make an accurate estimate, how to deal with suppliers, use pricing sheets, and modify labor units. Provides extensive labor unit tables and blank forms for your next electrical job.
272 pages, 8½ x 11, $19.00

Estimating Excavation

How to calculate the amount of dirt you'll have to move and the cost of owning and operating the machines you'll do it with. Detailed, step-by-step instructions on how to assign bid prices to each part of the job, including labor and equipment costs. Also, the best ways to set up an organized and logical estimating system, take off from contour maps, estimate quantities in irregular areas, and figure your overhead.
448 pages, 8½ x 11, $39.50

Estimating Framing Quantities

Gives you hundreds of time-saving estimating tips. Shows how to make thorough step-by-step estimates of all rough carpentry in residential and light commercial construction: ceilings, walls, floors, and roofs. Lots of illustrations showing lumber requirements, nail quantities, and practical estimating procedures. **285 pages, 5½ x 8½, $34.95**

Estimating Home Building Costs

Estimate every phase of residential construction from site costs to the profit margin you include in your bid. Shows how to keep track of manhours and make accurate labor cost estimates for footings, foundations, framing and sheathing finishes, electrical, plumbing, and more. Provides and explains sample cost estimate worksheets with complete instructions for each job phase. **320 pages, 5½ x 8½, $17.00**

Estimating with Microsoft *Excel*

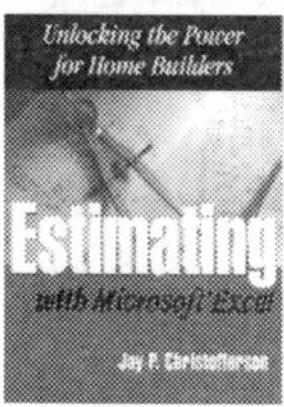

Most builders estimate with *Excel* because it's easy to learn, quick to use, and can be customized to your style of estimating. Here you'll find step-by-step how to create your own customized automated spreadsheet estimating program for use with *Excel*. You'll learn how to use the magic of Excel in creating detail sheets, cost breakdown summaries, and linking. You can even create your own macros. Includes a CD-ROM that illustrates examples in the book and provides you with templates you can use to set up your own estimating system.
148 pages, 8½ x 11, $49.95

Cost Records for Construction Estimating

How to organize and use cost information from jobs just completed to make more accurate estimates in the future. Explains how to keep the records you need to track costs for sitework, footings, foundations, framing, interior finish, siding and trim, masonry, and subcontract expense. Provides sample forms. **208 pages, 8½ x 11, $15.75**

Builder's Guide to Room Additions

How to tackle problems that are unique to additions, such as requirements for basement conversions, reinforcing ceiling joists for second-story conversions, handling problems in attic conversions, what's required for footings, foundations, and slabs, how to design the best bathroom for the space, and much more. Besides actual construction, you'll even find help in designing, planning, and estimating your room addition jobs. **352 pages, 8½ x 11, $27.25**

Greenbook Standard Specifications For Public Works Construction

Since 1967, eleven previous editions of the popular "Greenbook" have been used as the official specification, bidding and contract document for many cities, counties and public agencies throughout the West. New federal regulations mandate that all public construction use metric documentation. This complete reference, which meets this new requirement, provides uniform standards of quality and sound construction practice easily understood and used by engineers, public works officials, and contractors across the U.S. Includes hundreds of charts and tables.
480 pages, 8½ x 11, $59.95

Home Inspection Handbook

Every area you need to check in a home inspection — especially in older homes. Twenty complete inspection checklists: building site, foundation and basement, structural, bathrooms, chimneys and flues, ceilings, interior & exterior finishes, electrical, plumbing, HVAC, insects, vermin and decay, and more. Also includes information on starting and running your own home inspection business. **324 pages, 5½ x 8½, $24.95**

Steel-Frame House Construction

Framing with steel has obvious advantages over wood, yet building with steel requires new skills that can present challenges to the wood builder. This new book explains the secrets of steel framing techniques for building homes, whether pre-engineered or built stick by stick. It shows you the techniques, the tools, the materials, and how you can make it happen. Includes hundreds of photos and illustrations, plus a CD-ROM with steel framing details. **328 pages, 8½ x 11, $39.75**

Managing the Small Construction Business

Overcome your share of business hassles by learning how 50 small contractors handled similar problems in their businesses. Here you'll learn how they handle bidding, unit pricing, contract clauses, change orders, job-site safety, quality control, overhead and markup, managing subs, scheduling systems, cost-plus contracts, pricing small jobs, insurance repair, finding solutions to conflicts, and much more.
243 pages, 8½ x 11, $34.95

National Building Cost Manual

Square foot costs for residential, commercial, industrial, and farm buildings. Quickly work up a reliable budget estimate based on actual materials and design features, area, shape, wall height, number of floors, and support requirements. Includes all the important variables that can make any building unique from a cost standpoint.
240 pages, 8½ x 11, $23.00. Revised annually

Markup & Profit: A Contractor's Guide

In order to succeed in a construction business, you have to be able to price your jobs to cover all labor, material and overhead expenses, and make a decent profit. The problem is knowing what markup to use. You don't want to lose jobs because you charge too much, and you don't want to work for free because you've charged too little. If you know how to calculate markup, you can apply it to your job costs to find the right sales price for your work. This book gives you tried and tested formulas, with step-by-step instructions and easy-to-follow examples, so you can easily figure the markup that's right for your business. Includes a CD-ROM with forms and checklists for your use. **320 pages, 8½ x 11, $32.50**

National Electrical Estimator

This year's prices for installation of all common electrical work: conduit, wire, boxes, fixtures, switches, outlets, loadcenters, panelboards, raceway, duct, signal systems, and more. Provides material costs, manhours per unit, and total installed cost. Explains what you should know to estimate each part of an electrical system. Includes a CD-ROM with an electronic version of the book with *National Estimator*, a stand-alone *Windows*™ estimating program, plus an interactive multimedia video that shows how to use the disk to compile construction cost estimates.
520 pages, 8½ x 11, $47.75. Revised annually

Getting Financing & Developing Land

Developing land is a major leap for most builders — yet that's where the big money is made. This book gives you the practical knowledge you need to make that leap. Learn how to prepare a market study, select a building site, obtain financing, guide your plans through approval, then control your building costs so you can ensure yourself a good profit. Includes a CD-ROM with forms, checklists, and a sample business plan you can customize and use to help you sell your idea to lenders and investors. **232 pages, 8½ x 11, $39.00**

Residential Plumbing Inspection

You never know what you'll find in a plumbing inspection. Corrosion between pipes, illegal traps, and even unsafe shut-off valves can be found in crawl spaces and attics of homes that appear on the surface to be built perfectly to code. This unique book is filled with illustrations and photographs of typical plumbing code violations you're likely to encounter during inspections. Then it explains what should be done to bring them to code. It covers all the areas: drainage systems, vents and venting, fixtures, water distribution, fuel gas piping, water heaters, vents and combustion, private wells and pumps, solar heating systems, and more.
90 pages, 8½ x 11, $49.95. Published by ITA

Rough Framing Carpentry

If you'd like to make good money working outdoors as a framer, this is the book for you. Here you'll find shortcuts to laying out studs; speed cutting blocks, trimmers and plates by eye; quickly building and blocking rake walls; installing ceiling backing, ceiling joists, and truss joists; cutting and assembling hip trusses and California fills; arches and drop ceilings — all with production line procedures that save you time and help you make more money. Over 100 on-the-job photos of how to do it right and what can go wrong. **304 pages, 8½ x 11, $26.50**

Roofing Construction & Estimating

Installation, repair and estimating for nearly every type of roof covering available today in residential and commercial structures: asphalt shingles, roll roofing, wood shingles and shakes, clay tile, slate, metal, built-up, and elastomeric. Covers sheathing and underlayment techniques, as well as secrets for installing leakproof valleys. Many estimating tips help you minimize waste, as well as insure a profit on every job. Troubleshooting techniques help you identify the true source of most leaks. Over 300 large, clear illustrations help you find the answer to just about all your roofing questions. **432 pages, 8½ x 11, $35.00**

Residential Electrical Estimating

A fast, accurate pricing system proven on over 1000 residential jobs. Using the manhours provided, combined with material prices from your wholesaler, you quickly work up estimates based on degree of difficulty. These manhours come from a working electrical contractor's records — not some pricing agency. You'll find prices for every type of electrical job you're likely to estimate — from service entrances to ceiling fans.
320 pages, 8½ x 11, $29.00

Constructionary

A unique pocket-sized dictionary of up-to-date construction words and phrases in English-Spanish and Spanish-English. Here you'll find over 1000 construction terms and 70 commonly used on-the-job phrases. This dictionary includes phonetic pronunciation, tool section, commonly used sentences, and conversion tables.
170 pages, 4 x 7, $14.95. Published by ICBO

A Builder's Guide To Residential HVAC Systems

Here's the information you need to evaluate and specify heating, ventilation and air conditioning systems effectively, access and improve system quality, and apply and sell energy-efficient alternatives. It covers traditional and alternative duct systems; installation and operating costs, and gives current information on quality and energy-efficient design. Includes tables, charts, examples of duct layouts, and examples of trade language. **64 pages, 8½ x 11, $27.50**

Residential Structure & Framing

With this easy-to-understand guide you'll learn how to calculate loads, size joists and beams, and tackle many common structural problems facing residential contractors. It covers cantilevered floors, complex roof structures, tall window walls, and seismic and wind bracing. Plus, you'll learn field-proven production techniques for advanced wall, floor, and roof framing with both dimensional and engineered lumber. You'll find information on sizing joists and beams, framing with wood I-joists, supporting oversized dormers, unequal-pitched roofs, coffered ceilings, and more. Fully illustrated with lots of photos.
272 pages, 8½ x 11, $34.95. Published by JLC

Builder's Guide to Accounting Revised

Step-by-step, easy-to-follow guidelines for setting up and maintaining records for your building business. This practical, newly-revised guide to all accounting methods shows how to meet state and federal accounting requirements, explains the new depreciation rules, and describes how the Tax Reform Act can affect the way you keep records. Full of charts, diagrams, simple directions and examples, to help you keep track of where your money is going. Recommended reading for many state contractor's exams. **320 pages, 8½ x 11, $26.50**

Commercial Metal Stud Framing

Framing commercial jobs can be more lucrative than residential work. But most commercial jobs require some form of metal stud framing. This book teaches step-by-step, with hundreds of job site photos, high-speed metal stud framing in commercial construction. It describes the special tools you'll need and how to use them effectively, and the material and equipment you'll be working with. You'll find the shortcuts, tips and tricks-of-the-trade that take most steel frames years on the job to discover. Shows how to set up a crew to maintain a rhythm that will speed progress faster than any wood framing job. If you've framed with wood, this book will teach you how to be one of the few top-notch metal stud framers.
208 pages, 8½ x 11, $45.00

Contractor's Year-Round Tax Guide Revised

How to set up and run your construction business to minimize taxes: corporate tax strategy and how to use it to your advantage, and what you should be aware of in contracts with others. Covers tax shelters for builders, write-offs and investments that will reduce your taxes, accounting methods that are best for contractors, and what the I.R.S. allows and what it often questions. **192 pages, 8½ x 11, $26.50**

Carpentry Estimating

Simple, clear instructions to help you take off quantities and figure costs for all rough and finish carpentry. Provides checklists, manhour tables, and worksheets with calculation factors built in. Take the guesswork out of estimating carpentry so you get accurate labor and material costs every time. Includes FREE *National Estimator* software, with all the manhours in the book, plus the carpentry estimating worksheets in formats that work with *Word*, *Excel*, *WordPerfect*, and *MS Works* for *Windows*™. **336 pages, 8½ x 11, $35.50**

Wood-Frame House Construction

Step-by-step construction details, from the layout of the outer walls, excavation and formwork, to finish carpentry and painting. Contains all new, clear illustrations and explanations updated for construction in the '90s. Everything you need to know about framing, roofing, siding, interior finishings, floor covering and stairs — your complete book of wood-frame homebuilding. **320 pages, 8½ x 11, $25.50. Revised edition**

Electrician's Exam Preparation Guide

Need help in passing the apprentice, journeyman, or master electrician's exam? This is a book of questions and answers based on actual electrician's exams over the last few years. Almost a thousand multiple-choice questions — exactly the type you'll find on the exam — cover every area of electrical installation: electrical drawings, services and systems, transformers, capacitors, distribution equipment, branch circuits, feeders, calculations, measuring and testing, and more. It gives you the correct answer, an explanation, and where to find it in the latest *NEC*. Also tells how to apply for the test, how best to study, and what to expect on examination day.
352 pages, 8½ x 11, $32.00

Finish Carpenter's Manual

Everything you need to know to be a finish carpenter: assessing a job before you begin, and tricks of the trade from a master finish carpenter. Easy-to-follow instructions for installing doors and windows, ceiling treatments (including fancy beams, corbels, cornices and moldings), wall treatments (including wainscoting and sheet paneling), and the finishing touches of chair, picture, and plate rails. Specialized interior work includes cabinetry and built-ins, stair finish work, and closets. Also covers exterior trims and porches. Includes manhour tables for finish work, and hundreds of illustrations and photos. **208 pages, 8½ x 11, $22.50**

Kitchens & Baths

Practical, detailed solutions to every problem you're likely to approach when constructing or remodeling kitchens and baths. Over 50 articles from the *Journal of Light Construction* plus up-to-date reference material on codes, safety, space planning and construction methods for kitchens and bathrooms. Includes over 400 photos and illustrations. If you're looking for a complete reference on designing, remodeling, or constructing kitchens and baths, you should have this book. **256 pages, 8 x 11, $34.95**

Painter's Handbook

Loaded with "how-to" information you'll use every day to get professional results on any job: the best way to prepare a surface for painting or repainting; selecting and using the right materials and tools (including airless spray); tips for repainting kitchens, bathrooms, cabinets, eaves and porches; how to match and blend colors; why coatings fail and what to do about it. Lists 30 profitable specialties in the painting business.
320 pages, 8½ x 11, $21.25

Roof Framing

Shows how to frame any type of roof in common use today, even if you've never framed a roof before. Includes using a pocket calculator to figure any common, hip, valley, or jack rafter length in seconds. Over 400 illustrations cover every measurement and every cut on each type of roof: gable, hip, Dutch, Tudor, gambrel, shed, gazebo, and more.
480 pages, 5½ x 8½, $22.00

CD Estimator Heavy

CD Estimator Heavy has a complete 780-page heavy construction cost estimating volume for each of the 50 states. Select the cost database for the state where the work will be done. Includes thousands of cost estimates you won't find anywhere else, and in-depth coverage of demolition, hazardous materials remediation, tunneling, site utilities, precast concrete, structural framing, heavy timber construction, membrane waterproofing, industrial windows and doors, specialty finishes, built-in commercial and industrial equipment, and HVAC and electrical systems for commercial and industrial buildings. **CD Estimator Heavy is $69.00**

Simplified Guide to Construction Law

Here you'll find easy-to-read, paragraphed-sized samples of how the courts have viewed common areas of disagreement — and litigation — in the building industry. You'll read about legal problems that real builders have faced, and how the court ruled. This book will tell you what you need to know about contracts, torts, fraud, misrepresentation, warranty and strict liability, construction defects, indemnity, insurance, mechanics liens, bonds and bonding, statutes of limitation, arbitration, and more. These are simplified examples that illustrate not necessarily who is right and who is wrong — but who the law has sided with. **298 pages, 5½ x 8½, $29.95**

Remodeling Contractor's Handbook

Everything you need to know to make a remodeling business grow: identifying a market, inexpensive sales and advertising techniques that work, making accurate estimates, building a positive company image, training effective sales people, placing loans for customers, and bringing in profitable work to keep your company growing.
304 pages, 8½ x 11, $18.25

How to Sell Remodeling

Proven, effective sales methods for repair and remodeling contractors: finding qualified leads, making the sales call, identifying what your prospects really need, pricing the job, arranging financing, and closing the sale. Explains how to organize and staff a sales team, keep your crews busy, and much more. Includes blank forms, tables, and charts.
240 pages, 8½ x 11, $17.50

Vest Pocket Guide to HVAC Electricity

This handy guide will be a constant source of useful information for anyone working with electrical systems for heating, ventilating, refrigeration, and air conditioning. Includes essential tables and diagrams for calculating and installing electrical systems for powering and controlling motors, fans, heating elements, compressors, transformers and every electrical part of an HVAC system. **304 pages, 3½x 5½, $18.00**

Scheduling Residential Construction for Builders and Remodelers

Successful builders organize staff, coordinate, monitor, and control what is taking place at one or more jobsites. Too often this involves a lot of stress and sixteen-hour days. Learn scheduling techniques that take relatively little time to implement and streamline your workload. These techniques include manual and computerized scheduling. **116 pages, 8½ x 11, $29.00**

Craftsman Book Company
6058 Corte del Cedro
P.O. Box 6500
Carlsbad, CA 92018

☎ 24 hour order line
1-800-829-8123
Fax (760) 438-0398

Name ______________________

Company ______________________

Address ______________________

City/State/Zip ______________________

○ This is a residence

Total enclosed______________________(In California add 7.25% tax

We pay shipping when your check covers your order in full.

In A Hurry?

We accept phone orders charged to your

○ Visa, ○ MasterCard, ○ Discover or ○ American Express

Card#______________________

Exp.date______________Initials______________

Tax Deductible: Treasury regulations make these references tax deductible when used in your work. Save the canceled check or charge card statement as your receipt.

Order online http://www.craftsman-book.com

Free on the Internet! Download any of Craftsman's estimating costbooks for a 30-day free trial! http://costbook.com

10-Day Money Back Guarantee

- ○ 34.00 Basic Engineering for Builders
- ○ 38.00 Basic Lumber Engineering for Builders
- ○ 33.00 Basic Plumbing with Illustrations
- ○ 14.75 Blueprint Reading for the Building Trades
- ○ 74.95 BNI Public Works Costbook
- ○ 34.75 Build Smarter with Alternative Materials
- ○ 26.50 Builder's Guide to Accounting Revised
- ○ 27.50 A Builder's Guide to Residential HVAC Systems
- ○ 27.25 Builder's Guide to Room Additions
- ○ 35.00 Building Contractor's Exam Preparation Guide
- ○ 35.50 Carpentry Estimating w/FREE *National Estimator* carpentry estimating program on a CD-ROM.
- ○ 68.50 CD Estimator
- ○ 69.00 CD Estimator Heavy
- ○ 45.00 Commercial Metal Stud Framing
- ○ 17.75 Concrete and Formwork
- ○ 25.00 Concrete Construction & Estimating
- ○ 14.95 Constructionary
- ○ 39.50 Construction Estimating Reference Data w/FREE *National Estimator* on a CD-ROM.
- ○ 41.75 Construction Forms & Contracts with a CD-ROM for *Windows*™ and Macintosh.
- ○ 36.00 Contracting in All 50 States
- ○ 19.00 Contractor's Growth & Profit Guide
- ○ 44.50 Contractor's Guide to QuickBooks Pro 2000
- ○ 39.00 Contractor's Guide to the Building Code Revised
- ○ 26.00 Contractor's Index to the *UBC* — Paperback
- ○ 29.00 Contractor's Index to the *UBC* — Looseleaf
- ○ 59.95 Contractor's Legal Kit
- ○ 22.25 Contractor's Survival Manual
- ○ 26.50 Contractor's Year-Round Tax Guide Revised
- ○ 15.75 Cost Records for Construction Estimating
- ○ 36.00 Craftsman's Illustrated Dictionary of Construction Terms
- ○ 32.00 Electrician's Exam Preparation Guide
- ○ 19.00 Estimating Electrical Construction
- ○ 39.50 Estimating Excavation
- ○ 34.95 Estimating Framing Quantities
- ○ 17.00 Estimating Home Building Costs
- ○ 49.95 Estimating with Microsoft *Excel*
- ○ 22.50 Finish Carpenter's Manual
- ○ 39.00 Getting Financing & Developing Land
- ○ 59.95 Greenbook Standard Specifications for Public Works Construction
- ○ 24.95 Home Inspection Handbook
- ○ 17.50 How to Sell Remodeling
- ○ 28.50 How to Succeed w/Your Own Construction Business
- ○ 34.95 Kitchens & Baths
- ○ 34.95 Managing the Small Construction Business
- ○ 32.50 Markup & Profit: A Contractor's Guide
- ○ 44.75 National Building Assembly Costs
- ○ 23.00 National Building Cost Manual
- ○ 47.50 National Construction Estimator w/FREE *National Estimator* on a CD-ROM.
- ○ 47.75 National Electrical Estimator w/FREE *National Estimator* on a CD-ROM.
- ○ 48.00 National Painting Cost Estimator w/FREE *National Estimator* on a CD-ROM.
- ○ 49.50 National Renovation & Insurance Repair Estimator w/FREE *National Estimator* on a CD-ROM.
- ○ 21.25 Painter's Handbook
- ○ 29.00 Pipe & Excavation Contracting
- ○ 19.25 Planning Drain, Waste & Vent Systems
- ○ 29.00 Plumber's Exam Preparation Guide
- ○ 32.00 Plumber's Handbook Revised
- ○ 28.25 Plumbing & HVAC Manhour Estimates
- ○ 24.50 Professional Kitchen Design
- ○ 19.75 Profits in Buying & Renovating Homes
- ○ 18.25 Remodeling Contractor's Handbook
- ○ 33.50 Renovating & Restyling Older Homes
- ○ 29.00 Residential Electrical Estimating
- ○ 49.95 Residential Plumbing Inspection
- ○ 34.95 Residential Structure & Framing
- ○ 22.00 Roof Framing
- ○ 35.00 Roofing Construction & Estimating
- ○ 26.50 Rough Framing Carpentry
- ○ 29.00 Scheduling Residential Construction for Builders and Remodelers
- ○ 29.95 A Simplified Guide to Construction Law
- ○ 39.75 Steel-Frame House Construction
- ○ 32.50 Troubleshooting Guide to Residential Construction
- ○ 18.00 Vest Pocket Guide to HVAC Electricity
- ○ 25.50 Wood-Frame House Construction
- ○ 48.50 National Repair & Remodeling Estimator w/FREE *National Estimator* on a CD-ROM.
- ○ FREE Full Color Catalog

Prices subject to change without notice